DMU 0487241 01 0
WITHDRAWN
DE MONTFORT UNIVERSITY
LIBRARY

AF616289

Vision

The Approach of Biophysics and Neurosciences

Istituto Italiano per gli Studi Filosofici

Series on Biophysics and Biocybernetics

Coordinating Editor: Cloe Taddei-Ferretti

Vol. 1: Biophysics of Photoreception: Molecular and Phototransductive Events
edited by: C. Taddei-Ferretti

Vol. 2: Biocybernetics of Vision: Integrative Mechanisms and Cognitive Processes
edited by: C. Taddei-Ferretti

Vol. 3: High-Dilution Effects on Cells and Integrated Systems
edited by: C. Taddei-Ferretti and P. Marotta

Vol. 4: Macromolecular Interplay in Brain Associative Mechanisms
edited by: A. Neugebauer

Vol. 5: From Structure to Information in Sensory Systems
edited by: C. Taddei-Ferretti and C. Musio

Vol. 6: Downward Processes in the Perception Representation Mechanisms
edited by: C. Taddei-Ferretti and C. Musio

Vol. 7: Chaos and Noise in Biology and Medicine
edited by: M. Barbi and S. Chillemi

Vol. 8: Neuronal Bases and Psychological Aspects of Consciousness
edited by: C. Taddei-Ferretti and C. Musio

Forthcoming volumes:

Vol. 9: Neuronal Coding of Perceptual Systems
edited by: W. Backhaus

Vol. 10: Emotions, Qualia, Consciousness
edited by: A. Kaszniak

Vol. 12: Memory and Emotions
edited by:P. Calabrese and A. Neugebauer

ISTITUTO ITALIANO PER GLI STUDI FILOSOFICI
SERIES ON BIOPHYSICS AND BIOCYBERNETICS
Vol. 11 – Biophysics

Vision

The Approach of Biophysics and Neurosciences

Proceedings of the International School of Biophysics
Casamicciola, Napoli, Italy, 11–16 October 1999

Edited by

C. Musio

Istituto di Cibernetica, CNR, Arco Felice, Napoli, Italy

Published by

World Scientific Publishing Co. Pte. Ltd.

P O Box 128, Farrer Road, Singapore 912805

USA office: Suite 1B, 1060 Main Street, River Edge, NJ 07661

UK office: 57 Shelton Street, Covent Garden, London WC2H 9HE

British Library Cataloguing-in-Publication Data
A catalogue record for this book is available from the British Library.

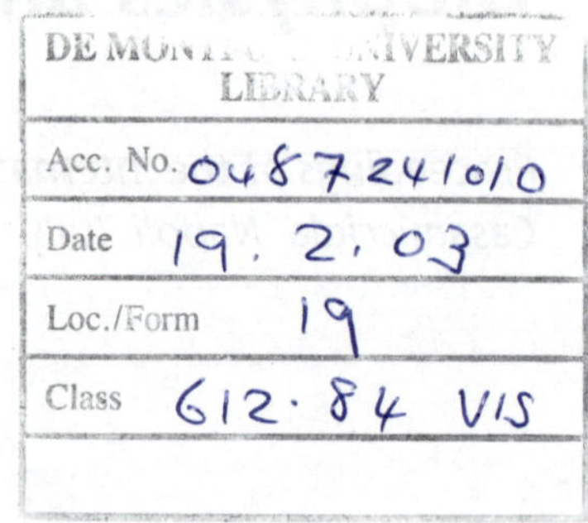

VISION
The Approach of Biophysics and Neurosciences

ISBN 981-02-4647-1

Printed in Singapore by Uto-Print

PREFACE

This is the eleventh volume of the Series on Biophysics and Biocybernetics promoted by the Istituto Italiano per gli Studi Filosofici.

It appears as the Proceedings of the sixth Course of the International School of Biophysics entitled "Vision: The Approach of Biophysics and Neurosciences", which was inaugurated at the site of the Istituto Italiano per gli Studi Filosofici, Palazzo Serra di Cassano, Naples, Italy, and was held at Casamicciola on the isle of Ischia, Italy, on October 11-16, 1999, under the direction of this volume's Editor.

The School is promoted and supported by the Istituto Italiano per gli Studi Filosofici, while the organization of the Course was carried on by the Istituto di Cibernetica of the Italian National Research Council (CNR), Arco Felice, Naples, Italy, under the auspices of the Italian Society of Pure and Applied Biophysics (SIBPA).

The Course sessions covered all latest aspects of vision, ranging from the "Molecular Level" to the "Computational and Cognitive Level" through the "Cellular Level" and the "Integrative Level".

Vision, in a general meaning, is conceivably the key-sense in both animal and vegetal kingdom. The research in this field is providing impressive results thanks to fast-growing theoretical and methodological advances. Overall, the approach of biophysics and neurosciences has proved to be greatly advantageous and of considerable heuristic value. In this direction, the present book provides an outline of most recent acquisitions reached in these fields. Visual mechanisms and processes are analysed and understood at several levels (molecular, cellular, integrative, computational and cognitive ones) through different theoretical tools and experimental methods applied to different living models (from protists to humans, *via* invertebrates and lower vertebrates).

I would like to thank the members of the Course Advisory Board for their fruitful advices and suggestions, as well as for their helpful cooperation also as reviewers of the participants' communication papers: J.E. Dowling (USA), A. Fiorentini (I), H.J. Karten (USA), L. Lagnado (UK), S.B. Laughlin (UK), C. Taddei-Ferretti (I).

I would also acknowledge the partial financial support of the Italian Society of Pure and Applied Biophysics (SIBPA) and of Eliografia Maria, Pozzuoli (NA), Italy. The precious help of the Istituto Italiano per gli Studi Filosofici has provided several deserving participants, especially those coming from needy countries, with substantial grants to attend the Course.

I wish to thank all scientists who agreed to lecture and contributed to the Course with their discussions: they favoured, toghether with all participants, a charming and friendly atmosphere in a highly stimulating scientific milieu.

Finally, I am grateful to the members of the local organizing committee, A. Cotugno, S. Santillo of the Istituto di Cibernetica, CNR, and the administrative advisor S. Aprile of the Istituto Italiano per gli Studi Filosofici; without their work the Course could not have been realized, nor its cordial atmosphere obtained. The precious and patient work together of S. Santillo in the preparation of this book is warmly acknowledged too.

Last but not least, I wish to express my gratitude to C. Taddei-Ferretti, the permanent director of the International Schools of Biophysics and Biocybernetics, who charged me with this demanding nevertheless honourable committment.

The beauty of the isle of Ischia and the courtesy of the staff of the Hotel Gran Paradiso at Casamicciola, where the Course was held, completed the pleasantness of the environment.

Carlo Musio

CONTENTS

COMPUTATIONAL AND COGNITIVE LEVEL

PARTICIPANTS

INTRODUCTORY LECTURE

THE OPTICS OF ANIMAL EYES

MICHAEL F LAND
Sussex Centre for Neuroscience
School of Biological Sciences, University of Sussex
Brighton BN1 9QG, United Kingdom

ABSTRACT

Eyes with well-developed optical systems evolved many times at the end of the Cambrian period, 500 million years ago. There are now about ten optically distinct mechanisms. These include pinholes, lenses of both multi-element and inhomogeneous construction, aspheric surfaces, concave mirrors, apposition compound eyes that employ a variety of lens types, and three kinds of superposition eye that utilize lenses, mirrors, or both. Because the number of physical solutions to the problem of forming an image is finite, convergent evolution has been very common. The best example is the inhomogeneous Matthiessen lens, which has evolved independently in the vertebrates, several times in the molluscs and annelids, and once in the crustaceans. Similar cases of convergence can also be found among compound eyes.

1. Introduction

1.1 The Evolution of Optical Mechanisms

There is a relatively small number of ways to produce an eye that gives a usable image, and most have been "discovered" more than once, thus giving rise to similar structures in unrelated animals. Thus when we trace the evolution of different kinds of eye, the greatest problem lies in deciding whether similarity in structure is due to evolutionary convergence or to common descent. Citing the most notorious example, the phylogenetically unrelated eyes of squid and fish are similar in a great many details, presumably because the logic of the production of large, camera-type eyes necessitates a spherical lens, iris, eye-muscles, etc. (Packard, 1972). By contrast, human and fish eyes are related by common descent, although optically they are rather different from each other. A superficial study of the eyes does not always allow such a distinction to be made, and lineages in eyes must be traced by either knowing the phylogeny of the animals in advance, or looking at other characters that are related less to optical "design principles." In the case of fish and cephalopods, the inverted and multilayered structure of the fish retina, compared with the simpler, noninverted retina of octopus and squid, demonstrates most clearly the unrelatedness of the eyes themselves.

Eye evolution has proceeded in two stages. In almost all the major animal groups, one finds simple eye-spots that consist of a small number of receptors in an open cup of screening pigment cells (Figure 1*a*). In an impressive analysis of the detailed structure, anatomic origins, and phylogenetic affinities of these eye-spots, Salvini-Plawen and Mayr (1977) concluded that such structures had evolved independently at least 40 times, and probably as many as 65 times. These eye-spots are useful in selecting a congenial environment, as they can tell an animal a certain amount about the distribution of light and dark in the surroundings. However, with only shadowing from the pigment cup to restrict the acceptance angle of individual receptors, the resolution is much too poor for the eye to detect predators or prey, or to be involved in pattern recognition or the control of locomotion. All these tasks require the eye to have an optical system that can restrict receptor acceptance angles to a few degrees or better. This second stage in eye evolution, the provision of a competent optical system, has occurred much less frequently than the first, in only six of the 33 metazoan *phyla* listed by Barnes (1987): the Cnidaria, Mollusca, Annelida, Onychophora, Arthropoda, and Chordata. These are, however, the most successful *phyla*, as they contribute about 96% of known species. Perhaps the attainment of optical "lift-off" has contributed to this success.

The most exciting feature of the later stages of optical evolution has been the diversity of mechanisms that have been tried out in various parts of the animal kindgom. At last count, there were ten optically distinct ways of producing images (Figures 1 and 2). These include nearly all those known from optical technology (the Fresnel lens and the zoom lens are two of the few exceptions that come to mind), plus several solutions involving array optics that have not really been invented. Some of these solutions, such as the spherical graded-index lens (Figure 1*e*), have evolved many times; others, such as the reflecting superposition eyes of shrimps and lobsters (Figure 2*j*), have probably only evolved once. Four or five of these mechanisms have only been discovered in the last 25 years, which is remarkable given that excellent anatomic descriptions of most of these eyes have been available since the 1900s or earlier.

As most textbooks continue to refer to "the lens eye" or "the compound eye," as though these represented the totality of optical types, it seems appropriate to provide a brief review of all known mechanisms of image formation in eyes. We concentrate here on the new mechanisms, but do not omit those that have been understood for much longer.

Thus, this chapter is mostly devoted to the mechanisms, capabilities, evolutionary origins, and affinities of the many kinds of "advanced" image forming eye. The conventional division of eyes into "simple," *i.e.* single-chambered or camera-like, and "compound" is retained, because the mechanisms involved really are very different and represent topologically "concave" and "convex" solutions to the problem of image formation (Goldsmith, 1990). Useful, supplementary accounts of optical mechanisms in the invertebrates are given by Land (1981), and in several chapters in Ali (1984). Nilsson (1989) gives an excellent account of compound eye optics and evolution. Walls (1942) still provides the best comparative account of vertebrate eye optics, but Hughes (1977) and Sivak (1988) offer important new perspectives.

2. Simple Eyes

2.1. Pit Eyes

These eye-spots are of interest here only because they must have provided the ancestors for optically more advanced eyes (Figure 1*a*). These eyes are typically less than 100 μm in diameter and contain from only 1 to about 100 receptors. They are found in all but five of the 33 metazoan *phyla.* They may be derived from ciliated ectodermal cells or, less commonly, from nonciliated ganglionic cells. The eyes may be "everse," *i.e.* the receptors are directed towards the light, and the nerve fibers pass through the back of the eye-cup. Or, they may be "inverse," *i.e.* the nerve fibers emerge from the front of the cup (Salvini-Plawen and Mayr, 1977). Burr (1984) has reviewed behaviours that these eyes can mediate. There are three ways to improve the performance of an eye-spot. An enlarged cup and reduced aperture produces a pinhole eye (Figure l*b*). The incorporation of a refractile structure into the eye sharpens the retinal image and thus improves directionality (Figure l*c*). And, the provision of a reflecting layer behind the receptors has two effects: First, it increases the amount of light available to the receptors; second, if the receptors move forward in the eye, it throws an image on them (Figure 1*f*). One can discern the beginnings of all these processes in the eye-spots of different invertebrate groups (reviews in Ali, 1984).

2.2. Pinholes

The only one good example of a pinhole eye is found in the ancient cephalopod mollusk *Nautilus* (Figure l*b*). A few other mollusks have what one might describe as "improved pits." In the abalone *Haliotis,* the eye-cup is 1 mm long with a 0.2 mm pupil, and perhaps 15,000 receptors (Messenger, 1981). The *Nautilus* eye, however, is quite different. Except for the absence of a lens, it is an advanced eye in all respects. It is large, almost l cm in diameter; it has an aperture that can be expanded from 0.4 to 2.8 mm; and it has extraocular muscles that mediate a response to gravity, thus stabilizing the eye against the rocking motion of the swimming animal (Hartline *et al.*, 1979). Optically, however, this is a poor eye. The point-spread function (blur circle) on the retina cannot be smaller than the pupil, which limits resolution to several degrees at best. Muntz and Raj (1984) used the animal's optomotor response to test resolution and found that the minimum effective grating period was 11-22.5°, which is worse than expected. The real problem with this eye is that a reduction of the pupil diameter to improve resolution means a serious loss of retinal illuminance, and vice versa. Even at full aperture, the image is six times dimmer than in the eye of an octopus or fish, and the resolution is awful. The real mystery is that the pinhole has been retained. Almost any lens-like structure, however crude, placed in the aperture would improve resolution, sensitivity, or both. Thus, it must remain an evolutionary conundrum that this simple modification has not occurred here, when it has so often elsewhere.

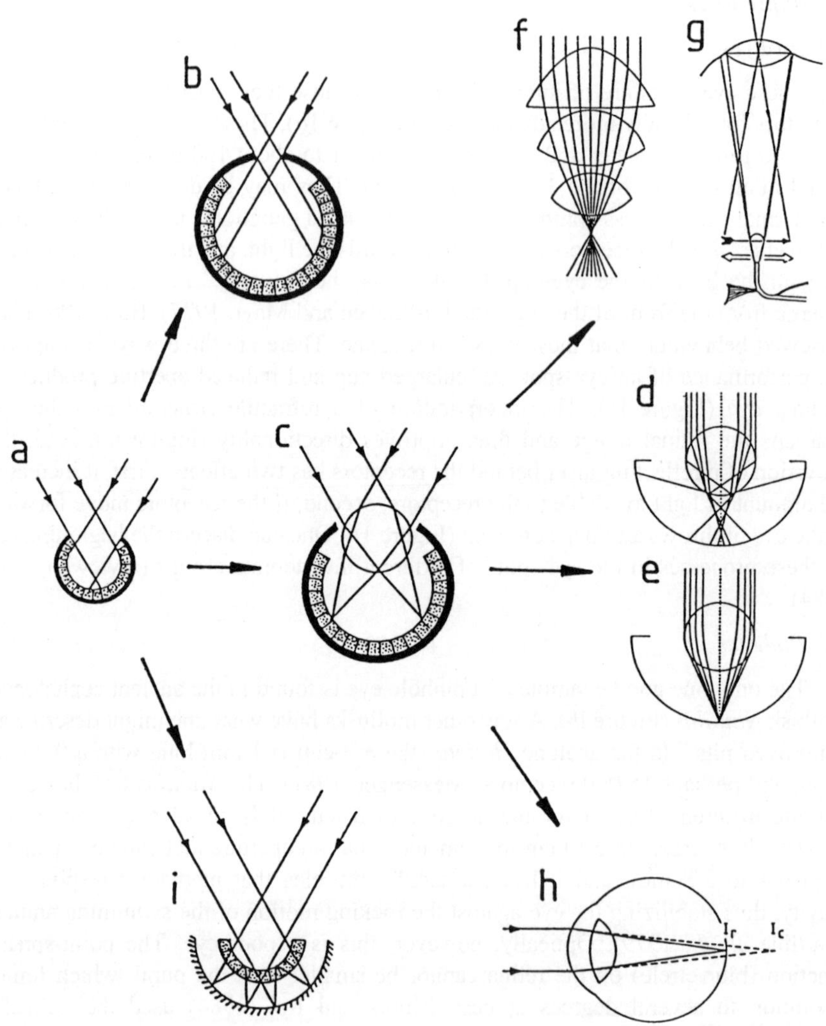

Figure 1. The evolution of single-chambered eyes. The arrows indicate developments, rather than specific evolutionary pathways. See text for details and references. Compiled from many sources. *(a)* Pit eye, common throughout the lower *phyla*. *(b)* Pinhole eye of *Haliotis* or *Nautilus*. *(c)* Eye with lens. *(d)* Homogeneous lens. *(e)* Inhomogeneous "Matthiessen" lens. *(f)* Multiple lens eye of male *Pontella*. *(g)* Two-lens eye of *Copilia*. Solid arrow shows image position; open arrow, the movement of the second lens. *(h)* Terrestrial eye of man with cornea and lens; *Ic*, image formed by cornea alone; *Ir*, final image on retina. *(i)* Mirror eye of the scallop *Pecten*.

2.3. Spherical Lenses

In aquatic animals, the most common optical system in single-chambered eyes is based on a spherical lens (Figures 1*c-e)*. Initially, such a lens would have arisen by an increase in the refractive index of the material within the eye-cup, brought about by the addition of protein or carbohydrate. Eyes with such undifferentiated (or "Fullmasse") lenses can still be found in some gastropod mollusks and annelids (see Land, 1981 for earlier references). However, such a lens can only reduce the diameter of the blur circle on the retina, not form a sharp image, because the focal length cannot be shortened enough to fit the eye. In more advanced lens eyes, the required reduction in focal length is achieved because the lens has a special inhomogeneous construction, with dense, high refractive index material in the center, and a gradient of decreasing density and refractive index toward the periphery. In 1877, Matthiessen discovered this gradient in fish lenses (see Pumphrey, 1961; Axelrod *et al.*, 1988). He was struck by the short focal length (about 2.5 radii, known as "Matthiessen's ratio"); if the lens were homogeneous, the refractive index would be 1.66, an unattainable value. In fact, the central refractive index is about 1.52, which falls to less than 1.4 at the periphery. The effect of the gradient is twofold. First, the focal length is reduced (and, concommitantly, the relative aperture increased) because light is continuously bent within the lens, not just at its surfaces. Second, with the correct gradient the lens can be made aplanatic, *i.e.* free from the spherical aberration, which makes homogeneous spherical lens virtually unusable (Pumphrey, 1961) (Figure 1*d* and *e*). The exact form of the gradient that permits this condition was not achieved theoretically until quite recently (Luneberg, 1944; Fletcher *et al.*, 1954), although Matthiessen had proposed a parabolic gradient that was very similar. It now seems that the gradient is not as smooth as was formerly supposed. Kröger *et al.* (1999) have shown that some fish lenses have multiple focal lengths, the function of which is to provide in-focus images at distances corresponding to the peak spectral sensitivities of the different cone types, thereby partially compensating for the lens' otherwise uncorrectable chromatic aberration.

By measuring the focal length, it is easy to tell whether a particular group of animals has "discovered" how to make the Matthiessen type of lens. If the focal length is around 2.5 radii, then the lens must have a gradient construction. A homogeneous lens with the same central refractive index would have a focal length of 4 radii. By this criterion, "Matthiessen" lenses have evolved at least eight times: in the fish, in the cephalopod molluscs (excluding *Nautilus),* at least four times in the gastropod molluscs (littorinids, strombids, heteropods, and some pulmonates), in the annelids (alciopid polychaetes), and once in the copepod crustaceans *(Labidocera).* Details are given in Land (1981; 1984a). The remarkable lens eyes of cubomedusan jellyfish (Piatigorsky *et al.*, 1989) are not included here, as their optical properties have not been examined. Interestingly, the above list does include all aquatic lens eyes of any size; none have homogeneous lenses. One can conclude that there is one right way of producing such lenses, and that natural selection always finds it. Matthiessen lenses are indeed of excellent optical quality, as they offer high resolution with high light-gathering power. Their only residual defect is chromatic aberration, mentioned above. Lens construction accounts for one aspect of the remarkable convergence between fish and cephalopod eyes. The identity of Matthiessen's ratio in the two groups, itself a result of the refractive index of the dry material of the lens center, and the inevitable spherical symmetry of the image effectively dictate the eyes' shape and proportions. The presence of eye muscles can be explained from the need to stabilize the image. This need grows with image quality, if that quality is not to be compromised by blur. Similarly, the need for an accommodation mechanism is determined by eye size, in the same way that focusing becomes more critical for camera lenses as the focal length increases. Thus, many of the convergent features that seem so remarkable (Packard, 1972) are inevitable, given a particular type and size of eye.

2.4. Multiple Lenses

Among aquatic eyes (Figure l*f* and *g*), there are alternatives to the single spherical lens, but they are certainly not common. Two of the most interesting are found in copepod crustaceans, in which they are derived from parts of the single median eye. In *Pontella,* the lens is a triplet (Figure 1*f*); two elements are actually outside the eye in the animal's rostrum, and a third element is close to the retina of only six (!) receptors (Land, 1984a). The eyes are sexually dimorphic - the females only have a doublet - and the animals themselves are conspicuously marked in blue and silver, which suggests a role for the eyes in the recognition of species and potential mates. Optically, the intriguing feature of the eyes is the first surface, which is parabolic. Ray tracing shows that this configuration can correct the spherical aberration of the other five interfaces in the optical system to provide a point image. This seems to be an interesting alternative solution, as an aspheric surface achieves the same result as the inhomogeneous optics of the Matthiessen lens. Another copepod, *Copilia,* has fascinated biologists for more than a century. Its eyes are constructed strangely (Figures l*g* and 3*b*), and they move to and fro in the longitudinal plane, thus scanning the water in front of the animal (Exner, 1891). Each eye has two lenses that are arranged like a telescope: A large, long focal length “objective” lens forms an image on or close to a second, short focal length “eyepiece” lens immediately in front of the cluster of five to seven photoreceptors (Gregory, 1991). The second lens and receptors move together as a unit during scanning. The function of this astonishing system is still not well understood.

2.5. Corneal Refraction

In our own eyes, two thirds of the optical power lies in the cornea (Figure l*h*). The lens, which is entirely responsible for image formation in our aquatic ancestors, is now mainly concerned with adjustments of focus. The use of a curved air/tissue interface for image formation is limited to terrestrial animals, and is actually a rather uncommon optical mechanism. Apart from the land vertebrates, the only other large group to use corneal refraction are the spiders (Land, 1985a), whose eyesight can be remarkably acute. Williams and McIntyre (1980) estimate that the interreceptor angle in the jumping spider *Portia* is only 2.5 arc min. Considering the size of the animal (l cm), this compares quite favorably with 0.5 arc min in the human fovea. The larvae of some insects also have simple eyes that form an image by using the cornea; the most impressive are the eyes of tiger beetle larvae *(Cicindela),* in which the interreceptor angle is about 1.8°. This is quite comparable in performance to the compound eyes of the adults that supplant them (Land, 1985b). The dorsal ocelli of adult insects are of the same general design, but are profoundly out of focus. They are concerned with stabilizing flight relative to the sky, and not with imaging.

For an eye of the corneal type to realize its maximum possible (diffraction limited) acuity, it must be corrected for spherical aberration. There are two ways this might be done: The cornea itself might be aspherical, as the surface that directs all parallel rays to a single point is not spherical, but elliptical; alternatively, an inhomogeneous lens might be used to produce the correction. According to Millodot and Sivak (1979), the cornea of the human eye is aspheric and thus corrected; the lens corrects itself by being inhomogeneous. The penalty of an aspheric correction is that the eye loses its radial symmetry, and thus has one "good" axis and reduced resolution elsewhere. Where all-round vision is needed, it may be better to go for the other solution. In the rat eye, which has a nearly spherical cornea, the lens is in fact overcorrected for spherical aberration, thus compensating for the cornea (Chaudhuri *et al.*, 1983). One further trick that seems to obtain a little more resolution from the eye is the inclusion of a negative lens, which is formed from the retinal surface, into the fovea, immediately in front of the receptors. This produces a system with telephoto properties and a locally enlarged image. Snyder and Miller (1978) first described this arrangement in an eagle, in which the eye's focal length effectively increased by 50%; a similar mechanism also occurs in some jumping spider eyes (Williams and McIntyre, 1980).

The transition from lens-based to cornea-based optics, which accompanies the evolution of terrestrial life, must have involved a weakening of the power of the lens as the cornea became effective, much as happens today during metamorphosis in frogs and toads. Greater problems arise when an animal needs to operate effectively in both media at the same time. There seem to be two solutions. One solution is to retain all the power in the lens and have a flat cornea without power in either medium. This is approximately the situation in penguins and seals (Sivak, 1988). An interesting variant of this occurs in porpoises *(Phocoena),* in which the cornea retains some power by having different inner and outer radii and an internal refractive index gradient. When the porpoise focuses in air, the cornea is flattened further. The alternative to a nearly flat cornea is to provide the lens with huge powers of accommodation. This occurs in some diving birds, in which the powerful ciliary muscle squeezes the lens into, and partly through, the rigid iris, thus deforming the front surface into a locally very high curvature. In diving mergansers, this mechanism can produce as much as 80 diopters of accommodation, compared with 3-6 diopters in nondiving ducks (Sivak *et al.*, 1985).

2.6. Concave Reflectors

Small eye-spots, in which the pigment cup is overlaid by a multilayer mirror, are found in some rotifers, platyhelminthes, and copepod crustaceans (see Ali, 1984). However, none of these eyes are large enough to form usable images. In scallops *(Pecten)* and their relatives, the situation is different (Figure 1*i*). They have up to 100 respectable-sized (l mm) eyes around the edge of the mantle, each of which contains a "lens," a two-layered retina, and a reflecting tapetum. If one looks into the eye through the pupil, a bright inverted image is visible. Its location indicates that it could only have been formed by the concave reflector, not by the weak, low refractive index lens (Land, 1965). The image visible to an observer is indeed the same one the animal sees. It falls onto the distal layer of the retina, where there are receptors that give "off"-responses. Thus, the animal sees moving objects - and shuts - as the image crosses successive receptors. These eyes represent an evolutionary line that is apparently quite unrelated to other molluscan eyes (Salvini-Plawen and Mayr, 1977, Figure 8). The only other large eye that uses a mirror as an imaging device - rather than just a light-path doubler, as in the tapetum of a cat's eye - is in the deep-water ostracod crustacean *Gigantocypris.* These large (1 cm) animals have a pair of parabolic reflectors that focus light onto blob-like retinae at their foci. The resolution is probably very poor, but the light-gathering power is enormous, with a calculated F-number of 0.25 (Land, 1984a).

3. Compound Eyes

In the last 30 years, we have seen a great revival of interest in compound eyes, with the discovery of three new optical types (reflecting and parabolic superposition and afocal apposition), the reinstatement of a fourth (refracting superposition), and rediscovery and naming of a fifth (neural superposition). In fact, the only type of compound eye to have avoided recent reappraisal is the classical apposition eye of diurnal insects and crustaceans, in which the erect image in the eye as a whole is built up from the elementary contributions of all the separate ommatidia (Figure 2*b*). Even that mechanism, proposed in the 1826 "mosaic theory" of Johannes Müller, came close to eclipse in the mid-nineteenth century and had to be revived by Sigmund Exner in his great monograph on compound eye optics (Exner, 1891). Exner, the undisputed father of the subject, made two major discoveries, which we discuss below: the lens cylinder and the principle of superposition imagery.

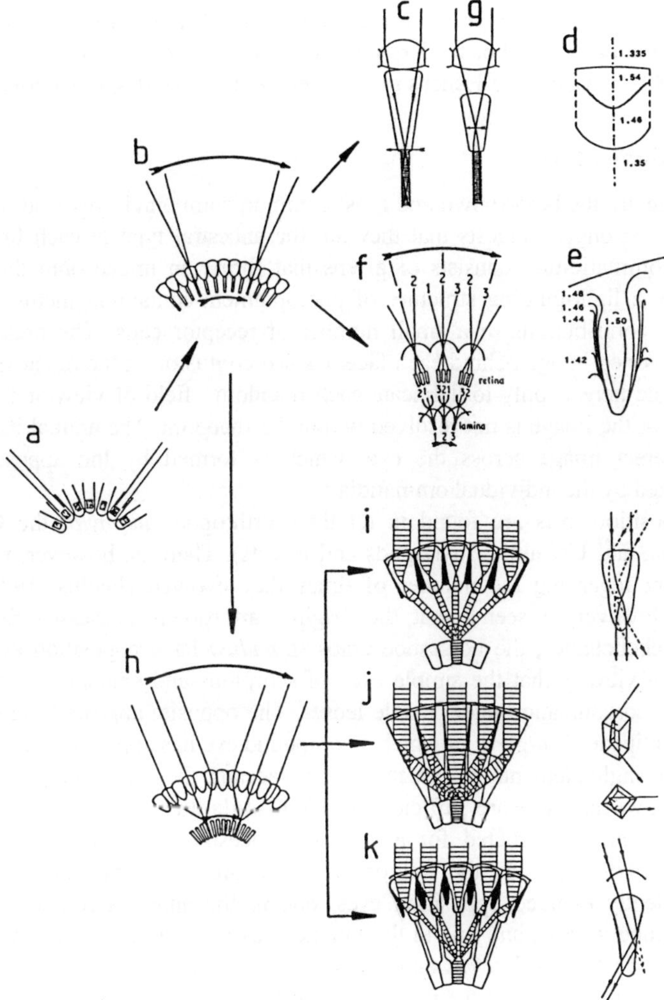

Figure 2. The evolution of compound eyes. Arrows indicate developments, rather than specific evolutionary pathways, which are more complex. For further details and references see text. Compiled from many sources. *(a)* Hypothetical ancestor with receptors in pigmented tubes. *(b)* Apposition eye. *(c)* Focal apposition ommatidium with image at rhabdom tip. *(d)* Multi-interface lens *(Notonecta)*. *(e)* Lens-cylinder *(Limulus)*; numbers in *d* and *e* are refractive indices. *(f)* Neural superposition in a dipteran fly; the numbers indicate the receptors and laminar structures that view the same directions in space. *(g)* Afocal apposition optics with intermediate image and collimated exit beam. *(h)* Superposition eye with deep-lying image. *(i)* Refracting superposition; inset shows axial and oblique ray paths. *(j)* Reflecting superposition; inset shows two views of ray paths through mirror box. *(k)* Parabolic superposition *(Macropipus)*; inset shows focused beam recollimated by a parabolic mirror.

As seems to be the fate of ideas about compound eye function, they also came close to abandonment in the 1960s (see Land, 1981; Nilsson, 1989), but survived the challenge undamaged. For readers interested in the history of the subject, Hardie's new (1989) translation of Exner's monograph, which includes a modern appendix, is a feast.

3.1. Apposition Eyes

These are the best-known and most common compound eyes, and their relative simplicity strongly suggests that they are the ancestral type in each lineage. Each unit, or ommatidium, consists of a lens that forms an image onto the tip of the rhabdom, a light-guiding structure of photopigment-containing membrane formed from the contributions of a small number of receptor cells. The presence of the small, inverted image behind each facet caused confusion in the nineteenth century, but its role here is only to delineate each rhabdom's field of view and increase its brightness; the image is not resolved within the rhabdom. The animal itself sees the overall erect image across the eye, which is formed by the apposed "pixels" contributed by the individual ommatidia.

Apposition eyes are found in all three arthropod *subphyla*; the Chelicerata, Crustacea, and Uniramia (myriapods and insects). There is, however, no universal agreement regarding the number of times they evolved (Paulus, 1979; Nilsson, 1989). However, it seems that the original arthropods possessed faceted eyes. Among chelicerates, the horseshoe crabs *(Limulus)* have apposition eyes, and the prevailing view is that the simple eyes of scorpions and spiders are derived from these by reorganization under single lenses. The opposite appears to have occurred in the centipede *Scutigera,* in which a compound eye has apparently reevolved from scattered single elements. In insects and crustaceans, the compound eyes take many forms. However, there are sufficient detailed similarities in the way that individual ommatidia are constructed for a common ancestry to be a distinct possibility (Paulus, 1979). Outside the arthropods, there are two remarkable examples of independently evolved apposition eyes, one in the annelids (on the tentacles of sabellid tube worms) and one in the mantle eyes of bivalve molluscs of the family *Arcacae* (see Salvini-Plawen and Mayr, 1977; Nilsson, 1994). In both cases, the eyes' function is to detect the movements of predators. In some of the tube worms, the eyes are little more than collections of pigmented tubes with receptors at the bottom. This was probably how compound eyes originated in the mainstream of the Arthropoda (Figure 2*a*).

The image in each ommatidium may be produced in three different ways. In terrestrial insects, the curved cornea nearly always forms the image (Figure 2*c*). This mechanism is not available underwater; the alternatives are the use of other lens surfaces *(Notonecta,* Schwind, 1980), or a lens with a variable refractive index *(Limulus,* Exner, 1891) (Figure 2*d* and *e*). Exner discovered the latter mechanism, which has affinities with the Matthiessen lens. He described it as a lens cylinder and showed that such a cylinder would form an image if the gradient of refractive index fell in an approximately parabolic fashion from the axis to the circumference. Eighty years later, interference microscopy made it possible to confirm Exner's farsighted conjecture in *Limulus* (Land, 1979) and in the superposition eyes of many species (Figure 2*i*, see below).

The most serious limitation to the resolving power of apposition eyes, and of compound eyes in general, is diffraction (Mallock, 1894; Snyder, 1979). Image quality depends on lens diameter; the smaller the lens, the more blurred the image. The half-width of the diffraction image of a point source is given by ë/*D* radians. Thus, for green light (ë = 0.5 μm) and a lens diameter *D* of 25 μm, the diffraction image is 1.1° wide. The minimum angle that separates ommatidial axes cannot usefully be much smaller than this, which severely limits the quality of compound eye vision. By comparison, humans resolve 100 times better, as they have a single lens and a daylight pupil 2.5 mm in diameter. An improvement in the resolution of a compound eye requires an increase in both the sizes and number of the facets, which quickly results in structures of absurd dimensions. This is beautifully illustrated in Kirschfeld (1976).

3.2. Neural Superposition Eyes

In the dipteran flies, there is a variant of the apposition eye in which the elements (rhabdomeres) that comprise the rhabdom are not fused, but separated from each other (Figure 2*f*). In these insects, each inverted image is really resolved by the seven receptive elements in the focal plane, which raises all the problems of how the many inverted images are put together to form the overall erect image. The solution to this was first proposed by Vigier in 1908, and rediscovered and proved by Kirschfeld (1967). The angle between the visual directions of the rhabdomeres within an ommatidium is the same as that between the ommatidial axes themselves, so that the six eccentric rhabdomeres in one ommatidium all have fields of view that coincide with the central rhabdomeres in adjacent ommatidia. Beneath the retina, the axons of all the retinula cells that view the same direction (eight, as the central rhabdomere is double) from seven adjacent ommatidia, collect up into the same "cartridge" in the lamina, after an impressively complicated piece of neural rewiring. Therefore, there is no difference between these eyes and ordinary apposition eyes at the level of the lamina. Dipterans thereby gain a sevenfold increase in the effective size of the photon signal and do not have to sacrifice resolution by increasing rhabdom size and, hence, acceptance angle. Thus, the flies have about an extra 15 minutes of useful vision at dawn and sunset.

3.3. Afocal Apposition

Butterflies have apposition eyes, but with an unusual construction (Figure 2*g*). The cornea forms an image, just as in the eye of a bee or grasshopper. Unlike those insects, however, the image is not at the rhabdom tip, but at the front focus of a second lens contained (as a lens cylinder) in the crystalline cone. This lens, which is of very short focal length, then recollimates the light, so that it emerges into the rhabdom as a parallel beam, not a focused spot (Nilsson *et al.*, 1988). This construction is basically the same as in a two-lens astronomical telescope, with an angular magnification of about x 6. It is considered "afocal" because there is no external focus, in contrast to the "focal" arrangement of an ordinary apposition eye (Figure 2*c*). As far as the resolution of the eye is concerned, the acceptance angle of each ommatidium is not determined by the angular subtense of the rhabdom tip, but by the critical angle for total internal reflection, which is set by the refractive index of the rhabdom itself. In practice, the situation is a little more complicated because the narrowness of the rhabdoms (ca. 2 μm) means that waveguide effects are important (Nilsson *et al.*, 1988). Overall, the performance of the afocal apposition eye is marginally better than its focal equivalent (van Hateren and Nilsson, 1987).

The afocal apposition eye is an important link between the apposition and superposition types, which we discuss next. It can be derived from an ordinary apposition eye by assuming that the second lens arises as a waveguide "funnel," which improves the transfer of light into the rhabdom. According to van Hateren and Nilsson (1987), such a structure can evolve into a lens without impediment. Once the second lens is present, and the system is afocal, it can further evolve into a superposition eye by an increase in the focal length of the second lens and a sinking of the retina to a more proximal position (Figure 2*h*).This type of transformation has apparently occurred several times in both the Lepidoptera and Coleoptera.

3.4. Refracting Superposition Eyes

In the eyes of many nocturnal insects and crustaceans, the rhabdom tips are not immediately behind the facet lenses, as they are in apposition eyes, but lie much deeper, with a zone of clear material that separates them from the optics (Figure 2*i*). Exner (1891) demonstrated that in the eye of the male european glowworm *Lampyris*, a real, erect image is formed at the level of the retina. This image is produced by the superposition of rays from many elements across the eye surface. Exner also showed that such imagery is not possible if the optical elements behave as simple lenses. However, a single image will be produced by the array if each element behaves as a two-lens telescope that inverts the light path, but (unlike afocal apposition) has little actual magnification. A problem with this mechanism seemed to be that in *Lampyris,* and in other eyes of this type, the optical elements do not have sufficient optical power in their curved surfaces to function as telescopes. Exner's solution was again to postulate the presence of lens cylinder optics (see Apposition Eyes, above). These lens cylinders differ from those of *Limulus* however, as they are twice the length, with a focus in the middle, not at the tip. Each half of the structure then behaves like one lens of a telescope, and overall the system becomes an afocal inverter, with a parallel output beam. Disbelief in both lens cylinders and superposition optics arose during the 1960s, and the modern reinstatement of Exner's ideas followed accurate refractive index measurements in the early 1970s (Kunze, 1979).

The feature crucial to the optical performance of all types of superposition eye is the accuracy with which the beams from each telescopic element coincide at the deep focus. In spite of an historic belief that there cannot be perfect coincidence, we now know that the superposition is so good in some diurnal moths that the eye operates at the diffraction limit for a single facet, which means that optically these eyes are as acute as equivalent apposition eyes (Land, 1984b). By having a large effective pupil and large receptors, superposition eyes gain a 100-fold, or even a 1000-fold, increase in sensitivity; hence, their popularity in dim-light situations. McIntyre and colleagues have published a particularly fine series of studies, which explores all these issues, on the design of scarab beetle eyes (McIntyre and Caveney, 1998).

In principle, a refracting superposition eye can be constructed using a single facet lens to give a single erect image, in the manner of a telescope used to project an image. Nilsson and Modlin (1994) found such an eye in a mysid shrimp, where a conventional refracting superposition eye has embedded in it a small high acuity region where the image is produced by a single giant crystalline cone.

3.5. Reflecting Superposition Eyes

Exner (1891) was actually wrong once. He thought that the eyes of long-bodied decapod crustaceans (shrimps, crayfish, lobsters) had superposition eyes of the refracting kind discussed above. However, attempts in the 1950s and 1960s to demonstrate lenses or lens cylinders in these eyes failed. Instead, these studies, which found square, homogeneous, low refractive-index, box-like structures, caused considerable confusion because no optical function could be ascribed to such elements. Thus, shrimps were blind, for about 20 years. This serious problem was resolved by Vogt (1980), who studied crayfish eyes. He discovered that the ray-bending was not done by lenses, but by mirrors in the walls of each "box." A comparison of Figures 2*i* and 2*j* shows that both telescopes and mirrors have the ability to invert the direction of a beam of light, so both can give rise to a superposition image. In many ways, the mirror solution seems more straightforward than the complicated telescope arrangement. This, however, is only true for the rather idealized case of Figure 2*j,* which illustrates rays in a section along a perfect row of mirrors in the center of the eye. Most rays away from the eye's plane of symmetry do not encounter a single mirror, but are reflected from two sides of the mirror "box" that makes up each optical element. There are, then, two important questions: What is the fate of these doubly reflected rays? Do all initially parallel rays reach a common focus? Here, the square arrangement of the facet array - almost unique to the decapod crustaceans - turns out to be crucial. Image formation is only possible if most rays encounter a "corner-reflector."

Consider first a simpler arrangement for producing a point image by reflection. This consists of a series of concentric "saucer rims," each angled to direct rays to a common focus; Figure 2*j* would then be any radial section through this array. The problem here is that such a stack has a single axis, and only rays nearly parallel to that axis form an image; other rays are reflected chaotically around the stack. The alternative is to replace the single reflecting strips with an array of mirror-pairs set at right angles. This substitution is possible because rays reflected from a corner go through two right angles and leave in a plane parallel to the incident rays (Figure 2*j*, inset). In other words, the rays behave almost as though they had encountered a single mirror at normal incidence, as in the saucer rim array. The beauty of the corner-reflector arrangement is that the orientation of each mirror pair is no longer important, unlike the situation in the single mirror array. Thus, the structure as a whole no longer has a single axis and can be used to make a wide-angle eye (Vogt, 1980; Land, 1881a). Clearly, this mirror-box design only works with right-angle corners and not hexagons, which accounts for the square facets. Various other features of these eyes are important for their function. The mirror boxes must be the right depth, about twice the width, so that most rays are reflected from two faces, but not more. Rays that pass straight through are intercepted by the unsilvered "tail" of the mirror boxes, whose refractive index decreases proximally to provide the appropriate critical angle for reflexion (Vogt, 1980). Finally, there is a weak lens in the cornea of the crayfish. This lens "pre-focuses" the light that enters the mirror box, thus giving a narrower beam at the retina (Bryceson, 1981). All these features provide an image comparable in quality to that produced by refracting superposition optics (Nilsson, 1989).

Reflecting superposition eyes, which are only found in the decapod crustaceans, presumably evolved within that group back in the Cambrian. The nearest relatives of the decapods, the euphausiids (krill), have refracting superposition eyes. The larval stages of decapod shrimps have apposition eyes with hexagonal facets, which change at metamorphosis into superposition eyes with square facets (Nilsson, 1989). Presumably, this transformation would have been no more difficult in evolution than in ontogeny. Interestingly, most of the true crabs (Brachyura), normally regarded as "advanced" decapods, have retained the apposition eyes into adult life. Undoubtedly, this reflects the crabs' littoral or semiterrestrial environment, in which light levels are high compared with the benthic or pelagic environment of shrimps and lobsters.

3.6. Parabolic Superposition Eyes

This final type of eye is the most recently discovered (Nilsson, 1988) and the most difficult to understand. From an evolutionary viewpoint, it is also the most interesting because it has some characteristics of apposition eyes, as well as both other types of superposition eye (Figure 2*k*). It was first discovered in a swimming crab *(Macropipus = Portunus).* Each optical element consists of a corneal lens, which on its own focuses light close to the proximal tip of the crystalline cone, as in an apposition eye. Rays parallel to the axis of the cone enter a light-guiding structure that links the cone to the deep-lying rhabdom. Oblique rays, however, encounter the side of the cone, which has a reflecting coating and a parabolic profile. The effect of this mirror surface is to recollimate the partially focused rays, so that they emerge as a parallel beam that crosses the eye's clear-zone, as in other superposition eyes. This relatively straightforward mechanism is complicated because rays in the orthogonal plane (perpendicular to the page) encounter rather different optics. For these rays, the cone behaves as a cylindrical lens, thus creating a focus on the surface of the parabolic mirror. It then recollimates the rays on their reverse passage through the cone. This mechanism has more in common with refracting superposition. Thus, this eye uses lenses and mirrors in both apposition and superposition configurations and it would be the ideal ancestor of most kinds of compound eye. Sadly, the evidence is against this, as all the eyes of this kind discovered to date are from the brachyuran crabs or the anomuran hermit crabs, neither of which is an ancestral group to other crustaceans (Nilsson, 1989). However, this eye does demonstrate the possibility of mixing mirrors and lenses, thus providing a viable link between the refracting and reflecting superposition types. This is important because such transitions do appear to have occurred. The shrimp *Gennadas,* for example, has a perfectly good refracting superposition eye, whereas its ancestors presumably had reflecting optics as in other shrimps (Nilsson, 1990).

Acknowledgments

This chapter is an abridged and updated version of a review that first appeared in the *Annual Review of Neuroscience* (Land and Fernald, 1992).

References

Ali, M.A., ed. (1984) *Photoreception and Vision in Invertebrates.* New York: Plenum.

Axelrod, D., D. Lerner and P.J. Sands (1988) "Refractive index within the lens of a goldfish determined from the paths of thin laser beams", *Vision Res.* 28: 57-65.

Barnes, R.D. (1987) *Invertebrate Zoology*, (5th ed.) Philadelphia: Saunders.

Bryceson, K. (1981) "Focusing of light by corneal lenses in a reflecting superposition eye", *J. Exp. Biol.* 90:347-50.

Burr, A.H. (1984) "Photomovement behavior in simple invertebrates", in: *Photoreception and Vision in Invertebrates*, M.A. Ali, ed., New York: Plenum, pp. 179-215.

Chaudhuri, A., P.E. Hallett and J.A. Parker (1983) "Aspheric curvatures, refractive indices and chromatic aberration for the rat eye", *Vision Res.* 23:1351-63.

Exner, S. (1891) *The Physiology of the Compound Eyes of Insects and Crustaceans*, Transl. R.C. Hardie, (1989) Berlin: Springer, (From German).

Gregory, R.L. (1991) "Origins of eyes - with speculations on scanning eyes", in: *Vision and Visual Dysfunction, Vol 2*, J.R. Cronly-Dillon and R.L. Gregory, eds, Basingstoke: Macmillan, pp. 52-59.

Goldsmith, T.H. (1990) "Optimization, constraint, and history in the evolution of eyes", *Quart. Rev. Biol.* 65: 281-322.

Hartline, P.H., A.C. Hurley and G.D. Lange (1979) "Eye stabilization by statocyst mediated oculomotor reflex in *Nautilus*", *J. Comp. Physiol.* 132:117-28.

Hughes, A. (1977) "The topography of vision in mammals", in: *Handbook of Sensory Physiology Vol. VII/5*, F. Crescitelli, ed., Berlin: Springer, pp. 613-756.

Kirschfeld, K. (1967) "Die Projektion der optischen Umwelt auf der Raster der Rhabdomere im Komplexauge von *Musca*", *Exp. Brain Res.* 3:248-70.

Kirschfeld, K. (1976) "The resolution of lens and compound eyes", in: *Neural Principles in Vision,* F. Zettler and R. Weiler, eds, Berlin: Springer, pp. 354-70.

Kröger, R.H.H., M.C.W. Campbell, R.D. Fernald and H.-J. Wagner (1999) "Multifocal lenses compensate for chromatic defocus in vertebrate eyes", *J. Comp. Physiol. A* 184:361-369

Kunze, P. (1979) "Apposition and superposition eyes", in: *Handbook of Sensory Physiology Vol. VII/6A*, H.-J. Autrum, ed., Berlin: Springer, pp. 441-502.

Land, M.F. (1965) "Image formation by a concave reflector in the eye of the scallop, *Pecten maximus*", *J. Physiol. (London)* 179:138-153.

Land, M.F. (1979) "The optical mechanism of the eye of *Limulus*", *Nature* 280:396-397.

Land, M.F. (1981) "Optics and vision in invertebrates", in: *Handbook of Sensory Physiology Vol. VII/6B*, H.-J. Autrum, ed., Berlin: Springer, pp. 471-592.

Land, M.F. (1984a) "Crustacea", in: *Photoreception and Vision in Invertebrates*, M.A. Ali, ed., New York: Plenum, pp. 401-38.

Land, M.F. (1984b) "The resolving power of diurnal superposition eyes measured with an ophthalmoscope", *J. Comp. Physiol. A* 154:515-33.

Land, M.F. (1985a) "The morphology and optics of spider eyes", in: *Neurobiology of Arachnids*, F.G. Barth, ed., Berlin: Springer, pp. 53-78.

Land, M.F. (1985b) "Optics of insect eyes", in: *Comprehensive Insect Physiology, Biochemistry and Pharmacology Vol. 6*, G.A. Kerkut and L.I. Gilbert, eds, Oxford: Pergamon, pp. 225-75.

Land, M.F. and R.D. Fernald (1992) "The evolution of eyes", *Ann. Rev. Neurosci.* 15:1-29.

Luneberg, R.K. (1944) *The Mathematical Theory of Optics*, PhD thesis Brown Univ., Providence, RI. Republished (1964) Berkeley: Univ. Calif. Press.

Mallock, A. (1894) "Insect sight and the defining power of composite eyes", *Proc. R. Soc. London Ser. B* 55:85-90.

McIntyre, P. and S. Caveney (1998) "Superposition optics and the time of flight of onitine dung beetles", *J. Comp. Physiol A* 183:45-60.

Messenger, J.B. (1981) "Comparative physiology of vision in molluscs", in: *Handbook of Sensory Physiology Vol. VII/6C*, H.-J. Autrum, ed, Berlin: Springer, pp. 93-200.

Millodot, M. and J. Sivak (1979) "Contribution of the cornea and lens to the spherical aberration of the eye", *Vision Res.* 19:685-687.

Muntz, W.R.A. and U. Raj (1984) "On the visual system of *Nautilus pompilius*", *J. Exp. Biol.* 109:253-63.

Nilsson, D.-E. (1988) "A new type of imaging optics in compound eyes", *Nature* 332:76-78.

Nilsson, D.-E. (1989) "Optics and evolution of the compound eye". in: *Facets of Vision,* ed., D.G. Stavenga and R.C. Hardie, eds, Berlin: Springer, pp. 30-73.

Nilsson, D.-E. (1990) "Three unexpected cases of refracting superposition eyes in crustaceans", *J. Comp. Physiol. A* 167:71-78.

Nilsson, D.-E. (1994) "Eyes as optical alarm systems in fan worms and ark clams", *Phil. Trans. R. Soc. Lond. B* 346:195-212.

Nilsson, D.-E., M.F. Land, J. Howard (1988) "Optics of the butterfly eye", *J. Comp. Physiol. A* 162:341-66.

Nilsson, D.-E., R.F. Modlin (1994) "A mysid shrimp carrying a pair of binoculars", *J. Exp. Biol.* 189:213-236.

Packard, A. (1972) "Cephalopods and fish: the limits of convergence", *Biol. Rev.* 47:241-307.

Paulus, H.F. (1979) "Eye structure and the monophyly of the Arthropoda", in: *Arthropod Phylogeny,* A.P. Gupta, ed., New York: Van Nostrand Reinhold, pp. 299383.

Piatigorsky, J., J. Horwitz, T. Kuwabara and C.E. Cutress (1989). "The cellular eye lens and crystallins of cubomedusan jellyfish", *J. Comp. Physiol. A* 164:577-87.

Pumphrey, R.J. (1961) "Concerning vision", in: *The Cell and the Organism*, J.A. Ramsay and V.B. Wigglesworth, eds, Cambridge: Cambridge Univ. Press, pp. 193-208.

Salvini-Plawen, L.V. and E. Mayr (1977) "On the evolution of photoreceptors and eyes", *Evol. Biol.* 10:207-63.

Schwind, R. (1980) "Geometrical optics of the *Notonecta* eye: adaptations to optical environment and way of life", *J. Comp. Physiol. A* 140:59-68.

Sherk, T.E. (1978) "Development of the compound eyes of dragonflies (Odonata). III. Adult compound eyes", *J. Exp. Zool.* 203:61-80.

Sivak, J.G. (1988) "Optics of amphibious eyes in vertebrates", in: *Sensory Biology of Aquatic Animals*, J. Atema, R.R. Fay, A.N. Popper and W.N. Tavolga, New York: Springer, pp. 466-485.

Snyder, A.W. (1979) "Physics of vision in compound eyes", in: *Handbook of*

Sensory Physiology Vol. VII/6A, H.-J. Autrum, ed., Berlin: Springer, pp. 225-313.
Snyder, A.W. and W.H. Miller (1978) "Telephoto lens system of falconiform eyes", *Nature* 275:127-129.
Van Hateren, J. H. and D.-E. Nilsson (1987) "Butterfly optics exceed the theoretical limits of conventional apposition eyes", *Biol. Cybern.* 57:159-68.
Vogt, K. (1980) "Die Spiegeloptik des Flusskrebsauges. The optical system of the crayfish eye", *J. Comp. Physiol.* 135:1-19.
Walls, G. L. (1942) *The Vertebrate Eye and its Adaptive Radiation*. Bloomington Hills: Cranbrook Inst. reprinted (1963), New York: Hafner.
Williams, D.S. and P. McIntyre (1980) "The principal eyes of a jumping spider have a telephoto component", *Nature* 288:578-80.

MOLECULAR LEVEL

RHODOPSIN-LIKE PROTEINS: THE UNIVERSAL AND PROBABLY UNIQUE PROTEINS FOR VISION

PAOLO GUALTIERI
Istituto di Biofisica, CNR, Via S. Lorenzo 26, 56127 Pisa, Italy

ABSTRACT

Rhodopsin-like proteins that range from bacteriorhodopsin, the light-transducing protein of the purple membrane of *Halobacterium halobium*, through sensory rhodopsins, the light-sensing proteins in the archeabacteria, towards rhodopsin, the protein responsible for the conversion of light into an optic nervous impulse, have significantly different biological roles. Nevertheless, natural selection has converged on very similar design for all these kind of proteins. As consequence, the combination of serendipity and natural selection has produced a family of proteins with characteristics near to the optimum for light detection and therefore for many linear and non-linear optical applications.

1. Introduction

According to Mayr's words (Mayr, 1982), living organisms are made up of macromolecules having extraordinary characteristics. Many of these macromolecules are so specific and unique for their ability to carry on a particular function, as rhodopsin does in the photoreceptive process, to be present in animal and plant kingdoms every time this specific function is demanded. Therefore, image-forming eyes present only in mollusks, arthropods and vertebrates, and no-imaging forming eyes (photoreceptor) found especially in unicellular species, plants and fungi, although anatomically very different, would employ similar visual transducers, namely rhodopsins, 7- transmembrane domain receptors with retinal as chromophore.

What is so special about this light absorbing group? First, retinal-opsin complex has an intense absorption band whose maximum can be shifted into the visible region of the spectrum, over the entire range from 340 nm to 640 nm, almost nanometer by nanometer (for example see the table 1 in Kusmic and Gualtieri, 2000, where it is shown that the distribution in freshwater fish photoreceptors covers an incredible gamut of wavelength, *i.e.* from UV to the red part of the visible spectrum).

Second, light isomerizes retinal very efficiently and rapidly, and this phenomenon occurs in less than one picosecond, with quite high efficiency (the quantum yield for converting a photon into a nerve impulse, for example, is about 67%). Moreover, its rate of isomerization in the dark is very low, about once in a thousand years. No other compound in nature comes close to matching the extremely high isomerization rate of retinal, due to a barrierless excited state potential surface (Barlow *et al.*, 1993), and the extremely high signal-to-noise ratio of retinal (Aho *et al.*, 1988).

Third, remarkable structural changes (movements of single α-helices) are produced by isomerization of retinal. Light is converted into atomic motion of sufficient magnitude to trigger a signal reliably and reproducibly (Spudich *et al.*, 1995).

Fourth, retinal is derived from β-carotene, a precursor with a very broad biological distribution (Stryer, 1988). In the following we will examine the most known photoreceptive proteins: Bacteriorhodopsin and Rhodopsin.

2. Structure and function

Bacteriorhodopsin (MW ≈ 26 kDa) is the light transducing protein of *H. halobium*. It is located in the purple membrane of this bacterium. The chromophore is all-*trans* retinal covalently bound via protonated Shiff base-linkage to the Lysine 216 of helix 7. The chromophore spans the intrahelical region and thus has potential interaction with almost all the seven helices. Purple membrane, which contains bacteriorhodopsin in a lipid matrix (protein:lipid, 1:10), is produced by the bacterium when the concentration of oxygen in the water becomes too low to sustain the generation of ATP via oxidative phosphorylation (Bogomolni and Spudich, 1991). Upon absorption of light, bacteriorhodopsin converts from a dark-adapted state to a light-adapted state. Subsequent absorption of a photon by the latter state generates a photocycle that pumps protons across the membrane, with a net transport from the inside (cytoplasm) to the outside (medium) of the membrane (Bogomolni and Spudich, 1991). The resulting pH gradient ($\Delta pH \approx 0.2$) generate a protonmotive force used by the bacterium to synthesize ATP. The primary photochemical event involves an all-*trans* to 13-*cis* photoisomerization. Primary counterion (Arg-82, Asp-212, Tyr-185, or possibly Asp 85) can change during photocycle (Kusnetzow *et al.*, 1999).

Rhodopsin (MW ≈ 40 kDa) is the protein responsible for generating an optic nervous impulse in the visual receptors of the three phyla possessing image-resolving eyes: mollusks, arthropods and vertebrates. The generation of a nerve impulse following rhodopsin excitation involves a complex series of reactions. The photo-excited visual pigment activates the GTP-binding protein trasducin, which in turn stimulates cGMP phospodiesterase. This enzymes hydrolyzes cGMP, allowing cGMP-gated cationic channel in the surface membrane to close, hyperpolarize the cell, and modulate transmitter release at the synaptic terminal (Baylor, 1996). In this case, the chromophore 11-*cis* retinal isomerizes to 11-*trans*. As in bacterio-opsin, this chromophore spans an intrahelice region and is linked to Lys 296. Candidates to be the primary counterion are: Asp83, Glu-113, Glu-122, Glu-134 (Birge, 1990).

3. The photocycle of bacteriorhodopsin

At ambient temperature and under low-light conditions, the membrane of *H. halobium* contains a mixture of two proteins, dark-adapted bacteriorhodopsin with 13-cis retinal as chromophore, and light-adapted bacteriorhodopsin, bR with all-trans retinal as chromophore. The latter protein undergoes a photocycle sequence (for detailed scheme see Birge, 1990).

The phototrasformation of bR to K is the primary event, which involves the all-*trans* 13-*cis* isomerization. K intermediate stores about 16 Kcal*mol^{-1}. The key thermal intermediate in the photocycle is M, because the formation of this intermediate coincides with the pumping of proton (Lipson and Horovitz, 1991).

4. The photobleaching sequence of rhodopsin

At ambient temperature and neutral pH, rhodopsin undergoes a photobleaching sequence (for detailed scheme see Birge, 1990). A key difference between bacteriorhodopsin and rhodopsin is that the final reaction in the photobleaching process of rhodopsin involves the expulsion of the isomerized chromophore from the binding protein site.

This denaturation precludes a rapid recycling of the protein, because an enzyme (retinyl-ester isomerase) is required to reisomerize the all-*trans* chromophore to 11-*cis*-retinal prior to regeneration of the protein. The total process takes about 20 minutes and the time differential is utilized *in vivo* for light adaptation (Knoles and Dartnall, 1977).

5. UV-Vis and two photons spectroscopy

One photon and two photons absorption spectra of bacteriorhodopsin and rhodopsin are shown in figure 1 (main spectroscopic data of bacteriorhodopsin is presented in table 1).

Table 1. Photophysical and photochemical properties of bR and purple membrane

Photochromism	All-*tran*/13-*cis* isomerization Reversible protonation and deprotonation of the Schiff base linkage
Molar absorption coefficients	ε bR = 66,0000 l mol^{-1} cm^{-1} ε K = 64,0000 l mol^{-1} cm^{-1} ε M = 43,0000 l mol^{-1} cm^{-1}
Quantum efficiencies	Φ (bR->M) = 64% Φ (M->bR) = 64%
Time scales	bR->K 3 ps bR->M 50 μsec M->bR 100 nsec M->bR 100 msec (thermal)
Refractive index of membrane Refractive index changes	1.47<n<1.57 0.0001<Δn<0.001

The primary photoproduct spectra (bathorhodopsin and K) were recorded out at 77K where the two products are stable. All the pigment spectra display two

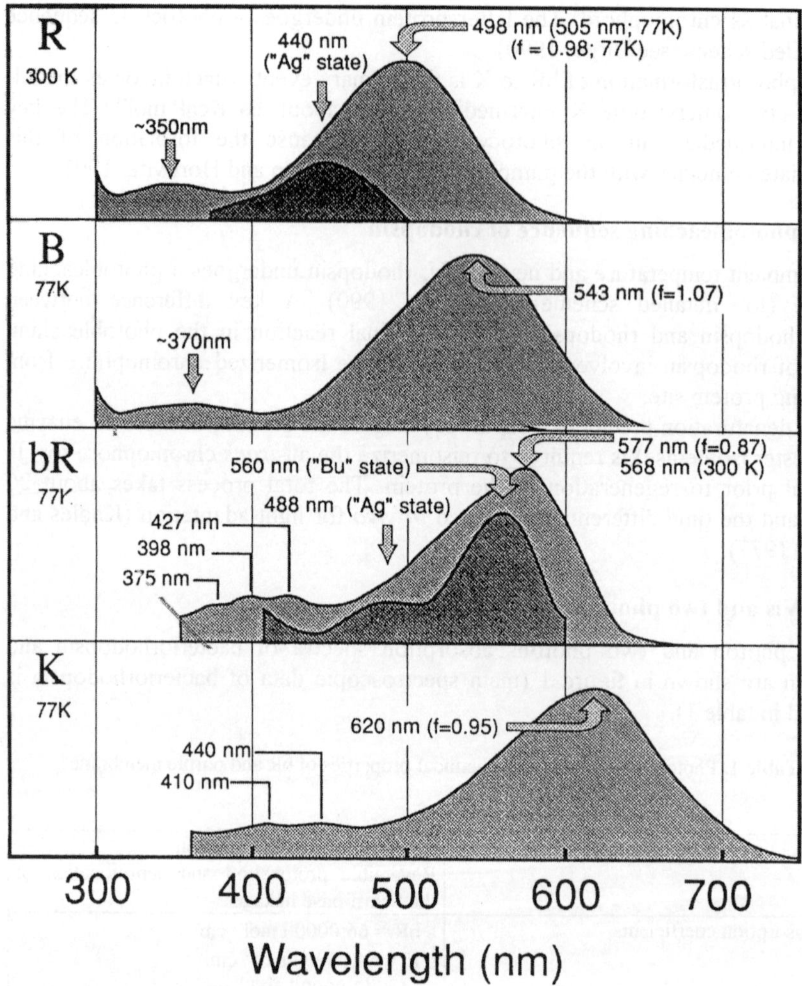

Figure 1. One-photon and two-photon absorption spectra of rhodopsin R, bathorhodopsin (B), light-adapted bacteriorhodopsin (bR), and K. The two-photon spectra are the darker insets. (Redrawn from Birge, 1990).

absorption bands, a strong unresolved λ_{max} band in the green-red region and a weaker band in the blue region.. There are three long-lying π,π^* electronic states, in both isolated and protein-bound chromophores, $^1B_u{}^{*+}$, $^1A_g{}^{*+}$, $^1A_g{}^{*-}$. The first state is a strongly allowed in all the polyenes; the last one is a forbidden state. Two-photon spectroscopy inverts the selection rules, hence, the $^1B_u{}^{*+}$ state is now forbidden and the $^1A_g{}^{*-}$ is strongly allowed (Birge, 1990).

6. Noise problem, a practical case

As we have seen, intense selective pressure has optimized many of the functional characteristics of photoreceptors. These include low thermal noise, efficient photon capture, rapid activation and inactivation, and variable gain control. Many of these characteristics are embodied within the rhodopsin-like molecules and make these proteins more efficient than other possible candidate proteins for light detection. They will be emphasized here.

A dark-adapted human can detect a flash of light that produces only 5-7 photoisomerizations, corresponding to an energy of 6×10^{-22} kcal. This very low dark-noise value is linked to the unusual and very high pK of the retinal chromophore inside the rhodopsin-like proteins (Steinberg *et al.*, 1993). By requiring an ensemble of approximately 500 rods to produce 5-7 isomerizations within a several hundred millisecond integration time, the retina works as a coincidence detector. This strategy is necessary because the threshold for signal detection is limited by noise derived from thermal activation of rhodopsin, most likely as a result of thermal isomerization of retinal (Birge, 1990). Not surprisingly, the rate of thermal isomerization of rhodopsin is very low, approximately 1 per 1000 years (10^{10} sec) at 37 °C (Baylor *et al.* ,1993). This corresponds to an energy barrier of 22 kcal/mol. Photons of 500 nm (the wavelength of maximal absorption of rhodopsin) carry 57 kcal/mol, resulting in a 40% efficient utilization of available energy in setting the signal-to-noise ratio. The low thermal noise allows for a large number of rhodopsins per cell. A human rod has approximately $4*10^7$ rhodopsins, and the rods of some amphibian have 10^7 rhodopsins.

Despite the large barrier to thermal activation, isomerization is extremely efficient: its quantum efficiency is 0.67 from 11-*cis* retinal to all-*trans* when retinal is part of rhodopsin. Approximately 20% of photons at a wavelength of 500 nm that strike the human retina lead to a transduction event, an efficiency comparable to that of the best photomultiplier tubes.

Let us try to analyze the dark noise effect in the photoreceptive system of *Euglena,* to determine the threshold and the contrast this cell can detect. We can calculate the Signal to Noise Ratio (SNR) of the *Euglena* photoreceptor in the case it uses rhodopsins or flavoproteins.

Assuming 10^7 is the number of rhodopsin-like molecules in the photoreceptor of *Euglena* (Gualtieri, 1993), a false signal from the photoreceptor arises every 1000 sec.

The rate of the ground state electron transfer process of a flavoprotein is about 10^2 sec^{-1} (Stryer, 1988). This rate has to be higher than the rate of oxidation of the reduced flavoproteins by means of molecular oxygen which is 10^{-3} sec^{-1} (Stryer, 1988). If we suppose that our specialized photoreceptive flavoprotein uses some special tricks, as altering distances between electron donor and acceptor, we can decrease its ground state transfer rate to the limit represented by oxidation rate, namely 10^{-2} sec^{-1}. In the darkness, assuming 10^7 is the number of these proteins in the photoreceptor of *Euglena*, a false signal will arise from the photoreceptor every 10 μsec. Therefore, during *E. gracilis* photoreceptor exposition (0.2 sec), and in the case of a flavoprotein photoreceptor, the thermal noise corresponds to 20,000 photons (Root Mean Squared, RMS equal to 450), whereas in the case of a rhodopsin photoreceptor the thermal noise effect is almost zero.

We can now calculate the absolute threshold of *Euglena* photoreception and how large a photon flux is required in order to avoid false alarm. Knowing the area of the photoreceptor, the effective wavelength, and the light intensity, the energy necessary (and the number of photons) to produce a response can be calculated. The lowest light intensity impinging on a glass slide and capable of inducing a phototactic response is about 10 mW/m^2 of blue light (Lenci and Colombetti, 1978). This energy corresponds to 10^4 photons/ (μm^2 sec). Assuming the area of the photoreceptor equal to 1 μm^2, an exposition time of 0.2 sec and a transmission factor of 0.5 (in this approximation we take into account cell rotation time, loss of 4% of incident energy due to reflection, and a scattering factor), 10^3 photons impinge on the photoreceptor during its exposition.

Considering a 1% absorption for a flavin photoreceptor, only 10 photons are absorbed. This is not yet the signal. The signal results from the difference between these 10 absorbed photons, and the fraction of photons absorbed when the stigma screens the photoreceptor. Considering a 50% screening effect (Gualtieri, 1991), the true signal is 5 photons. Flavoprotein photoreceptor should have a thermal noise 4000 times greater than the measured threshold for *Euglena* photoreception.

It is often stated or implied in discussions on detection systems that the threshold of detectability of a signal occurs when the signal is equal to the noise. This is a somewhat misleading statement, since in this case, 50% of the observation will be false alarm. To avoid false alarm, the "real" signal to be detected must exceed the level of noise by a factor depending on the statistical fluctuation of the noise (Rose, 1977). The signal to be detected in our example is the difference between the average number of photons impinging on the photoreceptor during light exposure and during stigma screening. In order to find a signal value for which we are reasonably confident of not mistaking a noise fluctuation for the real signal, in the case of a flavoprotein photoreceptor with 10^7 elements, the signal must have an amplitude 5 times larger than the RMS noise (Rose, 1977). This means that at least 22,250 photons are required for a signal to be discriminated as such, that the incremental threshold is about 2,225 photons, and that *Euglena* could "consciously" detect and therefore respond only to 50 W/m^2 of blue light.

Obviously, rhodopsin photoreceptor would perfectly function as a result of its negligible noise.

7. Technically attractive molecular function of rhodopsin-like proteins

We can analyze the different aspects of the molecular function of bR that may find technical applications (table 1). First, bR shows a photoelectric response that occurs on a time scale of less than 5 picosecond. The molecular origin is the initial charge displacement in the molecule after the absorption of a photon. This charge separation is caused by the movement of the positively charged Schiff base away from its negatively charge counterion(s) located in the opsin backbone. The photovoltage and the concomitant photocurrent could be utilized in photoreceptor devices and opto-electronic detectors (Zeisel and Hampp, 1995).

As we have seen, bR is a photochromic protein. Light absorption leads to reversible changes of its absorption spectrum. The largest spectral shift is observed between bR and M state; the color changes takes place on time scale of several milliseconds. This property and the linked refractive index changes, could be use for optical information storage and processing (Birge 1995).

Moreover, bR is a light-driven proton pump. Each single photocycle results in the traslocation of a proton over a distance of about 5 nm. Due to the high quantum efficiency (Φ about 64%) only 1 or 2 photons are needed for proton translocation. The light-driven proton pump can be used for the direct conversion of sun light into chemical energy.

For all these application the property of bR in the purple membrane are essential. Only in this form bR does show its outstanding stability and reversibility, *i.e.* negligible photochemical side reactions, stability under air and light exposition, and capability of 10^6 switchings between the two photochromic states (Zeisel and Hampp, 1995).

References

Aho, A.C., K. Donner, C. Hyden, L.O. Holsen and T. Reuter (1988) "Low retinal noise in animals with low body temperature allows high visual sensitivity", *Nature* 334:348-350.

Barlow, D. (1996) "How photon starts vision", *Proc. Natl. Acad. Sci.* 93:560-565.

Barlow, R.B., R.R. Birge, E. Kaplan, and J.R. Tallen (1993) "On the molecular origin of photoreceptor noise", *Nature* 366:64-66.

Birge, R.R. (1990) "Nature of the primary photochemical events in rhodopsin and bacteriorhodopsin", *Biochem. Biophys. Acta* 1016:293-327.

Birge, R.R. (1995) "Protein-Based Computers", *Sci. Amer.* March:66-71.

Bogomolni, R.A. and J.L. Spudich (1991) "Archaebacterial rhodopsin: Sensory and energy transducing membrane proteins", in: *Sensory Reception and Signal Transduction*, J.L. Spudich and B.H. Satir, eds, New York: Wiley-Liss, pp. 1-65.

Gualtieri, P. (1991) "Microspectroscopy of photoreceptor pigments in flagellated *algae*", *Critic. Rev. Plant Sci.* 9:475-495.

Gualtieri, P. (1993) "*Euglena gracilis*: Is the photoreception enigma solved?", *J. Photochem. Photobiol.* 19:3-14.

Knoles, A. and H.J.A. Dartnall (1977) "Property of visual pigment related to their structure", in: *The Eye, Vol. 2B, Photobiology of Vision,* H. Davson, ed., London: Academic Press, pp.175-246.

Kusmic, C. and P. Gualtieri (2000) "Morphology and spectral sensitivities of retinal and extraretinal photoreceptors in freshwater teleosts", *Micron* 31:183-200.

Kusnetzow, A., D.L. Singh, C.H. Martin, I.J. Barani and R.R. Birge (1999) "Nature of the chromophore binding site of bacteriorhodopsin: The potential role of Arg-82 as principal counterion", *Biophys. J.* 76:2370-2389.

Lenci, F. and G. Colombetti (1978) "Photobehavior of microorganisms: A biophysical approach", *Ann. Rev. Biophys. Bioeng.* 7:341-361.

Lipson, E.D. and B.A. Horowitz (1991) "Photosensory reception and trasduction", in: *Sensory Reception and Signal Transduction*, J.L. Spudich and B.H. Satir eds, New York: Wiley-Liss, pp.1-65.

Mayr, E. (1982) *The Growth of Biological Thought. Diversity, Evolution and Inheritance*, Cambridge (MA), London: The Belknap Press of Harvard University Press.

Rose, A. (1977) *Vision, Human and Electronic*, London: Plenum Press.

Spudich, J.L., D.N. Zacks and R.A. Bogomolni (1995) "Microbial sensory rhodopsins: Photochemistry and function", *Israel J. Chem.* 35:495-513.

Steinberg, G.M., M. Ottolenghi and M. Sheves (1993) "The pK of the protonated Schiff base of bovine rhodopsin", *Biophys. J.* 64:1499-1502.

Stryer, L. (1988) *Biochimistry*, New York: Freeman.

Zeisel, D. and N. Hampp. (1995) "Realization of molecular otical devices using bacteriorhodopsin", *FED J.* 6:33-45.

PHOTORECEPTION BEFORE MEN

PAOLO GUALTIERI
Istituto di Biofisica, CNR, Via S. Lorenzo 26, 56127 Pisa, Italy

ABSTRACT

This chapter deals with photoreception strategies of some of our oldest ancestors belonging to the three super kingdoms of Eubacteria, Archaebacteria, and Eukaryotes. For each organism, the main characteristics of its photobehaviour and some of the experiments that have strengthened the presence of rhodopsin-like proteins have been outlined.

1. Introduction

Cells and organisms obtain nearly all their information about the environment around them from receptors. Detection of the outside became an essential strategy for life and evolution to be and to continue to be possible: if you don't want to be cut off from your surroundings and deprived of defense against them, you need structures that can inform you on what is going on out there. Cell membrane being the frontier of the organism, photic, chemical and mechanical stimuli reach it first. Membrane proteins are cell surface receptors that enable a cell to communicate with the external environment. Most of these sensory receptors have sensitivities at the limit of their respective stimulus energy: photoreceptors are activated by few photons, hair cells by stretching of the size of one hydrogen atom, odor receptor cells by one or few odorant molecules. Sensory cells can discriminate and extract different types of information carried by a stimulus and encode them in an electrical or chemical response. As we have seen in the other chapter, signal-to-noise ratio (SNR) always represents the limiting factor in each sensory process.

Light is a major environmental signal controlling most life processes; hence, light perception is absolutely necessary for organisms whose life rely on light. Because light is composed of discrete photons, organisms evolved structures for catching and counting photons. In the three super kingdoms of Archaebacteria, Eubacteria and Eukaryotes (Muller, 1997) structures have evolved for capturing photons. We will describe their different or constant features.

2. Overview

2.1. Eubacteria

Leptolyngbya is a thin (< 3μm) filamentous cyanobacteria that show a very marked photobehavior (Anagnostidis and Komarek, 1988). Microscope observation reveals that the entire tricomes of this filamentous cyanobacterium move as a screw. If we allowed these cells to move in a semi-solid medium in a Petri dish with a source of light located in a precise spot, the trichomes grow from the mother colony fragments toward the light source. Microscope observation of the dish reveals that the trichomes move in parallel rows. By moving the dish to achieve a change in the light direction at a right angle to the original direction, a subsequent deviation of the growth direction is elicited, which reflects the light direction change. Therefore, these photosynthetic bacteria use light to grow toward optimal light intensities, with a sort of oriented movement with respect to the stimulus direction. These red *Leptolyngbia* possess an orange spot at the tip of the apical cell of every trichome, which resembles the eyespot or stigma of carotenoid-rich lipid globules almost invariably present in phototactic flagellated algae (Foster and Smyth, 1980). Electron microscopy revealed that this orange tip is characterized by osmiophilic globules of about 50 nm in diameter arranged in a peripheral cap extending 2-3 μm from the apex and with a possible layered pattern. These globules are smaller than those present in eukaryotic algae whose size is about 0.10 - 0.30 μm (Melkonian and Robenek, 1984).

Micro-spectrophotometric analysis of the tip of the apical cell of *Leptolyngbya* trichomes revealed a complex absorption spectrum. Two principal bands, centered at 456 nm and 504 nm respectively, were identified. Rhodopsin-like proteins (λ_{max} at 504 nm) could be located inside the plasma membrane of the tip of the apical cell, whereas carotenoids inside the globules (λ_{max} at 456 nm) would have the screening role. In the presence of hydroxylamine (NH_2OH), photo-orientation capability of cyanobacteria was completely impaired. The trichomes progressively lost their guidance mechanism and their growth pattern changed from parallel to more and more disordered. This photo-orientation impairment was reflected in the spectroscopic characteristics of the tip. In this case the spectrum did not show the band with λ_{max} at 504 nm, whereas the band with λ_{max} at 456 nm was present without significant difference with respect to the spectrum of untreated cells.

Our findings show that, from an evolutionary point of view, *Leptolyngbya* may possess the first complex photoreceptive system so far described in Bacteria, Archaea, and Eucarya (Albertano *et al.*, 2000).

2.2. Archaebacteria

Halobacterium halobium is an archaebacterium 2-3 μm in length with a diameter of about 0.5 μm, adapted to media of extreme salinity (4M NaCl), which shows both phototactic and chemotactic behavior. The motility of *H. halobium* depends on two flagellar bundles emerging from the cell poles. The motility behavior consists of a number of almost straight runs, separated by pauses and/or reversal. This motility behavior is influenced by light, more precisely by changes in illumination intensity, since cells rapidly adapt to light of constant intensity. Typically, flashes or positive steps of blue-green light make reversal in the swimming direction more frequent, whereas the opposite occurs with red-orange light. Using terms relevant to chemotaxis studies, we may refer to red-orange light as an attractant and to blue-green light as a repellent. The reversal of bacterium movement is due to a change in the direction of rotation of flagellar bundles from clockwise to counterclockwise (or vice versa).

Light stimuli are sensed through retinal-containing proteins, referred to as sensory rhodopsins, located in the plasma membrane. Six sensory rhodopsins are known in halopilic Archaebacteria. The best characterized is sensory rhodopsin I (SR-I) a color sensitive receptor that relays attractant and repellent photosignals to a tightly bound transducer protein HtrI (halobacterial transducer for sensory rhodopsin I) homologous to methyl accepting eubacterial chemotaxic transducer (for example Tsr, the serine chemotaxis receptor/transducer in *Escherichia coli*). A second protein, SR-II, or phoborhodopsin, in *H. salinarum*, mediates repellent phototaxis in a different spectral range with respect to SR-I. A third protein, the pSR-II, has been shown in *Natronobacterium pharaonis* to be a repellent receptor with photochemistry similar to that of SR-II. The other three proteins are SG1-SR-I, (*Halobacterium* sp., strain SG1), Slow SR-I (halobacterial strains), and vSR-II (*Haloarcula vallismortis*) are so named because of the similarity of their spectra and photocycles to those of SR-I and SR-II, respectively. For further details see the review of Spudich *et al.* (1995), and the references therein.

2.3. Eukaria

The unicellular flagellate *Euglena* dwells in natural shallow ponds, and uses sunlight as a source of energy and information. Its chloroplasts are the energy-supplying devices, whereas a simple but sophisticated system is used as light detector. This system consists of a locomotory flagellum, a stigma and a photoreceptor. As the cell rotates while swimming, the stigma comes between the light source and the photoreceptor, thus modulating the light that reaches it, and regulating the steering of the locomotory flagellum. The photoreceptor, a three-dimensional proteic crystal, is composed of a stack of crystalline sheets with a regular organization of component proteins (Walne *et al.,* 1998). In 1989 Gualtieri (Gualtieri *et al.*, 1989) and subsequently in 1992 Crescitelli (James *et al.*,1992) measured the absorption spectrum of a single *Euglena* photoreceptor suggesting a pigment based on a rhodopsin-like protein. Successive experiments on inhibition of both the formation of the photoreceptor and cellular photo-orientation by means of hydroxylamine (Barsanti *et al.*, 1993) and nicotine (Barsanti *et al.*, 1992) showed that retinal is the chromophore present in the photoreceptor. The extraction of retinal from intact cells and from photoreceptors isolated from demembranated cells has further strengthened the hypothesis of a rhodopsin-like protein (Gualtieri *et al.*, 1992). The estimated number of proteins, on the basis of the number of retinal molecules and of photoreceptor integrated optical density, is about 10^7 molecules. In 1997 Barsanti reported the presence of a photochromic pigment in the photoreceptor of *E. gracilis,* which undergoes repeated and reversible fluorescence changes with a determinate kinetics. Therefore, the photoreceptor possesses optical bistability, i.e., the non-fluorescent parent form (first conformer) of its molecules upon photo-excitation generates a fluorescent stable intermediate (second conformer) that can be photochemically driven back to the parent form (Barsanti *et al.*, 1997). The quantum yields of the forward and reverse reactions are almost the same and close to unit. At present, the protein of the photoreceptor, which shows an absorption spectrum with a λ_{max} at 510 nm, and a molecular weight of 27 Kda, is under purification (Gualtieri *et al*, in preparation).

Plants are autotrophic and dynamic organisms that grow and extend into the surrounding environment. Hence, survival strategies have evolved to ensure adaptation through effective and rapid responses to changes of this environment. Developmental plasticity is one major strategy in which environmental signals are used by plants as stimuli to respond and exploit changing circumstances. Certain critical features of the plant lifestyle are relevant to understand molecular signalling in plants. Probably the primary one is the lack of obvious movements. Coordination between tissues and cells is much less obvious in plants than in animals, and plant also lack the pronounced differentiation between sensory and responding cells and tissues found in mammals. Some plant tissues, such as stomata and pulvini, do show limited movement, and signaling between the major part of plants can also be detected, but parallels with nerve-muscle coordination or long range coordination by hormones are not easily made. The signals - between root and shoot, for example - are usually a complex mixture reflecting the major function of these tissues and the different environments in which they grow. Plant growth and development may be considered the equivalent of movement. Plants "grow" rather than "move" into their local environment, and changes in form due to growth, can be considered for a plant the equivalent of "behaviour". Growing plants are permanently embriogenic, and many plant cells in the mature plant body remain unspecialized, accounting for their remarkable regenerative capabilities. The developmental plasticity of higher plants necessitates an accurate sensing and integration of the incoming direction of enviromental signals such as light, gravity, wind, minerals, water, gases and soil structure. Each growing part has a semi-autonomous capability both to perceive and to act on the information that impinge upon it. Indeed individual plant cells (e.g. guard cells) can both directly sense and respond to environmental stimuli. This external information must be transduced and integrated with internal signals to generate optimal growth and developmental patterns. As well as other organisms plants obtain nearly all their information about the environment from receptors. Cell membrane being the frontier of the organism, and photic, chemical and mechanical stimuli reaching it first, it represents the best and most effective location for receptors, which are membrane proteins that enable a cell to communicate with its external environment. Different types of information carried by a stimulus can be discriminated by sensory cells, which then encode them in an electrical or chemical response.

Since light is a major environmental signal controlling plant growth and development, light detection is absolutely necessary for plant life. Plants are caplable of sensing light accurately throughout the whole spectrum of sunlight (290 nm - 800 nm). As far as we know, three different sensor pigment categories are present in higher plants: the red-far red light absorbing phytochrome, the blue light absorbing pigment(s) (cryptochrome), and the UV absorbing pigment(s) (Lercari 1991). Each pigment has a specified action and a specified spectral range of absorption, but co-actions between them seem highly probable (Fernbach and Mohr 1990; Mohr and Drumm-Herrel 1991). There has been considerable progress recently in understanding plant responses to blue light, mediated rather by the cryptochrome than by the blue light-absorbing bands of phytochrome (Briggs and Short 1991). Although blue light regulates many aspect of growth and development in higher plants, a blue light photoreceptor has not yet been characterized. Candidate photoreceptive molecules, such as flavins (with pterin cofactor) are wide spread believed to act as photoreceptive proteins. Although the proposition that flavins could function as a near-UV/visible-light detector dates back to fifty years ago, at present a reliable biochemical identification of flavin photoreceptors lacks, and the identification of photoreceptor protein content is principally based on action spectroscopy. However, Cashmore and his group, have recently reported mutant plants of *Arabidopsis* lacking CRY1 and CRY2 genes, which show no evidence of first-positive curvature, but still retain second-positive curvature. In their opinion, the CRY1 and CRY2 genes might regulate the expression of a protein kinase with a putative redox-sensing domain (Ahmad and Cashomore, 1993; Ahmad *et al.*, 1998). They strongly believe that these genes are the long-sought blue light receptor and they named it cryptochrome (Cashomore et al., 1999). More recently Briggs *et al.* claimed that these two genes cannot serve as primary photoreceptor for phototropism, though they may modulate the magnitude of the phototropic response by acting downstream in the phototropism signal-transduction patway. According to this author the photoreceptor for phototropism is the Phototropin, the protein purified by his research group (Briggs *et al.,* 1999).

Researches carried out in the last 10 years have shown that many components of an animal-type phosphoinositide ($InsP_3$) signalling system are present in higher plants (Drobak 1991). It has been suggested that this signalling system may function in several important transduction events in plant cells, and evidence is accumulating that supports this view (Blatt *et al.* 1990; Trewavas and Gilroy 1991; Berridge 1993). The formation of $InsP_3$ is the focal point of a pathway initiated by a family of G protein-coupled receptors (Fein *et al.*, 1984). The G protein-coupled receptors identified so far are all characterized by seven membrane spanning domains connected by extracellular and intracellular loops, (Dohlman *et al.,* 1991; Berridge 1993). These G proteins have been identified in higher plants, and associated to photoreception in the monocotyledonous plant *Lemna paucicostata*

(Hasunuma *et al.* 1987) and to blue light photoreception in the dicotyledonous *Pisum sativum* (Warpeha *et al.* 1992). It is a consolidate belief that reception roles (chemoreception, mechanoreception, photoreception) are assigned to seven membrane spanning domain proteins, (Shepherd 1991). The recent discover that G proteins, always coupled to a 7 transmembrane domain receptor protein, are present in plants coupled with the inositol trisphosphate as second messenger, together with the newly collected data on the presence of all-trans retinal in tomato (*Lycopersicum esculentum*) leaves, suggest that plant photoreceptors should use rhodopsins for blue light photoreception (Lorenzi *et al.*, 1994). This finding is fundamental since what has been identified is the ligand of a G protein-coupled receptor and allows us to suggest rhodopsin-like proteins as the blue light photoreceptor in higher plant. Moreover, the finding of an all-trans retinal rhodopsin in Chlorophyceae (Deininger *et al.*, 1996) is very important since Chlorophyceae are located at the first divergence in the branch of plant kingdom (Margulis 1993); therefore, it can be postulated that the rhodopsin gene could have been passed on to higher plants by discendent evolution.

We can conclude that a rhodopsin-like protein could be responsible of blue light sensing and the two previously reported cases of flavin-pterin identification could be cases of photoregulation and not of photoreception.

As we have seen, photoreceptors are present in all the three kingdoms, with enormous structural variations, although the nature of the photoreceptor pigment is highly conserved, probably because of the stringent conditions it has to satisfy (see the other chapter by the same author). Martin and coworkers (1986) identified opsin-like genes in a wide variety of vertebrate, invertebrate and unicellular species, and speculated that a primitive photopigment gene first evolved in unicellular organisms (i.e. Archaebacteria and Algae) and was passed on, with some degree of sequence conservation, to the wide variety of present-day species. Rhodopsin could have originated as an early experiment in photosynthesis, a role presently retained only in Archaebacteria. The acquisition of a sensory role through association with a G protein could also have occurred early, since G proteins, found in yeast, are homologous with the universally distributed elongation factor of protein synthesis. If rhodopsin originated as early as this interpretation suggests, we could reasonably expect it or its homologues to be found in any group of living organisms that manifest photobehavior. Unfortunately, this can be considered a certainty only in Archaebacteria and Animalia, since data available on the other phyla is few and contradictory. But, as we have just reported, indications on the presence of rhodopsin-like proteins in the three kingdoms are emerging, hence, rhodopsin-like proteins sense blue light from Archaebacteria (Oesterhelt and Stoekenius 1971) through Algae (Foster *et al.* 1984; Gualtieri 1993), and Invertebrates (Wald 1968) toward Vertebrates (Wald 1964). Moreover, in Fungi such as *Phycomyces,* Lipson (Chen *et al.*, 1993) indicated a rhodopsin-like protein as sensory pigment mediating the cryptic subliminal-light effect, and in yeast, such as *Neurospora Crassa*, Spudich identified a rhodopsin-like protein with a photocycle analogous to the sensory rhodopsin of *Chlamydomonas* (Spudich, personal communication).

All-*trans* retinal rhodopsins are thought to be the ancient type of rhodopsins (Hegemann *et al.* 1991; Kreimer *et al.* 1991), whereas in higher animals a new type of rhodopsin containing 11-*cis* retinal is used for visual process. All-*trans* retinal serves as the chromophoric group, and apparently a 13 *trans*- to *cis*-isomerization is responsible for rhodopsins activation. Only in vertebrate rhodopsin, light excitation leads to dissociation (photolysis) of the chromophore from the opsin apoprotein. On the other hand, when all-trans retinal rhodopsin becomes excited, isomerization of all-*trans* retinal triggers a photobiological reaction which does not end with chromophore release, since the ligand remains bound to the opsin. Therefore, when these rhodopsins are excited by light they are converted to thermostable metarhodopsins, which can be photoreverted to the originary form. Thus, this kind of rhodopsin-like proteins operate as a photochromic switch, analog to other plant photomorphogenic sensor phytochrome. The presence of these two photochromic switches in higher plants is likely to be central to sensory transduction of solar radiation in different environments.

References

Ahamad, M., J.A. Jarillo, O. Smirnova and A.R. Cashmore (1998) "Cryptochrome blue-light photoreceptors of *Arabidopsis* implicated in phototropism", *Nature* 392:720-723.

Ahmad, M. and A.R Cashmore (1993) "*HY4* gene of *A. thaliana* encodes a protein with characteristics of a blue-light photoreceptor", *Nature* 366:162-166.

Albertano, P., L. Barsanti, V. Passarelli and P. Gualtieri (2000) "A complex photoreceptive structure in the cyanobacterium *Leptolyngbya sp.*", *Micron* 31:27-34.

Anagnostidis, K. and J. Komarek (1988) "Modern approach to the classification system of cyanophytes. 3 – Oscillatoriales", *Arch. Hydrobiol. Suppl.* 80:327-472.

Barsanti, L., V. Passarelli, P. Lenzi, P.L. Walne, I.R. Dunlap and P. Gualtieri (1993) "Effects of hydroxylamine, digitonin and Triton X-100 on photoreceptor (paraflagellar swelling) and photoreception of *Euglena gracilis*", *Vision Res.* 33:2043-2050.

Barsanti, L., V. Passarelli, P.L. Walne and P. Gualtieri (1997) "*In vivo* photocycle of the *Euglena gracilis* photoreceptor", *Biophys. J.* 72:545-553.

Barsanti, L., V. Passarelli, P. Lenzi and P. Gualtieri (1992) "Elimination of photoreceptor (paraflagellar swelling) and the photoreception in *Euglena gracilis* by means of the carotenoid biosynthesis inhibitor nicotine", *J.*

Photochem. Photobiol. 13(2):135-144.

Berridge, M.J. (1993) "Inositoltriphospate and calcium signalling", *Nature* 361:315-325.

Blatt, M.R., G. Thiel and D.R. Trentham (1990) "Reversible activation of K^+ channels of *Vicia* stomatal guard cell following the photolysis of caged inositol (1,4,5) triphosphate", *Nature* 346:766-769.

Briggs, W.R. and T.W. Short (1991) "The trasduction of light signal in Plants: Response to blue light", in: *Phytochrome Proprieties and Biological Action*, B. Thomas and C.B. Johnson, eds, Berlin: Springer Verlag, pp. 289-301.

Briggs, W.R., J.M. Christie, M. Salomon and M. Olney (1999) "Phototropin (nph1) is a new and different kind of plant photoreceptor", *European Symposium on Photomorphogenesis*, Berlin.

Cashomore, A.R., J.A. Jarillo, Y. Wu and D. Liu (1999) "Cryptochromes: Blue light receptors for plants and animals", *Science* 284:760-765.

Chen, Xi-Yin, Y.-Q. Xiong and E. Lipson (1993) "Action spectrum for subliminal ligt control of adaptation in *Phycomyces* phototropism", *Photochem. Photobiol.* 58(3):425-431.

Deininger, W., P. Kroger, U. Hegemann, F. Lottspeich and P. Hegemann (1996) "Chamyrhodopsin represents a new type of sensory photoreceptor", *EMBO J.* 14:5849-5858.

Dohlman, H.G., J. Thorner, M.C. Caron and R.J. Lefkowitz (1991) "Model system for the study of seven-transmembrane segment receptors", *Annu. Rev. Biochem.* 60:653-688.

Drobak, B.K. (1991) "Plant signal perception and transduction: The role of the phosphoinositide system", *Essay in Biochem.* 26:27-37.

Fein, A., R. Payne, D.W. Corson, M.J. Berridge and R.F. Irvine (1984) "Photoreceptor excitation and adaptation by inositol 1,4,5-triphospate", *Nature* 311:157-160.

Fernbah, E. and H. Mohr (1990) "Coaction of blue/ultraviolet: A light and light absorbed by phytochrome in controlling growth of pine (*Pinus syltestris L.*) seedlings", *Planta* 180:212-216.

Foster, K.W. and R.D. Smyth (1980) "Light antenna in phototactic algae", *Microbiol. Review* 44:572-630.

Foster, K.W., J. Saranak, N. Patel, G. Zarilli, M. Okabe, T. Kline and K. Nakanishi (1984) "A rhodopsin is the functional photoreceptor for fototaxis in the unicellular eukaryote *Chlamydomonas*", *Nature* 311:576-579.

Gualtieri, P. (1993) "*Euglena gracilis*: Is the photoreception enigma solved? ", *J. Photochem. Photobiol.* 19:3-14.

Gualtieri, P., L. Barsanti and V. Passarelli (1989) "Absorption spectrum of a single isolated paraflagellar swelling of *Euglena gracilis*", *Biochem. et Biophys. Acta.* 993:293-296.

Gualtieri, P., P. Pelosi, V. Passarelli and L. Barsanti (1992) "Identification of a rhodopsin photoreceptor in *Euglena gracilis*", *Biochem. Biophys. Acta* 1117:55-59.

Hasunuma, K., K. Furukawa, K. Funadera, M. Kubota and M. Watanabe (1987) "Partial characterization and light-induced regulation of GTP-binding proteins in *Lemna paucicostata*", *Photochem. Photobiol.* 46:531-535.

Hegemann, P., W. Gartner and R. Uhl (1991) "All-trans retinal constitutes the function chromophore in *Chlamidomonas* rhodopsin", *Biophys. J.* 60:1477-1489.

James, T.W., F. Crescitelli, E.R. Loew and W.N. McFarland (1992) "The eyespot of *Euglena gracilis*: a microspectrophotometric study", *Vision Res.* 32:1583-1591.

Kreimer, G., F.J. Marner, U. Brohsonn and M. Melkonian (1991) "Identification of 11-cis retinal and all-trans retinal in the photoreceptive organelle of a flagellate green alga", *FEBS Letters* 293:49-52.

Lercari, B. (1991) "Photomorphogenic response to UV light: Involvement of phytochrome and UV photoreceptors", in: *Photobiological Techniques*, D.P. Valenzeno, ed., New York: Plenum Press, pp. 231-248.

Lorenzi, R., N. Ceccarelli, B. Lercari, P. Gualtieri (1994) "Retinal identification in higher plants: Is a rhodopsin-like protein the blue light receptor?", *Phytochemistry* 36:599-600.

Margulis, L. (1993) *Symbiosis in Cell Evolution*, New York: Freeman and Co.

Martin, R.L., C. Wood, W. Baehr and M. Applebury (1986) "Visual pigment homologies revealed by DNA hybridization", *Science* 232:1266-1269.

Melkonian, M. and H. Robenek (1984) "The eyespot apparatus of flagellate algae: A critical review", *Progr. Phycol. Res.* 3:193-268.

Mohr, H. and H. Drumm-Herrel (1991) "Mode of coaction between phytochrome and blue/uv photoreceptor", in: *Photobiology*, E. Riklis, ed., Plenum Press, New York, pp. 445-453.

Müller, W.E.G. (1997) "Evolution of Protozoa to Metazoa", *Theory Biosci.* 116:145-168.

Oesterhelt, D. and W. Stoekenius (1971) "Rhodopsin-like protein from the purple membrane of *Halobacterium halobium*", *Nature* 233:149-152.

Shepherd, G.M. (1991) "Sensory transduction: Entering the mainstream of membrane signalling", *Cell* 67:845-851.

Spudich, J.l., D.N. Zacks and R. Bogomolni (1995) "Microbial sensory rhodopsins: Photochemistry and function", *Israel J. Chem.* 35:495-513.

Trewavas, A. and S. Gilroy (1991) "Signal transduction in plant cell", *Trends Genetics* 7(11):356-361.

Wald, G. (1964) "The receptors of human color vision", *Science* 145:1007-1010.

Wald, G. (1968) "Single and multiple visual system in arthropods", *J. Gen.*

Physiol. 51:125-132.

Walne, P.L., V. Passarelli, L. Barsanti and P. Gualtieri (1998) "Rhodopsin: A photopigment for phototaxis in *Euglena gracilis*", *Crit. Rev. Plant Sci.* 17:569-574.

Warpeha, K.M.F., L.S. Kaufman and W.R. Briggs (1992) "A flavoprotein may mediate the blue light-activated binding of guanosine 5'-triphospate to isolated plasma membranes of *Pisum sativum L.*", *Photochem. Photobiol.* 55:595-603.

THE MOLECULAR DESIGN OF A VISUAL CASCADE: MOLECULAR STAGES OF PHOTOTRANSDUCTION IN *DROSOPHILA*

R. PAULSEN, M. BÄHNER, A. HUBER, M. SCHILLO, S. SCHULZ, R. WOTTRICH and J. BENTROP
Department of Cell- and Neurobiology, University of Karlsruhe, 76128 Karlsruhe, Germany

ABSTRACT

The *Drosophila* visual cascade follows principles which are valid for numerous other signaling pathways. The initial step, light absorption, occurs at rhodopsins, prototypical members of the seven helix transmembrane receptor families. In the *Drosophila* compound eye, light is perceived by as much as 5 rhodopsins, which are differentially expressed in a distinct cell pattern. After transformation of rhodopsin molecules into their activated state (metarhodopsin), the cascades operating in different subtypes of photoreceptor cells converge to a singular pathway. The heterotrimeric G protein activated by metarhodopsin is characterized by its Gα subunit, which belongs to the phospholipase C-activating Gq family. Its γ subunit identifies the G protein complex as a visual G protein. Two arrestin isoforms, both implicated as regulators of visual pigment deactivation, interact with each of the five rhodopsins. A phospholipase C acts as central target enzyme of the cascade generating the messengers inositol-1,4,5-trisphosphate IP_3 and diacylglycerol (DAG). It has been proposed that IP_3 liberates calcium from internal stores, while DAG may serve as a source for the generation of long chain unsaturated fatty acids. Both calcium and free fatty acids have been implicated in the activation of TRP and TRPL, the ion channels that compose the light-activated conductance.

1. Introduction

The evolutionary design of visual cascades follows principles, which are valid for a considerable number of other signaling pathways, in particular of the G protein coupled receptor (GPCR) pathways activated by hormones and neurotransmitters. The process of visual transduction is usually compartmentalized to a well-elaborated membrane system, which is optimized for photon capture and for the generation of amplified, efficiently controlled, and highly time-resolved visual responses. Taking advantage of this remarkable compartmentalization, research directed to investigate the biochemical and molecular basis of phototransduction has run ahead of the studies in other fields of transmembrane signaling in many aspects.

As a result of the systematic use of genetic strategies in combination with biochemical and molecular physiological approaches, the phototransduction pathway operating in the compound eye of the fruitfly *Drosophila* emerged as a widely recognized model for the study of signal transduction by GPCRs. These studies have provided a detailed insight not only into the phototransduction pathway but also into other aspects of transmembrane signaling, *e.g.* receptor trafficking, spatial organization of signaling pathways and inherited retinal degeneration. Several reviews and lectures focus on different aspects of phototransduction and related cellular processes in *Drosophila*. They mirror the remarkable improvements in understanding the molecular basis of phototransduction gained during the last decade: Ranganathan *et al.*, 1995; Pak, 1995; Zuker, 1996; Scott and Zuker, 1997, 1988; O'Day *et al.*, 1997; O'Tousa, 1997; Montell, 1998; Bentrop, 1998.

The following section introduces structural features of the *Drosophila* eye, as they are required for understanding the function of key components of the phototransduction cascade, which is described in molecular terms in subsequent chapters. The molecular structure of the supramolecular INAD signaling complex and the consequences of the spatial organization provided by the modular PDZ domain protein INAD are topics of the second paper (Paulsen *et al.*, this volume).

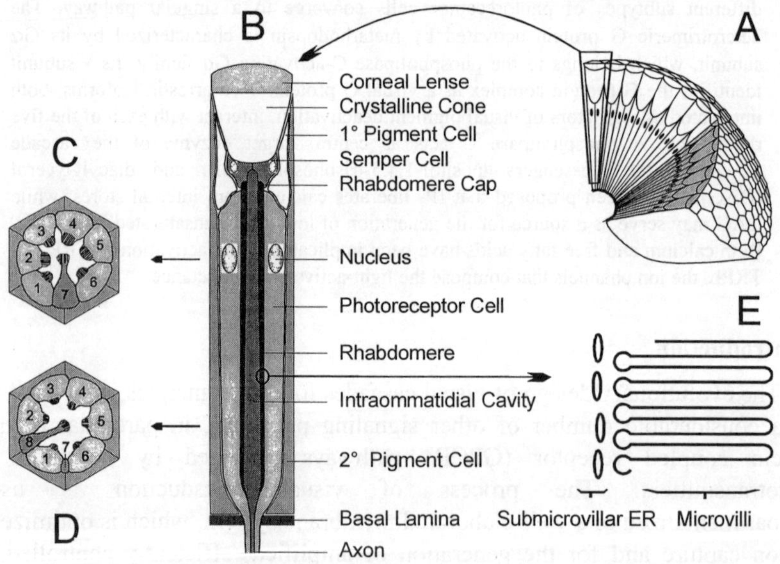

Figure 1. The *Drosophila* eye. Schematic drawing of a compound eye (A), a longitudinal section (B), cross-sections taken at the levels indicated by arrows (C, D), and a detail of the rhabdomere (E).

2. Structure of the *Drosophila* Compound Eye

The compound eye of *Drosophila* is composed of 750 ommatidia or single (unit) eyes (Fig. 1). Each ommatidium contains eight photoreceptors (retinula cells). The phototransduction machinery synthesized in the photoreceptors is targeted to a specialization of the plasma membrane, numerous microvilli projecting from the side of the cell, which faces the intraommatidial cavity. This cavity is filled by a prominent extracellular matrix to which the microvilli adhere (Paulsen, 1984).

Together the microvilli form a light-guiding structure, the rhabdomere. In many respects rhabdomeres are a functional equivalent of the outer segment of vertebrate photoreceptors. According to the position of their rhabdomeres within the ommatidium, the eight photoreceptors can be divided into three different classes, six outer photoreceptors R1-6 and two central cells, R7 and R8 (Fig. 1, see also Fig. 3). R7 and R8 form a fused rhabdomere, which is located centrally in the intraommatidial cavity, with the rhabdomere of R7 on top of the rhabdomere of R8. With respect to photon capture photoreceptors R1-6 fall into one group as they all express the same rhodopsin, Rh1 (O'Tousa *et al.*, 1985, Zuker *et al.*, 1985). Recent studies have revealed that the central photoreceptors of *Drosophila* express four different rhodopsins in non-overlapping subsets (Fig. 3). In a precisely coordinated fashion, the R7 and R8 cells of a single ommatidium may either express the "odd" rhodopsin pair Rh3/Rh5 or the "even" pair Rh4/Rh6 (Chou *et al.*, 1999).

In general, the morphology of the compound eyes of larger dipteran flies, *e.g.* the housefly (*Musca*) or the blowfly (*Calliphora*), is the same as in *Drosophila*. The number of ommatidia forming a single compound eye, however, is considerably higher (up to 5300 in a male *Calliphora*), and the individual ommatidium may be four times as long as an ommatidium of *Drosophila* (up to 340 μm in *Calliphora*). Photochemical and biochemical studies requiring the isolation of high amounts of photoreceptor membranes are, therefore, often performed with the eyes of these larger flies.

3. A Short View on the Basic Design of the Visual Cascade in *Drosophila*

As schematically depicted in Fig. 2, the initial step, light absorption, occurs at rhodopsins. Rhodopsins are prototypical members of the seven-helix transmembrane receptor family, which activate a signal transduction cascade via a heterotrimeric G protein. Upon photon capture, the 11-*cis* form of the chromophore (3-OH retinaldehyde, Vogt, 1983) of an individual rhodopsin molecule is isomerized to the all-*trans* configuration, and rhodopsin (P) is transformed into an active metarhodopsin state (M). Formation of M activates the visual G protein which in turn activates a phospholipase Cβ (PLC), the central target enzyme of this signaling cascade. The function of the messengers, which are generated by the PLC-catalyzed hydrolysis of the membrane lipid phosphatidylinositol-4,5-bisphosphate (PIP_2), namely inositol trisphosphate (IP_3) and diacylglycerol (DAG), is not yet exactly understood. IP_3 is proposed to liberate calcium from internal stores (for a recent evaluation see: Cook and Minke,

1999), while DAG may serve as activator of an eye-specific protein kinase C (ePKC) (Huber *et al.*, 1998), and as a source for the generation of long chain poly-unsaturated fatty acids (Chyb *et al.*, 1999). Both, calcium depletion of internal stores and poly-unsaturated fatty acids, have been implicated in the activation of TRP and TRPL, two cation-selective channels that are essential components of the light-activated membrane conductance. The key components of the phototransduction cascade operating downstream of the G protein are assembled into a supramolecular signaling complex, which provides the basis for highly time-resolved and specific signaling.

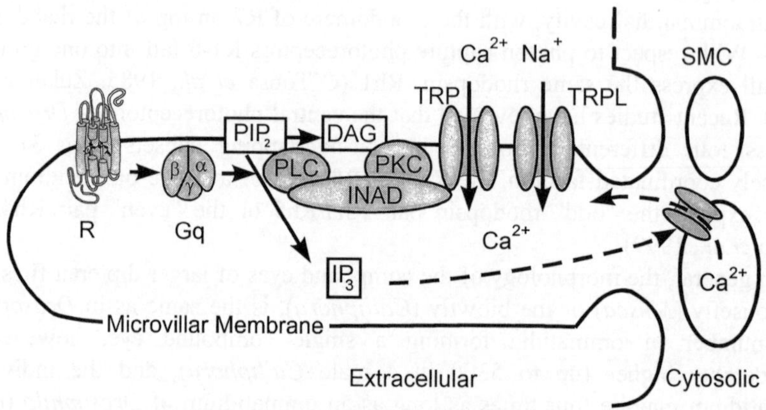

Figure 2. Hypothetical scheme of the signal transduction cascade operating in the *Drosophila* photoreceptor cell. R, rhodopsin; Gq, visual G protein, PLC, phospholipase Cβ; PKC, protein kinase C; SMC, submicrovillar cisternae. For further abbreviations and details see text.

4. Rhodopsin and the Rhodopsin Cycle

The five rhodopsins expressed in the *Drosophila* compound eye detect light ranging from the UV to the red spectral range (Fig. 3). In a distinct difference to the vertebrate transduction cascade, each rhodopsin, after catching a photon, is converted from its P-state via several short-lived intermediates, into a relatively long-lived M-state. This M-state represents the active rhodopsin state. The absorbance coefficient of M exceeds that of P by a factor of 1.5 to 2.0 (Fig. 3).

The high thermal stability of metarhodopsin allows the rhodopsins to be photoconverted back from their M-states into the corresponding P-states (Hamdorf *et al.*, 1973; Hillman *et al.*, 1983). This mechanism is of particular physiological significance as it eliminates the need for a costly chemical (enzymatic) regeneration of the P-state, as is required for vertebrate rhodopsins, which become "bleached" and not "coloured" by light.

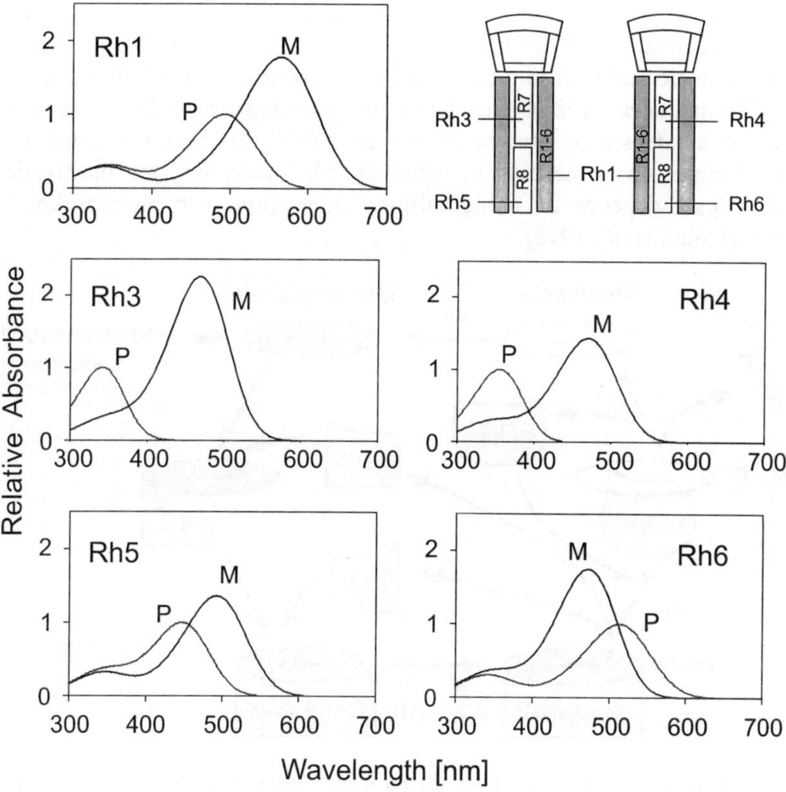

Figure 3. Expression pattern and absorbance spectra of the five different visual pigments expressed in the *Drosophila* compound eye. Shown are the absorbance spectra of rhodopsins (P) and metarhodopsins (M). The upper right panel indicates the expression pattern.

The photoregeneration mechanism evolved in the fly photoreceptor allows to retain high P-state levels and, thus, a high probability for light absorption even under light conditions which would completely abolish phototransduction in a human eye, for example exposure to direct sun light. The spectral absorbance characteristics separate the metarhodopsins into two groups. All metarhodopsins peak in the blue spectral range, apart from M of Rh1 that absorbs maximally at 568 nm. As M of Rh1 absorbs outside the shield provided by the screening pigments present in photoreceptors and accessory pigment cells of a fly ommatidium, the concentration of P in photoreceptors R1-6 is retained close to a level of 100 % at almost any light condition (Stavenga *et al.*, 1973).

The photoregeneration mechanism however does not eliminate the need for an efficient deactivation of the long-lived M-state which under certain conditions may accumulate and lead to a prolonged depolarization (Hamdorf and Razmjoo, 1977; Hillman *et al.*, 1983). The rhodopsin cycle depicted in Fig. 4 summarizes the sequential steps currently known to occur after P has been converted into the active M-state. As is the case in vertebrate photoreceptors, the inactivation of metarhodopsin depends on a high-affinity interaction with proteins known as arrestins (Dolph *et al.*, 1993).

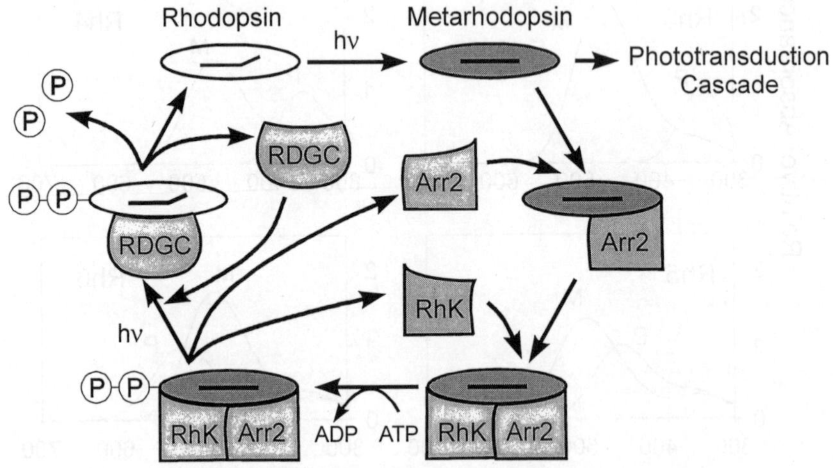

Figure 4. Activation and inactivation of *Drosophila* rhodopsin. hν, photon capture; Arr2, arrestin 2; RhK, rhodopsin kinase; RDGC, Retinal degeneration C protein (= rhodopsin phosphatase). For details see text.

There are, however, distinct differences. Firstly, phosphorylation is not a prerequisite for high affinity binding of arrestin to metarhodopsin (Plangger *et al.*, 1994). It appears rather that binding of arrestin favours the recruitment of a rhodopsin kinase that binds to the M-arrestin complex and phosphorylates M at C-terminal phosphorylation sites (Bentrop *et al.*, 1993), similarly as has been shown for the recruitment of a tyrosin protein kinase (Src) to the β-adrenergic receptor (Luttrell *et al.*, 1999).

Secondly, phosphorylation of metarhodopsin has not yet been shown to be a functionally important step in the deactivation of invertebrate rhodopsin. Thirdly, the functional relevance of two arrestin isoforms expressed in the *Drosophila* photoreceptor cell has not yet been elucidated.

According to the current state of knowledge, the conversion of any fly rhodopsin into its active M-state activates the same visual G protein that couples rhodopsin activation to downstream steps of the visual cascade. Thus, phototransduction, which is initiated at five different rhodopsins, converges to a singular pathway after G protein activation. This implies that some functional sites within the transmembrane region must be different to support the differences in spectral tuning of rhodopsins, while sites exposed to the cytosol which interact with the G protein and also with the arrestin isoforms, rhodopsin kinase and rhodopsin phosphatase, should be highly conserved.

This conclusion is supported by the finding that the ectopic expression of Rh1 to Rh6 in a *Drosophila* null mutant for Rh1 (*ninaE*) fully rescues the phototransduction of R1-6 cells (Feiler *et al.*, 1988, 1992; Salcedo *et al.*, 1999).

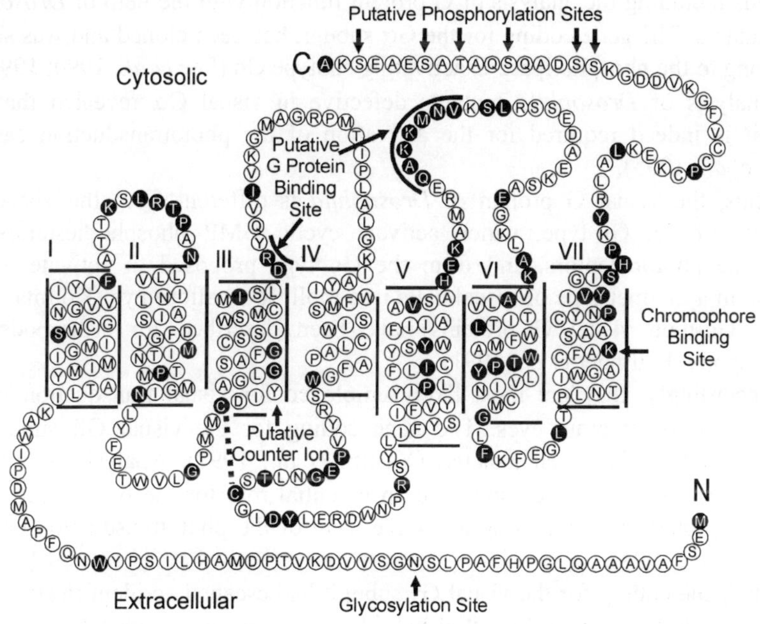

Figure 5. 2D-structural model of *Drosophila* Rh1 rhodopsin. Filled circles indicate amino acids conserved in all *Drosophila* opsins. Boxes, transmembrane domains.

Figure 5 shows a 2D-structural model for *Drosophila* Rh1. The model highlights putative interaction domains derived from comparing amino acid sequences of the six identified *Drosophila* rhodopsins, including Rh2, which is expressed in non-ocular photoreceptors. Amino acids conserved in these rhodopsins indicate putative interaction domains in all cytoplasmic loops. Amino acids proposed to be involved in G protein interaction are marked. There are also conserved features at the extracellular side which may contribute to the stabilization of rhodopsin's 3-dimensional structure by a disulfide bridge or indicate interaction with an extracellular protein. Conserved motifs are furthermore found in the seven transmembrane domains, for example the chromophore binding site in transmembrane domain 7.

5. The Visual G Protein

The trimeric visual G protein (Gαβγ) which reports rhodopsin activation to downstream members of the visual cascade has been characterized by biochemical methods, *e.g.* measuring light-activated GTPase activity (Blumenfeld *et al.*, 1985) or GTPγS binding (Devary *et al.*, 1987), as well as by molecular genetical methods including the analysis of G protein function with the help of *Drosophila* eye mutants. The gene coding for the Gα subunit has been cloned and was shown to belong to the phospholipase C-activating subtype Gq (Lee *et al.*, 1990; 1994).

Analysis of *Drosophila* mutants defective in visual Gα revealed that this subunit is indeed required for the activation of the phototransduction cascade (Scott *et al.*, 1995).

Thus, the visual G protein of *Drosophila* is different from the visual Gα subunits of the Gt type, which activate cyclic GMP phosphodiesterases of vertebrate photoreceptors, and from the Go type proposed to activate guanyl cyclase in a distinct type of visual cells in a mollusc (scallop) eye (Kojima *et al.*, 1997). Gq-type visual G proteins have been described for arthropods and cephalopods (Pottinger *et al.*, 1991).

Accordingly, they are most likely employed for visual transduction in the majority of invertebrate eyes. The gene coding for the visual Gβ subunit of *Drosophila* has also been isolated (Yarfitz *et al.*, 1991). Analysis of mutants defective in this subunit demonstrate an essential role for the βγ subunit of the visual G protein in terminating an active state of the phototransduction cascade (Dolph *et al.*, 1994).

The gene coding for the visual Gγ subunit had escaped an identification for a long time. However, by differential hybridization screens, we recently succeeded in the isolation of a novel Gγ subunit (Schulz *et al.*, 1999). This Gγ subunit is specifically expressed in the photoreceptor cells. It is associated with the rhabdomeric photoreceptor membrane where it forms a heterodimeric complex with the visual Gβ subunit. Amino acid sequence comparisons of the Gγ subunits of visual and non-visual G proteins revealed that the newly isolated visual Gγ subunit of *Drosophila* is more closely related to visual Gγ homologs from vertebrate eyes than to most of the non-visual Gγ subunits (Fig. 6).

Thus, its γ subunit characterizes a G protein as a member of a visual cascade irrespectively of the photoreceptor type - rhabdomeric or ciliated - and independently of the enzyme activated by the respective Gα subunit.

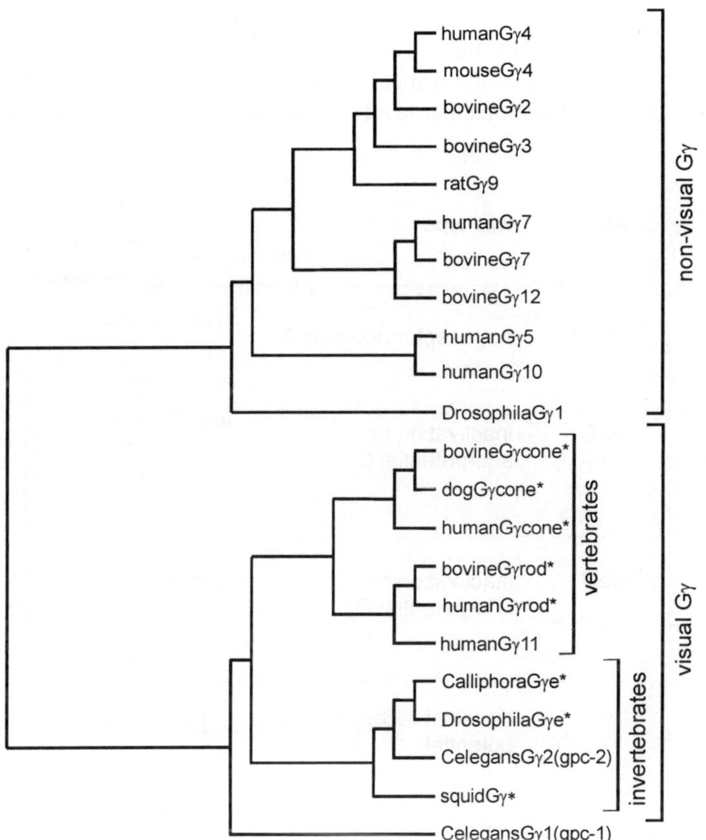

Figure 6. Evolutionary relationship of selected Gγ subunits. Asterisks mark the visual Gγ subunits. Accession numbers for sequences used: bovineGγ2, P16874; bovineGγ3, P29798; bovineGγ7, P30671; bovineGγ12, Q28024; bovineGγcone, P50154; bovineGγrod, P02698; CelegansGγ1(gpc-1), CAA91806; CelegansGγ2(gpc-2), AAC78236; dogGγcone, AAC98924; *Drosophila*Gγ1, P38040; humanGγ4, P50150; humanGγ5, P30670; humanGγ7, AAC32595; humanGγ10, P50151; humanGγ11, P50152; humanGγcone, O14610; humanGγrod, Q08447; mouseGγ4, P50153; ratGγ9, P43426; squidGγ, Q01821.

6. Proteins of the Phototransduction Pathway Downstream of the Visual G Protein

Members of the visual cascade downstream of the visual G protein have been identified primarily by mutagenesis screens which resulted in the isolation of mutants defective in certain aspects of phototransduction. Figure 7 shows examples for the phenotypes *norpA* (*no receptor potential A*), *inaC* (*inactivation no afterpotential C*) and *trp* (*transient receptor potential*) in which the physiological response to light, recorded as an electroretinogram (ERG), is affected by the mutation of the respective gene. These genes have all been cloned and characterized at the molecular level as described in the following paragraphs.

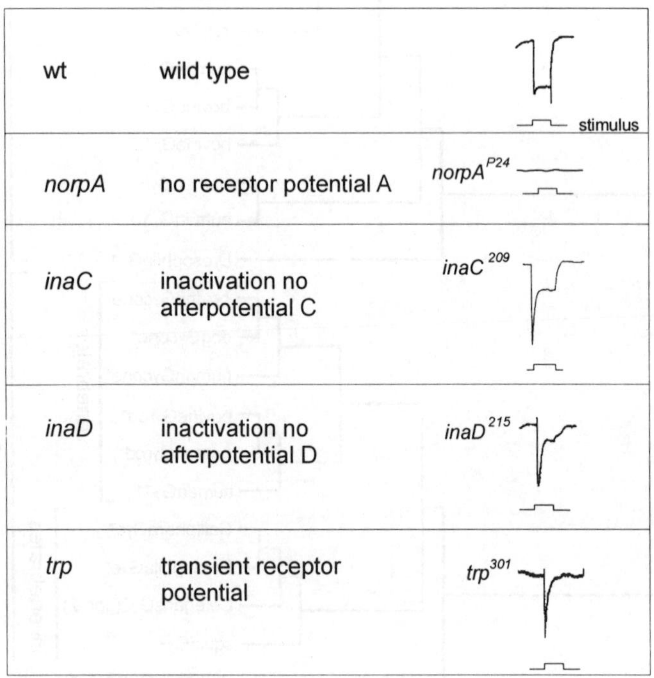

Figure 7. Electroretinograms of *Drosophila* eye mutants *norpA, inaD, inaC, trp*. After Peretz *et al.*, 1994 and Tsunoda *et al.*, 1997.

6.1. Phospholipase C, a Key Protein of the Visual Cascade

Phospholipase C has been identified as a key enzyme of the phototransduction cascade. The major insight into this crucial role was gained by the cloning and characterization of the *norpA* gene of *Drosophila* (Bloomquist *et al.*, 1988). *norpA* encodes the structural gene for a phosphatidylinositol-4,5-bisphosphate (PIP_2)-hydrolyzing phospholipase subtype Cβ (PLCβ). In *norpA* mutants, phototransduction is completely abolished, rendering the mutant fly blind (Pak *et al.*, 1970; Hotta and Benzer, 1970). The defect in the light response can be rescued by germline transformation of an intact *norpA* gene (McKay *et al.*, 1995). This strongly suggests that the PLC-catalyzed formation of the intracellular messengers diacylglycerol (DAG) and/or inositol-1,4,5-trisphosphate (IP_3) from PIP_2 is an absolute requirement for visual excitation to proceed.

Phosphatidyl inositol-specific lipases C of the subclass PLCβ show a domain structure which comprises a pleckstrin homology (PH) domain, an EF hand domain, the catalytic region composed of X and Y domains, a C2 domain, and a putative G protein-binding region in the C-terminal portion of the protein (Fig. 8). The X and Y domains are highly conserved between different subtypes of PLCs. By determining the crystal structure of the PLCδ subtype, they were shown to form the catalytic core (Essen *et al.*, 1996). The function of the PH domain in photoreceptor-specific PLCβ isoforms is not yet clear. In addition to binding PIP_2 and IP_3, this domain was implicated in a putative regulation of PLCβ by G protein βγ subunits and protein kinase C. The C2 domain, which is also present in Ca^{2+}-dependent members of the protein kinase C family, is most likely responsible for the calcium dependency of the enzymatic activity, which is common to all PLC subclasses. The function of another Ca^{2+}-binding domain, the EF hand motif located between the PH domain and the X domain has yet to be clarified. In addition, as indicated in Fig. 8, PLCβ isoforms contain an extended C-terminal region that has been identified as the binding site for Gqα (Singer *et al.*, 1997). Finally, a FCA motif located at the C-terminus of the *norpA*-encoded PLCβ serves as the binding site for the PDZ domain protein INAD (Shieh *et al.*, 1997).

With respect to PLCβ employed in phototransduction, activation by Gqα has been demonstrated for the isoenzymes expressed in squid photoreceptors (Mitchell *et al.*, 1995, Suzuki *et al.*, 1995). Direct biochemical evidence indicating that the active state of visual Gqα of *Drosophila* activates PLCβ has been missing so far, but the genetic experiments strongly suggest that this interaction takes place. Taken together, the domain structure of PLCβ suggests that PLC in the phototransduction process becomes activated by a visual Gqα protein and is subject to an extensive regulation by calcium.

6.2. The Eye-Specific Protein Kinase C (ePKC)

Members of the protein kinase C (PKC) family phosphorylate serine or threonine residues of their substrate proteins. Conventional PKCs (subtypes α,β,γ) are activated by calcium, diacylglycerol and phorbolesters. One member of this family, encoded by the *inaC* locus (*inactivation no afterpotential C*) has been shown to be expressed exclusively in the photoreceptor cells of fly eyes (ePKC) (Schaeffer *et al.*, 1989; Smith *et al.*, 1991; Huber *et al.*, 1996, 1998). Sequence alignment of ePKC with functionally well characterized vertebrate PKCs reveals the presence of characteristic, conserved regions of this family of protein kinases. As indicated in Fig. 8, there are four conserved regions (C1 to C4) which are separated by variable regions (V1 to V5) of limited sequence homology. A regulatory domain (comprising V1 to C2) and a catalytic domain (C3 to V5) are separated by the V3 linker region which is susceptible to hydrolytic cleavage by proteolytic enzymes such as trypsin or calpain (Hug and Sarre, 1993). The C1 regulatory domain is a cysteine-rich region which is responsible for the binding of diacylglycerol and phorbolesters. Calcium sensitivity is probably mediated by the C2 domain, which is also present in other calcium-binding proteins, *e.g.* PLCs, phospholipases A2 or synaptotagmin (Brose *et al.*, 1995). This domain is absent in calcium-independent PKC isoenzymes. The C3 domain contains the consensus sequence for an ATP binding motif, $xGxGx_2Gx_{16}Kx$ (x representing any amino acid), which is found in almost all protein kinases (Taylor *et al.*, 1990). The C4 domain comprises a substrate binding region as well as a phosphate transfer region with the highly conserved element $RDLx_nDFG_nAPE$, a signature of protein kinases. ePKC is a major component of the INAD signaling complex. It is linked to a PDZ domain of INAD through the C-terminal S/TXI motif (Fig. 8) (Adamski *et al.*, 1998).

6.3. TRP and TRPL, Two Candidates for Light-Activated Channels

The TRP protein family constitutes a novel class of ion channels implicated in calcium signaling. Prototypical members of this class have been firstly identified in the visual system of *Drosophila,* where they were shown to represent essential components of the light-activated conductance (Hardie & Minke, 1992, 1993; Niemeyer *et al.*, 1996).

Two structurally related channel proteins are encoded by the genes *transient receptor potential* (*trp*) (see Fig. 8) and *transient receptor potential like* (*trpl*) (Montell & Rubin, 1989; Phillips *et al.*, 1992). Both channel subunits share structural features which are also detected in vertebrate voltage-gated calcium channels, *e.g.* six putative transmembrane domains S1 to S6 (Fig. 8). The highest sequence similarity exists between the transmembrane domains S4, S5, and S6 which include the putative pore-forming region (between S5 and S6). S4 of TRP and TRPL, however, lacks a positive charge that acts as voltage sensor in voltage gated channels (Phillips *et al.*, 1992).

Common features of the TRP and TRPL structure are three ankyrin repeats at the intracellularly located N-terminal region, which are supposed to be involved in protein-protein interactions. Calmodulin binding sites, one in TRP, two in TRPL, are located in the respective carboxy-terminal regions. While the function of the calmodulin binding site in TRP is not clear, deletion of this site in TRPL alters the calcium dependency of the portion of the light-activated conductance which is carried by TRPL.

This indicates that calcium affects the inactivation of TRPL via these calmodulin binding sites (Scott *et al.*, 1997). Furthermore, a S/TXV-motif, located near the C-terminus of TRP (Fig. 8) has been identified as anchoring site for TRP to the INAD signaling complex (Shieh & Zhu, 1996; Tsunoda *et al.*, 1997).

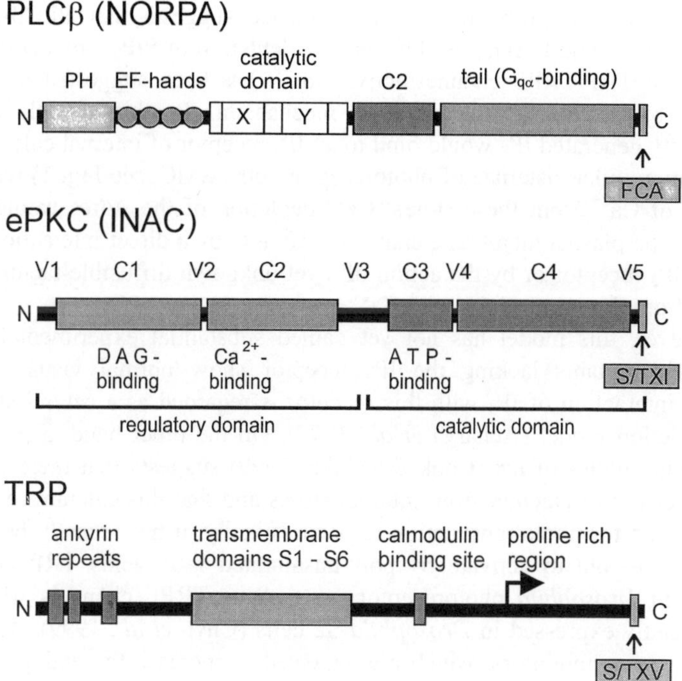

Figure 8. Domain structure of PLCβ, ePKC and TRP.

Several questions central to the understanding of the gating mechanism of the photoreceptor channel proteins TRP and TRPL are not yet solved or are discussed highly controversely (see Scott and Zuker, 1998; Montell, 1998).

Firstly, why, after all, in fly photoreceptors evolved a phototransduction mechanism that results in the activation of two types of cation channels is one of the important questions which remains to be solved. While studies with wild-type and mutant flies provided evidence that TRP and TRPL function as distinct light-activated channels of the photoreceptor cell (Niemeyer *et al.*, 1996; Reuss *et al.*, 1997), heterologously expressed TRP and TRPL appear to form heteromultimers (Gillo *et al.*, 1996; Xu *et al.*, 1997). Immunoprecipitation studies and quantification of TRP and TRPL (Xu *et al.*, 1997, Paulsen *et al.*, 1999) suggest that at least part of the TRP and TRPL subunits present in the photoreceptor membrane form heteromultimers. However, since TRP may be present in the membrane in a more than ten-fold excess to TRPL, most TRP monomers are likely to form homomultimeric TRP channels.

Secondly, the gating mechanism of the TRP and TRPL calcium channels is the most troubling problem. Current evidence suggests that PLCβ-catalyzed formation of IP_3 and DAG, possibly also the depletion of PIP_2, are a *conditio sine qua non* for TRP/TRPL channel activation. It has been suggested that TRP is activated by a store-operated gating mechanism. In this model the light-dependently generated IP_3 would bind to an IP_3 receptor of internal calcium stores (the submicrovillar cisternae of photoreceptor cells, SMC, see Fig. 1) resulting in a release of Ca^{2+} from these stores. Ca^{2+} depletion of the stores would then be sensed by the plasma membrane channel TRP, *e.g.* by a direct interaction of TRP with the IP_3 receptor or by the action of a yet unknown diffusible factor which is released from the stores along with Ca^{2+}.

However, this model has not yet gained substantial experimental support. *Drosophila* mutants lacking the IP_3 receptor show normal visual responses although interaction of IP_3 with this receptor is regarded as a central step in the store depletion model (Acharya *et al.*, 1997). On the other hand, a recent study with mutant alleles of *trp* (Cook & Minke, 1999) suggests that there is a light-induced release of calcium from internal stores and that this calcium release is at least required for phototransduction to occur. Finally, it has recently been shown that it is possible to activate by poly-unsaturated fatty acids TRP and TRPL channels in *Drosophila* photoreceptors as well as TRPL channels which were recombinantly expressed in *Drosophila* S2 cells (Chyb *et al.*, 1999). Application of lipoxygenase inhibitors, which are expected to increase the endogenous fatty acid content of photoreceptors, may also lead to the gating of TRP and TRPL. Accordingly, the formation of DAG by the PLCβ-catalyzed hydrolysis of PIP_2 may be the step in phototransduction that provides the substrate for the formation of poly-unsaturated fatty acids, which may represent the longtime-searched-for internal messengers of fly phototransduction. Despite a wealth of information which has been obtained by molecular genetical, biochemical and physiological studies, the internal transmitter problem in fly phototransduction has not yet been solved in detail. This may result from the assembly of the phototransduction cascade into supramolecular, multimeric signaling complexes, transducisomes (Tsunoda *et al.*, 1997) or signaling webs (Montell, 1998), which became obvious only recently. The biochemical methods required to study the role of putative messengers which activate and regulate this network are not yet at hand.

Acknowledgements

Original work reported in this review is supported by grants of the Deutsche Forschungsgemeinschaft (Pa 274/4-5) and the European Union (BMH4-VCT97-2341).

References

Acharya, J.K., K. Jalink, R.W. Hardy, V. Hartenstein and C.S. Zuker (1997) "InsP3 receptor is essential for growth and differentiation but not for vision in *Drosophila*", *Neuron* 18:881-887.

Adamski, F.M., M.Y. Zhu, F. Bahiraei and B.H. Shieh (1998) "Interaction of eye protein kinase C and INAD in *Drosophila*. Localization of binding domains and electrophysiological characterization of a loss of association in transgenic flies", *J. Biol. Chem.* 273:17713-17719.

Bentrop, J. (1998) "Rhodopsin mutations as the cause of retinal degeneration. Classification of degeneration phenotypes in the model system *Drosophila melanogaster*", *Acta Anat. (Basel)* 162:85-94.

Bentrop, J., A. Plangger and R. Paulsen (1993) "An arrestin homolog of blowfly photoreceptors stimulates visual-pigment phosphorylation by activating a membrane-associated protein kinase", *Eur. J. Biochem.* 216:67-73.

Bloomquist, B.T., R.D. Shortridge, S. Schneuwly, M. Perdew, C. Montell, H. Steller, G. Rubin and W.L. Pak (1988) "Isolation of a putative phospholipase C gene of *Drosophila*, norpA, and its role in phototransduction", *Cell* 54:723-733.

Blumenfeld, A., J. Erusalimsky, O. Heichal, Z. Selinger and B. Minke (1985) "Light-activated guanosinetriphosphatase in *Musca* eye membranes resembles the prolonged depolarizing afterpotential in photoreceptor cells", *Proc. Natl. Acad. Sci. U.S.A.* 82:7116-7120.

Brose, N., K. Hofmann, Y. Hata and T.C. Südhof (1995) "Mammalian homologues of Caenorhabditis elegans unc-13 gene define novel family of C2-domain proteins", *J. Biol. Chem.* 270:25273-25280.

Chou, W.H., A. Huber, J. Bentrop, S. Schulz, K. Schwab, L.V. Chadwell, R. Paulsen and S.G. Britt (1999) "Patterning of the R7 and R8 photoreceptor cells of *Drosophila*: evidence for induced and default cell-fate specification", *Development* 126:607-616.

Chyb, S., P. Raghu and R.C. Hardie (1999) "Poly-unsaturated fatty acids activate the *Drosophila* light-sensitive channels TRP and TRPL", *Nature* 397:255-259.

Cook, B. and B. Minke (1999) "TRP and calcium stores in *Drosophila* phototransduction", *Cell Calcium* 25:161-171.

Devary, O., O. Heichal, A. Blumenfeld, D. Cassel, E. Suss, S. Barash, C.T. Rubinstein, B. Minke and Z. Selinger (1987) "Coupling of photoexcited rhodopsin to inositol phospholipid hydrolysis in fly photoreceptors", *Proc. Natl. Acad. Sci . U.S.A.* 84:6939-6943.

Dolph, P.J., R. Ranganathan, N.J. Colley, R.W. Hardy, M. Socolich and C.S. Zuker (1993) "Arrestin function in inactivation of G protein-coupled receptor rhodopsin in vivo", *Science* 260:1910-1916.

Dolph, P.J., S.H. Man, S. Yarfitz, N.J. Colley, J.R. Deer, M. Spencer, J.B. Hurley and C.S. Zuker (1994) "An eye-specific G beta subunit essential for termination of the phototransduction cascade", *Nature* 370:59-61.

Essen, L.O., O. Perisic, R. Cheung, M. Katan and R.L. Williams (1996) "Crystal structure of a mammalian phosphoinositide-specific phospholipase C delta", *Nature* 380:595-602.

Feiler, R., W.A. Harris, K. Kirschfeld, C. Wehrhahn and C.S. Zuker (1988) "Targeted misexpression of a *Drosophila* opsin gene leads to altered visual function", *Nature* 333:737-741.

Feiler, R., R. Bjornson, K. Kirschfeld, D. Mismer, G.M. Rubin, D.P. Smith, M. Socolich and C.S. Zuker (1992) "Ectopic expression of ultraviolet-rhodopsins in the blue photoreceptor cells of *Drosophila*: visual physiology and photochemistry of transgenic animals", *J. Neurosci.* 12:3862-3868.

Gillo, B., I. Chorna, H. Cohen, B. Cook, I. Manistersky, M. Chorev, A. Arnon, J.A. Pollock, Z. Selinger and B. Minke (1996) "Coexpression of *Drosophila* TRP and TRP-like proteins in *Xenopus* oocytes reconstitutes capacitative Ca^{2+} entry", *Proc. Natl. Acad. Sci. U.S.A.* 93:14146-14151.

Hamdorf, K. and S. Razmjoo (1977) "The prolonged depolarizing afterpotential and its contribution to the understanding of photoreceptor function", *Biophys. Struct. Mech.* 3:163-170.

Hamdorf, K., R. Paulsen and J. Schwemer (1973) "Photoregeneration and sensitivity control of photoreceptors in invertebrates", in: *Biochemistry and physiology of visual pigments*, H. Langer, ed., Berlin, Heidelberg, New York: Springer Verlag, pp. 155-166.

Hardie, R.C. and B. Minke (1992) "The trp gene is essential for a light-activated Ca^{2+} channel in *Drosophila* photoreceptors", *Neuron* 8:643-651.

Hardie, R.C., A. Peretz, E. Suss-Toby, A. Rom-Glas, S.A. Bishop, Z. Selinger and B. Minke (1993) "Protein kinase C is required for light adaptation in *Drosophila* photoreceptors", *Nature* 363:634-637.

Hillman, P., S. Hochstein and B. Minke (1983) "Transduction in invertebrate photoreceptors: role of pigment bistability", *Physiol. Rev.* 63:668-772.

Hotta, Y. and S. Benzer (1970) "Genetic dissection of the *Drosophila* nervous system by means of mosaics", *Proc. Natl. Acad. Sci. U.S.A.* 67:1156-1163.

Huber, A., P. Sander and R. Paulsen (1996) "Phosphorylation of the InaD gene product, a photoreceptor membrane protein required for recovery of visual excitation", *J. Biol. Chem.* 271:11710-11717.

Huber, A., P. Sander, M. Bähner and R. Paulsen (1998) "The TRP Ca^{2+} channel assembled in a signaling complex by the PDZ domain protein INAD is phosphorylated through the interaction with protein kinase C (ePKC)", *FEBS Lett.* 425:317-322.

Hug, H. and T.F. Sarre (1993) "Protein kinase C isoenzymes: divergence in signal transduction?", *Biochem. J.* 291:329-343.

Kojima, D., A. Terakita, T. Ishikawa, Y. Tsukahara, A. Maeda and Y. Shichida (1997) "A novel Go-mediated phototransduction cascade in scallop visual cells", *J. Biol. Chem.* 272:22979-22982.

Lee, Y.J., M.B. Dobbs, M.L. Verardi and D.R. Hyde (1990) "dgq: a *Drosophila* gene encoding a visual system-specific G alpha molecule", *Neuron* 5:889-898.

Lee, Y.J., S. Shah, E. Suzuki, T. Zars, P.M. O'Day and D.R. Hyde (1994) "The *Drosophila* dgq gene encodes a G alpha protein that mediates phototransduction", *Neuron* 13:1143-1157.

Luttrell, L.M., S.S. Ferguson, Y. Daaka, W.E. Miller, S. Maudsley, R.G. Della, F. Lin, H. Kawakatsu, K. Owada, D.K. Luttrell, M.G. Caron and R.J. Lefkowitz (1999) "Beta-arrestin-dependent formation of beta2 adrenergic receptor-Src protein kinase complexes", *Science* 283:655-661.

McKay, R.R., D.M. Chen, K. Miller, S. Kim, W.S. Stark and R.D. Shortridge (1995) "Phospholipase C rescues visual defect in norpA mutant of *Drosophila* melanogaster", *J. Biol. Chem.* 270:13271-13276.

Mitchell, J., J. Gutierrez and J.K. Northup (1995) "Purification, characterization, and partial amino acid sequence of a G protein-activated phospholipase C from squid photoreceptors", *J. Biol. Chem.* 270:854-859.

Montell, C. (1998) "TRP trapped in fly signaling web", *Curr. Opin. Neurobiol.* 8:389-397.

Montell, C. and G.M. Rubin (1989) "Molecular characterization of the *Drosophila* trp locus: a putative integral membrane protein required for phototransduction", *Neuron* 2:1313-1323.

Niemeyer, B.A., E. Suzuki, K. Scott, K. Jalink and C.S. Zuker (1996) "The *Drosophila* light-activated conductance is composed of the two channels TRP and TRPL", *Cell* 85:651-659.

O'Day, P.M., J. Bacigalupo, C. Vergara and J.E. Haab (1997) "Current issues in invertebrate phototransduction. Second messengers and ion conductances", *Mol. Neurobiol.* 15:41-63.

O'Tousa, J.E. (1997) "Normal physiology and retinal degeneration in the *Drosophila* visual system", *Prog. Ret. Eye Res.* 16:691-703.

O'Tousa, J.E., W. Baehr, R.L. Martin, J. Hirsh, W.L. Pak and M.L. Applebury (1985) "The *Drosophila* ninaE gene encodes an opsin", *Cell* 40:839-850.

Pak, W.L. (1995) "*Drosophila* in vision research. The Friedenwald Lecture", *Invest. Ophthalmol. Vis. Sci.* 36:2340-2357.

Pak, W.L., J. Grossfield and K.S. Arnold (1970) "Mutants of the visual pathway of *Drosophila melanogaster*", *Nature* 227:518-520.

Paulsen, R. (1984) "Spectral characteristics of isolated blowfly rhabdoms", *J. Comp. Physiol. [A.]* 155:47-55.

Paulsen, R., M. Bähner and A. Huber (1999) "The PDZ assembled "transducisome", of microvillar photoreceptors: the TRP/TRPL problem", *Eur. J. Physiol.* in press.

Peretz, A., C. Sandler, K. Kirschfeld, R.C. Hardie and B. Minke (1994) "Genetic dissection of light-induced Ca^{2+} influx into *Drosophila* photoreceptors", *J. Gen. Physiol.* 104:1057-1077.

Phillips, A.M., A. Bull and L.E. Kelly (1992) "Identification of a *Drosophila* gene encoding a calmodulin-binding protein with homology to the trp phototransduction gene", *Neuron* 8:631-642.

Plangger, A., D. Malicki, M. Whitney and R. Paulsen (1994) "Mechanism of arrestin 2 function in rhabdomeric photoreceptors", *J. Biol. Chem.* 269:26969-26975.

Pottinger, J.D., N.J. Ryba, J.N. Keen and J.B. Findlay (1991) "The identification and purification of the heterotrimeric GTP-binding protein from squid (*Loligo forbesi*) photoreceptors", *Biochem.* . 279:323-326.

Ranganathan, R., D.M. Malicki and C.S. Zuker (1995) "Signal transduction in *Drosophila* photoreceptors", *Annu. Rev. Neurosci.* 18:283-317:283-317.

Reuss, H., M.H. Mojet, S. Chyb and R.C. Hardie (1997) "In vivo analysis of the *Drosophila* light-sensitive channels, TRP and TRPL", *Neuron* 19:1249-1259.

Salcedo, E., A. Huber, Henrich, L.V. Chadwell, W.H. Chou, R. Paulsen and S.G. Britt (1999) "Ectopic expression and physiological characterization of the R8 photoreceptor cell-specific Rh5 and Rh6 rhodopsins of *Drosophila*", *J. Neurosci.* in press.

Schaeffer, E., D. Smith, G. Mardon, W. Quinn and C. Zuker (1989) "Isolation and characterization of two new *Drosophila* protein kinase C genes, including one specifically expressed in photoreceptor cells", *Cell* 57:403-412.

Scott, K. and C. Zuker (1997) "Lights out: deactivation of the phototransduction cascade", *Trends. Biochem. Sci.* 22:350-354.

Scott, K. and C. Zuker (1998) "TRP, TRPL and trouble in photoreceptor cells", *Curr. Opin. Neurobiol.* 8:383-388.

Scott, K. and C.S. Zuker (1998) "Assembly of the *Drosophila* phototransduction cascade into a signalling complex shapes elementary responses", *Nature* 395:805-808.

Scott, K., Y. Sun, K. Beckingham and C.S. Zuker (1997) "Calmodulin regulation of *Drosophila* light-activated channels and receptor function mediates termination of the light response in vivo", *Cell* 91:375-383.

Scott, K., A. Becker, Y. Sun, R. Hardy and C. Zuker (1995) "Gq alpha protein function in vivo: genetic dissection of its role in photoreceptor cell physiology", *Neuron* 15:919-927.

Schulz, S., A. Huber, K. Schwab and R. Paulsen (1999) "A novel Gγ isolated from *Drosophila* constitutes a visual G protein γ subunit of the fly compound eye", *J. Biol. Chem.* 274:37605-37610.

Shieh, B.H. and M.Y. Zhu (1996) "Regulation of the TRP Ca^{2+} channel by INAD in *Drosophila* photoreceptors", *Neuron* 16:991-998.

Shieh, B.H., M.Y. Zhu, J.K. Lee, I.M. Kelly and F. Bahiraei (1997) "Association of INAD with NORPA is essential for controlled activation and deactivation of *Drosophila* phototransduction in vivo", *Proc. Natl. Acad. Sci. U.S.A.* 94:12682-12687.

Singer, W.D., H.A. Brown and P.C. Sternweis (1997) "Regulation of eukaryotic phosphatidylinositol-specific phospholipase C and phospholipase D", *Annu. Rev. Biochem.* 66:475-509.

Smith, D.P., R. Ranganathan, R.W. Hardy, J. Marx, T. Tsuchida and C.S. Zuker (1991) "Photoreceptor deactivation and retinal degeneration mediated by a photoreceptor-specific protein kinase C", *Science* 254:1478-1484.

Stavenga, D.G., A. Zantema and K.W. Kuiper (1973) "Rhodopsin processes and the function of pupil mechanism in flies", in: *Biochemistry and Physiology of Visual Pigments*, H. Langer, ed., Berlin, Heidelberg, New York: Springer Verlag, pp. 175-180.

Suzuki, T., A. Terakita, K. Narita, K. Nagai, Y. Tsukahara and Y. Kito (1995) "Squid photoreceptor phospholipase C is stimulated by membrane Gq alpha but not by soluble Gq alpha", *FEBS Lett.* 377:333-337.

Taylor, S.S., J.A. Buechler and W. Yonemoto (1990) "cAMP-dependent protein kinase: framework for a diverse family of regulatory enzymes", *Annu. Rev. Biochem.* 59:971-1005.

Tsunoda, S., J. Sierralta, Y. Sun, R. Bodner, E. Suzuki, A. Becker, M. Socolich and C.S. Zuker (1997) "A multivalent PDZ-domain protein assembles signalling complexes in a G- protein-coupled cascade", *Nature* 388:243-249.

Vogt, K. (1983) "Is the fly visual pigment a rhodopsin?", *Z. Naturforsch. [C.]* 38:329-333.

Xu, X.Z., H.S. Li, W.B. Guggino and C. Montell (1997) "Coassembly of TRP and TRPL produces a distinct store-operated conductance", *Cell* 89:1155-1164.

Yarfitz, S., G.A. Niemi, J.L. McConnell, C.L. Fitch and J.B. Hurley (1991) "A Gβ protein in the *Drosophila* compound eye is different from that in the brain", *Neuron* 7:429-438.

Zuker, C.S. (1996) "The biology of vision of *Drosophila*", *Proc. Natl. Acad. Sci. U.S.A.* 93:571-576.

Zuker, C.S., A.F. Cowman and G.M. Rubin (1985) "Isolation and structure of a rhodopsin gene from *D. melanogaster*" *Cell* 40:851-858.

THE MOLECULAR DESIGN OF A VISUAL CASCADE: ASSEMBLY OF THE *DROSOPHILA* PHOTOTRANSDUCTION PATHWAY INTO A SUPRAMOLECULAR SIGNALING COMPLEX

R. PAULSEN, M. BÄHNER, J. BENTROP, M. SCHILLO, S. SCHULZ and A. HUBER

Department of Cell- and Neurobiology, University of Karlsruhe, 76128 Karlsruhe, Germany

ABSTRACT

The PDZ domain protein INAD has emerged as the central organizer of a supramolecular protein complex in the microvillar photoreceptor membrane of fly photoreceptors. INAD is a modular scaffold protein which localizes the assembled proteins of the G protein coupled phototransduction cascade to the microvillar membrane of the polarized photoreceptor cell. INAD also provides the structural basis for interactions of the activated Gqα-subunit with its target enzyme phospholipase Cβ. Further functional interactions within the INAD signaling complex are revealed by protein kinase C-catalyzed phosphorylation of INAD and the light-sensitive Ca^{2+} channel TRP. The integration of phototransduction proteins into a signaling complex is of particular functional significance with respect to the specificity of the activation, control and time resolution of the phototransduction process.

1. Introduction

The assembly of signaling proteins into supramolecular complexes has been recognized as a principle holding for signaling pathways at synapses, for pathways controlling development and for visual transduction pathways (Fanning and Anderson, 1998, 1999; Pawson and Scott, 1997; Craven and Bredt, 1998; Gomperts, 1996). The respective signaling complexes constitute distinct functional entities (transducisomes) which may be organized to form higher order networks. The clustering of signaling proteins results in the spatial localization of a distinct pathway to a specific subcellular site, and thus provides the basis for a control of cross talk between different pathways and for enhancing the speed and specificity of the signaling process. Among the scaffold or anchoring proteins providing specialized sites for the clustering of signaling proteins, PDZ domain proteins constitute a rapidly growing family with a modular protein binding motif that is conserved from bacteria to man. The repeated, ca. 90 amino acid long sequences were designated as PDZ domains according to their occurrence in the post-synaptic density protein PSD-95 (**P**) (Cho *et al.*, 1992), the tumor suppressor protein disc large (Dlg) of *Drosophila* (**D**) (Woods and Bryant, 1991) and the *zonula occludens* protein ZO-1 (**Z**) (Willot *et al.*, 1993). The modular arrangement of binding sites may, according to the current state of art, comprise up to 13 PDZ domains (Ullmer *et al.*, 1998) in a single protein.

The concept of a signaling complex for phototransduction in the *Drosophila* eye has been derived from recent findings demonstrating that key proteins of the visual pathway are assembled into a supramolecular complex by the PDZ domain protein INAD (Huber *et al.*, 1996a, b; Shieh and Zhu, 1996, Chevesich *et al.*, 1997, Tsunoda *et al.*, 1997). The visual transduction cascade of *Drosophila* appears to be organized as a precisely adjusted network of interacting proteins with the INAD signaling complex as central core. In the rhabdomeral membrane of fly photoreceptors, the internal messenger-generating phospholipase Cβ (PLCβ), the eye specifically-expressed protein kinase C (ePKC) and the major light-activated Ca^{2+} channel TRP are key proteins of visual transduction (see previous chapter) which are directly linked to one of the five PDZ-domains of INAD. A further level of complexity is obtained by homomeric interactions between INAD molecules and by the binding of other proteins, for example calmodulin and the unconventional myosin NINAC (Chevesich *et al.*, 1997; Xu *et al.*, 1998; Wes *et al.*, 1999).

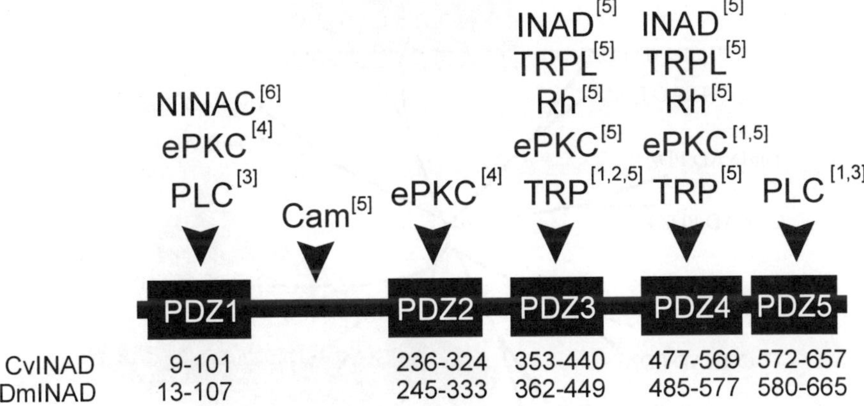

Figure 1. Domain structure of the PDZ domain protein INAD. The amino acid position of the 5 PDZ domains in *Calliphora* (Cv) and *Drosophila* (Dm) is indicated at the bottom. The binding sites for various INAD ligands are depicted by arrow heads as has been reported by [1] Tsunoda *et al.*, 1997, [2] Shieh and Zhu, 1997, [3] van Huizen *et al.*, 1998, [4] Adamski *et al.*, 1998, [5] Xu *et al.*, 1998, and [6] Wes *et al.*, 1999.

Functional studies show that the integration of phototransduction proteins into a supramolecular signaling complex is a prerequisite for the localization of the cascade members to the rhabdomeral membrane (Chevesich *et al.*, 1997, Tsunoda *et al.*, 1997), the generation of reliable single photon responses (Scott and Zuker, 1998) and for correct response termination (Shieh *et al.*, 1997; Adamski *et al.*,

1998, Wes *et al.*, 1999). Thus, contrary to the view that the interactions between proteins cooperating in this G protein coupled signaling pathway result primarily from random collisions of molecules rotating and diffusing freely in a fluid membrane, a sophisticated membrane organization provides the molecular framework for the generation of highly time-resolved and specific visual responses.

2. Molecular Structure of the INAD Signaling Complex

The *InaD* mutants of *Drosophila* were originally classified as inactivation-no afterpotential mutants (Pak *et al.*, 1970). The genes coding for the INAD protein of *Drosophila* and its *Calliphora* homologue have been isolated by subtractive hybridization and immunoscreening (Shieh and Niemeyer, 1995; Huber *et al.*, 1996c).

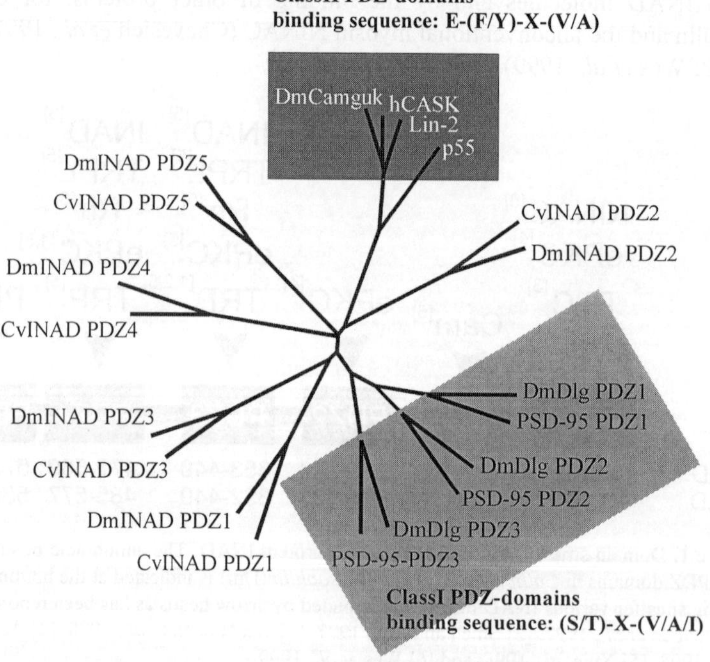

Figure 2. Phylogenetic relationship of selected PDZ domains. PDZ domains contained in *Calliphora* (Cv) and *Drosophila* (Dm) INAD, in the post synaptic density protein 95 (PSD-95), in the *C. elegans* protein Lin2, in the erythrocyte membrane protein p55, and in a *Drosophila* (DmCamguk) and a human (hCASK) PDZ protein with a region of homology to the enzyme calcium-calmodulin protein kinase II were aligned by multiple sequence alignment to generate the phylogramm.

The subsequent analysis of the INAD amino acid sequence revealed that INAD is a modular scaffold protein containing 5 PDZ domains (Fig. 1; Tsunoda *et al.*, 1997). In many cases, PDZ domains bind to specific sequence motifs located at or near the C-terminus of their target proteins. The structural basis of PDZ binding to a C-terminal S/TXV motif has been determined after crystallization of the third PDZ domain of PSD-95 in complex with a peptide ligand (Doyle *et al.*, 1996). The binding of target proteins to PDZ domains of a modular scaffold protein is selective to some extent. In Fig. 2 it is shown that PDZ domains located on different scaffold proteins may be structurally related in so far as they bind to similar C-terminal motifs. Two groups of PDZ domains can be distinguished which bind ligands via either of two possible C-terminal consensus sequences: E-(F/Y)-X-(V/A) or (S/T)-X-(V/A/I) (where X is any amino acid). Figure 2 also reveals that individual PDZ domains of proteins containing multiple PDZ domains, like INAD, can be more homologous to PDZ domains of other scaffold proteins than to their neighboring PDZ domains in the same molecule.

The affinity of target proteins to PDZ domains is high enough to isolate the supramolecular protein complexes by immunoprecipitation. Therefore, this method has become a widely used technique to characterize interactions between PDZ domain proteins and their target proteins. In fact immunoprecipitation, which was originally employed to separate INAD from other proteins of the photoreceptor membrane, has lead to the identification of the key members of the INAD signaling complex (Huber *et al.*, 1996a, b). In Fig. 3 it is shown that after extraction of proteins from eye membranes of *Drosophila* with a non-ionic detergent three other proteins, PLCβ, ePKC and TRP, of the phototransduction cascade co-precipitate with INAD. Quantification of the co-precipitated proteins shows that they occupy binding sites on INAD in an approximately equimolar ratio (Huber *et al.* 1996b) and thus should bind to three out of the 5 PDZ domains contained in INAD.

Although present in an equimolar ratio each key member of the INAD complex may bind to different PDZ domains of INAD. This is suggested by studies directed to identify the ligands of the five PDZ binding domains of INAD. As indicated in Fig. 1, apart from PDZ2 which appears to constitute a binding site only for ePKC, the remaining PDZ domains are able to bind various ligands (Xu *et al.*, 1998). Homomeric interactions between PDZ3 and/or PDZ4 of INAD were proposed to link multiple INAD signaling complexes into a supramolecular structure or a web of signaling proteins (Xu *et al.*, 1998). Additional proteins which include calmodulin, a second ion channel (TRPL), the unconventional myosin NINAC, and rhodopsin were also shown to interact with INAD. NINAC is proposed to interact with PDZ1 of INAD, while calmodulin appears to occupy a binding site which is located in the region between PDZ1 and PDZ2.

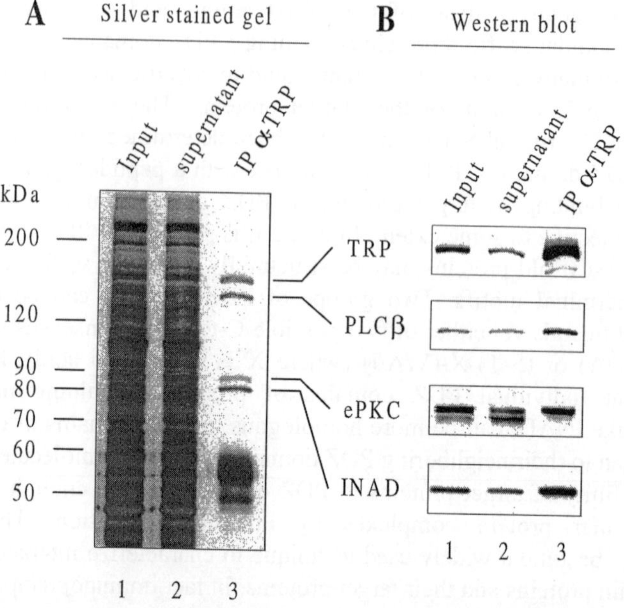

Figure 3. Co-immunoprecipitation of the INAD signaling complex. When a protein extract obtained from *Drosophila* head membranes is immunoprecipitated with anti-TRP antibodies, four major protein bands and the anti-TRP IgG are detected in the precipitate (panel A, lane 3). By Western blot analysis (panel B) the four protein bands are identified as the core components of the INAD signaling complex, TRP, PLCβ, ePKC, and INAD .

Rhodopsin was found to bind to PDZ3 and PDZ4. However, these latter proteins appear to be present at the signaling complex at a substoichiometric ratio and may bind to INAD transiently. They are therefore designated "associated components" as opposed to the core components INAD, TRP, PLCβ, and ePKC of the signaling complex (Table 1).

How the binding of ligands to PDZ domains of INAD is regulated is not yet known. It may be that binding of a ligand to one PDZ domain affects the binding properties of another PDZ domain thus causing a ligand to be exchanged for another one. Direct interactions between the proteins assembled by INAD may also contribute to variation in the binding pattern of PDZ domains of INAD. In addition, members of the INAD signaling complex might be displaced from or tethered to a distinct PDZ by protein phosphorylation. It has been shown that both, INAD and TRP, are putative substrates for ePKC-catalyzed phosphorylation (Huber *et al.*, 1996a, 1998). Phosphorylation of either a PDZ domain of INAD or of putative phosphorylation sites located in or near the C- terminal binding motif of the ligands may modify the binding affinity between a PDZ domain and a specific ligand. That phosphorylation can be part of a regulatory mechanism for binding to PDZ domains has already been demonstrated for the PDZ domain-mediated interaction of the receptor tyrosine kinase EphB3 and the Ras binding protein AF6 (Hock *et al.*, 1998).

Table 1. The INAD signaling complex is composed of major (core) and associated components.

Proposed INAD binding partner	Stoichiometry relative to INAD	Phenotype in INAD null mutant	
TRP	1:1	mislocalization; degradation in older flies	major (core) components
PLCβ	1:1	mislocalization; degradation in older flies	
ePKC	1:1	mislocalization; degradation in older flies	
TRPL	< 1:20	no detectable effect	Associated components
rhodopsin	n. d.	no detectable effect	
calmodulin	n. d.	no detectable effect	
ninaC	n. d.	no detectable effect	

3. Functional Relevance of the INAD Scaffold

Mutations in the *inaD* gene result in phenotypes with defects in phototransduction (see also Paulsen *et al.*, in this volume). Accordingly, the assembly of several key players of the visual cascade into the INAD signaling complex must be a prerequisite for phototransduction to proceed properly. In principle, the defects observed in INAD mutants could be explained by assuming that the scaffolding protein INAD primarily provides a means for the selective localization of phototransduction proteins at the microvillar (rhabdomeral) photoreceptor membrane. Immunocytochemical localization of the core components of the INAD signaling complex in fact shows that PLCβ, ePKC and TRP are mistargeted and become unstable in INAD null mutants of *Drosophila* (Tsunoda *et al.*, 1997), indicating that assembly of the signaling complex is also required to prevent uncontrolled internalization and degradation of the phototransduction proteins. Yet the importance of generating INAD-assembled signaling complexes may reach far beyond a simple protein sorting function. It is likely that the INAD scaffold serves as a site for the sequestration of phototransduction proteins to specialized microdomains of the microvillar membrane which represent functional entities that correspond to the coordinated activation of about 100 TRP channels triggered by the absorption of a single photon within the same microdomain (Scott and Zuker, 1998). The generation of such microdomains which autonomously perform all steps from photon absorption to channel opening would ensure short diffusion distances for messenger molecules and restrict cross-talk to other signaling pathways, thus providing the basis for highly time resolved and specific responses.

4. The INAD Signaling Complex, a Target for Light Activation of PLCβ

In order to provide direct evidence for a role of the INAD signaling complex in phototransduction signaling we investigated whether or not PLCβ tethered to PDZ domains of INAD interacts with the activated state of the visual G protein (Bähner *et al.*, 2000). The visual G protein was activated either by illuminating photoreceptor membranes in the presence of GTPγS with blue light or by adding AlF_4^- which activates G proteins independently of receptor activation (Sternweis and Gilman, 1982). Subsequently, we performed co-immunoprecipitation studies with detergent extracted membrane proteins. Analysis of the immunoprecipitates revealed that activated Gqα (Gqα-GTPγS) of the visual G protein binds to the INAD signaling complex with high affinity, no matter whether the activation is induced by light (Gqα-GTPγS) or mediated by AlF_4^-. Activated Gqα fails to bind to the INAD signaling complex isolated from a *norpA* mutant of *Drosophila* which does not express the PLCβ isoform involved in phototransduction. Thus, we are able to show that Gqα does not only interact with the INAD signaling complex but that PLCβ linked to INAD is the specific target for the light activation of the visual cascade. This strongly supports a concept in which the PDZ domain protein INAD not only serves to retain key players of phototransduction in the microvillar compartment of the photoreceptor cell but constitutes a functional element of the phototransduction mechanism.

5. Interactions between Members of the INAD Signaling Complex, Revealed by ePKC Catalyzed Protein Phosphorylation

A photoreceptor specifically expressed isoform of the Ca^{2+} and phospholipid-dependent protein kinase C (ePKC) is present in the INAD signaling complex in about equimolar amounts with INAD, PLCβ and the TRP channel protein (Fig. 1, Table 1). DAG, one of the messengers required for activation of ePKC is generated from the membrane lipid PIP_2 by the Gqα-activated PLCβ (Bloomquist *et al.*, 1988). Calcium ions which are, in addition to DAG, required for ePKC are provided by Ca^{2+} influx through the light sensitive TRP calcium channels (Hardie and Minke, 1992).

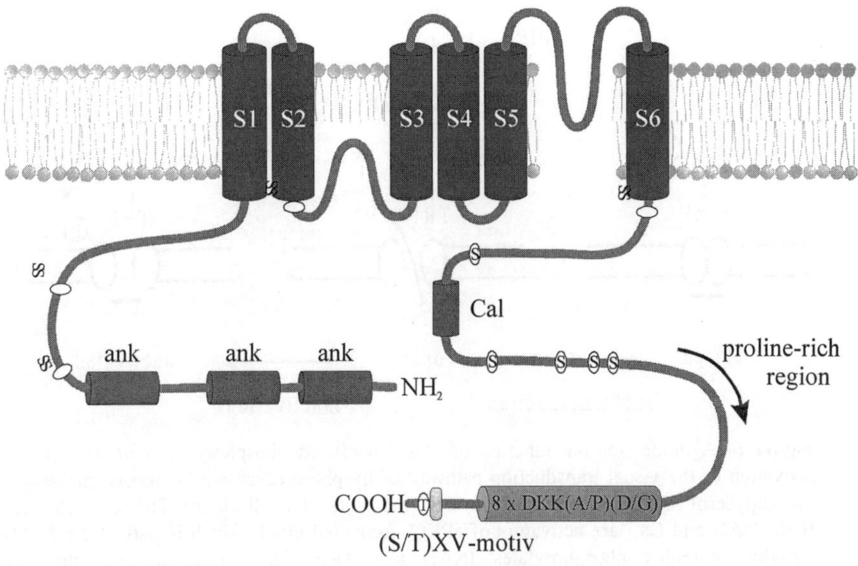

Figure 4. Model of the *Drosophila* TRP calcium channel. Conserved PKC phosphorylation sites are indicated by circles labeled with S (serine) or T (threonine). S1-S6, transmembrane domains; ank, ankyrin repeats; Cal, putative calmodulin binding site.

Thus the modular INAD scaffold allows the activation of ePKC to occur in close proximity to the light sensitive channels and also at the site where the second ePKC activator, DAG, is generated. If the integration of ePKC into the INAD signaling complex optimizes and specifies interactions between the components of the complex, the substrates for ePKC-catalyzed protein phosphorylation must be searched for among the members of the INAD signaling complex. Substrates of ePKC which we have identified so far are INAD and TRP but not PLCβ (Huber *et al.*, 1998). The TRP channel has been shown to become not only phosphorylated by ePKC in the isolated complex but also in intact photoreceptors (Huber *et al.*, 1998). Several potential sites for phosphorylation by ePKC are found in the amino acid sequence of TRP. Ten of these sites are conserved in TRP of *Drosophila* and *Calliphora* (Fig. 4; Huber *et al.*, 1996b). Conserved phosphorylation sites are found in functionally important regions, for example close to the second and sixth transmembrane domain of TRP or near the C-terminus, next to the PDZ target motif (S/TXV) of TRP. Since ePKC mutants (*inaC*) of *Drosophila* show a defect in the deactivation of the light response, *i.e.* the receptor potential evoked by a short light pulse is prolonged as compared to wild-type flies (Smith *et al.*, 1991; Hardie *et al.*, 1993), it is logical to assume that ePKC-catalyzed phosphorylation of TRP is a mechanism for channel inactivation.

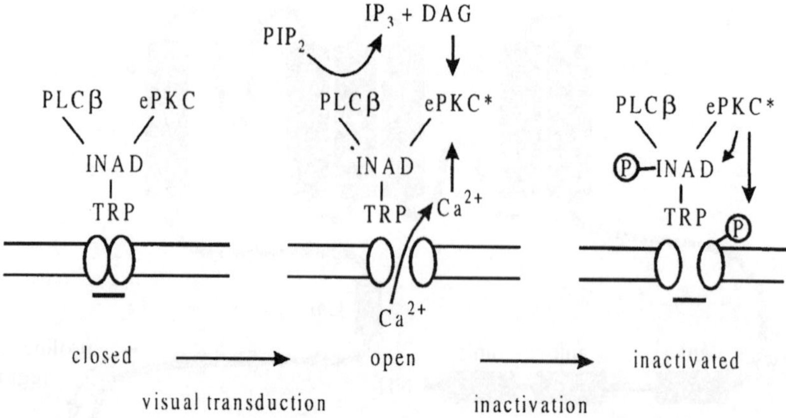

Figure 5. A model for the function of ePKC mediated phosphorylation of TRP. Upon activation of the visual transduction pathway of fly photoreceptors, the second messenger diacylglycerol (DAG) is generated and Ca^{2+} ions enter the cell via the TRP Ca^{2+} channel. Both, DAG and Ca^{2+} are activators of ePKC. Activated ePKC which is part of the INAD signaling complex phosphorylates INAD and TRP. The model suggests that the phosphorylation of TRP contributes to TRP channel inactivation and, thus, to the termination of the receptor potential. This role of TRP phosphorylation would be in line with findings showing that a defect in ePKC results in defective response inactivation (Smith *et al.*, 1991; Hardie *et al.*, 1993).

In the working model depicted in Fig. 5 it is assumed that TRP phosphorylation is a step in a negative feedback loop in which opening of TRP channels results in a local elevation of the calcium concentration. The rise in intracellular calcium then contributes to the activation of ePKC which in turn results in TRP phosphorylation and the subsequent closure of TRP channels. Phosphorylation of TRP and INAD might also be part of the mechanism that controls the binding of proteins to PDZ domains of INAD. Immunoprecipitation studies have so far provided no evidence for an effect of INAD and TRP phosphorylation on the overall composition of the INAD signaling complex (Huber *et al.*, 1998). One possibility is that the effects of protein phosphorylation are not detected under the *in vitro* conditions employed for quantification of the members of the INAD signaling complex. Another explanation for the lack of detectable effects of TRP and INAD phosphorylation on binding to INAD would be that phosphorylation only causes affinity shifts which lead to displacement of a protein ligand from one PDZ domain to another position on the INAD module. A phosphorylation-dependent redistribution of a protein from one PDZ domain to another would not affect the overall composition of the INAD complex although it might have an impact on the phototransducing properties of the complex.

6. A Model for the Role of the INAD Signaling Complex in Phototransduction

In a current model illustrating the protein-protein interactions mediated by the PDZ domains of INAD it is proposed that each microvillus is equipped with a signaling web of phototransducing proteins (Montell, 1998; Fanning and Anderson, 1999). Our quantification of the INAD associated proteins in relation to rhodopsin (Huber *et al.*, 1996b), however, indicates that there are just about 60 to 100 INAD signaling complexes present in each microvillus. This would not be sufficient to generate an extended web but still leaves possibilities for higher levels of membrane organization. This is taken into account in the model shown in Fig. 6 which summarizes interactions involving the INAD signaling complex. It is assumed that an essential part of the microvillar membrane is packed with largely immobile rhodopsin molecules which form a lattice for the efficient absorption of light quanta. INAD complexes may be spaced along the length of a microvillus according to the distribution of NINAC. NINAC is an unconventional myosin containing a domain homologous to the head region of myosin heavy chains which is linked to the INAD signaling complex (Wes *et al.*, 1999) and, may thus connect the signaling complex with the axial actin filament present in each microvillus. A core complex is proposed to consist of four TRP monomers as well as of a corresponding number of PDZ domain linked PLCβ and ePKC molecules (Huber *et al.*, 1996b, Bähner *et al.*, 2000). Core complexes may form larger structural units (transducisomes; Tsunoda *et al.*, 1997, 1998), for example via INAD-INAD interactions (Xu *et al.*, 1998). On the left side of Fig. 6 it is shown that a light-activated rhodopsin (metarhodopsin) interacts with the visual G protein. The activated G protein (Gqα-GTP) diffuses in the membrane to interact with PLCβ on the INAD signaling complex. The interaction of Gqα with INAD-linked PLCβ is indicated at the right side of the model. This interaction initiates a sequence of reactions involving all proteins assembled into the INAD signaling complex as well as phospholipid microdomains in close vicinity to the INAD signaling complexes. Interaction partners of G protein βγ subunits are not yet identified. A spatially restricted expansion of the activation signal to other core complexes, designated as an activated membrane patch, may result from the diffusion of the second messengers IP_3 and DAG generated by activation of PLCβ or from messengers derived from these components, for example long chain polyunsaturated fatty acids (Chyb *et al.*, 1999). Influx of calcium ions through activated TRP channels is thought to create a calcium microdomain. This rise in Ca^{2+} may activate ePKC (see above) and may regulate other components of the signaling complex via Ca^{2+} calmodulin. Calmodulin has been shown to bind to INAD, TRP, TRPL and NINAC (Chevesich *et al.*, 1997; Scott *et al.*, 1997; Porter *et al.*, 1993, 1995). This would generate a further level of calcium-dependent regulatory interactions between members of the INAD signaling complex.

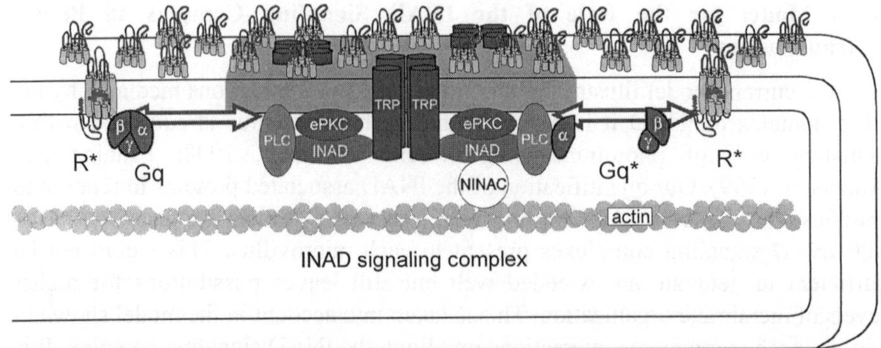

Figure 6. Molecular design of the visual transduction pathway. The model shows an INAD core complex, composed of 4 x (INAD + TRP + ePKC + PLCβ) and anchored to the actin cytoskeleton by the unconventional myosin NINAC. Along with rhodopsins and the visual heterotrimeric G protein molecules the signaling complexes are embedded into the rhabdomeral photoreceptor membrane. Recent evidence suggests that the Gqα subunit of the visual G protein functions as a molecular shuttle which transmits activation of the visual pigment rhodopsin to the INAD signaling complex. It is assumed that the transduction pathway terminates in the coordinated activation of a number of signaling core complexes which may be tethered together to form higher order complexes (transducisomes), and may constitute an activated membrane patch (darker membrane area). This coordinated activation results in a unitary electrical response known as a "quantum bump".

In conclusion, through a series of genetic, immunocytochemical and biochemical experiments it has been shown that the proteins involved in phototransduction by microvillar photoreceptors of *Drosophila* and closely related flies are assembled into supramolecular protein complexes. The assembly is mediated by PDZ domains of the modular scaffold protein INAD. The physical coupling of phototransduction proteins to INAD is required for retaining a distinct surface density of enzymes and ion channels within an individual microvillus. In addition, the clustering of phototransduction proteins is a prerequisite for a coordinated and highly specific activation and control of signal transducing enzymes in close proximity to the light-sensitive ion channels. The precise nature of multiple protein-protein interactions within the INAD signaling complex is still not resolved and thus remains an area of active investigation.

Acknowledgements

Original work reported in this review was supported by grants of the Deutsche Forschungsgemeinschaft (Pa 274/5-1) and the European Union (BMH4-VCT97-2341).

References

Adamski, F.M., M.Y. Zhu, F. Bahiraei and B.H. Shieh (1998) "Interaction of eye protein kinase C and INAD in *Drosophila*. Localization of binding domains and electrophysiological characterization of a loss of association in transgenic flies", *J. Biol.Chem.* 273:17713-17719.

Bähner, M., P. Sander, R. Paulsen and A. Huber (2000) "The visual G protein of fly photoreceptors interacts with the PDZ domain assembled INAD signaling complex via direct binding of activated Gαq to phospholipase Cβ", *J. Biol. Chem.* 275:2901-2904.

Bloomquist, B.T., R.D. Shortridge, S. Schneuwly, M. Perdew, C. Montell, H. Steller, G. Rubin and W.L. Pak, (1988) "Isolation of a putative phospholipase C gene of *Drosophila, norpA*, and its role in phototrans-duction", *Cell* 54:723-733.

Chevesich, J., A.J. Kreuz and C. Montell (1997) "Requirement for the PDZ domain protein, INAD, for localization of the TRP store-operated channel to a signaling complex", *Neuron* 18:95-105.

Cho, K.O., C.A. Hunt and M.B. Kennedy (1992) "The rat brain postsynaptic density fraction contains a homolog of the *Drosophila* discs-large tumor suppressor protein", *Neuron* 9:929-942.

Chyb, S., P. Raghu and R.C. Hardie (1999) "Polyunsaturated fatty acids activate the *Drosophila* light-sensitive channels TRP and TRPL", *Nature* 397:255-259.

Craven, S.E. and D.S. Bredt (1998) "PDZ proteins organize synaptic signaling pathways", *Cell* 93:495-498.

Doyle, D.A., A. Lee, J. Lewis, E. Kim, M. Sheng and R. MacKinnon (1996) "Crystal structures of a complexed and peptide-free membrane protein-binding domain: molecular basis of peptide recognition by PDZ", *Cell* 85:1067-1076.

Fanning, A.S. and J.M. Anderson (1998) "PDZ domains and the formation of protein networks at the plasma membrane", *Curr. Top. Microbiol. Immunol.* 228:209-233.

Fanning, A.S. and J.M. Anderson (1999) "Protein modules as organizers of membrane structure", *Curr.Opin.Cell Biol.* 11:432-439.

Gomperts, S.N. (1996) "Clustering membrane proteins: It's all coming together with the PSD- 95/SAP90 protein family", *Cell* 84:659-662.

Hardie, R.C. and B. Minke (1992) "The *trp* gene is essential for a light-activated Ca^{2+} channel in *Drosophila* photoreceptors" *Neuron* 8:643-651.

Hardie, R.C., A. Peretz, E. Suss-Toby, A. Rom-Glas, S.A. Bishop, Z. Selinger and B. Minke (1993) "Protein kinase C is required for light adaptation in *Drosophila* photoreceptors", *Nature* 363:634-637.

Hock, B., B. Bohme, T. Karn, T. Yamamoto, K. Kaibuchi, U. Holtrich, S. Holland, T. Pawson, H. Rubsamen-Waigmann and K. Strebhardt (1998) "PDZ-domain-mediated interaction of the Eph-related receptor tyrosine kinase EphB3 and the ras-binding protein AF6 depends on the kinase activity of the receptor", *Proc. Natl. Acad. Sci. U.S.A.* 95:9779-9784.

Huber, A., P. Sander and R. Paulsen (1996a) "Phosphorylation of the *InaD* gene product, a photoreceptor membrane protein required for recovery of visual excitation", *J. Biol. Chem.* 271:11710-11717.

Huber, A., P. Sander, A. Gobert, M. Bähner, R. Hermann and R. Paulsen (1996b) "The transient receptor potential protein (Trp), a putative store- operated Ca^{2+} channel essential for phosphoinositide-mediated photoreception, forms a signaling complex with NorpA, InaC and InaD", *EMBO J.* 15:7036-7045.

Huber, A., P. Sander, U. Wolfrum, C. Groell, G. Gerdon and R. Paulsen (1996c) "Isolation of genes encoding photoreceptor-specific proteins by immunoscreening with antibodies directed against purified blowfly rhabdoms", *J. Photochem. Photobiol. B* 35:69-76.

Huber, A., P. Sander, M. Bähner and R. Paulsen (1998) "The TRP Ca^{2+} channel assembled in a signaling complex by the PDZ domain protein INAD is phosphorylated through the interaction with protein kinase C (ePKC)", *FEBS Lett.* 425:317-322.

Montell, C. (1998) "TRP trapped in fly signaling web", *Curr. Opin. Neurobiol.* 8:389-397.

Pak, W.L., J. Grossfield and K.S. Arnold (1970) "Mutants of the visual pathway of *Drosophila melanogaster*", *Nature* 227:518-520.

Pawson, T. and J.D. Scott (1997) "Signaling through scaffold, anchoring, and adaptor proteins", *Science* 278:2075-2080.

Porter, J.A., M. Yu, S.K. Doberstein, T.D. Pollard and C. Montell (1993) "Dependence of calmodulin localization in the retina on the NINAC unconventional myosin", *Science* 262:1038-1042.

Porter, J.A., B. Minke and C. Montell (1995) "Calmodulin binding to *Drosophila* NinaC required for termination of phototransduction", *EMBO J.* 14:4450-4459.

Scott, K., Y. Sun, K. Beckingham and C.S. Zuker (1997) "Calmodulin regulation of *Drosophila* light-activated channels and receptor function mediates termination of the light response *in vivo*", *Cell* 91:375-383.

Scott, K. and C.S. Zuker (1998) "Assembly of the *Drosophila* phototransduction cascade into a signaling complex shapes elementary responses", *Nature* 395:805-808.

Shieh, B.H. and B. Niemeyer (1995) "A novel protein encoded by the *InaD* gene regulates recovery of visual transduction in *Drosophila*", *Neuron* 14:201-210.

Shieh, B.H. and M.Y. Zhu (1996) "Regulation of the TRP Ca^{2+} channel by INAD in *Drosophila* photoreceptors", *Neuron* 16:991-998.

Shieh, B.H., M.Y. Zhu, J.K. Lee, I.M. Kelly and F. Bahiraei (1997) "Association of INAD with NORPA is essential for controlled activation and deactivation of *Drosophila* phototransduction *in vivo*", *Proc. Natl. Acad. Sci. U.S.A.* 94:12682-12687.

Smith, D.P., R. Ranganathan, R.W. Hardy, J. Marx, T. Tsuchida and C.S. Zuker (1991) "Photoreceptor deactivation and retinal degeneration mediated by a photoreceptor-specific protein kinase C", *Science* 254:1478-1484.

Sternweis, P.C. and A.G. Gilman (1982) "Aluminum: a requirement for activation of the regulatory component of adenylate cyclase by fluoride", *Proc. Natl. Acad. Sci. U.S.A.* 79:4888-4891.

Tsunoda, S., J. Sierralta, Y. Sun, R. Bodner, E. Suzuki, A. Becker, M. Socolich and C.S. Zuker (1997) "A multivalent PDZ-domain protein assembles signaling complexes in a G- protein-coupled cascade", *Nature* 388:243-249.

Tsunoda, S., J. Sierralta and C.S. Zuker (1998) "Specificity in signaling pathways: assembly into multimolecular signaling complexes", *Curr. Opin. Genet. Dev.* 8:419-422.

Ullmer, C., K. Schmuck, A. Figge and H. Lubbert (1998) "Cloning and characterization of MUPP1, a novel PDZ domain protein", *FEBS Lett.* 424:63-68.

van Huizen, R., K. Miller, D.M. Chen, Y., Li, Z.C. Lai, R.W. Raab, W.S. Stark, R.D. Shortridge and M. Li (1998) "Two distantly positioned PDZ domains mediate multivalent INAD- phospholipase C interactions essential for G protein-coupled signaling", *EMBO J.* 17:2285-2297.

Wes, P.D., X.Z. Xu, H.S. Li, F. Chien, S.K. Doberstein and C. Montell (1999) "Termination of phototransduction requires binding of the NINAC myosin III and the PDZ protein INAD", *Nat. Neurosci.* 2:447-453.

Willott, E., M.S. Balda, A.S. Fanning, B. Jameson, C. Van Itallie and J.M. Anderson (1993) "The tight junction protein ZO-1 is homologous to the *Drosophila* discs-large tumor suppressor protein of septate junctions", *Proc. Natl. Acad. Sci. U.S.A.* 90:7834-7838.

Woods, D.F. and P.J. Bryant (1991). "The discs-large tumor suppressor gene of *Drosophila* encodes a guanylate kinase homolog localized at septate junctions", *Cell* 66:451-464.

Xu, X.Z., A. Choudhury, X. Li and C. Montell (1998) "Coordination of an array of signaling proteins through homo- and heteromeric interactions between PDZ domains and target proteins", *J. Cell Biol.* 142:545-555.

MOLECULAR CHANGES DURING PRIMARY VISUAL PATHWAY DEVELOPMENT

KENNETH L. MOYA*, ALVIN W. LYCKMAN[§] and ANNAMARIA CONFALONI[°]
*CNRS-CEA URA 2210 and INSERM U334, SHFJ, DRM DSV, Orsay, France
[§]Dept. of Brain and Cognitive Sciences, M. I. T., Cambridge, MA, USA
[°]Istituto Superiore di Sanità, Roma, Italy

ABSTRACT

Retinal ganglion cells form the essential link between the eye and the brain, conveying visual information transduced in the retina to higher visual centers for further processing. In visual animals, the primary visual pathway is marked by a high order of functional and anatomical organization. Information about developmental changes in the visual projection can contribute to understanding how global connectivity and the specificity of the synaptic connections are established.

The visual projection of the hamster has been extensively studied in terms of morphology and molecular changes during development.In the embryonic hamster, newly differentiated ganglion cells in the retina send out axons which fasciculate as they grow towards the optic disk. From there they leave the eye to form the optic nerve, decussate at the optic chiasm and grow along the surface of the diencephalon forming the optic tract.In the early postnatal hamster, axonal collaterals invade their target structures, the lateral geniculate body and superior colliculus, and then arborize forming synaptic neuropils. During this stage appropriate synaptic contacts are stabilized while extraneous contacts are eliminated and the cellular machinery for the regulated release of neurotransmitters is put in place. The use of *in vivo* molecular and cellular approaches has allowed us to characterize the program of protein expression during the rapid elongation of retinal axons through the optic nerve and to their targets in the brain. Elongating retinal ganglion cells express a specific set of proteins that includes growth cone motility proteins such as GAP-43 and the adhesion molecules L1 and NCAM. Also associated with retinal axon elongation are certain isoforms of the amyloid precursor protein, APP, and the related amyloid precursor-like protein 2, APLP2. At early developmental ages these proteins are localized along the axons and can be detected in growth cones. This phase of axon growth requires molecular mechanisms for growth cone motility, axon adhesion for fasciculated growth and guidance cues in order for the growing fibers to successfully reach the target. Function blocking experiments using either antisense oligonucleotides or antibodies suggest that APP and/or APLP2 are necessary retinal axon growth, at least in vitro. Experimental lesion studies *in vivo* showed that the regulation of NCAM and L1 expression was at least partly influenced by interactions in the target. Recent in vitro experiments suggest that L1 adhesion molecule expression in embryonic hamster retinal explants follows an intrinsic program, but the expression of L1 does not appear to be sufficient for retinal axon fasciculation in the presence of other extracellular cues. When retinal axons arborize in their target nuclei, they change their pattern of protein synthesis and axonal transport.Isoforms of APP, and an acidic 67kD protein are upregulated while elongation-associated proteins decline. Cellular interactions at this stage involve target recognition, axon branching, the initial steps for putting in place the molecular machinery for synaptic transmission, and the stabilization or elimination of synapses. Although the precise signals for the change from elongation to arborization are not yet known, experimental manipulations show that the expression of some retinal ganglion cell proteins depends on target interactions. Adult animals depend on the efficient and reliable processing of visual information. The pattern of proteins synthesized in mature retinal ganglion cells and transported to terminals is dominated by synaptic proteins, some of which form part of the synaptic vesicle docking complex. Other proteins may help stabilize synapses and the metabolic labeling studies show that certain proteins have a rapid turnover in the retinal terminals *in vivo*. The combination of anatomical and molecular approaches show that specific proteins are expressed in retinal ganglion cells at precise stages of development. These proteins provide the molecular basis for the growth of retinal axons to visual centers in the brain and the establishment of precise functional connections.

1. Introduction

Vision depends on the transfer of visual information from the site of its transduction to higher visual processing centers in the brain. The primary visual pathway provides the substrate for the transfer and processing of such visual information in vertebrates, and this pathway is marked by a high order of functional and anatomical organization. The pathway consists of retinal ganglion cells projecting their axons to the brain where they innervate the lateral geniculate nucleus (LGN) and superior colliculus (SC) (Fig. 1). In highly binocular animals such as primates, each eye projects to both sides of the brain while in rodents, the vast majority of axons from one eye cross at the optic chiasm and project to the contralateral side of the brain. We have been particularly interested in characterizing the cellular and molecular changes that accompany the formation of the visual pathway as one way to better understand the specificity of synaptic connections in the visual system and global connectivity in the brain.

2. Morphogenetic changes in the developing hamster visual pathway

The hamster has provided a useful animal model for studying morphological changes during development of the primary visual projection. Many of these changes have been well characterized and are summarized in Figure 2 (see: Frost *et al.*, 1979; Campbell *et al.*, 1984; Schneider *et al.*, 1985; Bhide and Frost, 1991; Jhaveri *et al.*, 1991; Jhaveri *et al.*, 1996). Starting at about embryonic day 11.5 (E11.5; date of birth = E15.5), the earliest born retinal ganglion cells grow out axons which are directed towards the optic disk. As the growing fibers encounter other retinal ganglion cell axons they fasciculate into bundles and upon leaving the eye, they form the optic nerve. The majority of retinal ganglion cell axons in the hamster cross at the chiasm and grow up the side of the diencephalon in the optic tract (see Fig. 1). The first retinal axons reach the surface of their principal targets, the LGN in the diencephalon and the SC in the mesencephalon at E14-E15; retinal axons of later born ganglion cells continue to arrive through the perinatal period up to about postnatal day 3 (P3). During this initial phase of axon growth, the retinal fibers elongate rapidly, are highly fasciculated and have large growth cones.

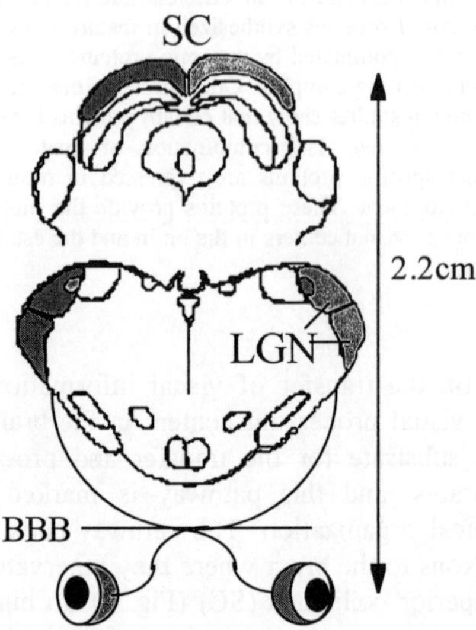

Figure 1. The hamster primary visual projection. SC, superior colliculus; LGN, lateral geniculate nucleus; BBB, blood-brain-barrier.

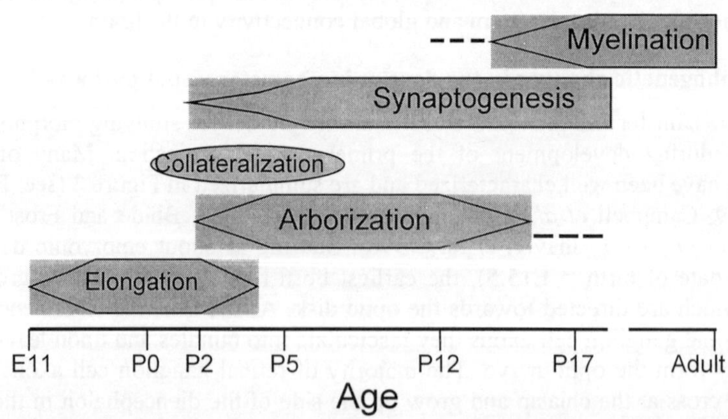

Figure 2. Major morphogenetic events in the developing hamster primary visual pathway.

After reaching the LGN or SC, retinal axons change their mode of growth. The axons advance much more slowly, are defasciculated and have smaller growth cones. Axons which will send a projection to the LGN are mostly situated in the superficial optic tract overlying the LGN. These axons emit a collateral branch from the trunk axon which penetrates into the neuropil of the LGN. The branch ramifies to form an immature terminal arbor. Retinal axon arborization occurs simultaneously in the SC with most taking place from P4-P14 when the eyes open and visually driven connections have been established. During the phase of terminal arborization the arbor size is refined, appropriate synaptic contacts are stabilized while extraneous contacts are eliminated. During the early arborization period (*i.e.*, P4-P7) the initially overlapping projection form the two eyes segregate into distinct territories in the hamster LGN (Fig. 1).

We were interested in characterizing the molecular changes that accompany these morphological changes in the developping primary visual pathway. Specifically, we used an in vivo cell biological approach to try and identify the proteins that were expressed by retinal ganglion cells and that were sent to the growing axon or terminal arbor. After identifying individual proteins synthesized and transported in developing retinal axons we used immunohistochemistry to examine the localization of certain proteins on the axons. Here we review some of the most prominent changes in protein expression and localization.

3. In vivo metabolic labeling

Retinal ganglion cells like most neurons, are highly polarized cells. The soma of these neurons are located in the ganglion cell layer of the retina with their dendrites in the inner plexiform layer, which may extend a hundred or more microns in rodents. The axons of the retinal ganglion cells projecting to the SC are about 2 cm long in the adult hamster. Of the thousands of proteins expressed by neurons, virtually all are synthesized in the soma, and thus, neurons require a specialized mechanism for delivering gene products from their site of synthesis to the appropriate cellular domain. Proteins destined for the axonal membrane are translated from their messenger RNAs in the endoplasmic reticulum, trafficked through the Golgi apparatus where they may be posttranslationally modified before being integrated into axonal transport vesicles. The membraneous organelles are then conveyed down the axons in the rapid phase of axonal tranport, propelled by ATP-hydrolyzing molecular motors of the kinesin family (Hammerschalg and Stone, 1982; see Hirokawa for recent review).

Residing on opposite sides of the blood-brain-barrier (Fig. 1), retinal ganglion cells and their distal axons are also metabolically isolated. The relative remoteness of soma and distal axon combined with metabolic isolation permits analysis of proteins transported down growing axons by labeling that occurs in the eye.

In vivo metabolic labeling of the primary visual pathway is carried out by introcular injection of radiolabeled amino acids which are taken up by cells in the eye and are incorporated into newly synthesized proteins (see Moya, 1998). A four hour interval between eye injection and sacrifice allows proteins newly synthesized in retinal ganglion cells to incorporate the radiolabeled amino acids and to be rapidly transported to the axon ending. The SC and LGN are then dissected out and the radiolabeled proteins analyzed by high resolution 2-dimensional gel electrophoresis. The gel contains all the proteins in the tissue *i.e.*, those synthesized locally in neurons and glia and those in projections from other brain regions. However, by analyzing the radiolabeled polypeptides we focus our attention on only those proteins which were synthesized in retinal ganglion cells at the time of the intraocular injection and which were then transported to the axon ending during the survival interval. In order to identify individual proteins we can use 2-D Western blotting in which antibodies against candidate proteins are reacted with a filter contining the electrophoretically separated proteins. For the developmental studies described here we held the eye injection/sacrifice interval constant and varied the age of the hamsters that we analyzed.

4. Molecular changes during the development of the primary visual pathway

Retinal ganglion cells express a specific set of proteins during the period of axon elongation and this is summarized in Figure 3. This phase of axon growth requires molecular mechanisms for growth cone motility, axon adhesion for fasciculated growth and the recognition of guidance cues in order for the growing fibers to successfully reach the target. In our studies, the highest levels of synthesis and axonal transport of the growth associated protein GAP-43, and the cellular adhesion proteins NCAM and L1 were observed at P2 (Moya *et al.*, 1988, 1992a; Lyckman *et al.*, in preparation). The levels of these proteins are greatly decreased at P5 or at P12. We also observed relatively high levels of higher molecular weight isoforms of the amyloid precursor protein (APP), and the related amyloid precursor-like protein 2 (APLP2) in neonatal retinal axons, however the developmental decrease for these two proteins was gradual and not as great (Moya *et al.*, 1994; Lyckman *et al.*, 1998).

Immunohistochemistry studies have revealed the localization of various proteins in elongating retinal axons. GAP-43 is abundant in retinal axon growth cones elongating in vitro and is present along the axons in the P0-P2 hamster optic tract (Moya, 1998; Moya *et al.*, 1989). At the time of arborization and when levels of GAP-43 synthesis and transport decrease, the localization of the protein in the optic tract changes from the axons to the neuropil and then is greatly diminished by the time the eyes open at P14. GAP-43 is a phosphoprotein, enriched in growth cones and localized to the inner face of the plasma membrane where it is thought to amplify signal transduction (see Benowitz and Routtenberg, 1997; Nakamura *et al.*, 1998).

The cell adhesion molecule L1 is distributed along retinal axons in vitro and in vivo when they elongate in fasciculated bundles (Lyckman *et al.*, in preparation).

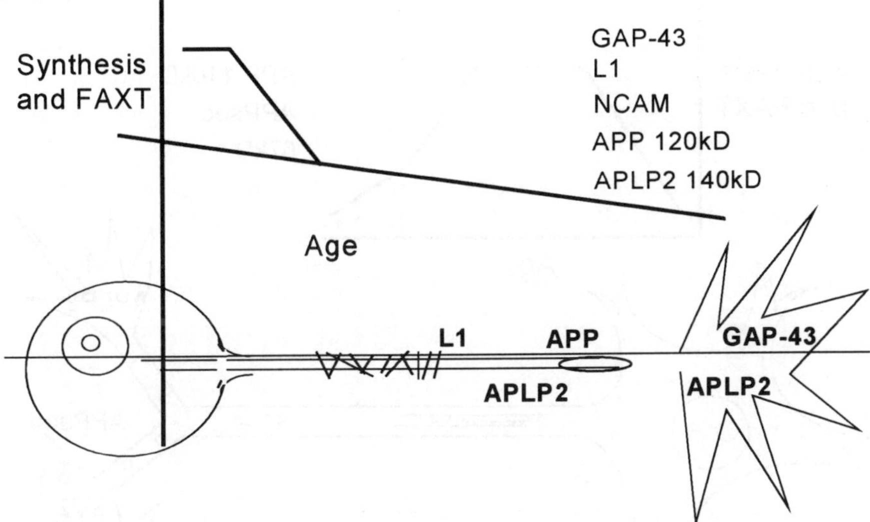

Figure 3. Protein expression during retinal ganglion cell axon elongation.

Biochemical characterization of the L1 transported in neonatal hamster retinal axons show that the protein is sulfated and carries a complex carbohydrate structure implicated in binding to laminin. L1 has documented adhesive functions and a high level of synthesis during axon elongation and its abundance along retinal axons at this stage suggests that the protein is involved in the fasciculated elongation of these axons. As the retinal axons defasciculate and arborize in the LGN and SC, there is a sharp decrease in L1 synthesis and axonal transport accompanied by a loss of L1 immunoreactivity on the optic axons.

In parallel with the transition from elongation to arborization, a new pattern of proteins is expressed in retinal ganglion cells and sent to the distal regions of their axons (Fig. 4). The synthesis and axonal transport of specific molecular weight isoforms of the amyloid precursor protein begin to increase at the end of retinal axon elongation and the early phases of terminal arborization (Moya *et al.*, 1994). A full-length transmembrane form and a C-terminal truncated form of APP remain at high levels during synaptogenesis and then their levels diminish towards the end of the period of arborization. The C-terminal truncated form of APP likely corresponds to APP secreted from the cell membrane and taken together, the results obtained by metabolic labeling suggest that the cellular mechanism for the secretion of APP from neurons in vivo is maturation-dependent and coincides with the formation of synaptic contacts. In addition to isoforms of APP, the developmental profile of a 67kD protein also increases at the time of synaptogenesis and then declines after the visual pathway matures (Moya *et al.*, 1988; 1992a).

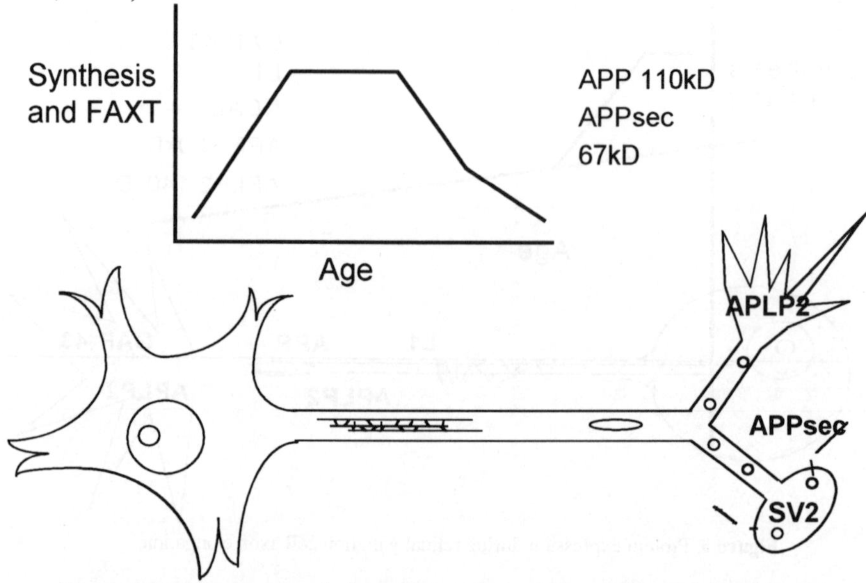

Figure 4. Protein expression during retinal axon arborization and synaptogenesis.

Experimental manipulations which disrupt the formation of normal retinofugal projections alters the developmental timecourse of protein expression. Obliteration of the superficial SC on the day of birth elminates one of the principal synaptic territories for retinal axons in the brain. The retinal axons respond to the loss of synaptic territory in the SC by forming dense patches of abnormal hyperinnervation in the LGN and by establishing a novel and long lasting visual projection in the lateral posterior nucleus in the thalamus (Moya *et al.*, 1990). We then examined the effects of the lesion and the formation of abnormal connections on the program of protein expression in retinal ganglion cells. The in vivo metabolic labeling experiments revealed that the developmental profile of some but not all proteins was altered (Moya *et al.*, 1992a). The decrease in L1 and NCAM which normally occurs after P2-P5 was delayed and levels of these protens remained elevated at P12 and P17 before diminishing to the low consititutive adult levels. A 67kD protein which increases during synaptogenesis and then decreases after the eyes open, remained significantly elevated at P17 in the animals with an early lesion of the SC. A 64kD protein which shows a steady increase in normal hamsters took longer to reach maximal levels in the lesioned group. The developmental timecourse of other proteins such as GAP-43, APP, APLP2 and

SNAP-25, however, were not affected by the early lesion and resultant abnormal connections. Taken together, the results show that during visual pathway formation the expression of some retinal ganglion cell proteins is independent of distal events and perhaps follows an intrinsic program of gene expression while the expression of other proteins destined for the growing axon in influenced by interactions in the target zone.

In the mature primary visual pathway, the prominent proteins synthesized in ganglion cells and transported in retinal axons are components of the synaptic terminal (Fig. 5). The most abundant are SNAP-25 and a syntaxin-like 67kD protein (Moya *et al.*, 1988; 1992a; Smirnova and Moya, unpublished results). We have also recently identified synaptobrevin as an axonally transported protein, however, further studies are necessary to characterize its developmental profile (Lyckman and Moya, unpublished results). SNAP-25, syntaxin and synaptobrevin interact to form the ternary SNARE complex which is essential for synaptic vesicle fusion and neurotransmitter release from the synapse (Hayashi *et al.*, 1995). In addition to these consituents of the SNARE docking complex, we have studied the localization of the major synaptic vesicle proteins, SV2 and rab3a, in retinal axon terminals (Confaloni *et al.*, 1997; Moya *et al.*, 1992b).

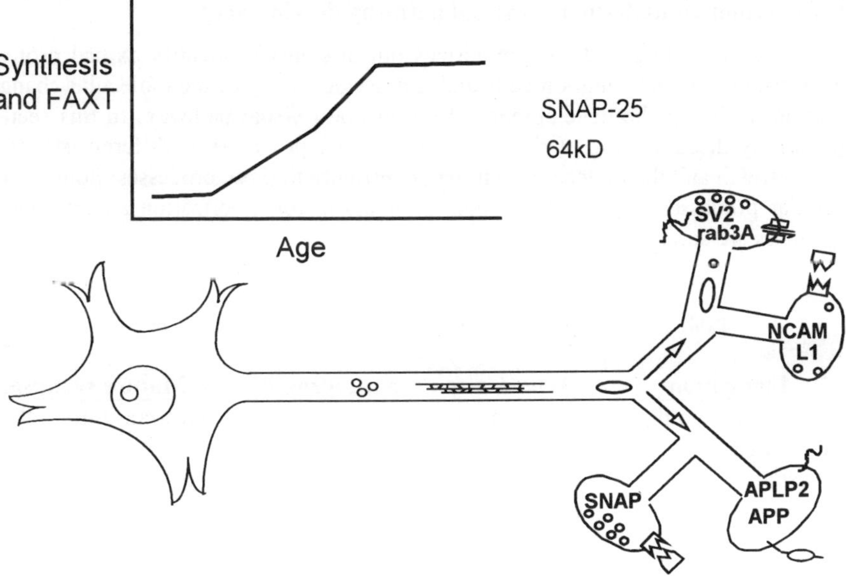

Figure 5. Protein expression in mature retinal ganglion cells with functional synapses.

SV2 is an ubiquitous synaptic vesicle protein with strong homology to a transporter molecule, but its function is still unknown (Feany *et al.*, 1992). SV2 is abundant in adult retinal terminals and during development its synaptic-like distribution coincides with early stages of synaptogenesis and precedes the emergence of fully functional synapses (Confaloni *et al.*, 1997; see Stettler *et al.*, 1996).

Rab3A is a small GTP-hydrolyzing protein of secretory vesicles that controls the formation or stability of the SNAP-25/synaptobrevin/ syntaxin SNARE complex (Johannes *et al.*, 1996).

Synaptic membrane glycoproteins sent to retinal terminals at relatively low, but constant levels include APP, APLP2, NCAM and L1 (Moya *et al.*, 1992a; 1994; Lyckman *et al.*, 1998 and in preparation). NCAM and L1 have been the object of extensive study and in addition to their role in axon adhesion, these membrane glycoproteins are also important for synaptic plasticity in hipocampal neurons (Lüthi *et al.*, 1994). As for the transmembrane glycoproteins APP and APLP2, the precise function of APP or APLP2 has yet to be established. However there is accumulating evidence for a role for APP in normal synaptic function (see related chapter this volume).

5. Molecular contributions to visual pathway development

The preceding section summarizes our attempts to identify axonal proteins expressed by retinal ganglion cells and it describes many of the molecular changes that occur during the development of the primary visual pathway. In this section we briefly discuss some of the important cellular processes at different stages of axon growth and the molecules that may contribute to these processes. Some of the cellular processes important for axon elongation, for arborization and in mature synapses are summarized in Table 1.

Table 1. Cellular processes during the formation of the retinal projection.

Elongation	Arborization/synaptogenesis	Mature synapses
growth cone motility	target recognition	neurotransmitter release
axon guidance	axon branching	vesicle cycling
adhesion	vesicle machinery	synaptic stability
	synapse elmination/stabilization	

Some of the proteins identified in our studies now have documented roles in the development of pathways in the brain. Disruption of the GAP-43 gene causes retinal axons to pause longer or halt completely at the optic chaism and results in the misrouting of retinal fibers into contralteral and ipsilateral optic tracts

(Strittmatter *et al.*, 1995; Stretevan and Kruger, 1998). This defect in retinal axon routing may be due to growth cone dysfunction since in vitro studies have reported that growth cones depleted of GAP-43 have an abnormal motility and adhere poorly to the substrate (Aigner and Caroni, 1995). Retinal axons that reach the LGN and SC of GAP-43 deficient mice do not form normal synaptic terminal zones and the topographical projection is abnormal (Zhu and Julien, 1999). Thus, while GAP-43 does not to appear to be essential for the differentiation of an axon by retinal ganglion cells nor for elongation *per se*, the protein is required for normal sorting of elongating retinal axons at the optic chaism where the extracellular environment of the midline provides instructive cues for axon path choice. The studies cited above also show that protein is important for normal retinal axon arborization and this coincides with the high level of synthesis and the presence of GAP-43 on retinal axons during the initial phase of arborization at P5 in the hamster.

L1 is essential for the correct development of specific neuronal pathways (see Hortsch, 1996). Mutations which disrupt L1 cause a loss of fibers and defects in decussation in the corticospinal tract which have been attibuted to a defect in cellular adhesion and axon fasciculation (Dahme *et al.*, 1997). More detailed analyses are necessary to determine whether the primary visual pathway might also have a high dependency on L1. In vitro studies using embryonic hamster retinal explants show that L1 is abundant on retinal axons even when the axons grow in a defasciculated manner (Moya, 1998; Lyckman *et al.*, in preparation). Thus, while L1 may be necessary for the fasciculation and normal growth of axons in certain pathways, the expression of L1 on axons by itself, is not always sufficient to induce and maintain fasciculated axon elongation.

The causative role of altered forms of APP in familial Alzheimer's disease and the synaptic loss associated with this pathology strongly implicate APP in the establishment or maintenance of synaptic contacts (see related chapter, this volume). Support for such a role in retinal axons is to be found in the results of our developmental studies in which the peak of synthesis and axonal transport of certain isoforms of APP are highly correlated with synaptogenesis (Moya *et al.*, 1994). Studies aimed at testing a direct contribution of APP to synaptogenesis using constitutive gene inactiviation, however, have been complicated by the expression of the closely homologous APLP2 which can functionally compensate for the loss of APP (van Koch *et al.*, 1997). APP may also contribute to earlier stages of neuronal differentiation as suggested by studies showing that the protein mediates NGF-induced neurite outgrowth and cell-substrate adhesion (Milward *et al.*, 1992; Schubert *et al.*, 1989; Breen *et al.*, 1991). Indeed, in vitro experiments in which antisense oligonucleotides specific for APP mRNA was administered to cultures of dissociated cortical neurons showed that neurite extension was inhibited when APP synthesis was blocked (Allinquant *et al.*, 1995). Recent preliminary studies have extended this finding to retinal axons in explant culture. When we applied antisense oligonucleotides specific for APP mRNA or an antibody directed against an extracellular region of APP retinal axon growth was reduced (Lyckman, Young and Moya, unpublished results). Although preliminary, these results suggest that blocking APP either at the level of its synthesis in retinal ganglion cells or at the level of the cell membrane, inhibits retinal axon growth.

Some of the proteins we have identified in mature retinal terminals have been shown in other systems to be required for the release of neurotransmitters. Disruption of the synaptic vesicle docking complex containing SNAP-25, synaptobrevin and syntaxin blocks secretion. For example, inactivation of SNAP-25 by selective cleavage with botulinum toxin A inhibits neurotransmitter release from synaptosomes, most likely by interferring with the docking of synaptic vesicles at the site of fusion (Blasi *et al.*, 1993). Studies with aplysia neurons have shown that mutant rab3 protein deficient for GTP-ase activity blocks neurotransmitter release (Johannes *et al.*, 1996). Subsequent analyses aimed at determining the step at which rab3 intervenes suggest that rab3 controls the formation or the stability of the SNARE complex which is necessary for vesicle fusion and regulated exocytosis (Johannes *et al.*, 1996).

The studies reviewed above describe changes in the most prominent proteins and clearly, further studies are needed to identify additional molecules present in developing retinal axons and to determine their function. On the basis of the results obtained to date, however, we can begin to establish a coherent view of the molecular changes and visual pathway development. Retinal ganglion cells synthesize L1, NCAM, APP, APLP2, and transport these proteins down their axons and to their growth cones which sample the surrounding environment. Some or all of these proteins may function as growth cone surface receptors to transduce guidance cues and other signals present in the surrounding environment. GAP-43 in the growth cone amplifies or modulates transduction of these signals. NCAM and L1 present along the trunk of the axons serve to maintain axons in fasciculated bundles reflecting these proteins' known role as adhesion molecules.

Upon reaching potential brain targets, the retinal axons must be able to recognize their appropriate target zone in order to stop their advance in the LGN or SC, or in the case of collateral branching, emit side branches from the correct segment of axon. At the time of this change in axon growth, retinal ganglion cells change their protein expression. Elongation-associated proteins such as GAP-43, L1 and NCAM are sharply reduced. Isoforms of APP increase as the retinal projection is refined, appropriate synapses are stabilized and others are eliminated. Some of the protein changes appear to follow an intrinsic program while other changes appear to be regulated by interactions in the target, although the nature of the cellular signals that initiate this transition remain unknown. An interesting question which remains open is whether the transition from elongation to synaptogenesis requires both the down regulation of elongation-associated proteins and the up regulation of synaptogenesis-associate proteins or whether, the expression of the molecules associated with arborization and synapse formation is sufficient to induce the change in growth.

As the branches become more complex and early boutons form, the molecular machinery for synaptic vesicle fusion and recycling must be made available to the newly formed terminals. The synthesis and transport of vesicle docking complex proteins is first detectable at the beginning of retinal axon arborization. Synthesis and axonal transport of SNAP-25 and the 64kD syntaxin-like protein continue to increase as synaptogenesis procedes and they remain at relatively high levels in the mature visual projection. As the terminal membrane turnsover, a high level of proteins involved in vesicle fusion and exocytosis as well as a lower constitutive flow of synaptic proteins (APP, APLP2, NCAM, L1) is required to maintain normal synaptic function.

In summary, the combination of anatomical and molecular approaches described here show that specific proteins are expressed in retinal ganglion cells at precise stages of development. These proteins provide the molecular basis for the growth of retinal axons to visual centers in the brain and the establishment of precise functional connections.

Acknowledgements

We thank L. Di Giamberardino for constant encouragement and support. Portions of the work reviewed here were supported by CNRS, INSERM, CEA and the EEC (BMH1-CT-94-8652).

References

Aigner, L. and P. Caroni (1995) "Absence of persistent spreading, branching, and adhesion in GAP-43-depleted growth cones", *J. Cell Biol.* 128:647-660.

Allinquant, B., P. Hantraye, P. Mailleux, K.L. Moya, C. Bouillot and A. Prochiantz (1995) "Down regulation of amyloid precursor protein inhibits neurite outgrowth in vitro", *J. Cell Biol.* 128:919-927.

Benowitz, L.I. and A. Routtenberg (1997) "GAP-43: An intrinsic determinant of neuronal development and plasticity", *Trends Neurosci.* 20:84-91.

Bhide, P.G. and D.O. Frost (1991) "Stages of growth of hamster retinofugal axons: Implications for developing pathways with multiple targets", *J. Neurosci.* 11:485-504.

Blasi, J., E.R. Chapman, E. Link, T. Binz, S. Yamasaki, P. De Camilli, T.C. Sudhof, H. Niemann and R. Jahn (1993) "Botulinum neurotoxin: A selectively cleaves the synaptic protein SNAP-25", *Nature* 365:160-163.

Breen, K.C., M. Bruce and B.H. Anderton (1991) "Beta amyloid precursor protein mediates neuronal cell-cell and cell-surface adhesion", *J. Neurosci. Res.* 28:90-100.

Campbell, G., K.-F. So and A.R. Lieberman (1984) "Normal post-natal development of retinogeniculate axons and terminals and identification of inappropriately-located transient synapses", *Neuroscience* 13:743-759.

Confaloni, A., A.W. Lyckman and K.L. Moya (1997) "Developmental shift of synaptic vesicle protein 2 (SV2) from axons to terminals in the primary visual projection of the hamster", *Neuroscience* 77:1225-1236.

Dahme, M., U. Bartsch, R. Martini, B. Anliker, M. Schachner and N. Mantei (1997) "Disruption of the mouse L1 gene leads to malformations of the nervous system", *Nat. Genet.* 17:346-349.

Feany, M.B., S. Lee, R.H. Edwards and K.M. Buckley (1992) "The synaptic vesicle protein SV2 is a novel type of transmembrane transporter", *Cell* 70:861-867.

Frost, D.O., K.-F. So and G.E. Schneider (1979) "Postnatal development of retinal projections in Syrian hamsters: A study using autoradiographic and anterograde degeneration techniques", *Neuroscience* 4:1649-1677.

Hammerschlag, R. and G.C. Stone (1982) "Membrane delivery by fast axonal transport", *Trends Neurosci.* 5:12.

Hayashi, T., H. Mahon, S. Yamasaki, T. Binz, T.C. Südhof and H. Nieman (1995) "Synaptic vesicle membrane fusion complex: action of clostidial neurotoxins on assembly", *EMBO J.* 13:5051-5061.

Hirokawa, N. (1993) "Axonal transport and the cytoskeleton", *Curr. Op. Neurobiol.* 3:724-731.

Hortsch, M. (1996) "The L1 family of neural cell adhesion molecules: Old proteins performing new tricks", *Neuron* 17:587-593.

Jhaveri, S., M.A. Edwards and G.E. Schneider (1991) "Initial stages of retinofugal axon development in the hamster: evidence for two distinct modes of growth", *Exp. Brain Res.* 87:371-382.

Jhaveri, S., R.S. Erzurumlu and G.E. Schneider (1996) "The optic tract in embryonic hamsters: Fasciculation, defasciculation, and other rearrangements of retinal axons", *Visual Neurosci.* 13:359-374.

Johannes, L., P.M. Llédo, P. Chameau, J.D. Vincent, J.P Henry and F. Darchen (1998) "Regulation of the Ca^{2+} sensitivit!y of exocytosis by Rab3a", *J. Neurochem.* 71:1127-33.

Johannes, L., L.F. Doussau, A. Clabecq, J.P. Henry, F. Darchen and B. Poulain (1996) "Evidence for a functional link between Rab3 and the SNARE complex", *J. Cell Sci.* 109:2875-84.

Lüthi, A., J.P. Laurent, A. Figurov, D. Muller and M. Schachner (1994) "Hippocampal long-term potentiation and neural cell adhesion molecules L1 and NCAM", *Nature* 372:777-779.

Lyckman, A.W., A. Confaloni, G. Thinikaran, S.S. Sisodia and K.L. Moya (1998) "Amyloid precursor superfamily protein postranslational processing and presynaptic turnover kinetics in the CNS *in vivo*", *J. Biol. Chem.* 273:11100-11106.

Milward, E.A., R. Papadopoulos, S.J. Fuller, R.D. Moir, D. Small, K. Beyreuther and C.L. Masters (1992) "The amyloid protein precursor of Alzheimer's disease is a mediator of the effects of nerve growth factor on neurite outgrowth", *Neuron* 9:129-137.

Moya, K.L. (1998) "Retinal ganglion cell axonal transport: Moving down the road to functional connections", in: *Development and Organization of the Retina, From Molecules to Function,* L.M. Chalupa and B. Finlay, eds, NATO Advanced Study Institute, New York, Plenum, pp. 259-274.

Moya, K.L., S. Jhaveri, L.I. Benowitz and G.E. Schneider (1988) "Changes in rapidly transported proteins in developing hamster retinofugal axons", *J. Neurosci.* 8:4445-4454.

Moya, K.L., S. Jhaveri, G.E. Schneider and L.I. Benowitz (1989) "Immunohistochemical localization of GAP-43 in the developing hamster retinofugal pathway", *J. Comp. Neurol.* 288:51-58.

Moya, K.L., L.I. Benowitz and G.E. Schneider (1990) "Abnormal retinal projections suppress GAP-43 in the diencephalon", *Brain Res.* 527:259-265.

Moya, K.L., L.I. Benowitz, B.A. Sabel and G.E. Schneider (1992a) "Changes in rapidly transported proteins associated with development of abnormal projections in the diencephalon", *Brain Res.* 586:265-272.

Moya, K.L., O. Stettler, A. Zahraoui, L. Di Giamberardino and B. Tavitian (1992b) "Rab3A is expressed in the primary visual system", *Neurosci. Abstr.* 18:1030.

Moya, K.L., L.I. Benowitz, G.E. Schneider and B. Allinquant (1994) "The amyloid precursor protein is developmentally regulated and correlated with synaptogenesis", *Dev. Biol.* 161:597-603.

Nakamura, F., P. Strittmatter and S.M. Strittmatter (1998) "GAP-43 augmentation of G protein-mediated signal transduction is regulated by both phosphorylation and palmitoylation", *J. Neurochem.* 70:983-92.

Schneider, G.E., S. Jhaveri, M.A. Edwards and K.-F. So (1985) "Regeneration, re-routing, and redistribution of axons after early lesions: Changes with age, and functional impact", in: *Recent Achievements in Restorative Neurology, Vol 1. Upper Motor Function and Dysfunction,* J. C. Eccles and M.R. Dimitrijevic, eds, Basel: Karger, pp. 291-310.

Schubert, D., W. Jin, T. Saitoh and G. Cole (1989) "The regulation of amyloid ß protein precursor secretion and its modulatory role in cell adhesion", *Neuron* 3:689-694.

Stettler, O., B. Tavitian and K.L. Moya (1996) "Differential synaptic vesicle protein expression in the barrel field of developing cortex", *J. Comp. Neurol.* 375:321-332.

Stretavan, D.W. and K. Kruger (1998) "Randomized retinal ganglion cell axon routing at the optic chiasm of GAP-43-deficient mice: association with midline recrossing and lack of normal ipsilateral axon turning", *J. Neurosci.* 18:10502-13.

Strittmatter, S.M., C. Frankhauser, P.L. Huang, H. Mashimo and M.C. Fishman (1995) "Neuronal pathfinding is abnormal in mice lacking the neuronal growth cone protein GAP-43", *Cell* 80:445-452.

von Koch, C.S., H. Zheng, H. Chen, M. Trumbauer, G. Thinakaran, L.H. van der Ploeg, D.L. Price and S.S. Sisodia (1997) "Generation of APLP2 KO mice and early postnatal lethality in APLP2/APP double KO mice", *Neurobiol. Aging* 18:661-669.

Zhu, Q. and J.P. Julien (1999) "A key role for GAP-43 in the retinotectal topographic organization", *Exp. Neurol.* 155:228-242.

METABOLISM OF A SYNAPTIC PROTEIN IN MATURE RETINAL TERMINALS *IN VIVO*: IMPLICATIONS FOR ALZHEIMER'S DISEASE

KENNETH L. MOYA*, ALVIN W. LYCKMAN[§] and ANNAMARIA CONFALONI[°]
**CNRS-CEA URA 2210 and INSERM U334, SHFJ, DRM DSV, Orsay, France*
[§]Dept. of Brain and Cognitive Sciences, M. I. T., Cambridge, MA, USA
[°]Istituto Superiore di Sanità, Roma, Italy

ABSTRACT

Normal function of retinal ganglion cell presynapses requires the constant replenishment of synaptic proteins. These molecules are synthesized in retinal ganglion cell bodies and transported down the nerve fibers to axon terminals.In the mature primary visual projection of the hamster, retinal axon terminals in the brain are separated from the soma of the ganglion cells by about two centimeters. Residing on opposite sides of the blood-brain-barrier, retinal ganglion cells and their terminals are also metabolically isolated. The relative remoteness of soma and terminal combined with metabolic isolation permits analysis of presynaptic brain proteins by labeling that occurs in the eye. In addition to furthering our understanding of the cell biology of retinal ganglion cell synapses, the primary visual pathway provides a CNS model system with which to study the metabolism of neuronal proteins destined for brain synapses in general. Molecular analysis of metabolically labeled proteins using high resolution two-dimensional gel electrophoresis showed that prominent among the proteins synthesized in retinal ganglion cells and transported down the axons to synaptic terminals is the amyloid precursor protein (APP) implicated in Alzheimer's disease (AD).

Alzheimer's disease is a neurodegenerative disease with severe and progressive memory loss and dementia. The neuropathological hallmarks of AD are the presence of senile plaques and neurofibrillary tangles. Senile plaques are deposits of amyloid in the brain parenchyma with a 42- amino acid peptide at the core of the plaques. This peptide, called the $\beta A4_{42}$ peptide, is a fragment of APP.A number of APP gene mutations have been described in families with a history of early onset AD. Neuronal APP is a 695 amino acid transmembrane glycoprotein implicated in cellular adhesion, however, the precise function of APP remains unknown. Neurofibrillary tangles are composed of paired helical filaments of aggregated tau protein. Tau is a microtubule associated protein which in normal neurons participates in microtubule stabilization. The tau protein found in neurofibrillary tangles from AD brain is hyperphosphorylated and recent cell biological studies have shown that abnormally phosphorylated tau aggregates into the paired helical filaments.

Based on these sets of data, two dominant models of AD etiology have emerged. In the *ßA4/senile plaque model*, the cause of the disease is an accumulation of amyloidergic $\beta A4_{42}$. These amyloid deposits are toxic for neurons and the subsequent neuronal loss accounts for the cognitive deficits seen inpatients with AD. Thus, in this model, the critical event leading to neural dysfunction is enhancement of processes that yield the $\beta A4_{42}$ peptide from APP.

In the *tau/neurofibrillary tangle model*, the cause of the disease is the accumulation of the neurofibrillary tangles in axons and cell bodies of neurons. These tangles cause neuronal cell death by blocking up or disrupting the cytoskeletal organization of the neurons and the neuronal loss leads to the cognitive deficits in AD. In this model, the critical event is the hyperphosphorylation of tau protein due to overactive protein kinase activity and/or underactive phosphatase activity.

Detailed anatomic studies have suggested, however, that synaptic loss is the most prominent feature of AD brain changes (DeKosky *et al.*, Ann Neurol. 27: 457-464, 1990) and that synaptic loss was the neuropathological change most highly correlated with the severity of dementia (Terry *et al.*, Ann. Neurol. 30:572-580, 1991). In these studies no correlation was noted between the number of senile plaques or the level of neuronal loss and dementia, and only a modest relationship was reported for the number of tangles and dementia.

The eye can be used as a window onto synaptic function in the brain. *In vivo* metabolic labeling studies of the primary visual projection have been used to study the APP that is destined for the retinal ganglion cell presynapse.APP is developmentallyregulated in the brain and the expression of some isoforms is correlated with synaptogenesis. The results also suggest that the mechanism for the cleavage and secretion of APP is maturation dependent and is correlated with the formation of synaptic contacts. The APP that arrives at the synaptic terminal *in vivo* is N- and O-glycosylated, contains sialic acid and is sulfated. These posttranslational modifications are consistent with the protein functioning as an adhesion molecule at the cell surface and we have suggested that APP acts at the synapse to maintain synaptic efficacy. The full-length transmembrane form of APP is cleaved and rapidly eliminated from the synapse *in vivo*. The half-life of APP at the synaptic terminal is 2-3 hours and the turnover of APP is not dependent on retinal ganglion cell activity.

These data obtained in the retinal projection lead us to propose *a model of Alzheimer's disease in which a loss of synaptic efficacy due to an alteration of APP metabolism is the precipitating cellular event in the disease process.* We hypothesize that APP plays a fundamental role in normal synaptic function, perhaps through interactions with other neuronal surface glycoproteins. In AD, subtle changes in APP metabolism could arise sporadically in cases of nonfamilial AD or from mutations in APP itself or proteins essential for APP processing such as the presenilins in familial forms of the disease. With a synaptic half-life of 2-3 hours, small perturbations in the rate of APP synthesis or its rate of elimination would result in rapid changes in the levels of APP at the nerve terminal with a subsequent loss of synaptic efficacy. Such a change in synaptic efficacy and associated loss of synapses would readily explain the early cognitive changes in the course of the disease which appear before the marked neuropathological changes. A continuing synaptic loss leads to neuronal dysfunction and eventually to the accumulation of amyloid plaques and neurofibrillary tangles resulting in frank neuronal loss and severe cognitive decline.

1. Introduction

Normal function of retinal ganglion cell presynapses requires the constant replenishment of synaptic proteins. These molecules are synthesized in retinal ganglion cell bodies and transported down the nerve fibers to axon terminals. In the mature primary visual projection of the hamster, retinal axon terminals in the brain are separated from the ganglion cell soma by about two centimeters (see Fig 1, preceding chapter). Residing on opposite sides of the blood-brain-barrier, retinal ganglion cells and their terminals are also metabolically isolated. The relative remoteness of soma and terminal combined with metabolic isolation permits analysis of presynaptic brain proteins by labeling that occurs in the eye. In addition to furthering our understanding of the cell biology of retinal ganglion cell synapses, the primary visual pathway provides a window onto the brain with which to study the metabolism of neuronal proteins destined for synapses in vivo.

In vivo metabolic labeling uses intraocular injection of radiolabeled amino acids followed by molecular analysis of metabolically labeled proteins with high resolution 2-dimensional gel electrophoresis. The amino acids injected into the eye are taken up by cells in the retina including the ganglion cells, and the amino acids are incorporated into proteins synthesized in the cells. Proteins destined for the synaptic terminal membrane are packaged into membranous organelles and axonally transported in the rapid phase axonal transport (Hammerschlag and Stone, 1982). For the analysis of synaptic protein turnover, we held the age of the animals constant (adult) and varied the intraocular injection/sacrifice interval. This allowed us to follow the arrival of newly synthesized, radiolabeled proteins at the retinal terminals and the subsequent clearance of these proteins from the terminals. The metabolic labeling approach combined with protein biochemical characterization can also provide useful information about the specific isoforms and the posttranslational modifications of proteins targetted to the axon terminal.

Metabolic labeling and 2-dimensional Western blotting have shown that the amyloid precursor protein (APP) is prominent among the proteins which were synthesized in hamster retinal ganglion cells and transported down the axon to synaptic terminals (Moya *et al.*, 1994a; Lyckman *et al.*, 1998). APP is a transmembrane glycoprotein which is preferentially expressed in brain (Weidemann *et al.*, 1989). The protein is coded for by a single gene on chromosome 21 in humans which can give rise to several different isoforms by alternative splicing of its mRNA. The neuronal form of APP is 695 amino acids in length and in vitro experiments implicate the protein in cell-cell and cell-substrate adhesion (Schubert *et al.*, 1989b; Breen *et al.*, 1991). Ultrastructural localization studies have shown that APP is present in the presynaptic terminal (Schubert *et al.*, 1991). Recent studies have reported that APP on the cell membrane interacts with Ga of the Go heterotrimeric complex raising the possiblity that the protein could function as a cell surface receptor whose interactions with an enodgenous ligand modulates G-protein signal transduction (Brouillet *et al.*, 1999).

2. Synthesis and turnover of APP in retinal terminals in vivo

Although only a single transcript is readily detectable in neurons, four dinstinct electrophoretic variants of APP arrive at retinal terminals (Lyckman *et al.*, 1998). Electrophoretic heterogeneity in APP has been attributed to variations in posttranslational modifications such as differential glycosylation (Weidemann *et al.*, 1989). The APP that arrives at the retinal axon terminal is glycosylated and lectin affinity chromatography showed that the carbohydrate domain on radiolabeled, axonally transported APP contains terminal D-glucose oligosaccharides (Moya *et al.*, 1994a). More recent characterization of axonally transported APP document that the protein is glycosylated on both asparagines and serine and/or threonine residues, N-linked and O-linked glycosylation, respectively (Lyckman *et al.*, 1998). APP sent to retinal terminals carries sialic acid and is sulfated, the latter modification most likely on tyrosine residues of the polypeptide back bone (Lyckman *et al.*, 1998; Moya *et al.*, 1994b; Schubert *et al.*, 1989a). These posttranslational modifications of the APP in retinal terminals, summarized in Figure 1, are consistent with the protein functioning as a cell surface adhesion molecule or receptor modulating signal transduction at retinal terminals.

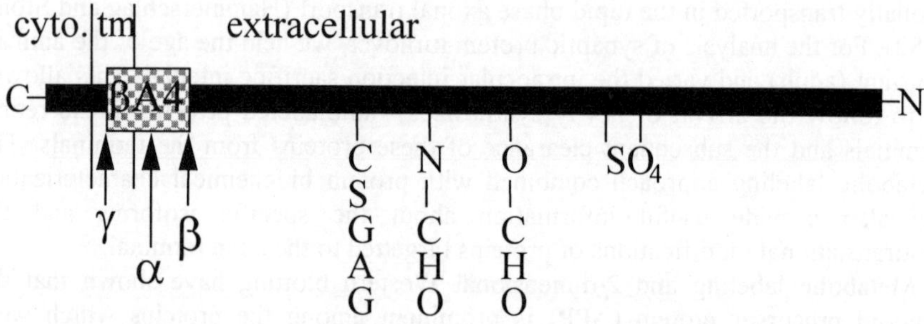

Figure 1. Summary of APP structure and posttranslational modifications. The cytoplasmic (cyto.), transmembrane domaine (tm) and extracellular regions are indicated. The relative position of sites for chondroitination, N- and O-linked glycosylation and sulfation are shown.

The synthesis and axonal transport of APP in hamster retinal ganglion cells is developmentally regulated. A 120kD full-length form of the protein decreases gradually with maturation (Moya *et al.*, 1994a) A 110kD transmembrane form and a soluble 100kD form lacking the C-terminal first increase during retinal axon arborization and then are diminished when functional visual connections have been formed. The 100kD, soluble form corresponds to secreted APP cleaved at the constitutive a secretase site (Fig. 1). The appearance of soluble APP at beginning of synaptogenesis suggests that the mechanism for APP secretion is age dependent and may require the formation of synaptic contacts. This is further supported by in vitro studies which show that in neurons which have not formed synaptic contacts, APP has a very short dwell time on the cell surface (Allinquant *et al.*, 1994). In these conditions, the immature neurons show no constitutive APP secretion nor calcium stimulated release of the protein (Allinquant *et al.*, 1994).

APP in retinal terminals has extremely rapid turnover kinetics in vivo. In our turnover studies in adult hamsters we quantitifed the levels of the radioactive signal for each of the APP forms at times ranging from 1hr to 3 days after intraocular injection (Lyckman *et al.*, 1998). By analyzing the radioactive signal we are able to follow the APP synthesized in identified CNS neurons and targetted to specfic presynapses, *i.e.* retinal ganglion cells and retinal terminals, respectively. The earliest time after in vivo metabolic labeling at which we could detect radiolabeled APP in the superior colliculus was 4hrs. This was also the time at which we measured the peak levels for the 3 full-length, transmembrane froms of APP. The levels of these APP forms in retinal terminals rapidly declined with a half-life of 3-4hrs depending on the APP form. The APP lacking the C-terminal and which was present in the soluble fraction reached its peak at 8hrs. The half-life of this form of APP was slightly longer, about 5 hrs.

The rapid turnover coupled with previous reports that APP turnover in brain slice preparations is linked to neuronal activity (Nitsch *et al.*, 1993) led us to directly test whether APP tunover in vivo depended on axonal activity. Tetrodotoxin is a sodium channnel blocker that inhibits action potentials including those of retinal ganglion cells. Quantitative assessment of metabolically labeled APP in retinal terminals after intraocular injection of tetrodotoxin showed no changes in APP levels demonstrating that the turnover of APP in retinal terminals in vivo is not dependent on axonal activity.

To summarize, APP has adhesive properties, is localized at synapses and can interact with asignal transduction complexe. Our data show that APP targeted to retinal terminals in vivo is sulfated, is glycosylated, carries sialic acid and is rapidly eliminated from the presynapse via a pathway that involves proteolytic cleavage but that does not depend on axonal activity. We propose that APP at the terminal plays a essential role in synaptic function, perhaps involving other membrane glycoproteins. The rapid turnover of APP may have important implications for Alzheimer's disease.

3. Alzheimer's disease

Alzheimer's disease is a neurodegenerative disease with severe and progressive memory loss and dementia. The neuropathological hallmarks of AD are the presence of senile plaques and neurofibrillary tangles (NFTs). Senile plaques are deposits of amyloid in the brain parenchyma with a 42-amino acid peptide at the core of the plaques. This peptide, called the ßA4$_{1-42}$ peptide, is a fragment of APP. A number of APP gene mutations have been described in families with a history of early onset AD emphasizing a central role for APP in the disease (see Selkoe, 1998). Neurofibrillary tangles are composed of paired helical filaments of aggregated tau protein. Tau is a microtubule associated protein which in normal neurons participates in microtubule stabilization. The tau protein found in neurofibrillary tangles from AD brain is hyperphosphorylated, abnormally glycosylated and causes disassembly of the microtubules (reviewed in Iqbal *et al.*, 1998).

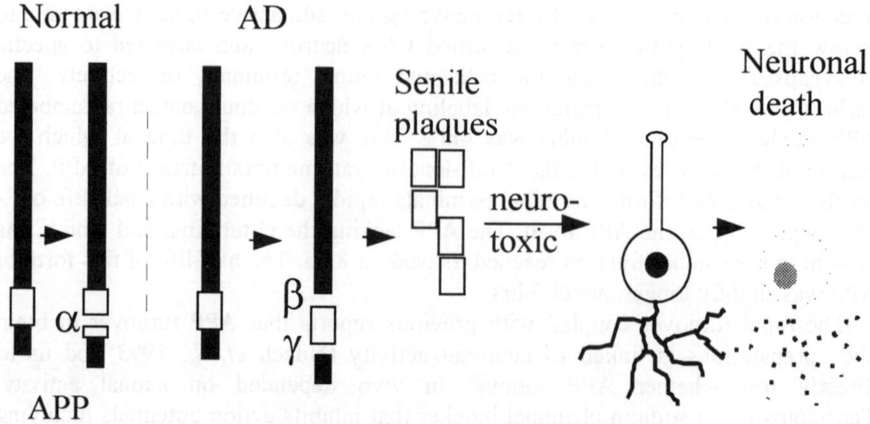

Figure 2. The ßAPP/senile plaque model of AD.

Based on these sets of findings, two dominant models of AD have emerged. In the *ßA4/senile plaque model of AD*, the cause of the disease is due to an accumulation of amyloidergic $ßA4_{1-42}$ (Fig 2). In the normal, nonpathological state, APP is constitutively cleaved by a secretase within the ßA4 peptide sequence (Fig. 2, left). This leads to the secretion of APP and prevents the formation of the ßA4 peptide. In the pathological state in the AD brain, the processing of APP is shifted from an a secretase pathway to one in which ß and g secretase events occur, cleaving APP at the N- and C-terminal limits of the $ßA4_{1-42}$ peptide, respectively (Fig. 2, right). The resulting $ßA4_{1-42}$ accumulates and aggregates, forming the core of amyloid plaques. These amyloid deposits are toxic for neurons and the subsequent neuronal loss accounts for the cognitive deficits seen in patients with AD. Thus, in this model, the precipitating event in terms of cellular dysfunction is abnormal cleavage of APP to yield the $ßA4_{1-42}$ peptide. In autosomal early onset AD, a number of mutations in the APP gene have been described. These mutations cluster at the various secretase cleavage sites and appear to shift the processing of APP from its consitutive secretion to the formation of $ßA4_{1-42}$ (reviewed in Hardy, 1997).

In the *tau/neurofibrillary tangle model*, the cause of the disease is the accumulation of neurofibrillary tangles in axons and cell bodies of neurons. Tau in the normal brain is a microtubule associated phosphoprotein which helps stabilize the neuronal cytoskeleton. In AD, overactive protein kinase activity and/or underactive phosphatase activity result in hyperphosphorylated tau. Tau is also abnormally glycosylated in the AD brain and together, these abnormal posttranslational modifications are thought to confer a propensity for tau to form paired helical filaments and to aggregate as interneuronal neurofibrillary tangles.

These tangles cause neuronal cell death by blocking up or disrupting the cytoskeletal organization of the neurons and the neuronal loss leads to the cognitive deficits in AD. In this model, the critical event is the hyperphosphorylation of tau protein.

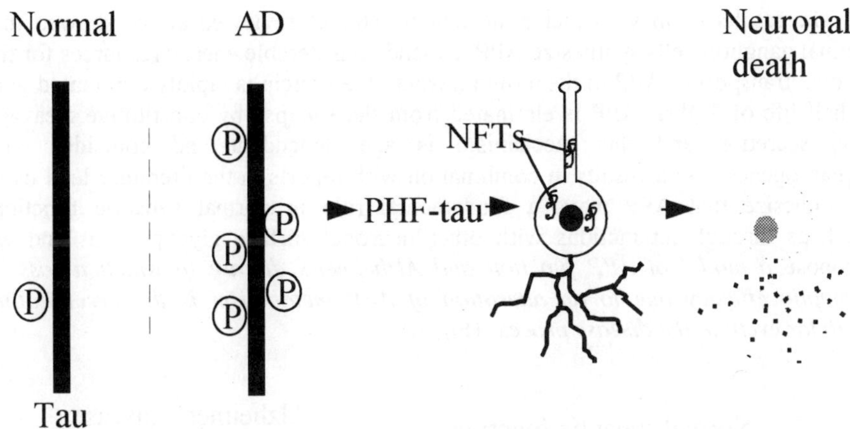

Figure 3. The tau/NFT model of AD. P, tau phosphorylation sites; PHF, paired helical filaments of tau; NFTs, neurofibrillary tangles.

Seemingly at odds with the two models above and their hypothesized causal role of senile plaques or neurofribrillary tangles on dimentia due to neuronal loss, is a growing body of literature suggesting that synaptic loss is a prominent feature of AD and that this may precede amyloid deposition and the appearance of neurofibrillary tangles (reviewed in DeKosky *et al.*, 1996). Electron microscopic quantitation of postmortem or biopsy brain tissue from confirmed or suspected AD patients show a significant loss of synapses (DeKosky *et al.*, 1990; Scheff *et al.*, 1990). Analysis of molecular markers of synapse density and function corroborate these findings (Terry *et al.*, 1991; Masliah *et al.*, 1994; Davidsson and Blennow, 1998). Furthermore, regression analysis has shown that synaptic loss was highly correlated with the severity of dementia (Terry *et al.*, 1991; Sze *et al.*, 1997). The former study used a stepwise linear regression model to examine the contribution of synapse loss, senile plaque density and NFT density to dementia and found that synapse loss was the pathological change most highly correlated with the severity of dementia (Terry *et al.*, 1991). No correlation was noted between the number of senile plaques or the level of neuronal loss and dementia, and only a modest relationship was reported for the number of tangles and dementia. Given the robust correlation between synaptic number and dementia in the study ($r=0.749$), over half of the variance in the cognitive changes could be explained by synaptic density.

Taken together such analyses provide increasing evidence that synaptic dysfunction and terminal loss may be the neuropathological change most responsable for the dementia in AD.

4. APP metabolism in retinal terminals: implications for AD?

The results of in vivo metabolic labeing studies reviewed above showed that retinal ganglion cells synthesize APP, expend considerable energy resources for the axonal transport of APP to the terminal where the protein is rapidly eliminated with a half-life of 3-4hrs. APP is elminated from the synapse by constitutive cleavage and secretion and this mechanism is age dependent and coincides with synaptogenesis. Our results in combination with reports in the literature lead us to hypothesize that APP plays a fundamental role in normal synaptic function, perhaps through interactions with other neuronal surface glycoproteins and we propose *a model of APP function and Alzheimer's disease in which a loss of synaptic efficacy due to an alteration of APP metabolism is the precipitating cellular event in the disease process* (Fig. 4).

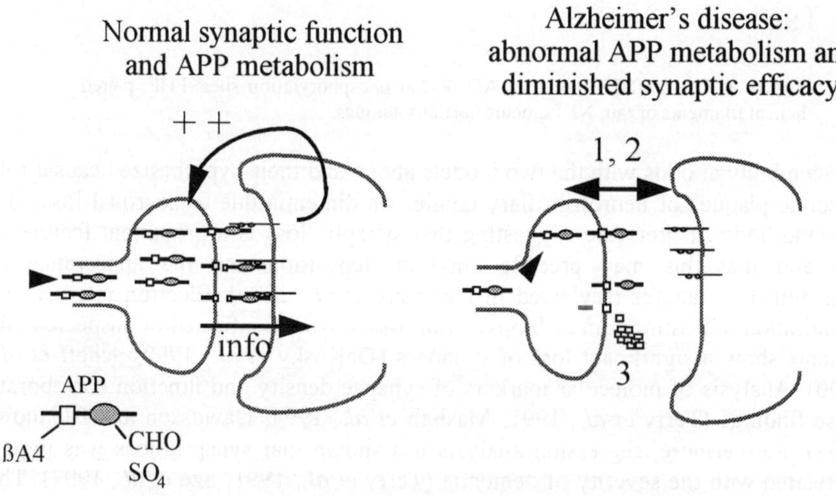

Figure 4. Synaptic efficacy model of APP function and AD.

In the normal brain, glycosylated and sulfated APP is rapidly transported to the terminal. APP reaches the terminal surface where it stabilizes pre- and postsynaptic membranes through adhesive interactions with other proteins (Fig 4, left). The interaction between APP and its molecular partner(s) (as yet unidentified) maintain synaptic integrity for the transmission of information along the neuronal network. One direct measure of synaptic integrity is its efficacy by which we mean the probablilty of a given postsynaptic response in the presence of a presynaptic signal. The synapse is reinforced (++, Fig 4, left) through Hebbian mechanisms by retrograde signals, thus increasing its stability. The intereaction between APP and other proteins at the synaptic surface also stabilizes the protein at the membrane where a secretase activity can proteolytically cleave APP and this also prevents possible accumualtion of $ßA4_{1-42}$ in a pathological state (Fig. 4, left). One indication of the importance of the balance of APP at the nerve terminal is the efficient cellular mechanism for APP homeostasis at the synapse even when afferent activity is blocked.

We propose that AD is caused by changes in APP metabolism. With a synaptic half-life of 3-4hrs, a small perturbation in the rate of APP synthesis or its rate of elimination would result in rapid changes in the levels of APP at the nerve terminal which we postulate would alter synaptic efficacy. In addition to a degradation in the fidelity of information transfer, a decrease in synaptic efficacy would also reduce retrograde reinforcement of the synapses rendering them less stable, eventually leading to their elmination. This progression of events fits well with the reports of synaptic loss in early stages of AD and can readily explain the cognitive changes in the course of the disease which appear before the marked neuropathological changes of plaque and NFT accumulation. A continuing synaptic loss leads to neuronal dysfunction. At this stage, the processing of APP in dysfunctional neurons shifts towards the ß and g secretases and the abnormal processing of tau protein leading to the accumulation of amyloid plaques and neurofibrillary tangles. The neurotoxic effects of ß amyloid and NFTs contribute to the frank neuronal loss and this compounds and accelerates the severe cognitive decline at later stages of AD.

It is now important to consider how a change in APP metabolism might occur. In familial AD due to mutations in the APP gene, the mutations cluster around the three proteolytic sites. At least one of the mutations near the a secretase site (the Flemish mutation) inhibits the effects of a secretase on APP. Reduced a secretase cleavage would increase the half-life of APP at the synaptic membrane, and thus alter the stability of the synapse. In our model the Flemish mutation in the absence of any compensatory reduction in the levels of APP synthesis and transport would increase synaptic levels of APP, further compounding the effects of an increased half-life on synaptic function. Mutations which increase ß and/or g secretase cleavage of APP such as the Swedish, London and Florida mutations would reduce the amount of APP available at the synaptic interface. In addition to mutations in the APP gene, mutations in the presenilin 1 and presenilin 2 genes cause autosomal dominant AD. It is interesting to note that the mutations in the presenilins are reported to augment g secretase processing of APP (see Hardy, 1997). As with APP mutations which increase increase ß and/or g secretase events, the presenilin mutations would also result in a reduction of APP at the synapse with similar consequences for synaptic function. The ever increasing array of transgenic mice being developed combined with in vivo metabolic labeling should render these predictions testable in the near future.

In sporadic AD, which represents the vast majority (86%) of cases (Baringa, 1995), possible causes of a change in APP metabolism are more speculative. However, with a half-life of 3-4hrs, any change in synthesis or axonal transport that reduces the levels of APP arriving at the terminals by 1% would decrease synaptic APP by about 3% per day. One possibility to consider is a global energy impariment. Since protein synthesis including posttranslational modification and axonal transport are energy dependent, an energy impairment at the level of the neuronal cell body and axon could reduce levels of synaptic proteins including APP delivered to the terminal over the long term, although APP would be particularly affected due to its rapid turnover. Evidence for the existence of a decrease in brain energy utilization in AD patients is accumulating (Chandrasekaran *et al.*, 1996). Future studies will be required, however, to determine if such changes in global brain metabolism can have a direct effect on synaptic APP levels.

Finally, our model generates a testable hypothesis concerning the biological function of APP, namely that the protein is essential for synaptic efficacy. While a transgenic approach may seem to provide a useful animal model, the effectivness of constitutive APP gene inactivation for the study of APP function is mitigated by the fact that the protein is one member of a family of proteins which share considerable structural similarity. Furhtermore, at least two members of the amyloid precurosor superfamily can finctionally comensate for each other (van Koch *et al.*, 1997). The primary visual projection provides an invaluable biological model with which to test this hypothesis in combination with an acute 'knockdown' of APP using antisense oligonucleotides injected intraocularly. A decrease of APP at retinal terminals in the brain can then be confirmed by metabolic labeling. Once conditions have been optimized, the effects of reducing APP at retinal presynapse on visually elicited reponses in the brain can be evaluated.

In summary, the primary visual pathway has provided a window onto the brain with which to analyze synaptic protein metabolism. The results have contributed to an alternative model for Alzheimer's disease which places the cause of the disease on synaptic dysfunction. In addition to furthering the discussion about AD, this model also generates hypotheses about the cell biology of normal synapses which can be tested in vivo in the visual system.

Acknowledgements

We thank L. Di Giamberardino for constant encouragement and support. Portions of the work reviewed here were supported by CNRS, INSERM, CEA and the EEC (BMH1-CT-94-8652).

References

Allinquant, B., K.L. Moya, C. Bouillot and A. Prochiantz (1994) "Amyloid precursor protein in differentiating neurons: distribution into two interrelated pools and association with the cytoskeleton", *J. Neurosci.* 14:6842-6854.

Barinaga, M (1995) "New Alzheimer's gene found", *Science* 268:1845-1846.

Breen, K.C., M.T. Bruce and B.H. Anderton (1991) "The beta amyloid precursor protein mediates cell-cell and cell-surface adhesion", *J. Neurosci. Res.* 28:90-100.

Brouillet, E., A. Trembleau, D. Galanaud, M. Volovitch, C. Bouillot, C. Valenza, A. Prochiantz and B. Allinquant (1999) "The amyloid precursor protein interacts with Go heterotrimeric protein within a cell compartment specialized in signal transduction", *J. Neurosci.* 19:1717-1727.

Chandrasekaran, K., K. Hatanpää, D.R. Brady and S.I. Rapoport (1996) "Evidence for physiological down-regulation of brain oxidative phsophorylation in Alzheimer's disease", *Exper. Neurol.* 142:80-88.

Davidsson, P. and K. Blennow (1998) "Neurochemical dissection of synaptic pathology in Alzheimer's disease", *Int. Psychogeriatr.* 10:11-23.

DeKosky, S.T. and S.W. Scheff (1990) "f8s24 synapse loss in frontal cortex biopsies in Alzheimer's disease: Correlation with cognitive severity", *Ann. Neurol.* 27:457-464.

DeKosky, S.T., S.W. Scheff and S.D. Styren (1996) "Structural correlates of cognition in dementia: Quantification and assessment of synapse change", *Neurodegeneration* 5:417-421.

Hammerschlag, R. and G.C. Stone (1982) "Membrane delivery by fast axonal transport", *Trends Neurosci.* 5:12.

Hardy, J. (1997) "Amyloid, the presenilins and Alzheimer's diseae", *Trends Neurosci.* 20:154-159.

Iqbal, K., A.C. Alonso, C.X. Gong, S. Khatoon, J.J. Pei, J.Z. Wang and I. Grundke-Iqbal (1998) "Mechanisms of neurofibrillary degeneration and the formation of neurofibrillary tangles", *J. Neural. Transm. Suppl.* 53:169-180.

Lyckman, A.W., A. Confaloni, G. Thinikaran, S.S. Sisodia and K.L. Moya (1998) "Amyloid precursor superfamily protein postranslational processing and presynaptic turnover kinetics in the CNS *in vivo*", *J. Biol. Chem.* 273:11100-11106.

Masliah, E., W.G. Honer, M. Mallory, M. Voigt, P. Kushner, L. Hansen and R. Terry (1994) "Topographical distribution of synaptic-associated proteins in the neuritic plaques of Alzheimer's disease hippocampus", *Acta Neuropathol (Berl)* 87:135-142.

Moya, K.L., L.I. Benowitz, G.E. Schneider and B. Allinquant (1994a) "The amyloid precursor protein is developmentally regulated and correlated with synaptogenesis", *Dev. Biol.* 161:597-603.

Moya, K.L., A. Confaloni and B. Allinquant (1994b) "In vivo neuronal synthesis and axonal transport of KPI-containing forms of the amyloid precursor protein", *J. Neurochem.* 63:1971-1974.

Nitsch, R.M., S.A. Farber, J.H. Growdon and R.J. Wurtman (1993) "Release of amyloid beta-protein precursor derivatives by electrical depolarization of rat hippocampal slices", *Proc. Nat. Acad. Sci. USA* 90:5191-5193.

Scheff, S.W., S.T. DeKosky and D.A. Price (1990) "Quantitative assessment of cortical synaptic density in Alzheimer's disease", *Neurobiol. Aging* 11:29-37.

Schubert, D., M. LaCorbiere, T. Saitoh and G. Cole (1989a) "Characterization of an amyloid ß precursor protein that binds heparin and contains tyrosine sulfate", *Proc. Natl. Acad. Sci. USA* 86:2066-2069.

Schubert, D., L.W. Jin, T. Saitoh and G. Cole (1989b) "The regulation of amyloid β protein precursor secretion and its modulatory role in cell adhesion", *Neuron* 3:689-694.

Schubert, W., R. Prior, A. Weidemann, H. Dircksen, G. Multhaup, C.L. Masters and K. Beyreuther (1991) "Localization of Alzheimer ßA4 amyloid precursor protein at central and peripheral synaptic sites", *Brain Res.* 563:184-194.

Selkoe, D.J. (1998) "The cell biology of beta-amyloid precursor protein and presenilin in Alzheimer's disease", *Trends Cell Biol.* 8:447-453.

Sze, C.I., J.C. Troncoso, C. Kawas, P. Mouton, D.L. Price and L.J. Martin (1999) "Loss of the presynaptic vesicle protein synaptophysin in hippocampus correlates with cognitive decline in Alzheimer disease", *J. Neuropathol. Exp. Neurol.* 56:933-944.

Terry, R.D., E. Masliah, D.P. Salmon, N. Butters, R. DeTeresa, R. Hill, L.A. Hansen and R. Katzman (1991) "Physical basis of cognitive alterations in Alzheimer's disease: Synapse loss is the major correlate of cogntive impairment", *Ann. Neurol.* 130:572-580.

von Koch, C.S., H. Zheng, H. Chen, M. Trumbauer, G. Thinakaran, L.H. van der Ploeg, D.L. Price and S.S. Sisodia (1997) "Generation of APLP2 KO mice and early postnatal lethality in APLP2/APP double KO mice", *Neurobiol. Aging* 18:661-669.

Weidemann, A., G. Konig, D. Bunke, P. Fischer, J.M. Salbaum and K. Beyreuther (1989) "Identification, biogenesis and localisation of precursors of Alzheimer's disease A4 amyloid protein", *Cell* 57:115-126.

SITE DIRECTED MUTAGENESIS OF PHOSPHORYLATION SITES IN THE C-TERMINAL REGION OF *DROSOPHILA* RH1 OPSIN

GEORGIA NEU, JOACHIM BENTROP,
KARIN SCHWAB and REINHARD PAULSEN
Department of Cell and Neurobiology, Institute of Zoology, University of Karlsruhe, Kornblumenstr. 13, 76128 Karlsruhe, Germany

ABSTRACT

A common structural feature of many G-protein coupled receptors (GPCRs) is the presence of phosphorylation sites within the C-terminal region. Protein phosphorylation sites are also located in the C-terminal tails of *Drosophila melanogaster* rhodopsins Rh1 to Rh6. Whereas the agonist-induced phosphorylation of many GPCRs has been shown to be the trigger for receptor inactivation and internalization, the function of these sites in invertebrate rhodopsins is completely elusive. In order to investigate the role of rhodopsin phosphorylation in *Drosophila* the phosphorylation sites were deleted by *in vitro* mutagenesis. Transgenic flies expressing the mutant Rh1 genes were generated in a Rh1-null background. Towards the analysis of these mutants, Western blot shows that the replacement of the respective phosphorylation site, serine or threonine, by alanine, does not affect Rh1 opsin expression. Immunocytochemistry reveals that the mutant rhodopsins are properly targeted to the rhabdomeric photoreceptor membrane. Consequently these phosphorylation sites are not essential for rhodopsin biosynthesis and targeting.

1. Introduction

G-protein coupled receptors comprise a class of receptor proteins that transduce extracellular signals to intracellular effector molecules through the activation of heterotrimeric G-proteins. In invertebrates, as in *e.g.* the fruit fly *Drosophila melanogaster,* light absorption by rhodopsin generates a thermostable metarhodopsin which in turn activates a photoreceptor specific G-protein (Paulsen and Bentrop, 1986). Details of the mechanism which terminates this step of the phototransduction cascade are still unknown. In vertebrate rhodopsin the uncoupling of the active metarhodopsin state from its G-protein is triggered by metarhodopsin phosphorylation followed by the binding of arrestin. This phosphorylation, which is catalyzed by a rhodopsin kinase, seems to be limited to the three serine and four threonine residues in the carboxy terminal region of rhodopsin (Thompson and Findlay, 1984; Palczewski *et al.*, 1991; Zhang *et al.*, 1997). Light dependent phosphorylation is also a feature of fly rhodopsin (Matsumoto and Pak, 1984; Paulsen and Bentrop, 1984; Bentrop and Paulsen, 1986; Bentrop, *et al.* 1993; Byk *et al.*, 1993; Plangger *et al.*, 1994). Although it has been shown that *Drosophila* rhodopsin is light-depentently phosphorylated, the exact sites of phosphorylation have however not yet been identified. In a distinct difference to the mechanism of the inactivation of vertebrate rhodopsin, phosphorylation of the C-terminus appears not to be an essential step of the mechanism of response termination (Vinós *et al.*, 1997). The current study addresses the question of the functional relevance of Rh1 rhodopsin phosphorylation in *Drosophila* by deletion of phosphorylation sites through site directed *in vitro* mutagenesis. Transformant flies which express the mutant rhodopsins were investigated for opsin expression and targeting.

2. Material and Methods

2.1. Generation of transgenic flies via P-element mediated germline transformation

In vitro mutagenesis was carried out by the Quik Change-Site-Directed-Mutagenesis. A single-stranded template was used, which consisted of a pGEM-vector containing a 5.5 kb DNA fragment encompassing the complete *Drosophila* Rh1 gene. Mutant Rh1 genes were cloned into the P-element transformation vector Yellow Carnegie 4. P-element-mediated germline transformation into host strain nina E^{ol17} was carried out, and transformants lines were made homozygous for the P-element insert. Flies were raised on a standard corn meal diet and were kept under a 12 h light / 12 h dark cycle.

2.2. Western-blot analysis

Eyes of mutant flies were inspected for opsin expression 1 day posteclosion. Membrane proteins from 100 compound eyes were prepared as described by Bentrop *et al.* (1997). Samples containing 12 μg of total protein were separated by SDS-PAGE according to Laemmli (1970). Immunoblots were carried out as described by Bentrop *et al.* (1997).

2.3. Immunolabeling of ultrathin sections

Eyes of transgenic flies were investigated 1 day post-eclosion. Immunolabeling of ultrathin sections was carried out according to Wolfrum (1995). Sections were examined with a Zeiss EM 912 electron microscope.

3. Results

To assess the functional importance of potential phosphorylation sites within the C-terminal domain of *Drosophila* Rh1 rhodopsin, we performed *in vitro* mutagenesis of the Rh1 gene that resulted in the substitution of serines and threonines for alanine. The five mutant Rh1 genes generated so far were designated Rh1 S357A, Rh1 S358A, Rh1 S362A, Rh1 T365A and Rh1 S367A, respectivley, to indicate the original amino acid, its position in the primary sequence and the amino acid introduced at this position by mutation.

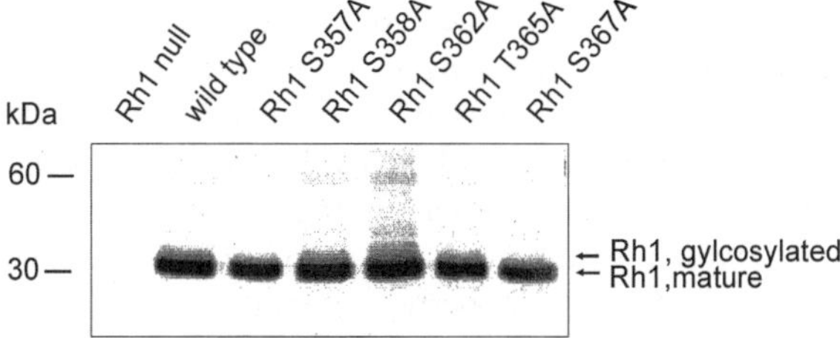

Figure 1. Rh1 opsin in wild-type and Rh1 mutant flies. Immunoblot obtained after separation by SDS-PAGE of protein extracts from eyes of the indicated fly strains. The blot was probed with antibodies directed against Rh1 opsin.

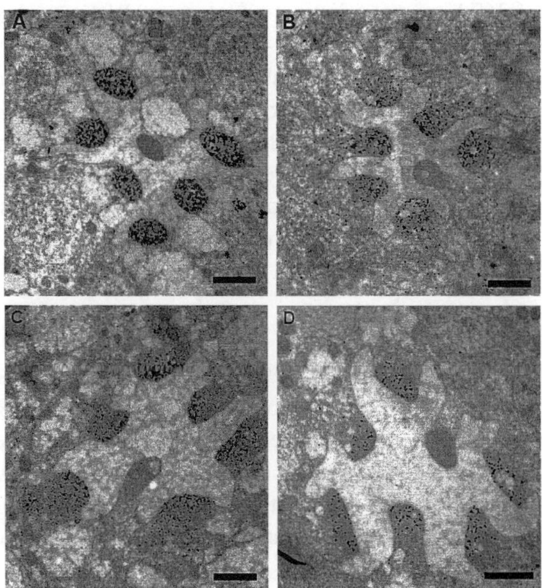

Figure 2. Immunocytologial localization of Rh1 opsin in wild-type and Rh1 mutant photoreceptor cells. Ommatidia of wild-type or Rh1 mutant flies were sectioned transversely at 1 day post-eclosion. Binding of antibodies directed against Rh1 opsin was visualized by immunogold-staining. Expression of mutant rhodopsin is restricted to cells R1 to R6. (A) wild-type, (B) Rh1 S362A, (C) Rh1 T365A, (D) Rh1 S367A. Scale bar = 2μm.

The mutant genes were introduced, by P-element transformation, into the germline of Rh1-null mutants, resulting in flies which only express the mutant rhodopsin gene in photoreceptor cells R1-6 (O'Tousa *et al.* 1985, Zuker *et al.* 1985, Huber *et al.* 1997). In studies directed to investigate structure / function relationship in Rh1, it has been observed that amino acid substitutions may produce rhodopsin which is mislocalized in the photoreceptor cell and / or degraded instead of beeing properly integrated within the microvillar photoreceptor membrane (Bentrop *et al.* 1997).

Thus, before the effect of rhodopsin phosphorylation can be investigated in Rh1 mutants, it had to first be assured that the mutant rhodopsin is expressed and that it is targeted to the correct cellular compartment. The opsin content of eye membrane was investigated by Western blot analysis using a polyclonal antiserum directed against a loop i3-peptide of Rh1 opsin.

As is shown in Figure 1, mutants Rh1 S357A, Rh1 S358A, Rh1 S362A, Rh1 T365A and Rh1 S367A express distinct amounts of Rh1 opsin protein with the same electrophoretic mobility as the wild-type opsin.

We also localized the opsin at the subcellular level by immunocytochemistry. For the three mutants tested so far, Rh1 S362A, Rh1 T365A and Rh1 S367A, we showed that opsin molecules are correctly targeted to the rhabdomeric microvilli of R1-6 photoreceptor cells. There are no signs of rhabdomeric membrane degradation visible in the cross sections through the eyes of transgenic flies (Figure 2 A to D) indicating that the mutant rhodopsins rescue the Rh1-null phenotype and do not induce rapid retinal degeneration.

4. Discussion

The C-terminal region of *Drosophila melanogaster* Rh1 opsin harbours potential phosphorylation sites, a common feature to many GPCRs. Vinós *et al.* (1997) presented evidence that transgenic *Drosophila* which express a Rh1 rhodopsin lacking the last 18 amino acids of the C-terminus have normal deactivation kinetics. These data were interpreted, such that phosphorylation sites located in these region of rhodopsin are not involved in the deactivation of *Drosophila* Rh1 rhodopsin.

In order to investigate the function of these phosphorylation sites we generated transgenic flies expressing Rh1 rhodopsin missing single putative phosphorylation sites. We showed, that the mutant flies synthesize distinct amounts of Rh1 opsin protein. In addition we were able to show by immunocytochemistry that the mutant opsins are correctly localized at the rhabdomeric photoreceptor membrane.

Consequently, the potential phosphorylation sites within the C-terminal domain of *Drosophila* Rh1 rhodopsin are not essential for rhodopsin biosynthesis and targeting. We conclude that these amino acids are not crucial structural determinants of binding domains, which the opsin molecule provides for interacting proteins involved in synthezising rhodopsin and transporting it to the rhabdomeral photoreceptor membrane. This will allow us to elucidate those aspects of rhodopsin function that may be affected by the elemination of these phosphorylation sites.

References

Bentrop, J. and R. Paulsen (1986) "Light modulated ADP-ribosylation, protein phosphorylation and protein binding in isolated fly photoreceptor membranes", *Eur. J. Biochem.* 161:61-67.

Bentrop, J., A. Plangger and R. Paulsen (1993) "An arrestin homolog of blowfly photoreceptors stimulates visual-pigment phosphorylation by activating a membrane-associated protein kinase", *Eur. J. Biochem.* 216:67-73.

Bentrop, J., K. Schwab, W.L. Pak and R. Paulsen (1997) "Site-directed mutagenesis of highly conserved amino acids in the first cytoplasmatic loop of *Drosophila* Rh1 opsin blocks rhodopsin synthesis in nascent state", *EMBO J.* 16:1600-1609.

Byk, T., M. Bar-Yaacov, Y.N. Doza, B. Minke and Z. Selinger (1993) "Regulatory arrestin cycle secures the fidelity and maintenance of the fly photoreceptor cell", *Proc. Natl. Acad. Sci. U.S.A.* 90:1907-1911.

Huber, A., S. Schulz, J. Bentrop, C. Groell, U. Wolfrum and R. Paulsen (1997) "Molecular cloning of Drosophila Rh6 rhodopsin: the visual pigment of a subset of R8 photoreceptor cells", *FEBS Lett.* 406:6-10.

Laemmli, U.K. (1970) "Cleavage of structural proteins during the assembly of the head of bacteriophage T4", *Nature* 227:682-685.

Matsumoto H. and W.L. Pak (1984) "Light-induced phosphorylation of retina-specific polypeptides of Drosophila in vivo", *Science* 223:184-186.

O´Tousa, J.E., W. Baehr, R.L. Martin, J. Hirsch, W.L. Pak and M.L. Applebury (1985) "The *Drosophila nina E* gene encodes an opsin", *Cell* 40:839-850.

Palczewski, K., J. Buczylko, M. Kaplan, A. Polans and C.W. Crabb (1991) "Mechanism of rhodopsin kinase activation", *J. Biol. Chem.* 266:12949-12955.

Paulsen, R. and J. Bentrop (1984) "Reversible phosphorylation of opsin induced by irradiation of blowfly retinae", *J. Comp. Physiol.* 155:39-45.

Paulsen, R. and J. Bentrop (1986) "Light-modulated biochemical events in fly photoreceptors", in: *Fortschritte der Zoologie, Bd. 33*, H.C. Lüttgau, ed., Stuttgart, New York: Gustav Fischer Verlag, pp. 299-319.

Plangger, A., D. Malicki, M. Whitney and R. Paulsen (1994) "Mechanism of arrestin 2 function in rhabdomeric photoreceptors", *J. Biol. Chem.* 269:26969-26975.

Thompson, P. and J. Findlay (1984) “Phosphorylation of bovine rhodopsin identification of the phosphorylated sites”, *Biochem. J.* 220:773-780.

Vinós, J., K. Jalink, R.W. Hardy, S.G. Britt and C. S. Zuker (1997) “A G-protein coupled receptor phosphatase required for rhodopsin function”, *Science* 277:687-690.

Wolfrum, U. (1995) “Centrin in the photoreceptor cells of mammalian retinae”, *Cell Motil. Cytoskeleton* 32:55-64.

Zhang, J., C.D. Sports, S. Osawa and E.R. Weiss (1997) “Rhodopsin phosphorylation sites and their role in arrestin binding”, *J. Biol. Chem.* 272:14762-14768.

Zuker, C., A.F. Cowman and G.M. Rubin (1985) “Isolation and structure of a rhodopsin gene from *Drosophila melanogaster*”, *Cell* 40:851-858.

ISOLATION OF NOVEL EYE-SPECIFICALLY EXPRESSED GENES BY DIFFERENTIAL HYBRIDIZATION OF A RETINAL cDNA LIBRARY OF *CALLIPHORA VICINA*

SIMONE SCHULZ, ARMIN HUBER, PHILIPP SANDER, AND REINHARD PAULSEN

Department of Cell- and Neurobiology, Institute of Zoology, University of Karlsruhe, Haid-und-Neu-Str. 9
76131 Karlsruhe, Germany

ABSTRACT

Visual transduction in the compound eye of flies is a well established model system to study G protein-coupled transduction pathways. It has been estimated that about 50 different gene products are dedicated to the functioning and regulation of phototransduction in *Drosophila melanogaster*. Almost all components involved in the main activating pathway of the phototransduction cascade have been cloned. Missing key components of the fly phototransduction cascade included, for example, until recently the visual Gγ subunit. The molecular cloning and characterization of the missing proteins would help to complete our knowledge about the fly visual transduction pathway. In order to isolate and characterize some of the not yet identified genes, which encode proteins of the photoreceptive membrane, we performed a substractive hybridization screen of a retinal cDNA library. In our screen we were able to isolate cDNA clones representing five different genes which are preferentially expressed in the retina. As determined by sequencing and subsequent database searches, four of these clones, Cv19, Cv25, CvGβe, and CvGγe, showed significant homologies to known genes, whereas Cv121 displayed no homology to genes published in public DNA sequence libraries.

1. Introduction

The phototransduction cascade of invertebrate photoreceptor cells depends on the precisely regulated interaction of a number of different photoreceptor-specific proteins: Upon light activation rhodopsin couples to a visual heterotrimeric G protein which activates a phospholipase Cβ (PLCβ) which in turn generates the second messenger inositol 1,4,5-trisphosphate and 1,2-diacylglycerol (Devary *et al.*, 1987; Selinger *et al.*, 1987). PLCβ itself and other signal transducing molecules, like the major light-activated Ca^{2+}-channel TRP and the eye-specific protein kinase C (ePKC) are assembled by the PDZ domain protein INAD into a signaling complex (*e.g.*, Huber *et al.*, 1996a; Shieh and Zhu, 1996; Tsunoda *et al.*, 1997; Chevesich *et al.*, 1997). So far, the exact mechanism for transmitting rhodopsin activation to the INAD signaling complex is not yet understood in detail. Part of the deficit in this knowledge is due to a lack of information about some missing key components of the phototransduction cascade. For example, the

Gγ subunit of the visual G protein had eluded identification at the molecular level until recently. The visual Gγ subunit has now been cloned (Schulz *et al.*, 1999) using the approach outlined in detail in the present paper. We performed a substractive hybridization screen using three different tissue-specific probes. In addition to the visual G protein γ subunit, this screen yielded four other genes preferentially expressed in the retina.

2. Materials and Methods

2.1. Fly stocks

Male *Calliphora vicina* Meig., chalky mutant, were raised at 25 °C in a 12 h light/12 h dark cycle. The *Drosophila* strains were raised on a standard corn meal diet and were kept under a 12 h light/12 h dark cycle.

2.2. Differential hybridization screening and Northern blot analysis

The differential hybridization screening of a *Calliphora* retinal cDNA library was carried out as described previously in Schulz *et al.* (1999). Northern blot analysis with digoxigenin-labeled antisense cRNA probes was performed as has been descibed (Huber *et al.*, 1994).

2.3. cDNA sequencing and sequence analysis

The nucleotide sequences of the isolated cDNA clones were determined with an automated sequencer (Alfexpress, Amersham Pharmacia Biotech) by the dideoxy chain termination method according to Sanger *et al.* (1977) using either Cy5-labeled vector-primers (Thermosequenase Kit, Amersham Pharmacia Biotech) or appropriate sequence-specific primers (Alfexpress Autoread Sequencing Kit, Amersham Pharmacia Biotech). Public internet resources of the National Center for Biotechnology Information (NCBI) were used to perform database searches for homologous proteins. Pairwise sequence alignments were performed using Vector NTI software (Informax). Amino acid sequence identities were calculated from pairwise sequence alignments.

3. Results

In order to isolate novel clones preferentially expressed in the fly eye we screened a *Calliphora* retinal cDNA library by differential hybridization with probes derived from mRNA of retinal tissue, of muscle tissue, and with a mixture of cDNAs encoding previously cloned retina-specifically expressed proteins (*Rh1, Rh6, arr1, arr2, dgq, inaD, inaC, trp, trpl, D19*). Each of the three probes was hybridized to one of three filter lifts of the replica-plated library (Figure 1).

Comparison of the positive signals on the three filter lifts allowed us to distinguish between non- and eye-specifically expressed clones and clones representing genes which are already known.

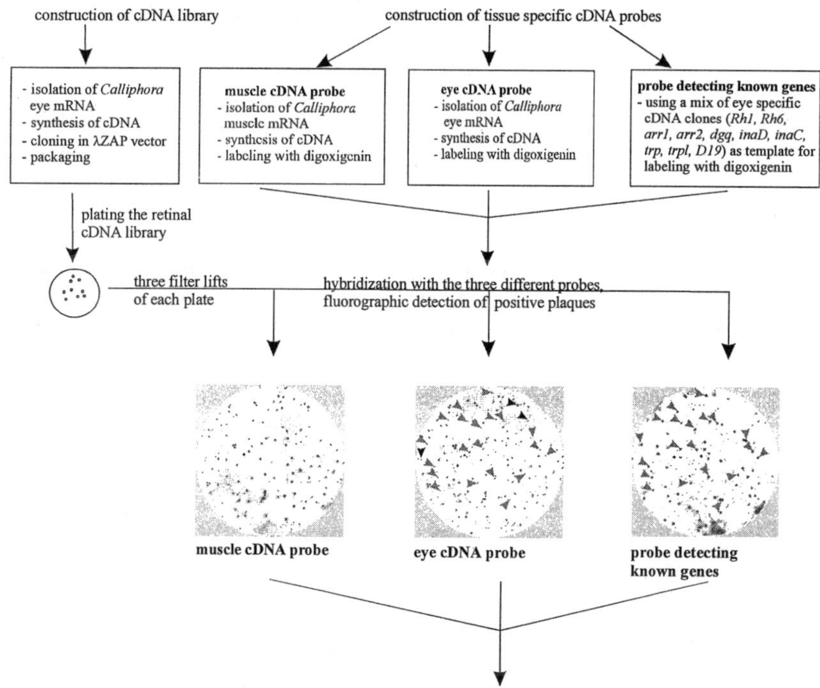

Figure 1. Strategy for the isolation of cDNA clones expressed preferentially in the retina. The gray arrowheads indicate retina-specific clones detected by the eye cDNA probe as well as by the probe detecting known genes, but not by the muscle probe. The black arrowheads indicate novel retina-specific clones detected by the eye cDNA probe only.

Out of 48000 clones screened, 125 clones hybridized with the eye-specific probe only. For further analysis we selected 17 clones which showed the strongest positive signals. As determined by sequencing and subsequent database searches, four of these clones, Cv19, Cv25, CvGβe, and CvGγe, showed homologies to known genes encoding the adult cuticle protein 1 precursor (Cv19), a homologue of a Y-box binding protein (Cv25), the *Calliphora* homologue of the already cloned DmGβe (CvGβe), and a novel G protein γ subunit (Gγe). Clone Cv121 displayed no significant homology to known genes. Table 1 summarizes the data obtained by Northern blot analyses of the different clones.

Table 1. Size and number of tissue-specifically expressed transcripts of the novel cDNA clones. The symbols indicate the expression-intensity of the different transcripts (+++, strong signal; +/-, weak signal; -, no signal; N.D., not determined).

		expressed in			
CDNA clone	transcript size	retina	brain	antenna	body
Cv19	1.0 kb	+++	+/-	-	-
Cv25	1.8 kb	+++	-	-	-
	2.5 kb	+/-	+/-	+/-	+++
Cv121	0.7 kb	+++	-	-	+/-
CvGγe	1.4 kb	+++	-	-	-
CvGβe	N.D.	N.D.	N.D.	N.D.	N.D.

In the absence of any molecular information about a visual Gγ subunit in the fly eye we concentrated our further studies on the isolated, novel G protein γ subunit. The *Drosophila* homologue DmGγe was isolated by a homology screen of a *Drosophila* head cDNA library with CvGγe. Furthermore, to complete our information about the G protein subunits in the fly compound eyes, we also cloned the α subunit of the visual G protein of *Calliphora* (CvGα_q) by a homology screen with DmGα_q. Comparison of the deduced amino acid sequences of CvGγe and DmGγe revealed that they are identical except for two amino acids at positions 26 and 43 (Figure 2C, Schulz *et al.*, 1999). Comparison of the spatial distribution of the DmGγe gene product with that of the visual DmGα_q and DmGβe by Western blot analysis and immunohistochemistry revealed a localization pattern for DmGγe which is indistinguishable from that of Gβe and Gα_q (Schulz *et al.*, 1999). Furthermore, immunoprecipitations with Gβe revealed that Gγe specifically co-precipitates with Gβe showing that the newly isolated Gγ subunit associates with Gβe of the visual G protein of *Drosophila* photoreceptors (Schulz *et al.*, 1999).

The *Calliphora* G protein subunits showed an amino acid identity of 95.5 % for CvGα_q, 89.3 % for CvGβe, and 97.2 % for CvGγe to the corresponding *Drosophila* proteins (Figure 2). Gα_q subunits from *Drosophila* and *Calliphora* show characteristics typical of other Gq proteins, including a putative cholera toxin ADP ribosylation site (Arg178), which may be the target for ADP-ribosylation in non-irradiated rhabdomeral membranes (Bentrop and Paulsen, 1986), and two N-terminal cysteine residues (Cys3, Cys4) which are putative sites for palmitoylation (Figure 2). A cysteine near the C-terminus which would be a putative site for pertussis toxin catalyzed ADP ribosylation is missing.

The structure of Gβ subunits is characterized by seven WD repeats which are also present in *Drosophila* and *Calliphora* Gβe with WD1, WD2, WD3, WD4, and WD7 being more than 95 % conserved between CvGβe and DmGβe (Figure 2B). The extremely high homology of the *Calliphora* and *Drosophila* G protein

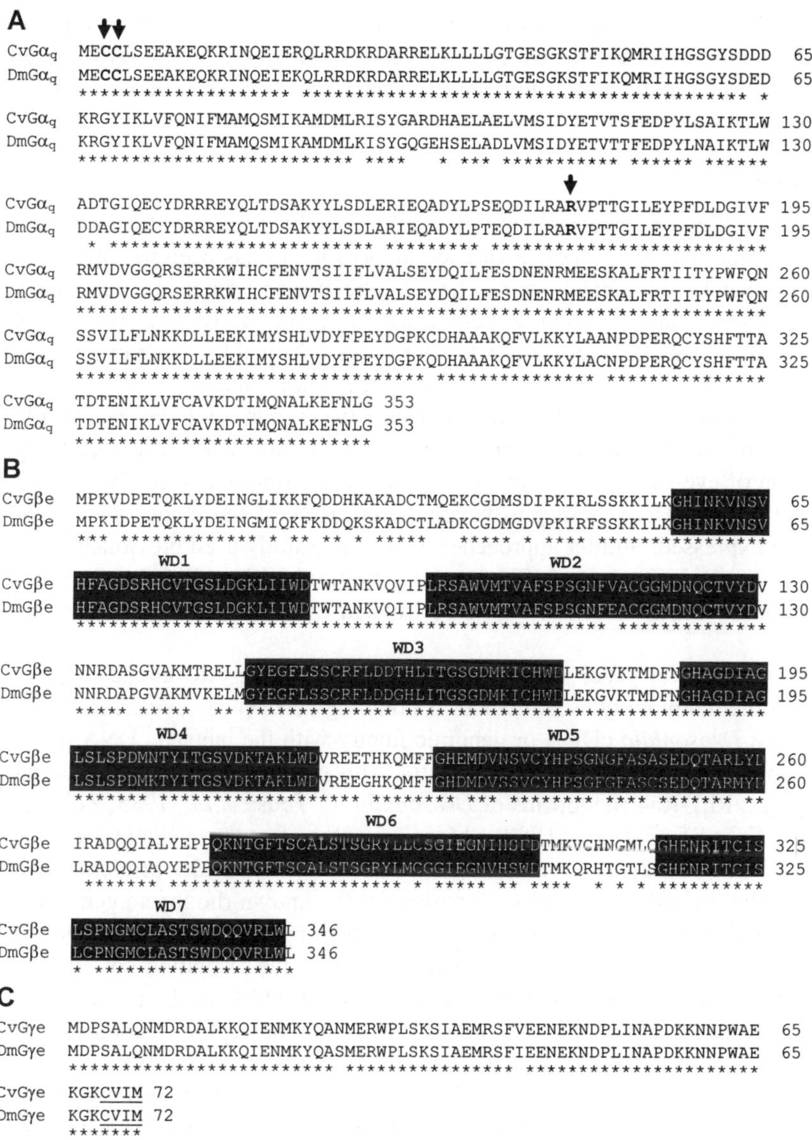

Figure 2. Amino acid alignment of the fly visual G protein subunits (A) $CvG\alpha_q$ and $DmG\alpha_q$, (B) CvGβe and DmGβe, and (C) CvGγe and DmGγe. The deduced amino acid sequences are shown in single letter code, identical amino acids are marked by an asterisk. Arrows in (A) indicate the putative N-terminal palmitoylation sites (C3, C4) and the putative cholera toxin ADP ribosylation site (R178). Black boxes in (B) mark the seven WD repeats of the β subunits. The farnesylation site in (C) is underlined.

subunits indicates the high degree of conservation in the mechanisms of phototransduction in these flies which is also evident when the homology of other phototranduction proteins of *Calliphora* is compared with that of *Drosophila* (Schulz *et al*., 1999).

4. Discussion

The identification and isolation of genes involved in the function and regulation of the phototransduction cascade in *Drosophila* has been performed by different molecular biological approaches. In the present paper we describe a differential hybridization screening method used to isolate novel eye-specifically expressed genes. The high degree of specificity was obtained by using two tissue-specific cDNA probes and a cDNA probe which detects already known visual genes. In this way we were able to distinguish between non- and eye-specifically expressed genes and we could discriminate already cloned genes. The specificity for selection of eye-specifically expressed genes was further enhanced by using a *Calliphora* retinal cDNA library in which about 10 % of the clones are retina-specifically expressed. Similar approaches were successfully used previously, *e.g.*, *arr1* (Hyde *et al*., 1990), *ninaA* (Shieh *et al*., 1989), *rh3* (Fryxell & Meyerowitz, 1987), or *inaD* (Shieh *et al*., 1995) were isolated by substractive hybridization screening.

Isolation of novel genes on the basis of their homology to already known genes is another commonly used method. A homology screen can be performed by hybridizing a *Drosophila* cDNA or genomic library with the labeled cDNA probe which is homologous to the gene searched for. This method yielded cDNA clones encoding *e.g.*, Rh1-Rh4 rhodopsins of *Drosophila* (O'Tousa *et al*., 1985; Zuker *et al*., 1985; Schaeffer *et al*., 1989). More recently, for the isolation of the *Drosophila* Rh5 rhodopsin gene, degenerated oligonucleotide primers were designed which hybridize to conserved regions of the known rhodopsin genes and were used to amplify cDNAs encoding for novel rhodopsins (Chou *et al*., 1996).

Immunoscreening represents another way for the isolation of tissue-specifically expressed novel genes. Huber *et al*. (1996b) used an antiserum directed against proteins of the rhabdomeral photoreceptor membrane of *Calliphora* for immunoscreening a *Calliphora* cDNA library. In this way they were able to isolate the homologues of *Drosophila trp*, *inaC*, and *inaD* as well as two novel genes which are preferentially expressed in the eye but have not yet been characterized in detail. By using P-element enhancer trap lines for retinal lacZ expression Wu *et al*. (1995) succeeded in isolating CDS, a CDP-diacylglycerol synthase, an enzyme required for the regeneration of the signaling molecule phosphatidylinositol-4,5-bisphosphate from phosphatic acid. These P-element enhancer trap lines possess an altered P-element carrying a lacZ gene fused to a minimal promoter. The P-element is mobilized to insert from a silent position in the genome to other sites in the genome nearby a tissue-specific (*e.g.*, retina-specific) enhancer. The tissue-specific expression can be detected by β-galactosidase activity staining and the role of the gene flanking the inserted P-element can be analyzed.

Finally, in the course of sequencing the whole genom of different organisms, bioinformatical methods are becoming an efficient and convenient tool to identify novel genes. Recently, Clyne *et al.* (1999) discovered olfactory receptor genes of *Drosophila* by analyzing the *Drosophila* genome database assuming that *Drosophila* olfactory receptors share structural similarities with known olfactory receptor genes. By a similar approach, Vosshall *et al.* (1999) isolated a *Drosophila* cDNA (*dor104*) encoding a putative odorant receptor by a difference cloning strategy. To isolate additional genes homologous to *dor104* they analyzed the database of the *Drosophila* genome project and were able to identify 11 encoded proteins with sequence similarity to the *dor104* sequence.

The screening methods described here represent effective tools for the isolation of novel genes encoding components of the phototransduction cascade of *Drosophila*. With the cloning of DmGγe one of the last missing key components directly involved in transmitting the visual signal has been characterized. It can be assumed that almost all components of the activation pathway of the *Drosophila* phototransduction cascade are now identified at the molecular level. Components not yet identified may comprise proteins required for the gating of the ion channels TRP and TRPL. There are evidences indicating that polyunsaturated fatty acids, such as arachidonic acid and linolenic acid may activate directly the light-sensitive channels TRP and TRPL (Chyb *et al.*, 1999). However, enzymes generating these messengers, *i.e.*, phospholipase A_2 or a lipoxigenase, have not been identified as phototransduction proteins so far. Furthermore, proteins regulating the inactivation of the phototransduction pathway, *e.g.*, a rhodopsin kinase, which phosphorylates metarhodopsin and may be involved in rhodopsin inactivation or internalization, has so far escaped identification at molecular level. The completion of the *Drosophila* genome project will help to close these gaps in our knowledge about the *Drosophila* visual transduction pathway.

Acknowledgements

This work is supported by funds provided by the European Union (BMH4-CT97-2341).

References

Bentrop, J. and R. Paulsen (1986) "Light-modulated ADP-ribosylation, protein phosphorylation and protein binding in isolated fly photoreceptor membranes", *Eur. J. Biochem.* 161:61-67.

Chevesich, J., A.J. Kreuz and C. Montell (1997) "Requirement for the PDZ domain protein, INAD, for localization of the TRP store-operated channel to a signaling complex", *Neuron* 18:95-105.

Chou, W.-H., K.J. Hall, D.B. Wilson, C.L. Wideman, S.M. Townson, L.V. Chadwell and S.G. Britt (1996) "Identification of a novel *Drosophila* opsin reveals specific patterning of the R7 and R8 photoreceptor cells", *Neuron* 17:1101-1115.

Chyb, S., P. Raghu and R. Hardie (1999) "Polyunsaturated fatty acids activate the *Drosophila* light-sensitive channels TRP and TRPL", *Nature* 397:255-259.

Clyne, P.J., S.J. Certel, M. de Bruyne, L. Zaslavsky, W.A. Johnson and J.R. Carlson (1999) "The odor specifities of a subset of olfactory receptor neurons are governed by Acj6, a POU-domain transcription factor", *Neuron* 22:339-347.

Devary, O., O. Heichal, A. Blumenfeld, D. Cassel, E. Suss, S. Barash, C.T. Rubinstein, B. Minke and Z. Selinger (1987) "Coupling of photoexcited rhodopsin to inositol phospholipid hydrolysis in fly photoreceptors", *Proc. Natl. Acad. Sci. U.S.A.* 84:6939-6943.

Fryxell, K.J and E.M. Meyerowitz (1987) "An opsin gene that is expressed only in the R7 photoreceptor cell of *Drosophila*", *EMBO J.* 6:443-451.

Huber, A., U. Wolfrum and R. Paulsen (1994) "Opsin maturation and targeting to rhabdomeral photoreceptor membranes requires the retinal chromophore", *Eur. J. Cell Biol.* 63:219-229.

Huber, A., P. Sander, A. Gobert, M. Baehner, R. Hermann and R. Paulsen (1996a) "The transient receptor potential protein (TRP), a putative store-operated Ca^{2+} channel essential for phosphoinositide-mediated photoreception, forms a signaling complex with NORPA, INAC and INAD", *EMBO J.* 15:7036-7045.

Huber, A., P. Sander, U. Wolfrum, C. Groell, G. Gerdon and R. Paulsen (1996b) "Isolation of genes encoding photoreceptor-specific proteins by immunoscreening with antibodies directed against purified blowfly rhabdoms", *J. Photochem. Photobiol B* 35:69-76.

Hyde, D.R., K.L. Mecklenburg, J.A. Pollock, T.S. Vihtelic and S. Benzer (1990) "Twenty *Drosophila* visual cDNA clones: One is a homolog of human arrestin", *Proc. Natl. Acad. Sci. USA* 87:1008-1012.

O'Tousa, J.E., W. Baehr, R.L. Martin, J. Hirsh, W.L. Pak and M.L. Applebury (1985) "The *Drosophila ninaE* gene encodes an opsin", *Cell* 40:839-850.

Sanger, F., S. Nicklen and A.R. Coulsen (1977) "DNA sequencing with chain-terminating inhibitors", *Proc. Natl. Acad. Sci. U.S.A.* 74:5463-5467.

Schaeffer, E., D. Smith, G. Mardon, W. Quinn and C. Zuker (1989) "Isolation and characterization of two new *Drosophila* protein kinase C genes, including one specifically expressed in photorceptor cells", *Cell* 57:403-412.

Schulz, S., A. Huber, K. Schwab and R. Paulsen (1999) "A novel Gγ isolated from *Drosophila* constitutes a visual G protein γ subunit of the fly compoud eye", *J. Biol. Chem.* 274:37605-37610.

Selinger, Z., O. Devary, A. Blumenfeld, O. Heichal, S. Barash and B. Minke (1987) "Light-dependent phospholipase C activity in Musca eye membranes and excitation of photoreceptor cells by inositol triphosphate and 2,3 diphosphoglycerate", *Prog. Clin. Biol. Res.* 249:169-178.

Shieh, B.-H., M.A. Stamnes, S. Seavello, G. L. Harris and C.S. Zuker (1989) "The *ninaA* gene required for visual transduction in *Drosophila* encodes a homologue of cyclosporin A-binding protein", *Nature* 338:67-70.

Shieh, B.-H. and B. Niemeyer (1995) "A novel protein encoded by the *InaD* gene regulates recovery of visual transduction in *Drosophila*", *Neuron* 14:201-210.

Shieh, B.-H. and M.Y. Zhu (1996) "Regulation of the TRP Ca^{2+} channel by INAD in *Drosophila* photoreceptors", *Neuron* 16:991-998.

Tsunoda, S., J. Sierralta, Y. Sun, R. Bodner, E. Suzuki, A. Becker, M. Socolich and C.S. Zuker (1997) "A multivalent PDZ-domain protein assembles signalling complexes in a G-protein-coupled cascade", *Nature* 388:243-249.

Vosshall, L.B., H. Amrein, P.S. Morozov, A. Rzhetsky and R. Axel (1999) "A spatial map of olfactory receptor expression in the *Drosophila* antenna", *Cell* 96:725-736.

Wu, L., B. Niemeyer, N. Colley, M. Socolich and C.S. Zuker (1995) "Regulation of PLC-mediated signalling in vivo by CDP-diacylglycerol synthase", *Nature* 373:216-222.

Zuker, C.S., A.F. Cowman and G.M. Rubin (1985) "Isolation and structure of a rhodopsin gene from *D. melanogaster*", *Cell* 40:851-858.

CELLULAR LEVEL

WHAT DO BUTTERFLIES "SEE" WITH THEIR GENITALIA? BIOLOGICAL FUNCTION OF THE GENITAL PHOTORECEPTORS OF THE SWALLOWTAIL BUTTERFLY, *PAPILIO XUTHUS*

KENTARO ARIKAWA

Graduate School of Integrated Science, Yokohama City University, Yokohama and PRESTO, Japan Science and Technology Corporation, Japan

ABSTRACT

Butterflies detect light by the genitalia. What sort of photoreceptor cells do the butterflies have on the genitalia? What are the photoreceptors for? In this chapter, I will give an overview of our studies on the butterfly genital photoreceptor system. I will start with the occurrence of the butterfly genital photoreceptors and their response characteristics. Then I will describe the anatomy, and finally I will discuss the biological function of the system.

1. Response characteristics

1.1. Occurrence of the genital photoreceptors

Extraocular photoreceptors (EOPs) have been found in various animals including arthropods (Yoshida, 1979). They are roughly divided into two categories. The first category of EOP is the found in the central nervous system (CNS-EOP). The crayfish caudal photoreceptor, a photoreceptive interneuron in the abdomen, is an extensively studied CNS-EOP (Wilkens, 1988). The second category of EOP is found outside the CNS as sensory neurons with the photoreceptive site located in the periphery of the animals. The existence of the peripheral EOP in arthropods was first conclusively demonstrated as the butterfly genital photoreceptors (Arikawa, *et al.*, 1980).

Figure 1. A mating pair of *Papilio xuthus*. Up, female; down, male.

The butterfly genital photoreceptor was discovered, rather accidentally, in the Japanese yellow swallowtail butterfly, *Papilio xuthus* (Fig. 1). I was working on a project on the neuronal mechanism of the host-plant selection by female *Papilio*, and thus I was analyzing input-output relations between neurons in the abdominal nervous system. In the course of analyzing mechanoreceptive inputs from the ovipositor, I encountered a sensory neuron in the nerve derived from the ovipositor actively producing spikes. The neuron was very active even when the mechanoreceptive hairs on the ovipositor were not stimulated. This unexpected neuronal activity obstructed the analysis I intended to perform. So I decided to take a rest, and turned off the illumination for the microscope. Strangely enough this caused the cessation of the spike activity. Surprisingly, the spikes immediately came back when I turned the light back on. Rather in an excited mood, I cut off the head, removed the thorax, and sectioned the abdomen. Some hours later, the remaining tiny piece of cuticle with a stump of nerve at the tip of the suction electrode still produced spikes in response to light flashes (Fig. 2). This was the moment I came to believe that *Papilio* "sees" with the genitalia.

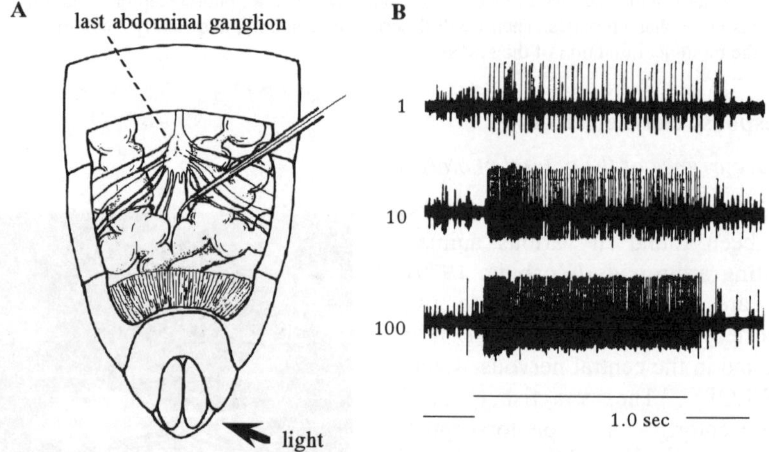

Figure 2. Photoreceptor responses. **A)** A schematic drawing of the abdominal nervous system of a female, fixed ventral side up. We used a suction electrode to pick up the nervous activity. **B)** Example of the sustained train of spikes of the photoreceptor recorded by the suction electrode in response to 1 sec light flashes of different intensities. Numbers on the left indicate relative intensities of stimulation.

1.2. Response characteristics

I first searched for similar effects in males, and I found that two out of six pairs of lateral nerves (N1-6), derived from the last abdominal ganglion (Fig. 2), contain the photoreceptor axon. Male and female photoreceptors appeared to be contained in equivalent nerves. The lateral nerves containing the photoreceptor axons are the posterior two pairs (N5 and N6). Later we investigated other lepidopteran species and revealed that the genital photoreceptors exist in all butterfly species tested, including skippers. However, we could not find them in moths, neither the diurnal nor nocturnal species (Arikawa and Aoki, 1982). Also the larvae of *Papilio xuthus* do not have the genital photoreceptor system: it develops in the late pupal stage (Miyako *et al.*, 1995), which suggests that it is used for adult-specific behaviors.

The photoreceptor produces a sustained train of spikes in response to a light flash (see Fig. 2). The photoreceptor is able to produce about 300 spikes per second at a maximum (Arikawa *et al.*, 1997).Using the spike frequency as a measure of the response intensity, we determined the spectral sensitivity of the photoreceptors. All four photoreceptor cells, both in males and females, appeared to be highly sensitive to light of the ultraviolet-blue (340-460 nm) wavelength region (Arikawa and Aoki, 1982).

1.3. Location of the photoreceptors

The precise location of the photoreceptors was studied also electrophysiologically. We localized the photoreceptive sites by scanning a small spot of light while recording the photoreceptor response from the nerve. Two pairs of photoreceptive sites (P1, P2) were found in both sexes (indicated by large arrows in Fig. 3). It appeared that the posterior-lateral nerve N6 contains the axon originating from the P1 site, whereas the N5 contains the axon from the P2 site.

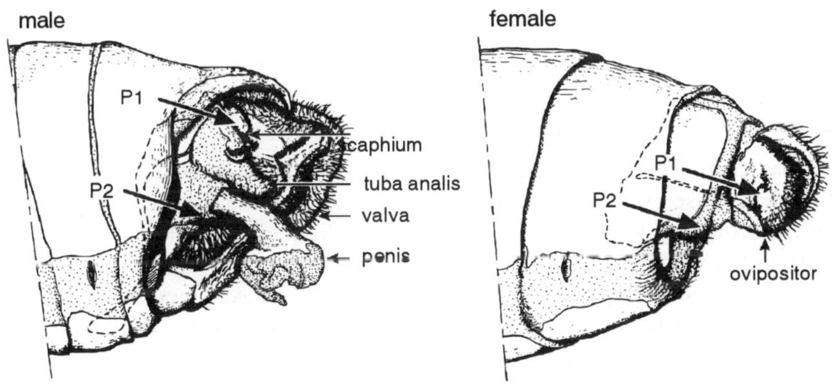

Figure 3. Location of two pairs of genital photoreceptors, P1 and P2. Side views of the abdominal tip.

In males, the P1 exists in the scaphium, the tanned sclerotization of the dorsal surface of the anal tube. The inner margin of the scaphium is clearly marked by a transparent patch of cuticle (Fig. 4A). When we cover the transparent region, the photoreceptor response disappears, indicating that the photosensitive structure is located immediately beneath or at least very close to the region (Fig. 3). The second pair, the P2, was found slightly ventral to the penis. The P2 region also appears to be transparent.

In females, the P1 was found on the lateral side of the ovipositor, or the papilla analis (Fig. 3). The ovipositor is a strongly tanned and hairy pair of lobes between which the anus and the oviduct open to the exterior. On the lateral side of each lobe, a characteristic concave structure exists, the inner surface of which is transparent and free of hairs (Fig. 4B). The photoreceptor response disappears when the transparent cuticle of the concave region is shielded, indicating again that the photoreceptive organelle exists around this region.

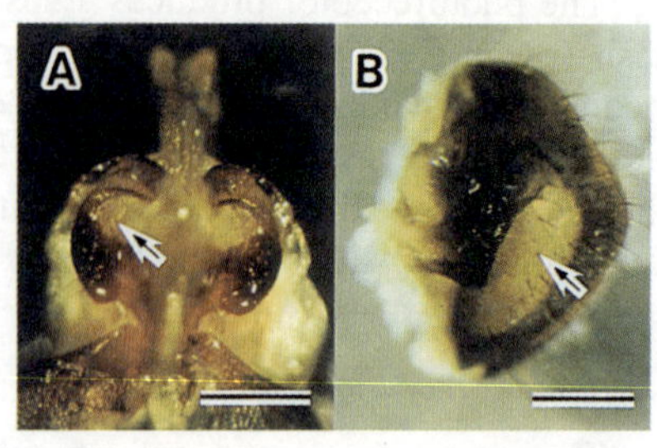

Figure. 4. Male scaphium **(A)** and female ovipositor **(B)**. Arrows: transparent cuticle. Scale = 500 μm.

The P2 region, which is slightly ventral to the ovipositor, is covered by yellow, hairy scales, but not tanned.

We carefully examined the surface structure of these transparent regions to see whethe there are any specialized structures for photoreception: we could not find any possible candidates, however.

2. Anatomy

2.1. Structure of the photoreceptors

The internal structure of the photoreceptive sites was studied by light and electron microscopy. We first prepared the photoreceptive sites for histology, and observed complete serial sections of the sites by light microscopy.

Both the scaphium and the ovipositor bear many hairs (Fig. 4). Most of them are mechanoreceptors, each of which has a small sensory neuron at the base. Among the numerous cell bodies of the small sensory neurons, we found a large (ca 30 x 40 μm) ovoid structure containing the cell body of a sensory neuron. The cell body in the ovoid structure tapers to form an axon, which extends via the N6 into the last abdominal ganglion where it arborizes and terminates (Arikawa and Aoki, 1982).

How does the ovoid structure look like? Does this really contain the photoreceptor? Light microscopy showed that the ovoid structure contains the cell body of a sensory neuron. It appeared by electron microscopy that about 30 % of the cross-sectional area of the ovoid structure is occupied by a very peculiar-looking organelle (Fig. 5). This organelle resembles a phaosome, a structure first found in the earthworm skin, which presumably functions as a photoreceptor (Roehlich *et al.*, 1970).

The phaosome found in the genitalia of *Papilio* consists of two components: membrane-enclosed electron-lucent areas and closely packed tubules of membranes. Serial sections revealed that the electron-lucent components are obliquely sectioned profiles of the processes that protrude from the distal side of the cell body. The tip of the distal processes bear tubular membranes. The diameter of the tubules ranges between 0.1 and 0.3 μm, which is larger than the diameter of the rhabdomeral microvilli of the compound eye photoreceptors (ca 0.07 μm). The variation in diameter is due to repeated bifurcation of the tubules (Miyako *et al.*, 1993).

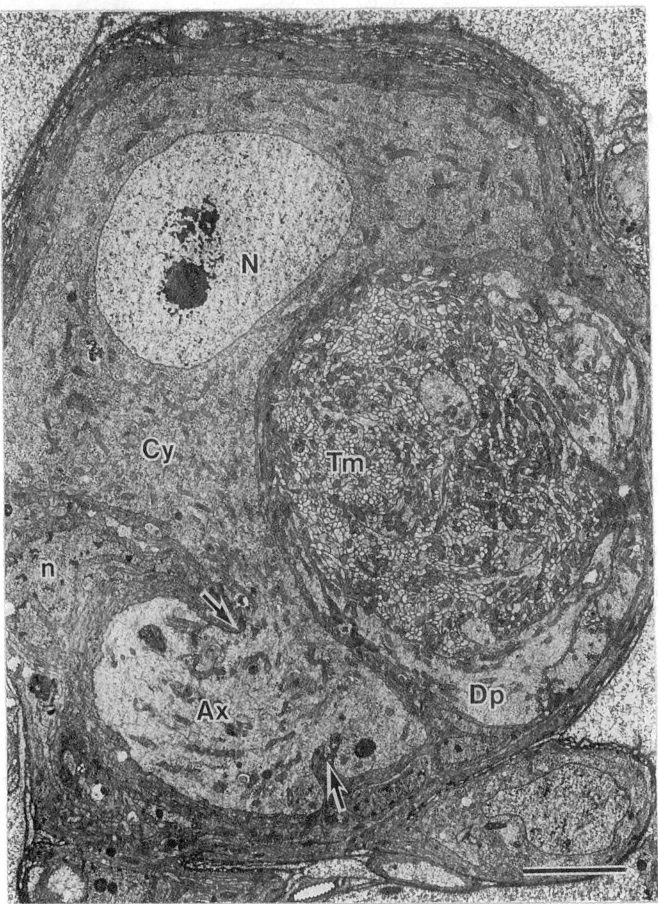

Figure 5. The whole appearance of the ovoid structure that contains the cell body of the sensory neuron in the female P1. The structure also contains a phaosome that consists of some distal processes (Dp) and tubular membranes (Tm). An axon (Ax) extends from the cell body. The axon hillock is marked by ar-rows. Cy, cell body cytoplasm; N, nucleus of the sensory neuron; n, nucleus of a glial cell. Scale = 5 μm.

3. Biological significance

Copulation in males

3.1. Valva-opening response

What is the genital photoreceptors for? In order to find some clue to answer this question, we observed the genitalia of an intact male under a dissecting microscope and then noticed an interesting phenomenon. In response to light stimulation of the genitalia, the male widely opened the valva, a pair of lobes for supporting female genitalia upon copulation (Fig. 6). The male closed the valva as the light was turned off. This response did not occur when the illumination light was red filtered so that it only contained wavelengths longer than 600 nm; these wavelengths are out of the range of the photoreceptor spectral sensitivity (Arikawa, 1993). As we assumed that the valva is most actively used upon mating, this finding strongly motivated further investigations into the possible involvement of the genital photoreceptors in the mating behavior (Arikawa *et al.*, 1996; Arikawa *et al.*, 1997).

Figure 6. Genitalia of a coupling pair. Taken from the ventral side. The male (down) supports the genitalia of the female (up) using the bilateral valva (arrowhead)

3.2. Copulation behavior

We hence constructed an outdoor cage (11 x 7 x 3 m), and performed a series of behavioral experiments. We first observed the mating behavior of intact *Papilio xuthus* in the cage (Fig. 7). We positioned a virgin female in the cage by attaching the dorsal cuticle of the thorax with beeswax to the bottom end of an insect pin whose upper end was fixed to a horizontal bar set 2 m above the ground. We subsequently released an intact male in the cage. It appeared that the mating behavior consists of at least six steps. It starts when a male finds a female and approaches her (step 1). The male gets into the 'ventor-to-ventor' position with the female, often opening the valva (step 2). The male then searches for the female's genitalia by touching her abdomen with his own genitalia exposed between the fully-opened valva (step 3), firmly clasps her genitalia by using the superuncus and scaphium (Fig. 7C). When the male has clasped the female's genitalia correctly (step 4), he inserts the penis, ejaculates, and then plugs the vagina with a sphragis by keeping the 'end-to-end' copulatory posture (step 5), after which the pair finally separates (step 6). Under the experimental condition, 66% intact males could copulate with the virgin females hanging in the cage (copulation success rate, CSR = 66 %).

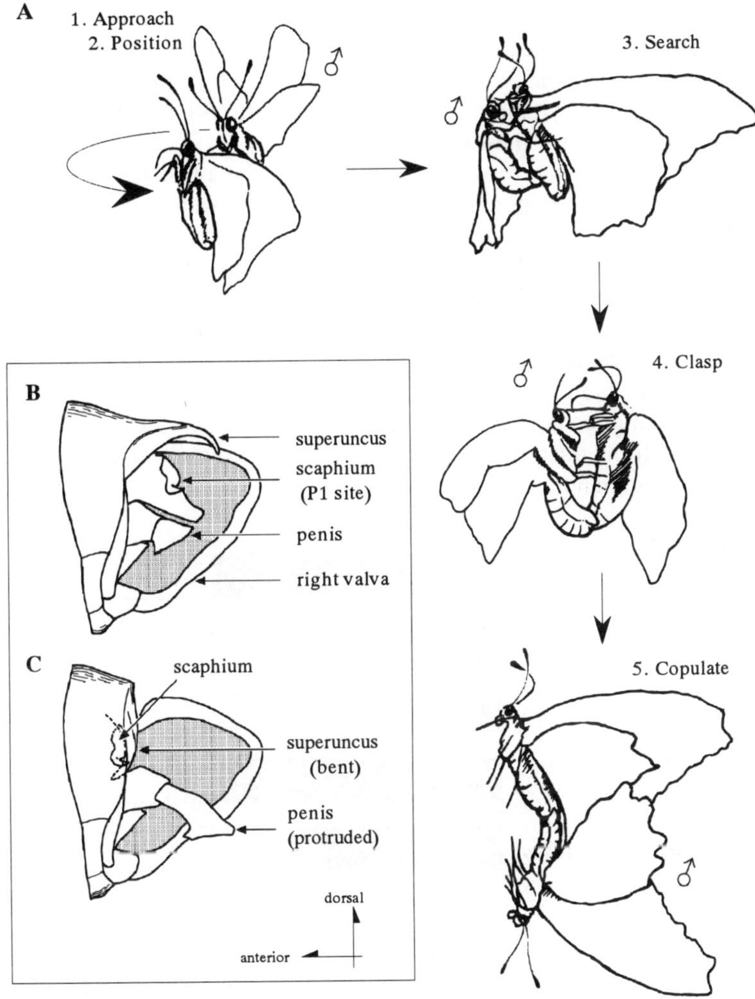

Figure 7. (A) Sequence of mating behavior of *Papilio xuthus* seen in the outdoor cage. The male was freely flying, whereas the female was mounted in the cage but could still flip their wings. **(B)** Left side view of male genitalia. Not copulating. Left valva was removed for clarity. **(C)** Genitalia of a copulating male (step 5) The female genitalia are omitted for clarity.

To see whether and how the genital photoreceptors are involved in the mating behavior, we next ablated the photoreceptor input in both males and females, and observed their behavior. We ablated the P1s by gently rubbing the P1 site with a fine heat-probe, or by painting with black mascara. As a control, we painted the P1 site with transparent mascara.

Figure 8 shows the results. We treated the P1s of males and let them mate with pinned intact females. The CSR of the P1-heat ablated males significantly decreased down to 28%. The decrease was also observed in the P1- black painted males (23%). The P1-clear painted males did not have any problems to mate with the intact females (CSR = 67 %). The results clearly indicate that the males somehow use the light signal from the genitalia for copulation. We also heat-ablated mechanoreceptors existing on the scaphium. The P1-ablation could have reduced the motivation to mate. However, the fraction of individuals that did not perform mating behavior, which was about 25% in intact males, remained constant in all the treatments. Clearly, the P1-ablated males did want to mate, but they simply could not.

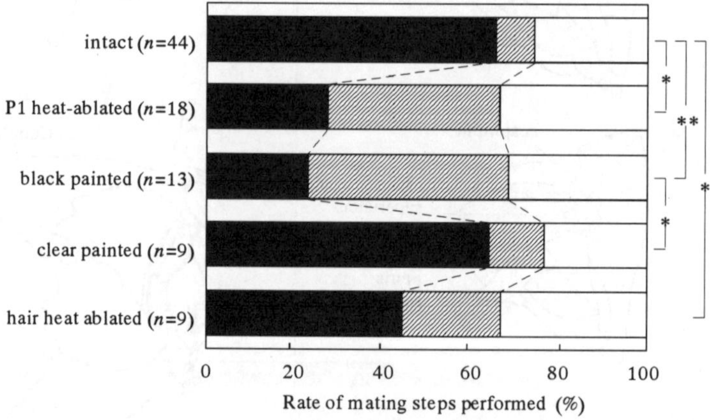

Figure 8. Effect of ablation of P1 and mechanoreceptive hairs on the mating behavior in males. Mann-Whitney *U*-test was used for analyzing statistical difference in CSR between treatments (*, $p > 0.01$; **, $p < 0.05$).

Next we treated the P1s of females and measured their CSR with intact males. However, the P1 ablation in females did not have any effect on the CSR. How do the P1s control mating behavior? To address this question, we investigated the P1 response in males during the mating behavior (Arikawa *et al.*, 1997). The P1 response was recorded with a suction electrode. Before the actual recording, we determined the relative positions of the genitalia of mating butterflies at different behavioral steps from several examples of videotaped copulation. These positions were mimicked by opening/closing the valva and placing an isolated female abdomen in various locations near the male fixed on the recording stage (Arikawa and Miyako-Shimazaki, 1996). The responses during clasp and copulation (steps 4 and 5) were recorded from males that were actually copulating.

Figure 9 summarizes the results. At step 2, when the male fully opens the valva, the P1 response increases (ca. 82 spikes per second at 2,000 lux). The response decreases to about 52 spikes per second while the male is searching for the female's genitalia (step 3), but it again drops to about 25 spikes per second (arrow) at step 4 when the male clasps the female's genitalia with the superuncus and scaphium.

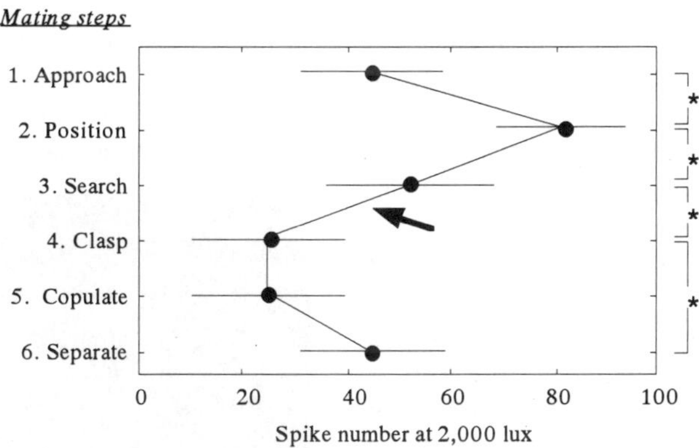

Figure 9. Simulation of the response of male P1 during the mating behavior. For each behavioral step, the spike number per a second elicited by a light flash of a maximal intensity (ca 2,000 lux) was plotted. A sharp decrease in response (arrow) was observed between the steps 3 and 4 when the male clasped female's genitalia by bending the superuncus.

The sharp drop in the P1 response, which occurs when the male correctly locates the female's genitalia, seems very important (Fig. 9). We therefore hypothesize that the sharp drop in the P1 response informs the male that the female's vagina is correctly positioned for penis insertion. The male P1 is located on the scaphium, which is used to clasp the female's genitalia together with the bent superuncus (Fig. 7C). Apparently, the P1s get dark when the mates properly couple. But in case the genitalia of both sexes are mal-aligned, the coupling would be incomplete, leaving some space through which light can enter, so that the P1 response continues. In such a case, the male releases the clasp and goes back to the search step (Fig. 7). The males with the P1 painted black never experience such a drop in response, as there is little P1 activity to begin with. These animals continue the genitalia search even when the female's genitalia are correctly aligned, until they finally give up.

The scaphium bears not only photoreceptors but also mechanoreceptive hairs. We also heat-ablated the mechanoreceptors on the scaphium of males, and observed their mating behavior. The CSR of the mechanoreceptor-ablated males reduces to 44%, indicating the necessity of mechanical sense for achieving copulation (Fig. 8).

This may explain why one third of P1 heat-ablated males can still copulate: males locate the females' genitalia by using the mechanical sense. We note that another pair of genital photoreceptors, the P2s, may also contribute to mating behavior, although this possibility has not yet been tested.

3.3. Oviposition behavior

P1 ablation in females did not have any effect on the CSR. Actually, the P1s of females are located on the lateral side of the ovipositor, which is usually half-covered with the hairy scale at the abdominal tip. The ovipositor is neither exposed to light nor tightly covered when coupling, so that it is rather difficult to assume that the function of the female P1s is involved in the mating behavior.

Then, what are the females' photoreceptors for? Upon light stimulation of the abdominal tip it appeared that the females sometimes push out the ovipositor. A similar movement of the ovipositor can be observed when the females are laying eggs (Fig. 10). The females of *Papilio* lay eggs on leaves of citrus plants, the food of the larvae, after they confirm the leaf "taste" by using the contact chemoreceptors on the forelegs.

Figure 10. A female *Papilio xuthus* laying an egg on a citrus leaf.

If they know that the leaf is correct, they curl the abdomen, push out the ovipositor, touch the leaf surface with it, and then deposit an egg on the leaf. The pushing-out of the ovipositor is similar to what happens upon light stimulation of the genitalia. Therefore, we hypothesized that the female photoreceptors are involved in the oviposition behavior.

Presumably, the mechanical sense of the ovipositor plays an important role in oviposition control (Yamaoka *et al.*, 1971). In fact, when we ablated the mechanoreceptors on the ovipositor of a female *Papilio* that was actively laying eggs, the female could not lay eggs anymore. The mechanoreceptor-ablated females push the ovipositor strongly against the leaf to locate the egg-laying site, but they cannot deposit an egg. This indicates that the mechanical input from the ovipositor informs females that the leaf is there.

What happens on the oviposition behavior if the P1s are removed? To test this, we heat-ablated the P1s of female as we did in males. Once an intact female curls the abdomen to lay an egg on the leaf, the female deposits an egg at the success rate of about 80 %; *i.e.* 8 eggs deposited per 10 abdomen curls. However, the success rate significantly dropped to only a few percent if the P1s are heat-ablated. The P1-ablated females strongly push the leaf with the ovipositor to try laying eggs, which is similar what happened to the mechanoreceptor-ablated females.

It is somewhat strange that the P1-ablated females touch the leaf with the ovipositor and still do not lay eggs, for their mechanoreceptors are intact. Probably the mechanical input from the ovipositor is only effective when the P1s are active. The P1 activity most likely tells the female that the ovipositor is sufficiently pushed out and is now ready to accept the mechanical input as the indicator of the leaf location (Arikawa *et al.* unpublished).

4. Concluding remarks

We now know what is the *Papilio* hindsight for. But, why do the butterflies have to use light in the reproductive behaviors? How does such an ability evolve?

A provocative hypothesis is that the photoreceptors prevent interspecific cross breeding, at least in males. The mating behavior of butterflies consists of many steps, in which almost all available sensory inputs are used for mate recognition. Coupling the genitalia is the final step. Butterfly genitalia are so complex, and differ so much between species even within the genus *Papilio*, that a precise lock-and-key relationship exists only between male and female of the same species. Thus a mechanical, or rather optical, coupling between a male and female of different species cannot be perfect. Measuring the light leaking through the space between mismatched genitalia would provide a useful and sensitive test of the appropriateness of a particular mating.

In the case of female oviposition, the use of light for detecting ovipositor location is somewhat reasonable. Of course they could have used mechanical sense, for example from muscle receptor organs, if any, in the ovipositor extensor muscles. In such a case however, the information could be wrong: positive decision will be made even though the exit of the ovipositor is blocked somehow. Light, on the other hand, is only available when the ovipositor is truly exposed to the open air. Thus the females are able to avoid loosing eggs at the very final step of oviposition, to which they have already invested a lot of energy.

Acknowledgments

I very much thank Kiyoshi Aoki, Eisuke Eguchi, Yumiko Miyako-Shimazaki, Daisuke Suyama, Takanori Fujii, Miho Sato, and Nobuhiro Takagi, for scientific contribution of the study. I thank Doekele Stavenga for critical reading of the manuscript. This work was supported by the Research grants from the Whitehall Foundation, the Uehara Memorial Foundation, the Sumitomo Foundation, the Novartis Foundation, Kanagawa Academy of Science and Technology, and the Ministry of Education, Science, and Culture of Japan.

References

Arikawa, K. (1993) "Valva-opening response induced by the light stimulation of the genital photoreceptors of male butterflies, *Naturwissenschaften* 80:326-328.

Arikawa, K. and K. Aoki (1982) "Response characteristics and occurrence of extraocular photoreceptors on lepidopteran genitalia", *J. Comp. Physiol. A* 148:483-489.

Arikawa, K., E. Eguchi, A. Yoshida and K. Aoki (1980) "Multiple extraocular photoreceptive areas on genitalia of butterfly, *Papilio xuthus*", *Nature* 288:700-702.

Arikawa, K. and Y. Miyako-Shimazaki (1996) "Combination of physiological and anatomical methods for studying extraocular photoreceptors on the genitalia of the butterfly, *Papilio xuthus*", *J. Neurosci. Meth.* 69:75-82.

Arikawa, K., D. Suyama and T. Fujii (1996) "Light on butterfly mating", *Nature* 382:119.

Arikawa, K., D. Suyama and T. Fujii (1997) "Hindsight by genitalia: Photo-guided copulation in butterflies", *J. Comp. Physiol. A* 180:295-299.

Miyako, Y., K. Arikawa and E. Eguchi (1993) "Ultrastructure of the extraocular photoreceptor in the genitalia of a butterfly, *Papilio xuthus*", *J. Comp. Neurol.* 327:458-468.

Miyako, Y., K. Arikawa and E. Eguchi (1995) "Morphogenesis of the photoreceptive site and development of the electrical responses in the butterfly genital photoreceptors during the pupal period", *J. Comp. Neurol.* 363:296-306.

Roehlich, P., B. Aros and S. Viregh (1970) "Fine structure of photoreceptor cells in the earthworm, *Lumbricus terrestris*", *Z. Zellforsch.* 104:345-357.

Wilkens, L.A. (1988) "The crayfish caudal photoreceptor: Advances and questions after the first half century", *Comp. Biochem. Physiol.* 91C:61-68.

Yamaoka, K., M. Hoshino and T. Hirao (1971) "Role of sensory hairs on the anal papillae in oviposition behaviour of *Bombyx mori*", *J. Insect Physiol.* 17:897-911.

Yoshida, M. (1979) "Extraocular photoreception", in: *Handbook of Sensory Physiology*, H. Autrum, ed., Berlin, Heidelberg, New York: Springer-Verlag, pp. 582-640.

COLOR VISION AND RETINAL RANDOMNESS OF THE JAPANESE YELLOW SWALLOWTAIL BUTTERFLY, *PAPILIO XUTHUS*

KENTARO ARIKAWA[1,2], MICHIYO KINOSHITA[1], JUNKO KITAMOTO[1] and DOEKELE G. STAVENGA[3]
[1]*Graduate School of Integrated Science, Yokohama City University, Japan*
[2]*PRESTO, Japan Science and Technology Corporation, Japan*
[3]*Department of Neurobiophysics, University of Groningen, The Netherlands*

ABSTRACT

Compound eyes consist of many unit eyes, called ommatidia. Since only recently, the ommatidia were commonly believed to be identical in both structure and physiological characteristics, at least within restricted eye regions. Although this view still may be correct for many insect species, our studies on the eyes of butterflies, performed during the last decade, have accumulated firm evidence that butterfly eyes often are a random mesh of different types of ommatidia. In this chapter, we will introduce our extensive studies on the Japanese yellow swallowtail butterfly, *Papilio xuthus*, with particular attention to the distribution of different types of spectral receptors. We will also refer to our recent behavioral studies on color vision and color constancy of *Papilio xuthus*.

1. The ommatidium of *Papilio*

1.1. Five types of spectral receptors

Butterflies are colorful animals, many of which feed on nectar offered by colorful flowers. Such flower-visiting behavior has attracted many researchers for studying the possible ability of butterflies to see color. Because color vision requires a certain set of spectral receptors in the retina, the identification of different types of spectral receptors is instrumental to the understanding of color vision. Therefore, many attempts have been made for identifying the spectral receptors that participate in the color vision system of animals. In insects, pioneering work was done on honeybees. There ultraviolet (UV) sensitive photoreceptors were first identified, which, together with the coexisting blue and green receptors, form the physiological basis of the trichromatic color vision system. Since then, many insects have been shown to have two to four types of spectral receptors (Menzel, 1979). In 1987, we found the first example of a retina furnished with at least five different types of spectral receptors in the Japanese yellow swallowtail butterfly, *Papilio xuthus*, by recording the photoreceptor potential from single photoreceptor cells, especially in the latero-frontal region of the compound eye (Arikawa *et al.*, 1987). The five different types of spectral receptors peak in the UV, violet, blue, green, and red wavelength regions, respectively (Fig. 1).

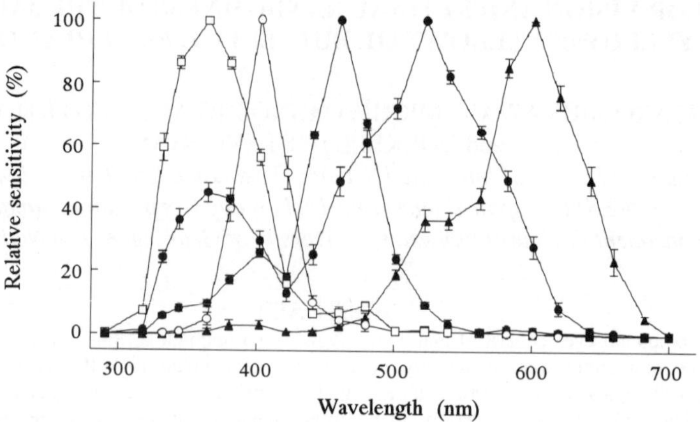

Figure 1. Spectral sensitivity curves of five types of spectral receptors identified by intracellular electrophysiology.

The sensitivity spectrum of a photoreceptor cell is basically determined by the absorption spectrum of the visual pigment expressed in the cell. In the most simple conception, sensitivity and absorption spectrum should match each other. The UV and the green receptors of *Papilio* conform rather well to this case. The UV receptor has a single sensitivity band, but the green receptors have two sensitivity bands; the secondary sensitivity peak in the UV corresponds to the so-called β-band of the visual pigment's absorption spectrum.

However, the *Papilio* retina also appears to contain photoreceptors whose sensitivity spectra distinctly differ from the absorption spectra predicted for visual pigments. For example, the sensitivity spectrum of the violet receptor, peaking at 400 nm, is considerably narrower than the absorption spectrum of a 400 nm-peaking visual pigment. Similarly, the sensitivity band of the red receptor, peaking at 600 nm, is narrower than the absorption band of a visual pigment, absorbing maximally at 600 nm. The narrower sensitivity spectra can be explained, at least in part, by filtering effects by screening pigments contained in the photoreceptor cells, as will be discussed later in this chapter (Arikawa *et al.*, 1999c). Furthermore, there is a special type of green receptor, which lacks the secondary peak in the UV (Bandai *et al.*, 1992). The underlying mechanism will also be discussed later (Arikawa *et al.*, 1999b).

1.2. Cellular organization of an ommatidium

A compound eye of *Papilio* consists of about 12,000 ommatidia. An ommatidium contains nine photoreceptor cells (R1-9, Fig. 2). Each of the photoreceptor cells bears closely packed and parallel microvilli forming a rhabdomere, where visual pigment molecules reside and absorb light.

The rhabdomeres of the nine photoreceptor cells construct a cylindrical rhabdom, whose diameter is about 2.5 μm, in the center of the ommatidium. Four (R1-4) out of nine photoreceptors are called distal photoreceptors, because they contribute their rhabdomeral microvilli to the rhabdom in the distal two-thirds of the ommatidium. R5-8 contribute to the rhabdom in the proximal one-third, so that they are called the proximal photoreceptors. R9, the basal photoreceptor, contributes a small rhabdomere at the base of the ommatidium.

We attempted to identify which of the five characterized types of spectral receptors corresponds to which of the R1-9 photoreceptors (Fig. 1). By applying intracellular recordings combining dye injection with measurements of the spectral as well as the polarization sensitivity, we were able to spectrally classify the R1-9 photoreceptor cells (Arikawa and Uchiyama, 1996, Bandai *et al.*, 1992).

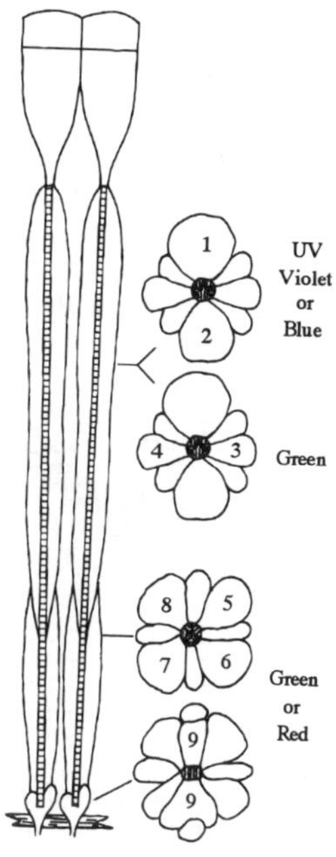

Figure 2. Cellular organization of the *Papilio* ommatidium. Each of the nine photoreceptors cells (1-9) were assigned to one of the five spectral receptor types.

The rhabdomeral photoreceptors bearing parallel and straight microvilli are known to be most sensitive to polarized light whose *e*-vector orientation is parallel to the microvillar longitudinal axis (Moody and Parriss, 1961).

In the *Papilio* ommatidium, the microvilli of R1-9 are basically parallel and straight. For example, the microvilli of the R1 and R2 are parallel to the animal's dorso-ventral (vertical = 0°) axis, so that the cells are maximally sensitive to polarized light oscillating in the vertical plane. Similarly, R3 and R4, whose microvilli are parallel to the antero-posterior axis, are maximally sensitive to polarized light oscillating horizontally.

Two examples of polarization sensitivity measurements are shown in Fig. 3. The polarization sensitivity of all encountered UV receptors peaks at 0°, indicating that the UV receptors are either R1 or R2. On the other hand, the polarization sensitivity curves of the green receptors peak at 90°, which is predicted for R3 and R4 photoreceptors. Recordings performed in this way in both the distal and proximal layers revealed that R1 and R2 are either of the UV, violet, or blue type, while R3 and R4 are of the green type, and R5-9 are of the green or red type (Fig. 2).

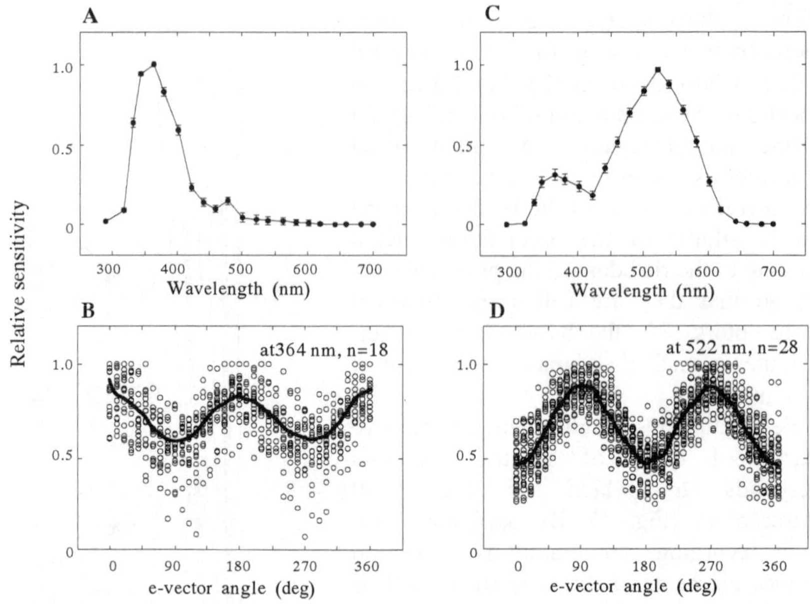

Figure 3. Example of the results of the polarization method. **(A)** UV receptor. **(B)** Polarization sensitivity of the UV receptor, peaking at 0 degree, indicating that the UV receptors are euther R1 or R2. **(C)** Green receptor. **(D)** Polarization sensoitivity of the green receptor, peaking at 90 degree, indicating that the green receptor are either R3 or R4.

1.3. Molecular biology of visual pigments

When photoreceptor cells have different sensitivity spectra it is generally accepted that these are caused by different visual pigments, *i.e.* visual pigments with differing absorption spectra. The basic structure of all visual pigment molecules so far identified, including those of arthropods appears to be identical. The protein moiety, the opsin, consists of about 350 amino acids, with seven transmembrane helices, and a retinal chromophore is attached to a lysine in the seventh helix (Applebury and Hargrave, 1986, Gaertner and Towner, 1995). The chromophore is the 11-*cis* form of either retinal, 3-4-dehydroretinal, 3-hydroxyretinal, or 4-hydroxyretinal, depending on species. For example, the chromophore of human visual pigments is retinal and some dragonfly species use both retinal and 3-hydroxyretinal. *Papilio* exclusively employs 3-hydroxyretinal (Seki *et al.*, 1987). This indicates that the difference in absorption spectra of multiple visual pigments in *Papilio*, if any, must be attributed to the difference in the structure of opsin.

To uncover these differences we initiated a molecular biological study with the aim to clone the cDNAs encoding opsins. So far we identified five cDNAs each encoding different opsins (PxRh1-5, Papilio xuthus Rhodopsin 1-5). Figure 4 shows a result of a phylogenetic analysis of insect opsins including the ones of *Papilio*, based on the deduced amino acid sequences. It appeared that PxRh1-3 belong to the class of long wavelength-absorbing visual pigments (Kitamoto *et al.*, 1998), whereas PxRh4 and PxRh5 are of the short wavelength-absorbing type (Kitamoto *et al.*, 1999).

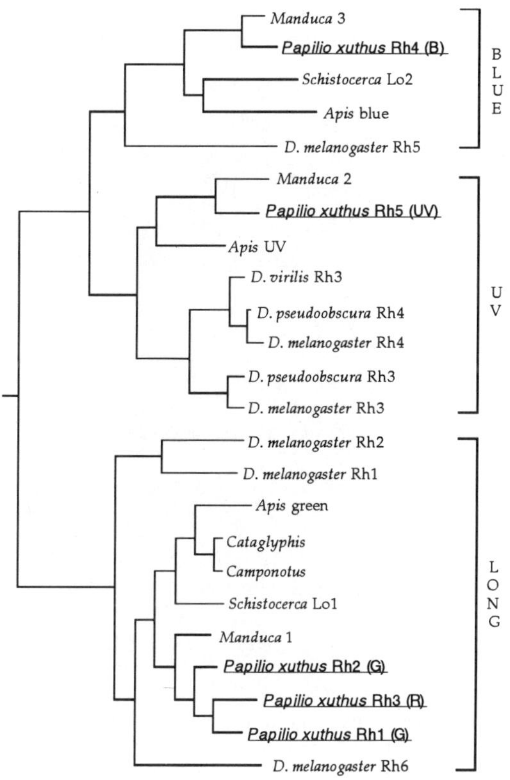

Figure 4. Phylogeny of insect opsin including the ones of *Papilio* (PxRh1-5). UV, ultraviolet; B, blue; G, green; R, red..

The histological distribution of the mRNAs of these opsins was studied by *in situ* hybridization. Figure 5 shows an example, where we labeled the retinal frozen sections with the probe detecting the PxRh2 mRNA.

The PxRh2 probe clearly hybridized to photoreceptor cells throughout the retina; *i.e.* from the dorsal to the proximal edge (Fig. 5A). In transverse sections, it appeared that the green sensitive distal photoreceptors, R3 and R4, were labeled in all ommatidia (Fig. 5B). Therefore, it is most likely that PxRh2 corresponds to the green-absorbing visual pigment. In the proximal layer, the PxRh2 probe labeled R5-8 in some ommatidia (Fig. 5C). Interestingly, the R5-8 in a single ommatidium are always labeled with the probe as a set, suggesting that the spectral property of R5-8 in a single ommatidium is identical. Considering the fact that the R5-8 are either green or red receptors (Fig. 2), the R5-8 labeled with the PxRh2 probe are most likely green sensitive.

The R5-8 not labeled with the PxRh2 probe were labeled, basically, with the PxRh3 probe instead, suggesting that PxRh3 corresponds to the visual pigment expressed in the red receptors (Fig. 5D).

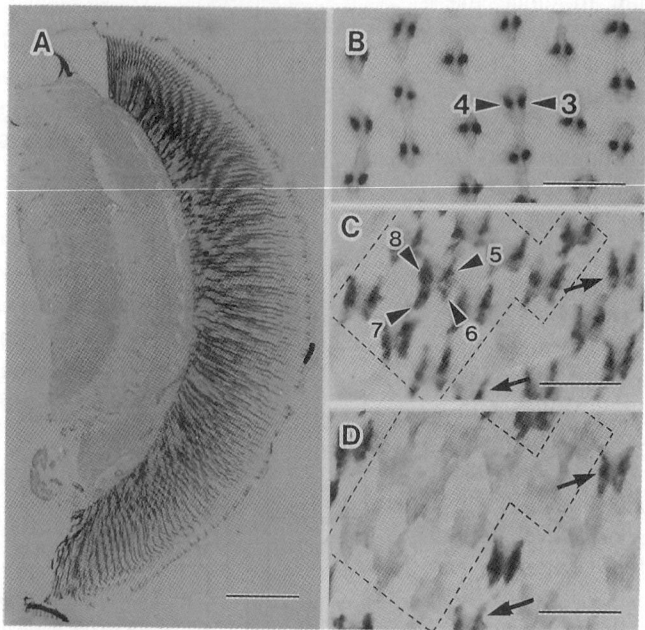

Figure 5. Histological distribution of PxRh2 and PxRh3 mRNAs by in situ hybridization. **(A)** Vertical section labeled with the PxRh2 probe. **(B)** Transverse section through the distal tier. PxRh2. **(C)** Transverse section through the proximal tier. PxRh2. **(D)** Transverse section through the proximal tier. PxRh3. Arrows in C and D indicate ommatidia containing R5-8 labeled both with the PxRh2 and PxRh3 probes. Scales = 250 μm (A), 25 μm (B-D).

Surprisingly, the R5-8 of about 18% of the ommatidia appeared to be labeled both the PxRh2 and PxRh3 probes. This strongly suggests that these photoreceptors simultaneously express visual pigments of the green and the red receptors (Figs 5C, D, arrows). Of course, the double expression of opsins could result in a broadened sensitivity spectrum. In fact, we recently found proximal photoreceptors with an abnormally broad sensitivity spectrum whose half-bandwidth was about 230 nm (Arikawa *et al.*, 1999a).

2. Ommatidial heterogeneity

2.1. Photoreceptor cell pigmentation

The findings described above clearly indicate that the ommatidia are spectrally heterogenous. For example, there are three spectral receptor types, UV, violet, and blue, assigned to only two anatomically identifiable photoreceptor cells, R1 and R2. Moreover, R5-8 in some ommatidia express PxRh2, whereas the R5-8 of other ommatidia bear PxRh3, or both. Then, how are the ommatidia different? How are the different ommatidia arranged in the retina?

To approach these questions, we first searched for histological relevance of the ommatidial heterogeneity. We carefully observed plastic sections of the retina stained with a Azur-II, but we could not find any clear difference between ommatidia. When we observed the sections before staining, rather accidentally, a surprising difference became evident. Some ommatidia bear yellow pigmentation around the rhabdom, whereas others have red pigmentation instead (Fig. 6).

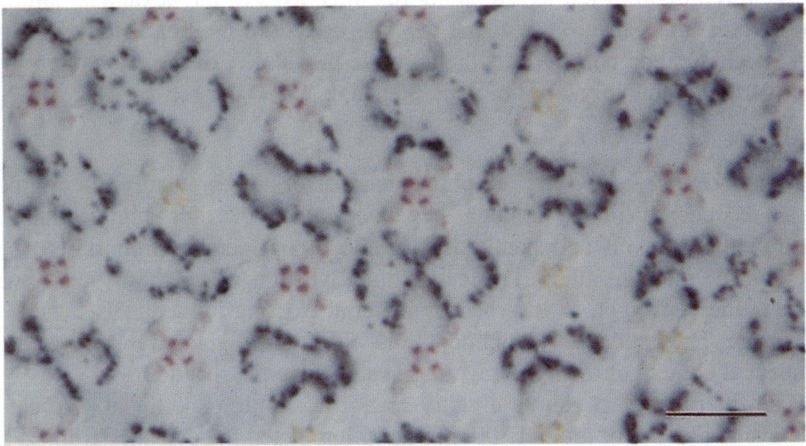

Figure 6. Transverse section of plastic-embedded specimen. Unstained. Red and yellow pigment around the rhabdom are evident. Scale = 20 μm.

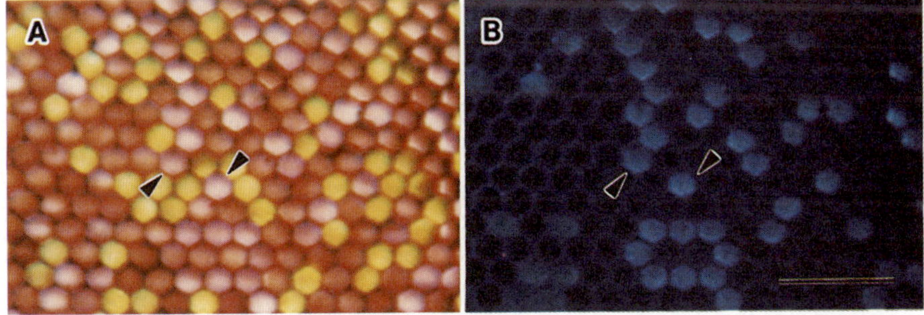

Figure 7. Slice of an eye cut at the depth of about 250 μm from the cornea. **(A)** Picture taken from the corneal side with white antidromic illumination. Yellow and red colors correspond to the yellow and red pigmentation around the rhabdom. **(B)** Autofluorescence of the same sample taken under UV epi-illumination. Less saturated red emit strong white fluorescence. Scale = 100 μm.

Extensive observations of serial transverse sections of ommatidia revealed that the pigment granules are contained in R3-8, and that the coloration of the pigment in these cells in one and the same ommatidium are identical. For clarity, we hereafter refer to the ommatidia with yellow and red pigmentation as yellow and red ommatidia, respectively. In the latero-frontal eye region, about 75% of ommatidia are of the red type, whereas the remaining 25% ommatidia are of the yellow type. Regardless of the pigmentation in R3-8, R1 and R2 have purple pigment in the distal tip. R9 is free from pigmentation.

We further investigated whether the array of differently colored ommatidia has any regularity. We counted the frequency of transition, for example from yellow to red or from red to red, along the three axes of the hexagonal lattice. It appeared that the transition frequency is independent of the frequency of the type of its neighbors and only reflects the absolute probability of the ommatidial type: they distribute randomly (Arikawa and Stavenga, 1997).

2.2. *Yellow and red filters*

What are the yellow and red pigmentations for? Theoretically, the pigments can act as spectral filters, although they exist outside the light-guiding rhabdom. When the light travels in a slender light-guide, a considerable proportion of light leaks outside the light-guide. The leaked light, which is called the boundary wave or the evanescent light, is absorbed by material existing immediately outside the light-guide. The yellow or red pigment thus will absorb light and therefore change the spectral content of the light traveling in and along the rhabdom.

The filtering effect can be directly seen in a simple experiment. We cut the fresh eye at a depth of around 300 mm and illuminated the eye slice from the cut surface with white light. Figure 7A is the picture of such a preparation taken from the corneal side, *i.e.* when viewing the transmitted light through the ommatidia. Because of the yellow and red filters, the ommatidia appear yellow or (more or less saturated) red.

To investigate any possible correlation between the spectral receptor types and the pigmentation, we recorded spectral sensitivities from single photoreceptors and marked the cells by injecting Lucifer yellow. Subsequently, we identified the pigmentation of the ommatidium to which the penetrated photoreceptor belonged by light microscopic histology. We found that the proximal R5-8 photoreceptors in the yellow ommatidia, without exception, appear to be green receptors. The R5-8 photoreceptors in the red ommatidia are always red receptors (Arikawa *et al.*, 1999c). The physiological function of the pigments is related to the tuning of spectral sensitivities of photoreceptors. As we mentioned above, the sensitivity spectrum of the 600 nm-peaking red receptors is narrow compared to the absorption spectrum of a 600 nm-absorbing visual pigment. This tuning can be explained by the filtering effect of the red pigmentation. Based on the anatomy of the *Papilio* ommatidium, we constructed an optical waveguide model for the *Papilio* rhabdom in order to predict the sensitivity spectrum of each photoreceptors.

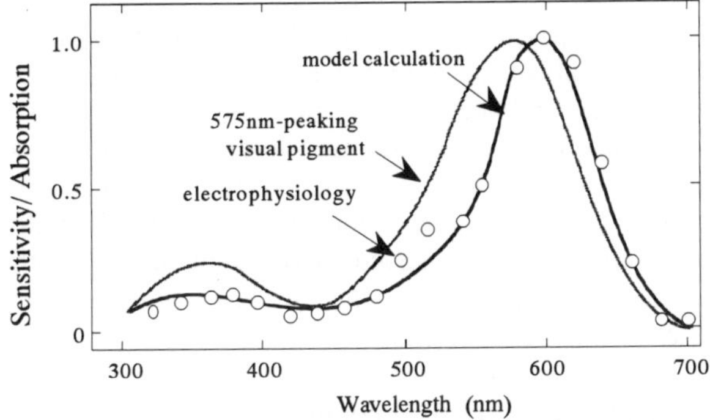

Figure 8. Tuning of the red receptor spectral sensitivity by red screening pigment around the rhabdom. Model calculation.

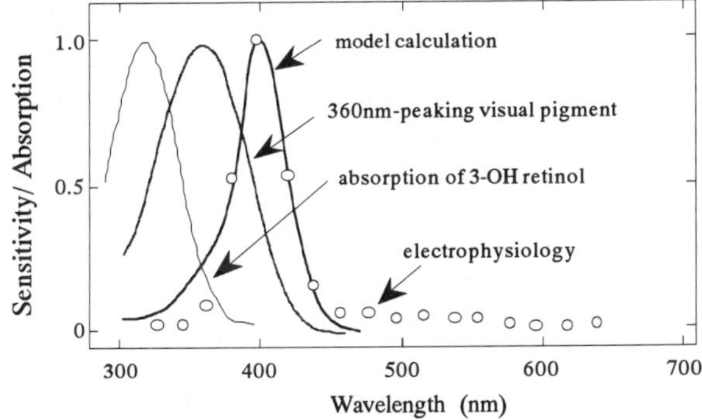

Figure 9. Generation of the violet receptor with the UV absorbing visual pigment and the filtering effect of 3-hydroxyretinol. Model calculation.

Figure 8 shows the sensitivity spectrum resulting from a spectral filter acting on the absorption spectrum of a visual pigment. When we put a 575 nm-absorbing visual pigment in the red ommatidium, the model predicted a spectrum close to the sensitivity spectrum of the red receptor. On the other hand, the effect of the yellow pigment is minor: the sensitivity spectrum of the green receptors was reproduced by putting a 515 nm-absorbing visual pigment in the yellow ommatidium. The details of the model are explained elsewhere (Arikawa *et al.*, 1999c).

2.3. UV-absorbing filter

Another interesting finding was made by epi-illumination fluorescence microscopy. When observed under UV excitation light, some ommatidia exhibit a strong, whitish fluorescence. In the latero-frontal region of the eye, about 30 % of ommatidia appear to be fluorescing. These fluorescing ommatidia are randomly distributed, and correspond to the less saturated red ommatidia in the eye slice (Fig. 7).

We investigated the origin and the physiological function of the fluorescence. An extensive HPLC analysis of retinoids in insect eyes by Seki *et al.* (1987) demonstrated that the extracts from the butterfly contain abundant 3-hydroxyretinol. Because retinols, including 3-hydroxyretinol, are known to emit whitish fluorescence under UV, we extracted the 3-hydroxyretinol from the *Papilio* retina, and compared the fluorescence spectrum with that of the ommatidial fluorescence. The two spectra closely match each other, suggesting that the ommatidial fluorescence is due to the 3-hydroxyretinol somehow concentrated in some ommatidia (Arikawa *et al.*, 1999b).

What is this 3-hydroxyretinol for? We hypothesized that it acts as an UV absorbing filter, and therefore incorporated the absorption spectrum of 3-hydroxyretinol in the optical waveguide model with 3-hydroxyretinol acting as a filter. Not surprisingly, the sensitivity spectra of the distal photoreceptors R1 and R2 having visual pigments peaking at 360 nm were considerably narrowed when the ommatidium contained a high concentration of 3-hydroxyretinol (Fig. 9). By adjusting the effective absorbance of the 3-hydroxyretinal a good match with the measured sensitivity spectra of the violet receptors was easily found. Presumably therefore the violet receptors are produced by the combination of a UV absorbing visual pigment and a UV absorbing filter. This hypothesis predicts that violet receptors must always be found in fluorescing ommatidia, and that the UV receptors in the non-fluorescing ommatidia.

We hence carried out an electrophysiological experiment, where we penetrated UV or violet receptors, injected Lucifer yellow in the cell and photographed the ommatidium containing the marked cell with an epi-fluorescence microscope. First we applied violet excitation to identify the ommatidium with the recorded cell and then we applied UV excitation to register the ommatidial fluorescence. Invariably, the violet receptors were found in the fluorescing ommatidia and the UV receptors in non-fluorescing ommatidia (Arikawa *et al.*, 1999b). Remarkably, the single-peaked green receptor was also found only in the fluorescing ommatidia. The UV-absorbing action of the 3-hydroxyretinol clearly eliminates the normally existing secondary sensitivity peak in the UV wavelength region (Arikawa *et al.*, 1999b).

3. Behavioral aspect of color vision

3.1. Color vision

The discovered spectral complexity of the butterfly retina strongly motivated us to study their color vision. All the behavioral experiments were carried out in an indoor cage (W x D x H = 80 x 60 x 45 cm). The visual stimuli were horizontally presented on the floor of the cage (Kinoshita *et al.*, 1999).

We used newly emerged females. We first tested whether they have innate preference to certain color. After 2 days of starvation, the naive females were released in the cage where four patches of different colors (blue, green, yellow, and red: 4-color pattern) were presented on the floor. We released only one butterfly at one time. The color of the patch on which the released naive butterfly visited for the first time was recorded as the innate preference of the individual. It appeared that the females of *Papilio xuthus* innately prefer red or yellow patches as the food source.

Next we trained butterflies in the cage to feed on a drop of sucrose solution put on a patch of a certain color (blue, green, yellow, or red). The patches of innately preferred colors, yellow and red, were very easily learned: after only a single training, the yellow- and red-trained butterflies significantly selected the patches of the trained colors from the 4-color pattern. Blue and green, which are not innately preferred, were also clearly learned after a few runs of the training. The observed relationship between the learning speed and the innate preference may be an interesting issue to approach the neuronal mechanism underlying learning and memory (Fig. 10).

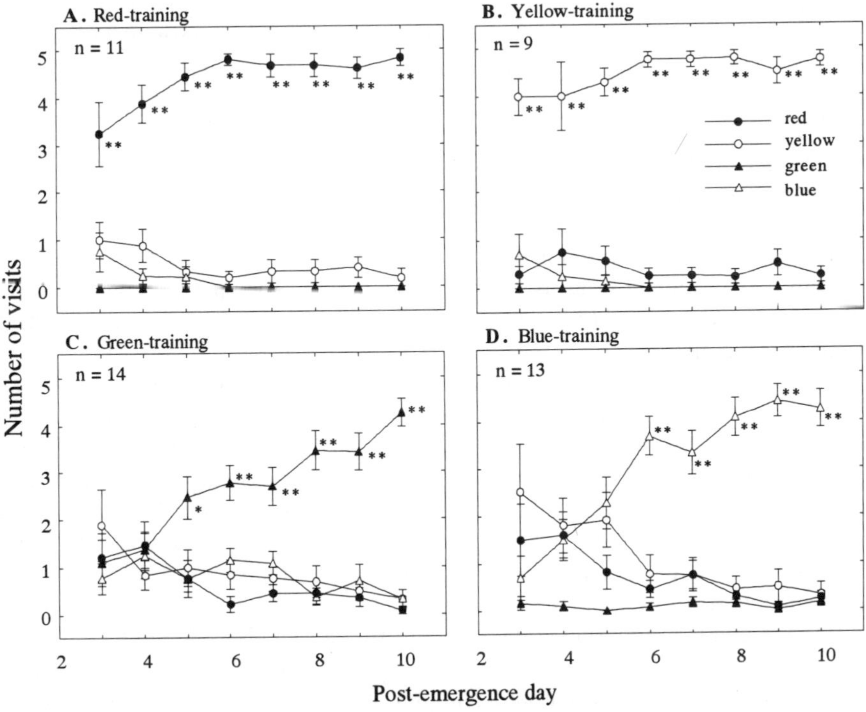

Figure 10. Effect of training on a disk of particular color. Asterisks indicate that the training color was selected significantly more often than any other color.

For a convincing demonstration of color vision, we must show that the butterflies do select the colored patch because of the chromatic content of the patches irrespective of their brightness. We therefore carried out two additional experiments. First, we presented a patch of the trained color with seven patches of different shades of greys, and tested whether the butterflies, trained to blue, yellow, green, or red, select the patch of the trained color. Indeed, the butterflies selected the patch of the trained color perfectly. Second, we presented the 4-color pattern with a neutral density filter placed of the disks. The neutral density filter was placed only on the disk of the trained color for the tested individuals. The trained butterflies successfully selected the patches of the trained color if their brightness was reduced about 50%.

Taken together, females of *Papilio xuthus* have color vision upon food search. Similarly, foraging males were also shown to use color vision.

3.2. Color constancy

Generally, the reliability of color vision is believed to be reinforced by the ability called color constancy. This enables animals to recognize an object's color, regardless of the spectral content of the illumination (Land, 1977). In fact, the butterflies have to search for flowers not only under the sunshine but also in shaded places, where the irradiation spectrum of the sun has been strongly biased. Here we tested whether the *Papilio* is also color constant (Kinoshita and Arikawa, 1999).

We used the same indoor cage used for the above experiment on color vision. The cage was illuminated with halogen lamps hanging from the ceiling above the cage. The emission of the halogen lamps at wavelengths shorter than 400 nm is negligible. Although *Papilio* is distinctly sensitive to UV (Fig. 1), we will refer to the unfiltered halogen light as "white" illumination. The spectral content of the illumination was changed by placing color filters (blue, green, yellow, and red) in front of the lamps.

Newly emerged butterflies were first trained to feed on a paper patch of a certain color (blue, yellow, or red) under white illumination in the cage. After confirming that the butterfly visited the patch of the trained color in the 4-color pattern, we changed the illumination from white to colored, using a filter of a certain color. In most cases, the butterflies correctly selected the patch of the trained color from the 4-color pattern under colored illumination. Some exceptions, where butterflies did not even fly, were observed under colored illumination of strong saturation.

Using a color Mondrian collage under differently colored illuminations is an approved method to demonstrate color constancy. Therefore, we also tested the trained butterflies on a Mondrian. The yellow- and red-trained butterflies could select the correct color from the Mondrian as they could with the 4-color pattern. However, in preliminary experiments, we noticed that it was necessary to train the butterflies to select the correct colour from a specialized 4-color training pattern on which four differently colored rectangles juxtaposed each other before they were willing to land on a Mondrian. Otherwise, most butterflies did not land on the appropriate patch in the Mondrian: they seemed trying hard to land on the correct color, but they appeared to be unsuccessful. Very likely, the butterflies learned how to land on a specific portion of the Mondrian-like pattern through the training process with the specialized 4-color pattern. Moreover, it appeared that such a training was effective for yellow and red, but not for blue. Actually, the blue-trained butterflies could not land on the blue patch in the specialized 4-color training pattern. The blue-trained butterflies could only land on the blue in the training pattern when a co-existing yellow patch, which was the brightest for butterflies, was covered with a neutral density filter of 40% transmittance. Apparently, the landing seemed to be inhibited by brighter patches of different colors. The underlying mechanism of this phenomenon is an open question.

Taken together, we conclude that the foraging *Papilio xuthus* has, at least to a certain extent, color constancy.

Acknowledgments

This work was supported by the Grants-in-Aid for Scientific Research from the Ministry of Education, Science, and Culture of Japan.

References

Applebury, M.L. and P.A. Hargrave (1986) "Molecular biology of the visual pigments", *Vision Res.* 26:1881-1895.

Arikawa, K., K. Inokuma and E. Eguchi (1987) "Pentachromatic visual system in a butterfly", *Naturwissenschaften* 74:297-298.

Arikawa, K., S. Mizuno, M. Kinoshita and D.G. Stavenga (1999a) "Two visual pigments simultaneously expressed in a sub-set of proximal photoreceptors of the butterfly, *Papilio xuthus*, cause an abnormally-broad spectral sensitivity", in: *Goettingen Neurobiology Report*, N. Elsner and U. Eisel, eds, Stuttgart, New York: Georg Thieme Verlag, pp. 409.

Arikawa, K., S. Mizuno, D.G.W. Scholten, M. Kinoshita, T. Seki, J. Kitamoto and D.G. Stvenga (1999b) "An ultraviolet absorbing pigment causes a narrow-band violet receptor and a single-peaked green receptor in the eye of the butterfly *Papilio*, *Vision Res.* 39:1-8.

Arikawa, K., D.G.W. Scholten, M. Kinoshita and D.G. Stavenga (1999c) "Tuning of photoreceptor spectral sensitivities by red and yellow pigments in the butterfly *Papilio xuthus*", *Zool. Sci.* 16:17-24.

Arikawa, K. and D.G. Stavenga (1997) "Random array of colour filters in the eyes of butterflies", *J. Exp. Biol.* 200:2501-2506.

Arikawa, K. and H. Uchiyama (1996) "Red receptors dominate the proximal tier of the retina in the butterfly *Papilio xuthus*", *J. Comp. Physiol. A* 178:55-61.

Bandai, K., K. Arikawa and E. Eguchi (1992) "Localization of spectral receptors in the ommatidium of butterfly compound eye determined by polarization sensitivity", *J. Comp. Physiol. A* 171:289-297.

Gaertner, W. and P. Towner (1995) "Invertebrate visual pigments", *Photochem. Photobiol.* 62:1-16.

Kinoshita, M. and K. Arikawa (1999) "Color constancy of the foraging swallowtail butterfly, *Papilio xuthus*", in: *Goettingen Neurobiology Report*, N. Elsner and U. Eisel, eds, Stuttgart, New York: Georg Thieme Verlag, pp. 422.

Kinoshita, M., N. Shimada and K. Arikawa (1999) "Colour vision of the foraging swallowtail butterfly *Papilio xuthus*", *J. Exp. Biol.* 202:95-102.

Kitamoto, J., K. Ozaki and K. Arikawa (1999) "Violet receptors in the retina of the butterfly *Papilio xuthus* contain UV-absorbing visual pigment", *Zool. Sci.* 16S:in press.

Kitamoto, J., K. Sakamoto, K. Ozaki, Y. Mishina and K. Arikawa (1998) "Two visual pigments in a single photoreceptor cell: Identification and histological localization of three mRNAs encoding visual pigment opsins in the retina of the butterfly *Papilio xuthus*", *J. Exp. Biol.* 201:1255-1261.

Land, E.H. (1977) "The retinex theory of color vision", *Sci. Am.* 237:108-128.

Menzel, R. (1979) "Spectral sensitivity and color vision in invertebrates", in: *Handbook of Sensory Physiology*, H. Autrum, ed., Berlin, Heidelberg, New York: Springer-Verlag, pp. 503-580.

Moody, M.F. and J.R. Parriss (1961) "The discrimination of polarized light by *Octopus*: A behavioral and morphological study", *Z. vergl. Physiol.* 44:268-291.

Seki, T., S. Fujishita, M. Ito, N. Matsuoka and K. Tsukida (1987) "Retinoid composition in the compound eyes of insects", *Exp. Biol.* 47:95-103.

PATCH-CLAMPING SOLITARY VISUAL CELLS TO UNDERSTAND THE CELLULAR MECHANISMS OF INVERTEBRATE PHOTOTRANSDUCTION

CARLO MUSIO
Istituto di Cibernetica del CNR, Via Toiano 6, I-80072 Arco Felice (NA), Italy
{carlom@biocib.cib.na.cnr.it}

ABSTRACT

The molecular and cellular mechanisms underlying phototransduction have been elucidated, recently, thanks to the "solitary photoreceptor approach" performed on both vertebrate and invertebrate photoreceptors. Despite the large amount of data available on vertebrate photoreceptors, electrophysiological studies of the cellular and molecular mechanisms on those of invertebrates are less comprehensive. In invertebrates, few electrophysiological works have been performed until now on single isolated visual cells, whereas there is a large data collection available on the processes underlying visual excitation in retinal and extra-retinal whole preparations. In recent years, the development of enzymatic dissociation protocol has provided useful cell model systems for the application of patch-clamp techniques to photoreceptor physiology. This paper will briefly survey the single photoreceptor models of invertebrates currently used to investigate the biophysical mechanisms of visual transduction.

1. Introduction

Unlike the vertebrate phototansductive machinery, the light transduction processes in invertebrate eyes are less understood even due to the high degree of complexity of their visual structures that do not allow a simple experimental approach (Land and Fernald, 1992; Fernald, 2000). In spite of the functional development of optical solutions, vertebrates share a substantially conserved structural scheme of eyes constituted by retinal ciliary photoreceptors, rods and cones (Rodieck, 1973). By contrast, invertebrates show a great variety of eyes and retinal structural patterns constituted by microvillar photoreceptors (with very few ciliary exceptions) (Eakin, 1972; Land, companion papers in this volume). Mainly in invertebrates, and in some lower vertebrates, extraretinal visual cells are arranged in simple structures like ocelli or are scattered (single or clustered) on dermal surface or are present in ganglia or regions belonging to the nervous system (Yoshida, 1979; Musio, 1997).

The two main evolutionary lineage of visual cells, ciliary and microvillar (rhabdomeric), have different functional properties of visual excitation, although in both the transduction mechanism is characterized by a G protein-coupled cascade mediated by a second messenger acting on the gating of light-dependent ion channels (see next section).

The easy exposure of light-sensitive surface afforded by vertebrate photoreceptors has facilitated electrophysiological study using different preparations such as whole retina (Dowling, 1987), single rod in intact retina (Baylor *et al.*, 1979), retinal slice (Werblin, 1978) as well as dissociated photoreceptors (Kaneko and Tachibana, 1986).

On the contrary, in invertebrates the cellular mechanisms of visual transduction have been largely carried out on whole-eye or semi-intact preparations (Dorlöchter and Stieve, 1997; Weckström and Hardie, 1995). These traditional preparations, although providing outstanding results by means of standard intracellular recordings, are not currently suitable (with rare exceptions) for the application of advanced gigaseal techniques because technical obstacles. These reside in morphological and functional features such as respectively the surrounding layer of glial cells or pigmented cells and cell-cell interactions.

Nowadays, the patch-clamp technique (Hamill *et al.*, 1981; Sakmann and Neher, 1995) represents a powerful tool to elucidate many aspects of phototransduction, in particular it provides a detailed account of the effector mechanisms and decisive proofs on the generation of the light response at single-channel level in both vertebrates (Matthews, 1987) and invertebrates (Bacigalupo *et al.*, 1986; Nagy and Stieve, 1990).

During the past few years, the problem of obtaining viable cells – suitable for patch-clamping – in which the light-sensitive membrane is clean and exposed, has been solved by the use of isolated photoreceptors. Viable single visual cells have been obtained thanks to the development of cell enzymatic dissociation protocols.

To this respect, solitary invertebrate photoreceptors represent a new suitable and attractive model to investigate and understand the cellular mechanisms and the biophysical processes of visual transduction.

2. Different phototransduction mechanisms in vertebrates and invertebrates

As previously reported, a common general scheme supervises the generation and the modulation of the visual response in ciliary and microvillar photoreceptors (Goldsmith, 1991). However, the expression of the structure-function relationship in vertebrate and invertebrate visual cells leads to different cellular mechanisms underlying phototransduction (Rayer *et al.*, 1990; Yarfitz and Hurley, 1994). Nevertheless, the early steps of the transduction cascade are notably conserved because the photopigments and G-proteins are substantially homologous.

Both vertebrate and invertebrate photoresponses begin with a light-induced isomerization of the photopigment rhodopsin (other visual pigment have been identified in some invertebrate species – Gartner and Tower, 1995) and a subsequent interaction with a G protein. A remarkable chemical amplification is involved in this process: ≈1,000 ionic channels per photo-excited pigment in vertebrate rods and 1,000 to 10,000 per absorbed photon in vertebrates and invertebrates respectively. This fact speaks in favour of the functional presence of a diffusible chemical effector (Bacigalupo and O'Day, 1996).

At present there is no doubt that cyclic GMP (cGMP) is the final messenger that gates light-sensitive channels in vertebrates (Fesenko *et al.*, 1985). It was found that cGMP opens about 3% light-dependent channels (cGMP-gated ion channels) in the dark, therefore the light stimulation activating a phosphodiesterase (PDE) hydrolize the cGMP to 5'GMP (Yau and Baylor, 1989). As final result, the cGMP-PDE, decreasing the cGMP level, closes the light-dependent channels producing a hyperpolarizing receptor potential due to a reduction or a termination of the Na^+ and Ca^{2+} influx (Koutalos *et al.*, in this volume).

Conversely, in invertebrates the presence of multiple second messenger signalling systems has been found and their role is still under debate (O'Day *et al.*, 1997). However, there is a general *consensus* that in rhabdomeric photoreceptors mainly a phosphoinositide (PI) pathway signalling system rules the visual excitation cascade. Upon light stimulation, the G-protein activates a phospholipase C (PLC) that generates a fast production of two intracellular messengers: cytosolyc inositol-1,4,5-trisphosphate (IP_3) and membrane lipid soluble diacylglycerol (DAG). These two messengers start parallel signalling parthways: 1) IP_3 triggers the release of Ca^{2+} from intracellular stores that causes a transient elevation of the intracellular Ca^{2+} concentration ($[Ca^{2+}]_i$), 2) DAG activates protein kinase C (PKC), a Ca^{2+}-dependent enzyme (Nagy, 1991). Thus Ca^{2+} has multiple effects on photoresponse and plays a key-role in the visual cascade.

The light-induced excitation terminates with the opening of light-sensitive channels that favours a cation influx and increases the membrane conductance leading to a depolarizing receptor potential. The identification of the "final" effector that gates the light-dependent channels is still controversial. In *Limulus* cGMP together with IP_3 seem to underly the activation of different components of the photocurrent (Nagy 1991), whereas a role of the IP_3-Ca^{2+} in the excitation and adaptation has been proposed too (Payne *et al.*, 1988). In *Drosophila*, during light response an increase of cytosolic calcium released by intracellular store has been demonstrated (Peretz *et al.*, 1994), although evidences on the Ca^{2+} acting on selective Ca^{2+}-store operated channels need confirmations. So far, the involvement of cGMP in the activation of light-dependent channels would be not excluded in this species (Bacigalupo *et al.*, 1995). However, a possible involvement of single component of the PI pathway is still far to be demonstrated. On the whole, in invertebrates the effectors of light-dependent channel-gating remain almost unsolved.

A different case is represented by the ciliary photoreceptors found in molluscs (McReynolds, 1976) in which the membrane conductance increasing produces a hyperpolarizing response and the light-sensitive channels are gated by cGMP (more details about these photoreceptors in section 4).

These facts depict a framework in which patch-clamp have proven as a unique tool of investigation. By this way, it can be reasonably argued that the utilisation of solitary photoreceptors has provided decisive and unambiguous data. The next sections will detail some of those findings surveying the most utilised solitary photoreceptor model systems among invertebrate species, just before a brief treatment on methods to isolate photoreceptors.

3. Getting solitary photoreceptors: Methodological issues

Apart from the invertebrate species used, the general scheme to obtain viable isolated photoreceptors suitable for patch-clamp recordings foresees four major steps (Figure 1). The procedure description (roughly based on a protocol used for crayfish photoreceptors – Musio, 1996) follows below.

3.1. Dissection

Eyes are removed from the animal by means of fine forceps and scissors, and placed in a physiological solution (PS). Once the eye has been removed, the retina and the first part of the eyestalk are forced out from the cornea under a stereomicroscope; after, the retina is roughly desheathed by forceps and the eyestalk is cut to make the preparation suitable for the dissociation procedures.

3.2. Dissociation

In different invertebrate species, several original protocols have been developed to isolate photoreceptors from retinas, eyes or whole heads (*e.g.*, Nasi, 1991a; Hardie, 1991; Nasi and Gomez, 1992b; Jinks *et al.*, 1993; Gomez and Nasi, 1994; Zhang *et al.*, 1994).

Generally, the basic plan consists of an enzymatic two-steps pre-treatment, followed by a mechanical dissociation of retinal tissues. In *Drosophila*, mechanical dissociation alone seems to be enough (Hardie, 1991; Ranganathan *et al.*, 1991) although several enzymes have been tested useful to obtain viable and clean cells (Ziemba *et al.*, 1995). The dissected eye is divided into some pieces and rinsed in PS. Retinal pieces are firstly incubated with collagenase or pronase or protease (at low dilution percentages depending on the animal species) to soften the connective tissue for a variable time (30-50 min), and then treated with suitably diluted enzymes, like trypsin or papain or dispase, for a variable time (15-30 min). Subsequently, the pre-treated pieces are mechanically triturated in PS by gentle repetitive sucking/expulsion with a fire-polished Pasteur pipette (1 mm tip diameter) or by lightly grasping with fine forceps. A finer dissociation could be performed on the reduced pieces by a gentle trituration with a Pasteur pipette having a fire-polished tip 0.5 mm in diameter.

3.3. Isolation

After the dissociation, the resulting cell suspension is composed by photoreceptors, glial, and pigmented cells. Selected photoreceptors now can be transferred, after several washings in PS, to a recording chamber or a plastic sterile Petri dish, rinsed with PS, for observation and electrophysiological recordings.

3.4. Attachment

The dish or the cover slip bottom of the recording chamber can be treated overnight with collagen and after for a variable times with an attachment factor such as concanavalyn A or poly-D-lysin or laminin to increase the adhesion of the dispersed isolated photoreceptors.

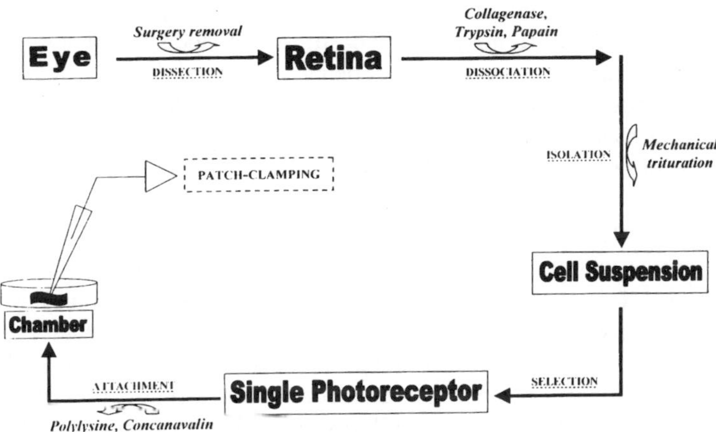

Figure 1. Schematic flow sheet showing the basic steps of photoreceptor cell handling from the whole eye, *in situ*, to the application of gigaseal techniques, *in vitro*. See text for details.

4. A look into isolated photoreceptors among invertebrate *phyla*

As already stated, the application of patch-clamp technique to solitary visual cells of invertebrate is quite recent lasting since about 10 years. During this period there has been a growing interest by some people engaged in photoreception research to find and test different photoreceptor model systems suitable for patch-clamping. Major attention was paid on those species in which a large amount of data was already gained with other classic preparations. For this reason *Mollusca* and *Arthropoda* species are currently the most utilised to test and develop dissociation protocols of visual cells. Furthermore, other different species can be chosen as they could provide the opportunity to study particular mechanisms not shared with other invertebrates and/or to bridge different visual processes by an evolutionary point of view. Hopefully, this will be a long story and now we are going to survey the first finished chapters.

4.1. Mollusca

An excellent series of three papers published by Enrico Nasi in the January 1991 issue of *The Journal of General Physiology* placed the solitary photoreceptors for the first time into the lime-light of invertebrate phototransduction research (Nasi, 1991a; 1991b; 1991c).

This study was carried out on rhabdomeric photoreceptors enzymatically dissociated from the retina of the file clam *Lima scabra* (Figure 2B). Isolated cells proved to be physiologically viable showing a classic depolarising light response (inward photocurrents graded with light intensity recorded in whole-cell clamp) due to an increase of conductance (Figure 3).

Furthermore the cells retained the functional properties measured with traditional intracellular techniques *in situ*. In *Lima* rhabdomeric photoreceptors a biphasic photocurrent were recorded being composed by two distinct light-dependent conductances. Each of two components had different latency, kinetics and light sensitivity (Nasi, 1991c).

Similar observations were fulfilled in another model, the isolated rhabdomeric photoreceptors of the scallop *Pecten irradians* (Figure 2D). In this animal, light-activated currents recordings from single channels of the rhabdomere showed two class of events with a conductance respectively of ≈47 and ≈18 pS presumably arising from different ion channels population that concur to the macroscopic light response (Nasi and Gomez, 1992a). The dominant conductance resembles that recorded in *Limulus* and it could be distinctive of other rhabdomeric photoreceptors. The presence of multicomponent events, observed also in *Drosophila* (see below) and *Limulus* (Deckert *et al.*, 1992), confirm the existence of multiple effector signalling systems in invertebrate visual cells.

In rhabdomeric photoreceptors of *Lima* and *Pecten* light-activated channels are selective for Na^+ primarily mediating the receptor potential while Ca^{2+} influx is poorly involved in the photocurrent (Gomez and Nasi, 1996). About the mechanisms of visual excitation, there are still no sharp indications on the identification of the chemical messenger and on the role of the PLC-IP_3-Ca^{2+} cascade although it has been demonstrated that internal Ca^{2+} plays an important role in the activation pathways and may modulate light-dependent channels (Gomez and Nasi, 1996).

Recently in *Lima*, an involvement of the DAG branch of the PLC cascade has been proposed in the activation of the photoresponse jointly with the IP_3/Ca^{2+} branch. In this case, the DAG could control the gating of the light-sensitive channels while the IP_3/Ca^{2+} should modulate the gain and the speed of the light response (Gomez and Nasi, 1998).

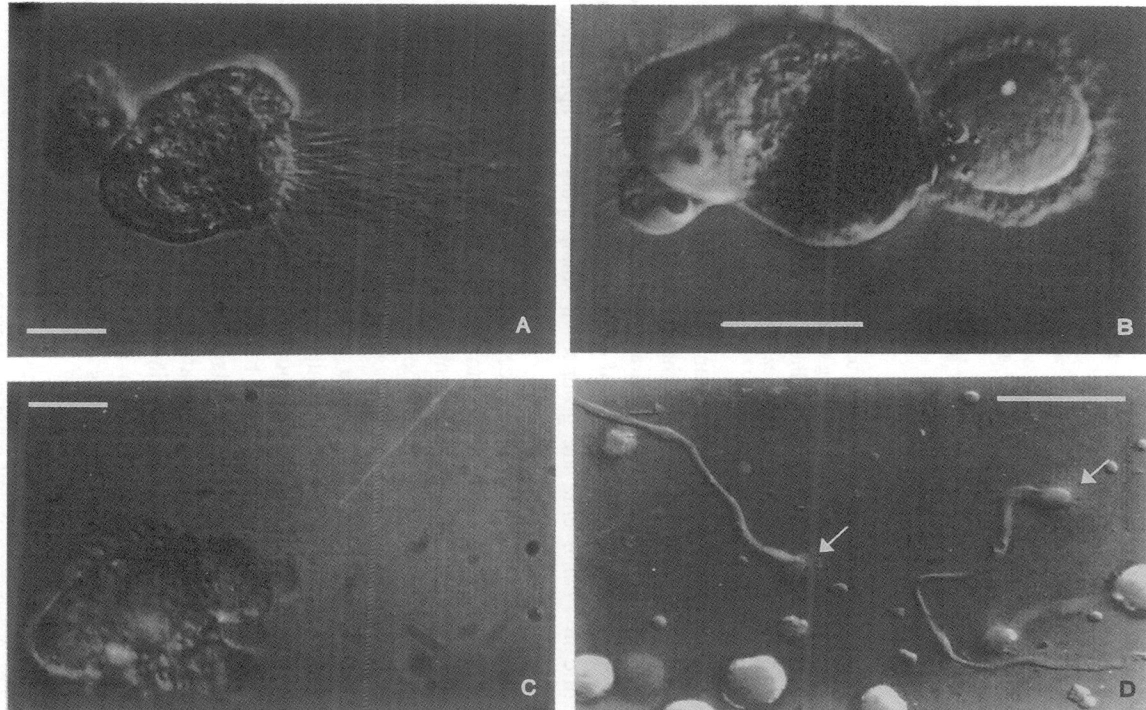

Figure 2. Nomarski DIC micrographs of solitary photoreceptors dissociated from the eyes of molluscs *Lima scabra* and *Pecten irradians*. (A) *Lima* ciliary cell (X100, oil-immersion) showing ciliary processes, bar 10μm; (B) *Lima* rhabdomeric photoreceptor (X100, oil-immersion), bar 20 μm; (C) *Pecten* ciliary cell (X40, water-immersion) displaying spherical appendages instead of fine cilia showed in (A), bar 10μm; (D) *Pecten* rhabdomeric photoreceptors showing a microvillar lobe (arrows), bar 20 μm. From Nasi, 1991a (B); Nasi and Gomez, 1992 (D); Gomez and Nasi, 1994 (A, C); *The Journal of General Physiology* by copyright permission of The Rockefeller University Press.

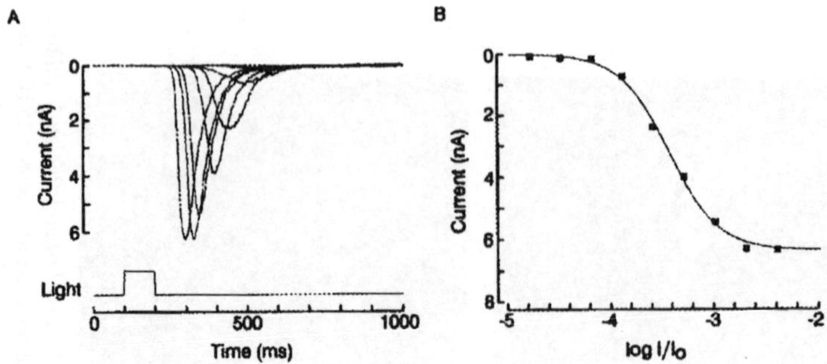

Figure 3. (A) Inward photocurrents recorded under whole-cell clamp in an isolated rhabdomeric photoreceptor from *Lima*. Holding potential –60 mV; light intensity was increased at 0.3 log increments. (B) Peak amplitude of the photoresponse plotted as a function of light attenuation. From Nasi and Gomez, 1998a, modified.

The molluscs *Lima* and *Pecten* represent an exceptional case of double retina having, in addition to a proximal layer of rhabdomeric photoreceptors, also a distal layer of ciliary cells (Dakin, 1910; Bell and Mpitsos, 1968). In *Lima* these cells have a spherical body with a bundle of ciliary processes 20-30 μm long (Figure 2A), while in *Pecten* cilia are grouped in small spherical appendages (Figure 2C).

In spite of their morphological differences, both *Lima* and *Pecten* ciliary photoreceptors show a hyperpolarising receptor potential upon photostimulation (outward photocurrents graded with light intensity recorded in whole-cell clamp, Figure 4A right), while, as reported above, light evokes a membrane depolarisation in the rhabdomeric proximal cells. For about twenty years, the hyperpolarising response has been believed a receptor potential on the basis of indirect considerations arising from intracellular recordings of intact retina (that can provide confounding factors) and/or extracellular recording from optic nerve (McReynolds, 1973). Definitive proofs have been obtained by Gomez and Nasi (1994) who study the physiology of hyperpolarising photoreceptors applying for the first time patch-clamp technique to enzymatically isolated cells.

Differences between ciliary and rabdhomeric photoreceptors in their electrophysiological behaviour are shown in Figure 4. Patch-clamp recordings have revealed the basic functional properties of ciliary photoreceptors (Gomez and Nasi, 1994):

1) light stimulation produces an outward current (graded with light intensity) accompanied by a decrease of cell input resistance;
2) the receptor potential is hyperpolarising;

3) the conductance underlying photocurrent is K^+ selective as the reversal potential is close to E_K and has Nerstian shifts with changes of $[K]_o$;
4) light-sensitive conductance increases with membrane depolarisation (outward rectification) and has an unitary value of ≈ 27 pS (notably lower than that of rhabdomeric cells).

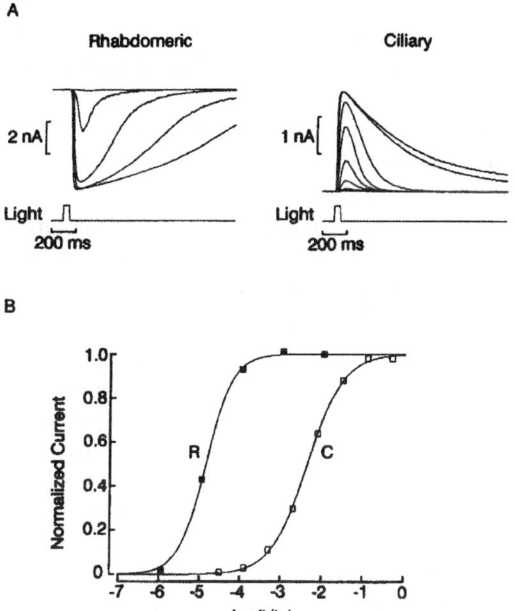

Figure 4. (A) Light-intensity series measured under voltage-clamp in isolated photoreceptors belonging to the two classes. The holding potential was set near the resting potential (-50 and -30 mV, respectively). (B) Comparison of the intensity-response curve for the two cells. From Nasi and Gomez, 1998b, modified.

The faster onset and the lower sensitivity of the photocurrent (Figure 4) suggest for ciliary photoreceptors a simpler transduction mechanism characterised by fewer biochemical processes in its cascade. Other results support this hypothesis: a) the photocurrent of ciliary cells does not show the biphasic behaviour typical of rhabdomeric reciprocal, b) light-sensitive conductance has an unitary value due to a single population of channels in ciliary cells, while multiple conductance components have been found in rhabdomeric cells.

Investigations on the nature of the second messenger in ciliary cells demonstrated that intracellular dialysis of cGMP induces a dose-dependent outward current with a concomitant increase of membrane conductance (Gomez and Nasi, 1995). On the other hand, manipulation of the IP_3/Ca^{2+} was ineffective on the activation of light-dependent channels (Gomez and Nasi, 1995). Furthermore, pharmacological blockers of the light-dependent conductance in vertebrates rods and cones have been tested effective in suppressing in reversible manner the photocurrent in *Pecten* ciliary cells. Taken together, these data indicate that ciliary photoreceptors utilise cGMP as internal messenger, like vertebrates rods and cones, whereas the IP_3/Ca^{2+} cascade is restricted to rhabdomeric photoreceptors.

In conclusion, light-sensitive channels of invertebrate ciliary photoreceptors resemble those of vertebrate visual cells (including also the parietal-eye photoreceptor, a ciliary extaretinal photoreceptors – Xiong *et al.*, in this volume), with the exception of different ion-selectivity, indicating that basic functional properties of ciliary photoreceptors are strongly conserved among different *phyla*. To this regard, these channels can bridge the gap between voltage-gated K^+ and cyclic-nucleotide-gated channels supporting the hypothesis of a common evolutionary origin of these two superfamilies of channel proteins (Yau, 1994; Gomez and Nasi, 1995; Nasi and Gomez, 1998b).

The studies on *Lima* and *Pecten* rhabdomeric and ciliary photoreceptors represent a striking amount of data collected on isolated visual cells in *Mollusca*, but investigations have been conducted also on other molluscan species.

Solitary photoreceptors of the squid *Loligo pealei* represent a useful model for applying gigaseal technique as the small size and the geometry of the cells has proved difficult the use of traditional microelectrodes techniques (Nasi and Gomez, 1992b). Whole-cell recordings showed outward currents upon increasing depolarising voltage steps in the dark and small inward current (up to 100 pA) graded with stimulus intensity upon light stimulation. Instead, perforated-patch recording revealed larger photocurrents (up to 1300 pA) indicating that isolated squid photoreceptors are more effective with non-invasive methods, only by which is possible to records photoresponse comparable to that of other rhabdomeric visual cells. Single-channel recordings were performed too, unfortunately without revealing any evaluation on the unitary conductance.

Two molluscan photoreceptor models are worthwhile to be treated in this survey, although they are not strictly dealing with research on phototransduction mechanisms while they concern photoreceptive aspects linked to circadian rhythms and cellular correlates of learning.

In *Aplysia*, dissociation of retinas provides many retinal cells classified as pacemaker neurones (see Section 5. below) and photoreceptors (PRs) that can be cultured single or clustered (Jacklet *et al.*, 1996). The largest and most noticeable is the microvillar or R-type. The R-type PRs responds to light with a stimulus-graded depolarisation that reaches 0 mV and reverses to a long-lasting hyperpolarisation with longer and more intense light pulse (Jacklet *et al.*, 1996). This activity resembles that recorded in PRs of intact retina confirming that cultured isolated photoreceptor retain their physiological activity and viability.

In *Hermissenda crassicornis* an enzymatically isolated eye preparation allows easy access to the three identified type B photoreceptors and very affordable single-channel recordings (Etchebarrigaray *et al.*, 1991). Light response and dark adaptation have been studied in freshly dissociated type A and type B photoreceptors in order to outline the functional differences among these cells without dealing with phototransduction processes. Transient Ca^{2+} and distinct inward rectifier currents have been described in both cell types and changes of their excitability have been proposed as correlates of classical conditioning (Yamoha *et al.*, 1998).

4.2. Arthropoda

Despite the great variety of species belonging to this phylum, in *Arthropoda* few single photoreceptor models have been developed and approached by patch-clamp, while the study of visual function has been restricted to other experimental preparations and methodologies. Just to mention, up to now arthropods are recognised as exceptional model systems for investigate vision either at behavioural and integrative levels (see Arikawa's and Land's companion papers in this volume) either at molecular (Tsunoda and Zuker, 1999; Paulsen' s group companion papers in this volume) and genetic ones (Pak, 1994).

Also in arthropods, the first reports of patch-clamp recordings applied to dissociated ommatidia date back to 1991 and are referred to the fruitfly *Drosophila* (Hardie, 1991; Ranganathan *et al.*, 1991), a species widely used in studies of phototransduction (Montell, 1999). A rich complexity emerges from phototransduction mechanisms in *Drosophila* thanks to the development of isolated cell preparations consisting of wild type (WT) and mutant photoreceptors. *Drosophila* mutants are a powerful tool to genetically dissect the phototransduction mechanisms identifying genes that encode multivalent component of signalling pathways (Tsunoda and Zuker, 1999; Paulsen's group companion papers in this volume).

Drosophila photoreceptors show the typical depolarising photoresponse of rhabdomeric visual cells and utilise as phototransductive cascade the PI signalling pathways. Light-induced currents (LIC) have been characterised and their selectivity to Ca^{2+} has been demonstrated, although light-sensitive channels readily permeate a variety of monovalent ions (Hardie, 1991; Ranganathan *et al.*, 1991). Investigations on isolated photoreceptor of the transient receptor potential (*trp*) mutant, impressively demonstrated a 10-fold reduction of Ca^{2+} permeability of the light sensitive conductance respect to the WT photoreceptor (Hardie and Minke, 1992; Peretz *et al.*, 1994). Thus the *trp* gene may encode the light-sensitive ion channels responsible for the Na^{+} and Ca^{2+} influx in *Drosophila* photoreceptors. A second putative channel gene, *trp*-like (*trpl*) has been proposed being involved in light-induced conductance. Although mutation has little effect on photoresponse, double mutants *trp;trpl* are completely unresponsive to light indicating that the residual response in *trp* mutant is carried by *trpl*-dependent channels (Niemeyer *et al.*, 1996). In conclusion, both *trp* and *trpl* constitute subunits of the *Drosophila* light-dependent channels (TRP and TRPL) and each can operate in the absence of the other (but the possibility that they form heteromultimeric channels in WT is not excluded) (Hardie, 1998a). These results are consistent with the demonstration that two separate light-activated conductances belonging to two classes of channels underlie the whole photocurrent (Hardie and Minke, 1994).

By contrast to the clear account about the functional properties of light-sensitive channels, the identification of the second messenger acting as chemical effector of the light excitation is still an enigma (Hardie, 1993).

Three main hypotheses of excitation are under debate (Hardie, 1998b):

1) the Ca^{2+} hypothesis, as proposed in *Limulus* (Payne *et al.*, 1988);
2) the multiple pathways, as proposed in *Limulus* in addition to the Ca^{2+} hypothesis (Deckert *et al.*, 1992);
3) the capacitive Ca^{2+} entry (Hardie and Minke, 1993).

Nowadays, despite the powerful experimental approach given by solitary photoreceptors, there are no direct evidences to confirm resolutely one of these hypotheses, although hypotheses 1 and 3 seem to have stronger indications than hypothesis 2. To this regard, genetic evidence in support of a role cGMP, as suggested by electrophysiological investigations (Bacigalupo *et al.*, 1996), is still lacking; however, an involvement of cyclic nucleotide in *Drosophila* photoreceptors has been proposed in the modulation of photoresponse (Chyb *et al.*, 1999b). Moreover, recently fatty acids such as linoleic acid have been proposed as an excitatory second messenger in *Drosophila* phototransduction (Chyb *et al.*, 1999a).

As shown so far, the outstanding data collected in *Drosophila* by groups of R. C. Hardie, B. Minke and C. S. Zuker have been obtained on dissociated ommatidia thanks to the combination of molecular biology and electrophysiological techniques.

In the horseshoe crab *Limulus*, the most extensively studied invertebrate species together with *Drosophila*, phototransduction mechanisms have been investigated on classic preparations (Dorlöchter and Stieve, 1997) and the use of solitary photoreceptors has not yet gained ground. Rhabdomeric photoreceptor cells have been enzymatically dissociated from the compound lateral eye and their functional properties were compared with those of cells in a cluster and *in situ* (Jinks *et al.*, 1993; Hanna *et al.*, 1993). Whole-cell recordings in all cell types showed spontaneous bumps in the dark and a depolarising receptor potentials upon light stimulation that has a slower onset in isolated cells respect to that of cells *in situ*. Single-channel recordings (performed on patch of both R- and A-segment, respectively the rhabdom and the arhabdomeral segment) showed no voltage-gated channel activity, while light-activated channels events were observed (only in R-segment membrane) showing inward currents of several conductances like those observed in ventral nerve photoreceptors (Bacigalupo *et al.*, 1986). In addition to lateral eye, in *Limulus* isolated ventral nerve photoreceptors have been dissociated from ventral optic nerves (Zhang *et al.*, 1994). According to the authors' evaluation this study was aimed just to develop a dissociation protocol in order to get a preparation offering viable cells and stable recordings for several hours. For these reasons light responses were recorded to make comparisons between different enzymatic assessment without dealing with phototransductive mechanisms.

In both cases, *Limulus* dissociated visual cells are viable and suitable for the study of processes involved in visual transduction by means patch-clamp recordings: in this view, they could represent a starting point for future experimentations.

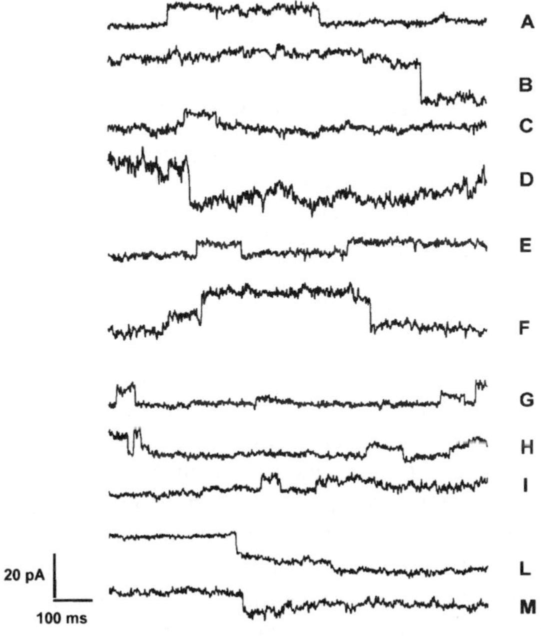

Figure 5. Spontaneous single channel recordings in solitary ommatidia from the eye of the crayfish *Orconectes limosus*. Holding potentials (V_H) of the cell-attached patches: +70 mV (A-I) and -80 mV (L-M). Negative currents induced depolarizing effects resulting in an opening probability due to outward currents (L-M); on the contrary positive currents were hyper-polarizing and inward (A-I). Channel openings are shown as upward (A-I) and downward (L-M) deflections. From Musio, 1996, modified.

To conclude the survey on solitary photoreceptors of arthropods, it could be worth of mention the first attempt to apply gigaseal technique to isolated photoreceptors of the crayfish (Musio, 1996). Viable ommatidia were dissociated from the eye of the crayfish *Orconectes limosus* and single-channel recordings were successfully made (Figure 5). A peculiar characteristic observed in all recordings was the presence of a long opening time and a large conductance from several channels (Figure 5A,B,D,E,F,L,M). In addition to this, several simultaneous opening activities were observed without the possibility to ascertain whether the examined activity was due to different channel types or to the same one but with different and complex kinetics (Figure 5F). Otherwise, in a particular case two or three different short opening probabilities were observed, due, presumably, to different kinds of channels (Figure 5G,H,I). In none of the tested patches were light-sensitive or light-stimulated channels found, although the portions of selected membranes appeared nearly pigmented. Concerning the voltage-gated nature, in this attempt different conductances were observed which made it difficult to assign identity characteristics. However on the basis of these initial data it could be speculated that the recorded channel activities belong to ion-sensitive and/or K^+ voltage-gated channels. Far from detailing any typical photoresponse, this preliminary work especially focussed on the suitability of the methods and their application to *Orconectes* ommatidia. However, the results presented demonstrated the suitability of *Orconectes* ommatidia for future investigation of photoreception biophysics, as well as their usefulness in studies of ionic channel activity.

5. Solitary photoreceptors and extraocular photoreception

This section will briefly deal with some examples of solitary photosensitive cells that do not exert the standard role of receptors in visual transduction but they are involved in extraocular photoreception or underlie other behavioural events (*e.g.*, circadian rhythms) strictly correlated and/or linked to light exposure.

It is known that the two eyes of *Aplysia* are inconspicuous, but each eye contains a precise circadian clock which produces a circadian rhythm (in the frequency of the compound action potentials of the optical nerve, CAP) controlling several animal rhythmic behaviours (Jacklet and Barnes, 1993). All cellular components of this activity are located within the isolated eye. In addition to photoreceptors (see Section 4.2. above), there are retinal photosensitive neurones (pacemaker neurones, PNs) that produce the CAP with the synchronous activity of their output axons (Jacklet and Barnes, 1993). Patch-clamp recordings performed on isolated PNs showed a spontaneous activity in darkness and responses to light stimulation by a bumpy depolarisation (corresponding under voltage clamp to increased inward current) and a spike train (Jacklet *et al.*, 1996). The photoresponse mechanism suggests that a second messenger cascade is present and that the photoreceptor responsible of the light induced phase shifts of the circadian clock is contained within PNs (Jacklet and Barnes, 1993).

In the related mollusc *Bulla*, basal retinal neurones (BRNs), similar to *Aplysia* PNs, generate the CAP activity paradigmatic of a circadian clock. BRNs have been dissociated and isolated in cell culture for patch-clamping and they expressed a circadian rhythm of membrane conductance changes (Michel *et al.*, 1993). In opposition to *Aplysia* PNs, they did not show spontaneous activity in darkness, so it is still questionable their role in producing a circadian rhythm.

An interesting case of extraocular photoreception is represented by the freshwater cnidarian *Hydra*. This animal shows a quantifiable behavioural photosensitivity (measured electrophysiologically as modulation of the animal's periodic behaviour) even though photoreceptor organs or individual photosensitive cells have not been identified (Taddei-Ferretti and Musio, 2000). A recent striking evidence speaks in favour of a photosensitiveness widely spread throughout the animal body and not restricted to specialised zones or cell aggregates. It is the demonstration, using polyclonal antibodies against squid opsin, of an immunofluorescence localisation of an opsin-like protein that is distributed on the whole animal's ectodermal surface (Musio *et al.*, 2000). As far as the identification of the cells bearing the phototransduction machinery, the electrophysiological screening of cell types of *Hydra* is unavoidable. By this way, gigaseals were obtained on enzymatically dissociated ectodermal epitheliomuscular cells and resting potential were recorded in whole-cell configuration (Santillo *et al.*, 1997), as first step in order to investigate in such cells and in other cell types the possible effect of light.

6. Concluding remarks

Far from to be comprehensive, the present overview has provided the current "state of the art" concerning the development of invertebrate solitary photoreceptor preparations suitable for the application of patch-lamp techniques. The several examples reported above have shown that the use of dissociated visual cells is unique in circumventing the hindrance of classic preparations that are inaccessible by advanced gigaseal techniques (see table below).

Advantages	Patch-clamp applications
Exposure of light-sensitive membranes (avoiding of cell-cell interactions)	whole-cell voltage clamp single-channel recording
Suitability for small cells	whole-cell voltage clamp single-channel recording
Homogeneous control of cytosolic milieu	perforated whole-cell clamp internal dialysis

To this regard, isolated-cells preparations for different types of primate retinal cells have been successfully used, overcoming the obstacle of a limited availability of primate retinas (Han *et al.*, 2000). Furthermore, it is possible to use data gained at single photoreceptor level for computer modelling of biophysical processes (Blackwell, 1999).

As pointed out by the achievements in vertebrate rods and cones, the understanding of cellular mechanisms of phototransduction cannot prescind from the study at ion channel level. This assumption has to be extended also to invertebrates, as confirmed by the following Roger Hardie's sentence (Hardie, 1993): "...further characterization of light-sensitive channels may help to unravel the enigma of invertebrate phototransduction".

Acknowledgements

This work was partially granted by the CNR Strategic Project "*Biological Sensors and Designing of Biosensors*". All my gratitude goes to my wife Angela for her tender patience in sustaining the weight of my intractableness during this adventure.

References

Bacigalupo, J. and P. O'Day (1996) "The second messenger for visual excitation in invertebrate phototransduction", *Biol. Res.* 29:319-324.

Bacigalupo, J., K. Chinn, J.E. Lisman (1986) "Ion channels activated by light in *Limulus* ventral photoreceptors", *J. Gen. Physiol.* 87:73-89.

Bacigalupo, J., D.M. Bautista, D.L. Brink, J.F. Hetzer and P. O'Day (1996) "Cyclic-GMP enhances light-induced excitation and induces membrane currents in *Drosophila* retinal photoreceptors", *J. Neurosci.* 15: 7196-7200.

Baylor, D.A., T.D. Lamb and K.-W. Yau (1979) "The membrane current of single rod outer segments", *J. Physiol.* 288:589-611.

Bell, A.L. and G.J. Mpitsos (1968) "Morphology of the eye of the flame fringe clam, *Lima scabra*", *Biol. Bull.* 135:414-415.

Blackwell, K.T. (1999) "Dynamics of the light-induced current in *Hermissenda*", *Neurocomput.* 26-27:61-67

Chyb, S., P. Raghu and R. Hardie (1999a) "Polyunsaturated fatty acids activate the *Drosophila* light-sensitive channels TRP and TRPL", *Nature* 397:255-259.

Chyb, S., W. Hevers, M. Forte, W.J. Wolfgang, Z. Selinger and R. Hardie (1999b) "Modulation of the light response by cAMP in *Drosophila* photoreceptors", *J. Neurosci.* 19:8799-8807.

Dakin, W.J. (1910) "The eye of *Pecten*", *Quart. J. Microsc. Sci.* 55:49-112.

Deckert, A., K. Nagy, C.S. Helrich and H. Stieve (1992) "Three components of the light induced current of the *Limulus* ventral photoreceptor", *J. Physiol.* 453:69-96.

Dorlöchter, M. and H. Stieve (1997) "The *Limulus* ventral photoreceptor: light response and the role of calcium in a classic preparation", *Prog. Neurobiol.* 53:451-515.

Dowling, J.E. (1987) *The Retina: An Approachable Part of the Brain*, Cambridge, MA: Harvard University Press (Belknap).

Eakin, R.M. (1972) "Structure of invertebrate photoreceptors", in: *Handbook of Sensory Physiology, Vol. 7 Part 1*, H.J. Dartnall, ed., Berlin: Springer, pp. 625-684.

Etchebarrigaray, R., P.L. Huddie and D.L. Alkon (1991) "Gigaohm single-channel recording from isolated *Hermissenda crassicornis* type B photoreceptors", *J. Exp. Biol.* 156:619-623.

Fernald, R.D. (2000) "Evolution of eyes", *Curr. Opin. Neurobiol.* 10:444-50.

Fesenko, E.E., S.S. Kolesnikov and A.L. Lyubarsky (1985) "Induction by cyclic GMP of cationic conductance in plasma membrane of retinal rod outer segment", *Nature* 313:310-313.

Gartner, W. and P. Towner P (1995) "Invertebrate visual pigments", *Photochem. Photobiol.* 62:1-16.

Goldsmith, T.H. (1991) "Photoreception and vision", in: *Comparative Animal Physiology: Neural and Integrative Animal Physiology* (4th edn), C.L. Prosser, ed., New York: Wiley-Liss, pp.171-245.

Gomez, M. and E. Nasi (1994) "The light-sensitive conductance of hyperpolarizing invertebrate photoreceptors: A patch-clamp study", *J. Gen. Physiol* 103:939-956.

Gomez, M. and E. Nasi (1995) "Activation of light-dependent K^+ channels in ciliary invertebrate photoreceptors involves cGMP but not IP_3/Ca^{2+} cascade", *Neuron* 15:607-618.

Gomez, M. and E. Nasi (1996) "Ion permeation through light-activated channels in rhabdomeric photoreceptors: Role of divalent cations", *J. Gen. Physiol.* 107:715-730.

Gomez, M. and E. Nasi (1997) "Antagonists of the cGMP-gated conductance of vertebrate rods block the photocurrent in scallop ciliary photoreceptors", *J. Physiol.* 500:367-378.

Gomez, M. and E. Nasi (1998) "Membrane current induced by protein kinase C activators in rhabdomeric photoreceptors: Implication for visual excitation", *J. Neurosci.* 18:5253-5263.

Hamill, O.P., A. Marty, E. Neher, B. Sakmann and F.J. Sigworth (1981) "Improved patch-clamp techniques for high resolution current recording from cells and cell-free membrane patches", *Pflügers Arch.* 391:85-100.

Han, Y., R.A. Jacoby and S.M. Wu (2000) "Morphological and electrophysiological properties of dissociated primate retinal cells", *Brain Res.* 875:175-186.

Hanna, W.J.B., E.C. Johnson, D. Chaves and G.H. Renninger (1993) "Photoreceptor cells dissociated from the compound lateral eye of the horseshoe crab, *Limulus polyphemus*, II: Function", *Visual Neurosci.* 10:609-620.

Hardie, R.C. (1991) "Whole-cell recordings of the light induced current in dissociated *Drosophila* photoreceptors: evidence for feedback by calcium permeating the light-sensitive channels", *Proc. R. Soc. Lond.* B 245:203-210.

Hardie, R.C. (1993) "Phototransduction: the invertebrate enigma", *Nature* 366:113-114.

Hardie, R.C. (1998a) "Properties of the light-sensitive channels in *Drosophila* photoreceptors", in: *From Structure to Information in Sensory Systems*, C. Taddei-Ferretti and C. Musio, eds, Singapore: World Scientific, pp. 383-397.

Hardie, R.C. (1998b) "Excitation in *Drosophila* photoreceptors: The role of Ca^{2+}", in: *From Structure to Information in Sensory Systems*, C. Taddei-Ferretti and C. Musio, eds, Singapore: World Scientific, pp. 398-411.

Hardie, R.C. and B. Minke (1992) "The *trp* gene is essential for light-activated Ca^{2+} channel in *Drosophila* photoreceptors", *Neuron* 8:643-651.

Hardie, R.C. and B. Minke (1993) "Novel Ca^{2+} channel underlying transduction in *Drosophila* photoreceptors: Implication for phosphoinositide-mediated Ca^{2+} mobilization", *Trends Neurosci.* 16:371-376.

Hardie, R.C. and B. Minke (1994) "Spontaneous activation of light-sensitive channels in *Drosophila* photoreceptors", *J. Gen. Physiol.* 103:389-407.

Jacklet, J. and S. Barnes (1993) "Photoresponsive pacemaker neurons from the dissociated retina of *Aplysia*", *NeuroReport* 5:209-212.

Jacklet, J., S. Barnes, A. Bulloch, K. Lukowiak and N. Syed (1996) "Rhythmic activities of isolated and clustered pacemaker neurons and photoreceptors of *Aplysia* retina in culture", *J. Neurobiol.* 31:16-28.

Jinks, R.N, W.J.B. Hanna, G.H. Renninger and S.C. Chamberlain (1993) "Photoreceptor cells dissociated from the compound lateral eye of the horseshoe crab, *Limulus polyphemus*, I: Structure and ultrastructure", *Visual Neurosci.* 10:597-607.

Kaneko, A. and M. Tachibana (1986) "Membrane properties of solitary retinal cells", *Prog. Retin. Res.* 5:125-146.

Land, M.F. and R.D. Fernald (1992) "The evolution of eyes", *Annu. Rev. Neurosci.* 15:1-29

Matthews, G. (1987) "Single-channel recordings demonstrate that cGMP opens the light-sensitive ion channel of the rod photoreceptor", *Proc. Natl. Acad. Sci. USA* 84:299-302.

McReynolds, J.S. (1976) "Hyperpolarizing photoreceptors in invertebrates", in: *Neural Principles in Vision*, F. Zettler and R. Weiler, eds, Berlin: Springer, pp. 394-409.

Michel, S., M. Geusz, J. Zaritsky and G. Block (1993) "Circadian rhythm in membrane conductance expressed in isolated neurons", *Science* 259:239-241.

Montell, C. (1999) "Visual transduction in *Drosophila*", *Annu. Rev. Cell Dev. Biol.* 15:231-268.

Musio, C. (1996) "Application of the patch-clamp technique to photoreceptor cells of the crayfish *Orconectes limosus*", *Ital. J. Zool.* 63:135-138.

Musio C. (1997) "Extraocular photosensitivity in invertebrates: a look into functional mechanisms and biophysical processes", in: *Biophysics of Photoreception: Molecular and Phototransductive Events*, C. Taddei-Ferretti, ed., Singapore: World Scientific, pp. 245-262.

Musio, C., S. Santillo, C. Taddei-Ferretti, L. Robles, L. Vismara, L. Barsanti and P. Gualtieri (2000) "First identification of a visual pigment in *Hydra*", *in preparation*.

Nagy, K. (1991) "Biophysical processes in invertebrate photoreceptors: Recent progress and a critical overview based on *Limulus* photoreceptors", *Quart. Rev. Biophys.* 24:165-226.

Nagy, K. and H. Stieve (1990) "Light-activated single channel currents in *Limulus* ventral nerve photoreceptors", *Eur. Biophys. J.* 18:221-224.

Nasi, E. (1991a) "Electrophysiological properties of isolated photoreceptors from the eye of *Lima scabra*", *J. Gen. Physiol.* 97:17-34

Nasi, E. (1991b) "Whole-cell clamp of dissociated photoreceptors from the eye of *Lima scabra*", *J. Gen. Physiol.* 97:35-54.

Nasi, E. (1991c) "Two light-dependent conductances in *Lima* rhabdomeric photoreceptors", *J. Gen. Physiol.* 97:55-72.

Nasi, E. and M. Gomez (1992a) "Light-activated ion channels in solitary photoreceptors of the scallop *Pecten irradians*", *J. Gen. Physiol.* 99:747-769.

Nasi, E. and M. Gomez M. (1992b) "Electrophysiological recordings in solitary photoreceptors from the retina of the squid *Loligo pealei*", *Visual. Neurosci.* 8:349-358.

Nasi E. and M. Gomez (1998a) "Visual excitation mechanisms in rhabdomeric photoreceptors: Multiple light-dependent channels, ion permeation and gating", in: *From Structure to Information in Sensory Systems*, C. Taddei-Ferretti and C. Musio, eds, Singapore: World Scientific, pp. 314-326.

Nasi, E. and M. Gomez (1998b) "Bridging the gap between vertebrate and invertebrate phototransduction: the light-activated conductance in hyperpolarizing photoreceptors", in: *From Structure to Information in Sensory Systems*, C. Taddei-Ferretti and C. Musio, eds, Singapore: World Scientific, pp. 327-340.

Niemeyer, B.A., E. Suzuki, K. Scott, K. Jalink and C.S. Zuker (1996) "The *Drosophila* light-activated conductance is composed of the two channels TRP and TRPL", *Cell* 85:651-659.

O'Day, P. and J. Bacigalupo (1997) "Current issues in invertebrate phototransduction", *Mol. Neurobiol.* 15:41-63.

Pak, W.L. (1994) "Retinal degeneration mutants of *Drosophila*", in: *Molecular Genetics of Inherited Eye*, A.F. Wright and B. Jay, eds, Chur, Switzerland: Harwood, pp. 29-52.

Payne, R., B. Walz, S. Levy and A. Fein (1988) "The localization of calcium release by inositol triphosphate in *Limulus* photoreceptors and its control by negative feedback", *Phil. Trans. Roy. Soc. Lond.* B 320:359-379.

Peretz, A., E. Suss-Tolby, A. Rom-Glas, A. Arnon, R. Payne and B. Minke (1994) "The light response of *Drosophila* photoreceptors is accompanied by an increase in cellular calcium: Effects of specific mutations", *Neuron* 12:1257-1267.

Ranganathan, R., G.L. Harris, C.F. Stevens and C.S. Zuker (1991) "A *Drosophila* mutant defective in extracellular calcium dependent photoreceptor deactivation and rapid desensitization", *Nature* 354:230-232.

Rayer, B., M. Naynert and H. Stieve (1990) "Phototransduction: different mechanisms in vertebrates and invertebrates", *J. Photochem. Photobiol. B: Biol.* 7:107-148.

Rodieck, R.W. (1973) *The Vertebrate Retina: Principle of Structure and Function*, San Francisco: Freeman.

Sakmann, B. and E. Neher (1995) *Single-Channel Recording* (2nd ed.), New York: Plenum.

Santillo, S., C. Taddei-Ferretti and M. Nobile (1997) "Resting potentials recorded in the whole-cell configuration from epithelial cells of *Hydra vulgaris*", *Ital. J. Zool.* 64:7-11.

Taddei-Ferretti, C. and C. Musio (2000) "Photobehaviour of *Hydra* (Cnidaria, Hydrozoa) and correlated mechanisms: A case of extraocular photosensitivity", *J. Photochem. Photobiol. B: Biol.* 55:88-101.

Tsunoda, S. and C.S. Zuker (1999) "The organization of INAD-signalling complexes by a multivalent PDZ domain protein in *Drosophila* photoreceptor cells ensures sensitivity and speed of signalling", *Cell Calcium* 26:165-171.

Weckström, M. and S.B. Laughlin (1995) "Visual ecology and voltage-gated ion channels in insect photoreceptors", *Trends Neurosci.* 18:17-21.

Werblin, F.S. (1978) "Transmission along and between rods in the tiger salamander retina", *J. Physiol.* 280:449-470.

Yamoha, E.N., L. Matzel and T. Crow (1998) "Expression of different types of inward rectifier currents confers specificity of light and dark response in type A and B photoreceptors of *Hermissenda*", *J. Neurosci.* 18:6501-6511.

Yarfitz, S. and J.B. Hurley (1994) "Transduction mechanisms in vertebrate and invertebrate photoreceptors", *J. Biol. Chem.* 269:14329-14332.

Yau, K.-W. (1994) "Cyclic-nucleotide-gated channels: An expanding new family of ion channels", *Proc. Natl. Acad. Sci. USA* 91:3481-3483.

Yau, K.-W. and D.A. Baylor (1989) "Cyclic CMP-activated conductance of retinal photoreceptor cells", *Annu. Rev. Neurosci.* 12:289-327.

Yoshida, M. (1979) "Extraocular photoreception", in: *Handbook of Sensory Physiology, Vol. 7 Part 6A*, H. Autrum, ed., Berlin: Springer, pp. 581-640.

Zhang, H.-J., R.N. Jinks, A.C. Wishart, B.A. Battelle, S.C. Chamberlain, W.H. Farenbach and L. Kass (1994) "An enzymatically enhanced recording technique for *Limulus* ventral photoreceptors: Physiology, biochemistry and morphology", *Visual Neurosci.* 11:41-52

Ziemba, S.E., S. Saks, Y. Janviriya and R.S. Stephenson (1995) "Dissociation of photoreceptors from whole heads of the fruit fly, *Drosophila melanogaster*", *Cell Tissue Res.* 280:473-477.

PHOTOTRANSDUCTION IN A DEPOLARIZING PHOTORECEPTOR OF VERTEBRATES

WEI-HONG XIONG, JOHN T. FINN*, EDUARDO SOLESSIO+ and KING-WAI YAU

Department of Neuroscience and Howard Hughes Medical Institute, Johns Hopkins University School of Medicine, Baltimore, MD 21205, USA

**CISAC, Stanford University, Stanford, CA 94305-6165, USA (present address)*

+Department of Ophthalmology, State University of New York Health Science Center, Syracuse, NY 13210, USA (present address)

ABSTRACT

Some lower vertebrates, such as lizard and fish, have a parietal (the "third") eye. This eye is developmentally related to the pineal gland. It resembles the lateral eyes in structure, but its retina has only photoreceptors and ganglion cells. The photoreceptors have an outer segment, but, unlike rods and cones, they depolarize to light under dark-adapted conditions, resulting from the opening of a non-selective cation channel.

Excised-patch recordings indicate that a cGMP-gated channel similar in properties to that in rods is present on the outer segment of the parietal-eye photoreceptor. A blocker of the cGMP-gated channel, L-*cis*-diltiazem, suppresses the light response when applied extracellularly, supporting that the opening of this channel indeed underlies the light response.

A membrane-permeant inhibitor of phosphodiesterase such as IBMX, which would elevate cGMP, induces a transient current very similar to the light response. When applied together, a light flash fails to induce any current additional to that produced by a saturating puff of IBMX, even when only a small percentage of the cGMP-gated channels is open. This result suggests that light elevates cGMP by inhibiting the phosphodiesterase and not by activating the guanylyl cyclase. It appears that the dark phosphodiesterase activity is driven by a stimulatory G protein, and light inhibits the phosphodiesterase through a separate, inhibitory G protein. This situation is reminiscent of the antagonistic control of adenylyl cyclase by G_s and G_i in cAMP signaling.

1. The Parietal Eye

Some lower vertebrates, such as lizard and fish, have a third eye called the parietal eye (Eakin, 1973). This eye, situated on the midline of the forehead, is developmentally derived from the diencephalon, the same part of the brain that gives rise to the pineal gland. The function of this eye is still unclear, but it has been proposed to serve as a dawn-dusk detector (Solessio and Engbretson, 1993). This eye is quite similar to the lateral eyes in structure, with a cornea, a lens, and a retina in the corresponding positions in the eye. However, the retina has only photoreceptors and ganglion cells, lacking bipolar, horizontal and amacrine cells (Eakin, 1973). There is also no pigment epithelium. The photoreceptors resemble rods and cones in morphology, including having an outer segment where light absorption and phototransduction presumably take place. Under electron microscopy, the outer segment resembles the cone outer segment in that it is tapered and the stack of membranous disks are continuous with the plasma membrane (Eakin, 1973). One clear difference between the parietal eye and the lateral eyes, however, has to do with the orientation of the retina, in that the photoreceptors in the parietal eye are centrally located and their outer segments project toward the front of the eye, i.e., facing the incident light.

2. The Light Response of Parietal-Eye Photoreceptors

Solessio and Engbretson (1993) first showed with intracellular recordings from isolated parietal-eye photoreceptors that these cells depolarize to light under dark-adapted conditions, regardless of stimulating wavelength. However, in the presence of a steady green background light, which would maintain a depolarizing response, blue light leads to a hyperpolarization. Thus, there is a chromatic antagonism within a single cell. It is not clear whether the chromatic responses are mediated by two distinct visual pigments or by two stable states of a single bistable pigment. The same workers have shown that the depolarizing response results from the opening of a non-selective cation channel (with a reversal potential near zero), though the underlying signaling mechanism is unknown.

3. Presence of a cGMP-Activated Channel

The cGMP-activated channel is a signature of the cGMP-mediated phototransduction pathway in rods and cones (for review, see Yau and Baylor, 1989; Koutalos and Yau, 1996). Thus, to test for the involvement of this pathway in phototransduction in the parietal-eye photoreceptor, a simple experiment would be to examine whether a cGMP-activated channel is present. With patch-clamp recordings from an excised patch of plasma membrane of the outer segment of these cells, such an ion channel was indeed identified (Finn *et al.*, 1997). This channel resembles the cGMP-activated channel of rod photoreceptors (see Finn *et al.*, 1996 for review) in many ways. First, it is much more sensitive to cGMP than to cAMP; the half-activation constant for cGMP is ca. 20 μM. Second, it is a cation channel that passes both monovalent and divalent cations, with a permeability ratio $P_{Ca}/P_{Na} \sim 10$. Third, it is blocked by the chemical L-*cis*-diltiazem. Fourth, Ca^{2+}-calmodulin negatively modulates the channel, reducing the current at low cGMP concentrations by about a factor of two.

The cGMP-activated channel is present on the outer segment with a density of ca. 140/μm^2 (Finn *et al.*, 1997), comparable to that for rods (Karpen *et al.*, 1992). The density elsewhere on the parietal-eye photoreceptor is ca. 100-fold lower. The selective presence of the channel on the outer segment (as is the case for rods) suggests that it is likely involved in phototransduction. The non-selectivity of the channel among cations is also consistent with the property of the light-activated channel described above. Thus, light probably leads to the opening of this cGMP-activated channel by elevating cGMP, opposite to the situation in rods and cones.

4. Phototransduction Pathway

The sensitivity of the cGMP-activated channel to L-*cis*-diltiazem provides a way to confirm whether the opening of this channel indeed underlies the light response (Xiong *et al.*, 1998). Dialyzing cGMP from a whole-cell recording pipette activated an inward current from a parietal-eye photoreceptor, and this current could be transiently inhibited by a puff of L-*cis*-diltiazem ejected at the cell from another pipette. The light response (recorded with perforated-patch recording using nystatin [Horn and Marty, 1988]) could be blocked in the same manner, supporting the above hypothesis.

Instead of intracellular dialysis, another way to elevate cGMP in a photoreceptor in darkness would be to use a phosphodiesterase inhibitor (Xiong *et al.*, 1998). Assuming that there is a basal turnover of cGMP in darkness consisting of synthesis by a guanylyl cyclase and hydrolysis by a phosphodiesterase, the inhibition of hydrolysis should elevate cGMP. Indeed, this is the case. When a puff of IBMX (a non-specific phosphodiesterase inhibitor) or zaprinast (a more specific inhibitor of cGMP-phosphodiesterase [Beavo, 1995; Gillespie and Beavo, 1989]), both highly membrane-permeant, was ejected at a photoreceptor extracellularly, a transient inward current was detected that resembled the light response in both time course and maximum amplitude. The IBMX-induced current is interpreted to indicate an accumulation of intracellular cGMP due to ceased hydrolysis and continuing cyclase activity. When a strong puff of IBMX and a light flash were applied in rapid succession, the light was unable to produce any additional current above that induced by IBMX (Xiong *et al.*, 1998). This lack of summation was observed even when most of the channels were still unopen. This observation indicates that the action of light is the same as that of IBMX, i.e., inhibiting the phosphodiesterase. If light activated the guanylyl cyclase, it would have produced a larger response in the presence of IBMX.

The signaling pathway thus seems to be that light activates a visual pigment, which, via a trimeric G protein (because the pigment is a seven-transmembrane-helix receptor), inhibits a cGMP-phosphodiesterase that is active in darkness.

5. Antagonistic Control of cGMP-Phosphodiesterase by Two G Proteins

One question is whether the dark phosphodiesterase activity in the parietal-eye photoreceptor represents spontaneous activity of the enzyme or is driven by a signal upstream, such as a stimulatory G protein. To address this question, a G protein activator or inhibitor could be dialyzed into a photoreceptor in darkness, and the effect on the phosphodiesterase activity examined (Xiong *et al.*, 1998).

When GTP was dialyzed together with cGMP into a photoreceptor from a whole-cell pipette, the inward current induced by cGMP in darkness was enhanced by illumination, supporting the idea that light inhibits dark phosphodiesterase activity. However, when GTPγS, a hydrolysis-resistant analog of GTP that "permanently" activates trimeric G proteins, was used instead, the effect of light on the dark current disappeared. The same was observed when $AlFl_4^-$, which also activates G proteins in the presence of GDP (Bigay *et al.*, 1985; Gilman, 1987), was used instead of GTPγS. In contrast, with dialysis of GDPβS, which inhibits trimeric G proteins, the dark current became very large, as if constant light were present. Together, these results suggest that the phosphodiesterase is active in darkness because it is driven by a stimulatory G protein. It is unlikely that this stimulatory G protein is the same as the one that mediates the inhibition of the enzyme by light, because this would require light to inhibit the stimulatory G protein, and there is no example of such an inhibition by any seven-transmembrane-helix receptor.

The most likely scenario is that two G proteins are involved in the control of the phosphodiesterase, one active in darkness and the other stimulated by light (Figure 1). The inhibitory G protein may act on the phosphodiesterase directly, or via inhibtion of the stimulatory G protein. In principle, a simpler scheme consisting of a mirror image of the phototranduction pathway in rods and cones (whereby all pigment is active in darkness to activate the phosphodiesterase, and light simply inactivates the pigment) can qualitatively explain all of the observations as well. However, this scheme fails in quantitative considerations, and has to be ruled out (Xiong *et al.*, 1998). This antagonistic control of the cGMP-phosphodiesterase is reminiscent of the G_s and G_i control of adenylyl cyclase in cAMP signaling (Sunahara *et al.*, 1996)

The question remains what signal, if any, drives the stimulatory G protein in darkness. From the work of Solessio and Engbretson (1993), a likely candidate would be a visual pigment (with peak absorption in the blue) distinct from that (with peak absorption in the green) driving the inhibitory G protein. It could also be a second active state of a single pigment.

6. Evolutionary Aspects

The parietal-eye photoreceptor represents the only known vertebrate photoreceptor that depolarizes to light under dark-adapted conditions. Among invertebrates, however, the great majority of known photoreceptors depolarize to light (Autrum, 1979). There are exceptions, however. One example is the hyperpolarizing photoreceptor in scallop, which now appears to use a cGMP signaling cascade for phototransduction (Gomez and Nasi, 1995). In this case, light elevates cGMP (though the underlying biochemistry is still unclear) as in the parietal-eye photoreceptor, but the cGMP activates a potassium channel, thus causing the cell to hyperpolarize. In molluscs such as *Onchidium verruculatum*, the neural ganglia have extraocular photoreceptors that likewise use a cGMP phototransduction cascade (Gotow *et al*., 1994).

Except for the molluscan extraocular photoreceptors, which do not have a recognizable photosensitive structure, all other photoreceptors that are now known to employ a cGMP cascade for phototransduction - whether vertebrate or invertebrate, depolarizing or hyperpolarizing - are ciliary photoreceptors, i.e., having a photosensitive structure derived from modified cilia. Thus, the unifying principle appears to be that all ciliary photoreceptors have evolved to use a cGMP signaling pathway for phototransduction, though there are variations with respect to the polarity of cGMP change and to the ion selectivity of the cGMP-activated channel. Most invertebrate photoreceptors, however, are rhabdomeric photoreceptors, with a photosensitive structure derived from modified microvilli (Autrum, 1979). These rhabdomeric photoreceptors include all known invertebrate depolarizing photoreceptors, such as in insects, molluscs and cephalopods. All evidence so far suggests that a phospholipase C pathway, and not a cGMP pathway, is central to phototransduction in these photoreceptors (Hardie and Minke, 1995; Ranganathan *et al*., 1995; Shin *et al*., 1993). Thus, there appears to be two major branches in photoreceptor evolution, each with a distinct motif of phototransduction.

7. Concluding Remarks

The parietal-eye photoreceptor has provided provocative information about the evolution of phototransduction mechanisms in the animal kingdom. At the same time, it gives new insight about how cGMP can be controlled in a signaling pathway. So far, all signaling pathways leading to a rise in cGMP have been known to involve the activation of a guanylyl cyclase – either a particulate cyclase activated by a ligand or a soluble cyclase activated by nitric oxide (Drewett and Garbers, 1994; Wedel and Garbers, 1997; Zhang and Snyder, 1995). In the parietal-eye photoreceptor, however, the rise in cGMP is caused by inhibition of the phosphodiesterase. Because the cellular cGMP level represents a balance between synthesis and hydrolysis, an up-regulation of synthesis is really equivalent in end result to a down-regulation of hydrolysis. It will be interesting to see how widespread this mode of control of cyclic nucleotides is among tissues.

Acknowledgments

The work described here has been supported in part by a grant from the US National Eye Institute.

References

Autrum, H. (1979) "Comparative Physiology and Evolution of Vision" in: *Invertebrates A: Invertebrate Photoreceptors*, Berlin: Springer-Verlag.

Beavo, J.A. (1995) "Cyclic nucleotide phosphodiesterases: functional implications of multiple isoforms", *Physiol. Rev.* 75:725-748.

Bigay, J., P. Deterre, C. Pfister, and M. Chabre (1985) "Fluoroaluminates activate transducin-GDP by mimicking the gamma-phosphate of GTP in its binding sites", *FEBS Lett.* 191:181-185.

Drewett, J.G. and D.L. Garbers (1994) "The family of guanylyl cyclase receptors and their ligands", *Endocr. Rev.* 15:135-162.

Eakin, R.M. (1973) *The Third Eye*, Berkeley: University of California Press.

Finn, J.T., M.E. Grunwald, and K.-W. Yau (1996) "Cyclic nucleotide-gated ion channels: an extended family with diverse functions", *Annu. Rev. Physiol.* 58:395-426.

Finn, J.T., E.C. Solessio, and K.-W. Yau (1997) "A cGMP-gated cation channel in depolarizing photoreceptors of the lizard parietal eye", *Nature* 385:815-819.

Gillespie, P.G. and J.A. Beavo (1989) "Inhibition and stimulation of photoreceptor phosphodiesterase by dipyridamole and M&B 22,948", *Mol. Pharmacol.* 36:773-781.

Gilman, A.G. (1987) "G proteins: transducers of receptor-generated signals", *Annu. Rev. Biochem.* 56:615-649.

Gomez, M. and E. Nasi (1995) "Activation of light-dependent K^+ channels in ciliary invertebrate photoreceptors involves cGMP but not the IP_3/Ca^{2+} cascade", *Neuron* 15:607-618.

Gotow, T., T. Nishi, and H. Kijima (1994) "Single K^+ channels closed by light and opened by cyclic GMP in molluscan extra-ocular photoreceptor cells", *Brain Res.* 662:268-272.

Hardie, R.C. and B. Minke (1995) "Phosphoinositide-mediated phototransduction in *Drosophila* photoreceptors: the role of Ca^{2+} and trp", *Cell Calcium* 18:256-274.

Horn, R. and A. Marty (1988) "Muscarinic activation of ionic currents measured by a new whole-cell recording method", *J. Gen. Physiol.* 92:145-159.

Karpen, J.W., D.A. Loney, and D.A. Baylor (1992) "Cyclic GMP-activated channels of salamander retinal rods: spatial distribution and variation of responsiveness", *J. Physiol. (Lond.)* 448:257-274.

Koutalos, Y. and K.-W. Yau (1996) "Regulation of sensitivity in vertebrate rod photoreceptors by calcium" *Trends in Neurosci.* 19:73-81.

Ranganathan, R., D.M. Malicki, and C.S. Zuker (1995) "Signal transduction in Drosophila photoreceptors", *Annu. Rev. Neurosci.* 18:283-317.

Shin, J., E.A. Richard, and J.E. Lisman (1993) "Ca^{2+} is an obligatory intermediate in the excitation cascade of limulus photoreceptors", *Neuron* 11:845-855.

Solessio, E. and G.A. Engbretson (1993) "Antagonistic chromatic mechanisms in photoreceptors of the parietal eye of lizards", *Nature* 364:442-445.

Sunahara, R.K., C.W. Dessauer, and A.G. Gilman (1996) "Complexity and diversity of mammalian adenylyl cyclases", *Annu. Rev. Pharmacol. Toxicol.* 36:461-480.

Wedel, B.J. and D.L. Garbers (1997) "New insights on the functions of the guanylyl cyclase receptors", *FEBS Lett.* 410:29-33.

Xiong, W.-H., E.C. Solessio, and K.-W. Yau (1998) "An unusual cGMP pathway underlying depolarizing light response of the vertebrate parietal-eye photoreceptor", *Nature Neurosci.* 1:359-365.

Yau, K.-W. and D.A. Baylor (1989) "Cyclic GMP-activated conductance of retinal photoreceptor cells", *Annu. Rev. Neurosci.* 12:289-327.

Zhang, J. and S.H. Snyder (1995) "Nitric oxide in the nervous system", *Annu. Rev. Pharmacol. Toxicol.* 35:213-233.

PHOTOTRANSDUCTION IN RETINAL RODS AND CONES

YIANNIS KOUTALOS*, KEI NAKATANI+, WEI-HONG XIONG
and KING-WAI YAU
*Department of Neuroscience and Howard Hughes Medical Institute,
Johns Hopkins University School of Medicine, Baltimore, MD 21205, USA*
**Department of Physiology and Biophysics, University of Colorado
School of Medicine, Denver, CO 80262, USA (present address)*
*+Institute for Biological Sciences, University of Tsukuba, Tsukuba,
Ibaraki, Japan (present address)*

ABSTRACT

Phototransduction in retinal rod and cone photoreceptors involves a cGMP signaling cascade. In darkness, cGMP in the photoreceptor outer segment opens a cGMP-gated channel and maintains a steady inward current carried mostly by Na^+ and Ca^{2+}. Light isomerizes rhodopsin, which, via the G protein transducin, stimulates a cGMP-phosphodiesterase to increase cGMP hydrolysis, hence leading to channel closure and a membrane hyperpolarization.

The closure of the cGMP-gated channel by light stops the Ca^{2+} influx, producing a decrease in intracellular Ca^{2+} concentration. This Ca^{2+} decrease triggers multiple negative-feedback effects: 1) an increase in guanylyl cyclase activity, 2) a decrease in the active lifetime of rhodopsin and hence a decrease in the light-stimulated phosphodiesterase activity, 3) a possible decrease in interaction between photoactivated rhodopsin and transducin, and 4) an increase in the affinity of the channel for cGMP. These feedbacks all tend to decrease the sensitivity of the photoreceptor to light, thus producing adaptation of the cell to steady illumination.

Rods and cones have similar phototransduction mechanisms, but why cones are less light-sensitive and have faster responses is still unclear. In this regard, it is worth noting that rods and cones have distinct, though highly homologous, isoforms of the various phototransduction proteins.

A phospholipase C signaling pathway is central to phototransduction in many known invertebrate photoreceptors, but it appears unimportant to rod and cone phototransduction, at least acutely.

1. General Properties of Rod and Cone Responses to Light

Unlike most other sensory receptors, retinal rods and cones hyperpolarize in response to illumination (Bortoff, 1964; Tomita, 1965). This hyperpolarization, triggered by light absorption in the cells' outer segment (where the visual pigment is situated), results from the closure of non-selective cation channels on the plasma membrane of the outer segment that are open in darkness (Hagins *et al.*, 1970; Tomita, 1970). In darkness, the open cation channels maintain a steady inward current and depolarization of the cell and, consequently, a steady release of neurotransmitter (glutamate) from the cell's presynaptic terminal (Byzov and Trifonov, 1968; Dowling and Ripps, 1973). The hyperpolarization produced by light decreases or completely stops this neurotransmitter release to second-order visual neurons (bipolar cells) in the retina. Early intracellular recordings with high-resistance microelectrodes have provided useful information about some of the basic properties of the rod and cone light responses (see, for example, Baylor and Fuortes, 1970; Baylor and Hodgkin, 1973). Subsequently, the suction-pipette recording method, which monitors the events in the outer segment more selectively by directly measuring the current through the light-sensitive channels, have provided finer details of the phototransduction process, especially in rods (Baylor *et al.*, 1979a). For example, it became known that the light-sensitive channels are uniformly distributed along the length of the rod outer segment; also, the saturated response of a rod corresponds to closure of all of the channels open in darkness (hence a complete cessation of the dark current, ca. 20-50 pA) (Baylor *et al.*, 1979a).

With flash illumination (i.e., brief light pulses), the suppression of the dark current is transient, with the relation between response peak amplitude and flash intensity described by the Michaelis equation $R = R_{max} [I/(I + \sigma)]$, where R is response peak amplitude, R_{max} is maximum (or saturated) response amplitude, I is flash intensity, and σ is the half-saturating flash intensity (Baylor *et al.*, 1979a). While useful, this relation is empirical, and does not convey the nature of the phototransduction mechanism. With increasing flash intensity, the response amplitude increases but at the same time the cell adapts to the light. This light adaptation begins to develop even before the flash response reaches peak. If the flash response is measured at sufficiently early times during the rising phase, before light adaptation occurs, the relation between response amplitude and flash intensity is steeper than the Michaelis relation, described instead by a saturating exponential function, $R = R_{max} [1 - \exp(-kI)]$, where k is a sensitivity parameter (Lamb *et al.*, 1981; Nakatani and Yau, 1988a). A simple interpretation of this relation is that the effect of an absorbed photon is local, consisting of the closure of all of the open channels within a small longitudinal region of the outer segment (Lamb *et al.*, 1981). While this scenario is somewhat idealized, there is evidence for a restricted spread of the single-photon effect (Baylor *et al.*, 1981; Matthews, 1986).

With continuous light, the response of a rod rises to a transient peak and then quickly relaxes to a plateau level (Baylor *et al.*, 1979b). This is again a sign of light adaptation. The relation between response plateau and light step intensity is quite shallow, being even less steep than the Michaelis relation (see, for example, Nakatani *et al.*, 1991).

Cones are much less sensitive to light than rods, with the half-saturating flash intensity (σ) ca. 25-100 times higher (Nakatani and Yau, 1989; Schnapf and McBurney, 1980). The response kinetics is faster by a factor of 2-4. They also adapt to light more effectively.

2. Response to a Single Photon

The response to a single absorbed photon is the fundamental building block of any response to brighter light of arbitrary duration. In rods, these single-photon effects can be observed individually with suction-pipette recording (Baylor *et al.*, 1979b). In dark-adapted conditions, each of these responses accounts for as much as 3% of the maximum response. These responses show very little variance in amplitude, form and kinetics from photon to photon. The mechanism underlying this stereotypy is a fundamental biophysical question (Rieke and Baylor, 1998; Whitlock and Lamb, 1999). The single-photon responses are also completely invariant with respective to the wavelength of light (Baylor *et al.*, 1979b). Thus, wavelength affects only the probability of absorption by the pigment, but not the downstream steps of signaling.

Each absorbed photon isomerizes a single visual pigment (rhodopsin) molecule. The ability to detect the response to a single absorbed photon therefore allows one to monitor the isomerization of a single rhodopsin molecule. Spontaneous isomerization events do occur in darkness, albeit infrequently, corresponding to ca. 1 event per minute for a toad rod (containing ca. 3 x 10^9 rhodopsin molecules) at 20°C (Baylor *et al.*, 1980). This rate suggests a half-life of rhodopsin due to spontaneous (thermal) isomerization in darkness of ca. 1,000 years. Thus, rhodopsin is very stable and rarely generates a false signal.

The response of a cone to a single photon is too small (25-100 times smaller) to be detected individually.

3. Mechanism of Phototransduction

The phototransduction mechanism is now quite well understood, thanks to a synergy of approaches from electrophysiology, biochemistry, cell biology, molecular biology and genetics (for reviews, see Baylor, 1987; Koutalos and Yau, 1996; Palczewski and Saari, 1997; Pugh and Lamb, 1993; Stryer, 1986; Yau, 1994). The two key second messengers involved in the process are cGMP and Ca^{2+}, the first important for photoexcitation and the second for light adaptation (Figure 1). The derived information is mostly about rods, but cones behave in a very similar manner (Yau, 1994).

In darkness, the cytoplasmic free concentration of cGMP in the rod outer segment is ca. one to a few μM (Nakatani and Yau, 1988b), maintained by a balance between synthesis by a guanylyl cyclase and hydrolysis by a cGMP-phosphodiesterase. The cGMP binds to and opens a non-selective, cGMP-activated cation channel (Fesenko *et al.*, 1985; Yau and Nakatani, 1985a). This open channel maintains the steady dark current mentioned above, carried by Na^+ (80%), Ca^{2+} (15%) and Mg^{2+} (5%) (Nakatani and Yau, 1988c). The Na^+ influx is balanced by an efflux via a Na/K ATPase situated on the cell's inner segment (adjacent to the outer segment). The Ca^{2+} influx is balanced by an efflux via a

Na/Ca,K exchanger situated also on the outer segment membrane (Cervetto *et al.*, 1989; Nakatani and Yau, 1988c). This exchanger employs the energy associated with the natural influx of Na^+ and efflux of K^+ down their respective electrochemical gradients to extrude Ca^{2+} against its electrochemical gradient. The pathway for Mg^{2+} efflux is still unknown.

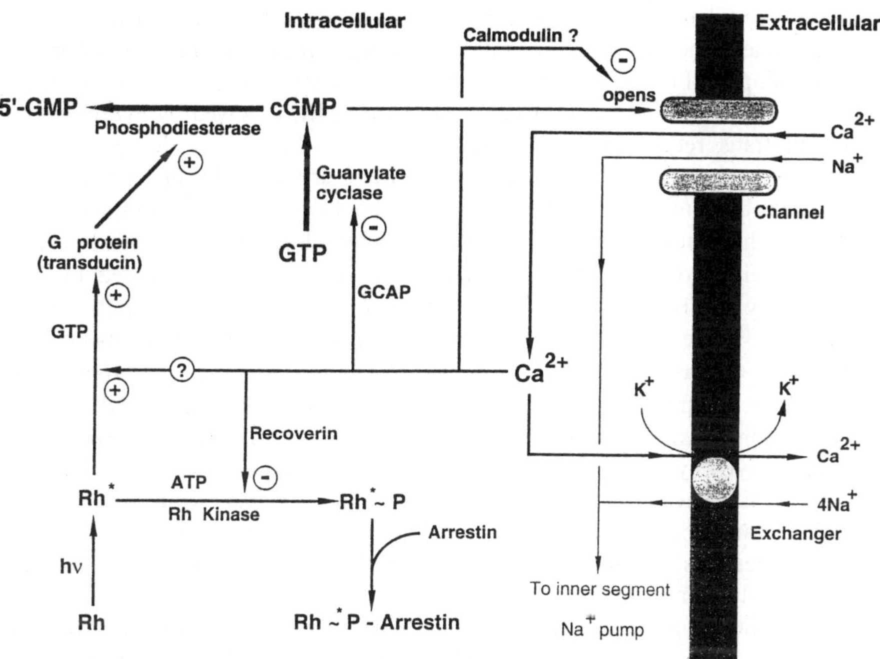

Figure 1. Phototransduction scheme in rods. Symbols: GCAP, guanylyl-cyclase-activating protein; hv, photon; Rh, rhodopsin; Rh*, photoactivated rhodopsin (metarhodopsin II); Rh*~P, phosphorylated form of Rh*; +, stimulation or positive modulation; -, inhibition or negative modulation. Question mark indicates that uncertainty exists. Mg^{2+} flux is not shown. (From Koutalos and Yau, 1996).

Upon absorption of a photon, the 11-cis retinal (chromophore moiety) in rhodopsin isomerizes into all-trans retinal. As a result, the protein moiety (opsin) of the pigment undergoes a number of spontaneous conformational changes. One of these intermediate conformations, metarhodopsin II, is formed within a millisecond of light absorption and is catalytically active. It catalyzes the exchange of GTP for GDP bound to the trimeric G protein transducin. As a result, the α-subunit of transducin (T_α), which binds GTP, dissociates from the βγ-subunits

($T_{\beta\gamma}$) and stimulates the activity of the cGMP-phosphodiesterase by removing the influence of the enzyme's inhibitory γ subunit (PDE_γ) on the catalytic $\alpha\beta$ subunits ($PDE_{\alpha\beta}$). $PDE_{\alpha\beta}$ hydrolyzes cGMP, lowering its cytoplasmic concentration and thus closing the cGMP-activated channels.

The recovery from light requires the termination of each of the active intermediates in the phototransduction cascade (see above reviews). The metarhodopsin II is phosphorylated within a second of its formation by rhodopsin kinase, a covalent modification which curtails some of its catalytic activity. At the same time, another protein called arrestin binds to the phosphorylated metarhodopsin II to cap its residual activity. The phosphorylated metarhodopsin II goes on to decay into the inactive metarhodopsin III, followed by hydrolysis into free all–trans retinal and opsin. At some stage during this decay, the opsin becomes dephosphorylated and loses the bound arrestin. The opsin recyles into rhodopsin by combining spontaneously with another 11-cis retinal molecule. The active T_α has endogenous GTPase activity, which hydrolyzes the bound GTP to GDP, whereby deactivating itself. This GTPase activity is promoted by PDE_γ (Arshavsky and Bownds, 1992), the substrate of T_α, as well as by another protein called RGS9 (He *et al.*, 1998), a member of the RGS family of proteins known to accelerate GTPase activity of G proteins. Finally, when T_α deactivates (and recycles by recombining with $T_{\beta\gamma}$), PDE_γ re-exerts its inhibition on $PDE_{\alpha\beta}$ to stop cGMP hydrolysis. The cGMP concentration returns to its dark level owing to on-going guanylyl cyclase activity, and the cGMP-activated channels re-open. During its active lifetime, a rhodopsin molecule is able to activate sequentially several hundred transducin molecules (Stryer, 1986), each of which goes on to activate a phosphodiesterase molecule. Thus, there is a high gain built into the phototransduction cascade.

It might be noted here that the total cGMP concentration in the rod outer segment is quite high, ca. 60 μM, although the free cGMP concentration is only one to a few μM (Nakatani and Yau, 1988b). Most of the cGMP is bound to high-affinity, non-catalytic binding sites on $PDE_{\alpha\beta}$, apparently to regulate $PDE_{\alpha\beta}$ functionally (Arshavsky and Bownds, 1992).

Over the years, there have been many reports of the presence of a light-activated phospholipase C pathway in the rod outer segment (see Xiong *et al.*, 1997). This signaling pathway is now found to be central to phototransduction in many invertebrates (Hardie and Minke, 1995; Ranganathan *et al.*, 1995; Shin *et al.*, 1993). Canonically, the phospholipase C pathway leads to IP_3 production and also the activation of protein kinase C. However, there is still no evidence for the presence of an IP_3 receptor in the rod outer segment (Peng *et al.*, 1991). At the same time, it does not appear that protein kinase C has any acute effect on phototransduction in rods (Xiong *et al.*, 1997).

4. Negative-Feedback Pathways Mediated by Ca^{2+}

As described above, there is in darkness a balance between Ca^{2+} influx through the cGMP-activated channels and efflux through a Na/Ca,K exchanger. Light closes the channels without affecting the activity of the exchanger. The resulting imbalance between Ca^{2+} influx and efflux leads to a decrease in the cytoplasmic free Ca^{2+} concentration (Nakatani and Yau, 1988c; Yau and Nakatani, 1985b). This Ca^{2+} decrease activates important negative-feedback control on phototransduction (see Koutalos and Yau, 1996 for review).

There are multiple targets of this negative feedback (Figure 1).

First, the guanylyl cyclase activity is sensitive to Ca^{2+}. This enzyme requires a protein called GCAP (guanylyl cyclase activating protein) for its activity, and GCAP is a Ca^{2+}-binding protein. In the Ca^{2+}-free form, GCAP activates guanylyl cyclase, but this ability is inhibited for the Ca^{2+}-bound form. In darkness, the free Ca^{2+} concentration in the outer segment is high enough (200-500 nM) to partially inhibit the guanylyl cyclase activity. When Ca^{2+} concentration decreases in the light, the guanylyl cyclase activity is disinhibited, thus diminishing the effect of the light-stimulated phosphodiesterase activity.

Second, rhodopsin kinase is inhibited by Ca^{2+}, apparently via the Ca^{2+}-binding protein recoverin. Thus, the phosphorylation of metarhodopsin II is faster at low than at high Ca^{2+}, meaning lower phosphodiesterase activity in the light, or a lower gain of phototransduction.

Third, there is indication that metarhodopsin II activates transducin less effectively at lower Ca^{2+}, with the same consequence as the feedback on rhodopsin kinase. Finally, the affinity of the cGMP-activated channel for cGMP is higher, making the channels more likely to open, at low than at high Ca^{2+}. This Ca^{2+} effect apparently involving calmodulin, a ubiquitous Ca^{2+}-binding protein. All of the above negative feedbacks serve to diminish the effect of illumination, resulting in adaptation of the cell to steady light (Matthews *et al.*, 1988; Nakatani and Yau, 1988a). The different feedback pathways do not contribute equally to light adaptation. From calculations based on experimental measurements (Koutalos *et al.*, 1995), the negative feedback on guanylyl cyclase is predominant at low light levels. The feedback on the light-stimulated phosphodiesterase becomes progressively important at high light levels. On the other hand, the feedback on the cGMP-activated channel does not appear to be very important at all light intensities.

The above negative control mediated by Ca^{2+} is important in darkness as well, by minimizing any spontaneous fluctuations in the synthesis or hydrolysis of cGMP. In this way, the background noise in phototransduction is reduced (Nakatani and Yau, 1988c; Yau, 1994).

5. Cone Phototransduction

It is mentioned above that cones have a very similar phototransduction mechanism as in rods. Cones are much less sensitive to light, and their responses have faster kinetics. What underlies these differences (which are likely to be mechanistically related) remains unclear. The outer segments of rods and cones have somewhat dissimilar geometries, though this does not appear to figure importantly in the differences in response properties (Nakatani and Yau, 1989). Interestingly, rods and cones have distinct isoforms of the various phototransduction proteins, such as pigment, transducin, phosphodiesterase, guanylyl cyclase, cGMP-activated channel, and others. It is quite likely that the different biochemistry associated with these protein isoforms is an important determinant of the functional differences (see Yau, 1994).

6. Important Functional Properties of the cGMP-Activated Channels

Unlike most ligand-activated channels (such as the nicotinic receptor channel and the glutamate channels), the rod and cone cGMP-activated channels do not show desensitization to the steady presence of cGMP (for review, see Yau and Baylor, 1989). While unusual, this property is critical for their function in phototransduction, by allowing the channels to stay open in darkness and maintain a dark current, and to be closed only by a decrease in cGMP concentration triggered by illumination.

Another important property of the rod and cone channels is that they do not have a high affinity for cGMP, with a half-activation constant, $K_{1/2}$, of ca. 50 μM cGMP. This $K_{1/2}$ value means that, at a cytoplasmic free cGMP concentration of one to a few μM in darkness, only ca. 1% of these channels are randomly open at any instant (Nakatani and Yau, 1988b). This low percentage does not ever increase under physiological conditions because light only closes channels. While it would seem wasteful for the cell to have a far greater number of channels on its plasma membrane than it ever engages in the open state, the situation is useful functionally. If the cell were to have far fewer channels on the membrane and keep all of them open, it would have to elevate the free cGMP concentration to a much higher level because of the asymptotic approach of any binding reaction to saturation. This scenario would be just as wasteful, because basal hydrolysis exists at all times and would increase with substrate concentration (Stryer, 1987). Instead of elevating cGMP concentration, an alternative way to compensate for the small number of channels would be to increase the channel's affinity for cGMP. In this case, however, cGMP would remain bound and the channel stay open for some time after the cytoplasmic cGMP level had decreased, thus adversely slowing the speed of phototransduction (Stryer, 1987; Yau and Baylor, 1989).

The rod and cone channels are permeable to both monovalent and divalent cations. Indeed, considering the significant fraction of inward dark current carried by divalent cations despite their low extracellular concentrations, the channels actually prefer Ca^{2+} to Na^{+} by 10:1 or more (Nakatani and Yau, 1988c). Estimates of permeability ratios based on reversal potential measurements have led to the same conclusion (Frings *et al.*, 1995; Picones and Korenbrot, 1995). In addition to the rod and cone channels, there is a highly homologous channel that mediates olfactory transduction in olfactory receptor neurons (Dhallan *et al.*, 1990; Nakamura and Gold, 1987). This olfactory cyclic-nucleotide-activated channel is likewise highly permeable to Ca^{2+} (Kurahashi and Shibuya, 1990). It now appears that these three channels are present in non-neural and other neural tissues besides sensory receptor neurons (for review, see Finn *et al.*, 1996). While the exact functions of these channels in other tissue locations remain unclear, it has been proposed that they serve as a pathway for Ca^{2+} influx (Kaupp, 1995).

Not only do divalent cations permeate through the cGMP-activated channels, they also partially block the passage of monovalent cations (Haynes *et al.*, 1986). In this respect, these channels resemble Ca^{2+} channels. This block by divalent cations is potent. In the case of the rod channel, for example, the divalent-cation block reduces the effective single-channel conductance by over 100-fold under physiological ionic conditions. This blockage has a purpose as well. It allows the cell to open more cGMP-activated channels for a given dark current, thus lowering the quantization noise associated with the steady dark current (Yau and Baylor, 1989). The noise associated with the divalent-cation blockage itself is fast and is filtered out by the slower membrane time constant of the photoreceptor and also by the slower kinetics of synaptic transfer from the photoreceptor to the retinal bipolar (postsynaptic) cell.

7. Molecular Structure of the cGMP-Activated Channels

The cGMP-activated channel in rods is the first example of an ion channel that is directly activated by cyclic nucleotides. As mentioned above, the corresponding channel on cones is a molecularly distinct species. The rod and cone channels, together with the highly homologous olfactory channel, form a family of ion channels that bear a distant relation in molecular structure to the Shaker superfamily of voltage-gated potassium channels (for review, see Finn *et al.*, 1996; Zagotta and Siegelbaum, 1996). Members of both families have six transmembrane domains, cytoplasmic N- and C-termini, and a pore region between transmembrane domains 5 and 6. A difference between the members of the two families is that cyclic-nucleotide-activated channels have, in addition, a consensus cyclic-nucleotide-binding site on the C-terminus. Transmembrane-domain 4 of the cyclic nucleotide-activated channels resemble the S4 domain (the molecular voltage sensor) of Shaker and voltage-activated channels by having evenly spaced positive-charged lysine or arginine residues. However, cyclic-nucleotide-activated channels cannot be activated by voltage alone. Like Shaker potassium channels, cyclic-nucleotide-activated channels are tetrameric complexes composed of 4 subunits (some combination of α- and β-subunit species). A lot is now known about the molecular structure and physiology of these channels (see above reviews).

8. Concluding Remarks

Remarkably detailed information has emerged about the phototransduction mechanism in retinal rods and cones. The functions of the various proteins involved in the process are well established, and genes coding for them are cloned. This information is being used successfully for understanding the etiology of many retinal diseases (see, for example, Rattner *et al.*, 2000 for review).

Acknowledgments

The above work has been supported in part by a grant from the US National Eye Institute.

References

Arshavsky, V.Y. and M.D. Bownds (1992) "Regulation of deactivation of photoreceptor G protein by its target enzyme and cGMP", *Nature* 357:416-417.

Baylor, D.A. (1987) "Photoreceptor signals and vision. Proctor lecture", *Invest. Ophthalmol. Vis. Sci.* 28:34-49.

Baylor, D.A. and M.G.F. Fuortes (1970) "Electrical responses of single cones in the retina of the turtle", *J. Physiol. (Lond.)* 207:77-92.

Baylor, D.A. and A.L. Hodgkin (1973) "Detection and resolution of visual stimuli by turtle photoreceptors", *J. Physiol. (Lond.)* 234:163-198.

Baylor, D.A., T.D. Lamb, and K.W. Yau (1979a) "Responses of retinal rods to single photons", *J. Physiol. (Lond.)* 288:613-634.

Baylor, D.A., T.D. Lamb, and K.W. Yau (1979b) "The membrane current of single rod outer segments", *J. Physiol. (Lond.)* 288:589-611.

Baylor, D.A., G. Matthews, and K.W. Yau (1980) "Two components of electrical dark noise in toad retinal rod outer segments", *J. Physiol. (Lond.)* 309:591-621.

Bortoff, A. (1964) "Localization of slow potential responses in the Necturus retina", *Vision Res.* 4:627-635.

Byzov, A.L. and J.A. Trifonov (1968) "The response to electric stimulation of horizontal cells in the carp retina", *Vision Res.* 8:817-822.

Cervetto, L., L. Lagnado, R.J. Perry, D.W. Robinson, and P.A. McNaughton (1989) "Extrusion of calcium from rod outer segments is driven by both sodium and potassium gradients", *Nature* 337:740-743.

Dhallan, R.S., K.W. Yau, K.A. Schrader, and R.R. Reed (1990) "Primary structure and functional expression of a cyclic nucleotide- activated channel from olfactory neurons", *Nature* 347:184-187.

Dowling, J.E. and H. Ripps (1973) "Effect of magnesium on horizontal cell activity in the skate retina", *Nature* 242:101-103.

Fesenko, E.E., S.S. Kolesnikov, and A.L. Lyubarsky (1985) "Induction by cyclic GMP of cationic conductance in plasma membrane of retinal rod outer segment", *Nature* 313:310-313.

Finn, J.T., M.E. Grunwald, and K.W. Yau (1996) "Cyclic nucleotide-gated ion channels: an extended family with diverse functions", *Annu. Rev. Physiol.* 58:395-426.

Frings, S., R. Seifert, M. Godde, and U.B. Kaupp (1995) "Profoundly different calcium permeation and blockage determine the specific function of distinct cyclic nucleotide-gated channels", *Neuron* 15:169-179.

Hagins, W.A., R.D. Penn, and S. Yoshikami (1970) "Dark current and photocurrent in retinal rods", *Biophys. J.* 10:380-412.

Hardie, R.C. and B. Minke (1995) "Phosphoinositide-mediated phototransduction in Drosophila photoreceptors: the role of Ca^{2+} and trp", *Cell Calcium* 18:256-274.

Haynes, L.W., A.R. Kay, and K.W. Yau (1986) "Single cyclic GMP-activated channel activity in excised patches of rod outer segment membrane", *Nature* 321:66-70.

He, W., C.W. Cowan, and T.G. Wensel (1998) "RGS9, a GTPase accelerator for phototransduction", *Neuron* 20:95-102.

Kaupp, U.B. (1995) "Family of cyclic nucleotide gated ion channels", *Curr. Opin. Neurobiol.* 5:434-442.

Koutalos, Y., K. Nakatani, and K.W. Yau (1995) "The cGMP-phosphodiesterase and its contribution to sensitivity regulation in retinal rods", *J. Gen. Physiol.* 106:891-921.

Koutalos, Y. and K.W. Yau (1996) "Regulation of sensitivity in vertebrate rod photoreceptors by calcium", *Trends Neurosci.* 19:73-81.

Kurahashi, T. and T. Shibuya (1990) "Ca^{2+}-dependent adaptive properties in the solitary olfactory receptor cell of the newt", *Brain Res.* 515:261-268.

Lamb, T.D., P.A. McNaughton, and K.W. Yau (1981) "Spatial spread of activation and background desensitization in toad rod outer segments", *J. Physiol. (Lond.)* 319:463-496.

Matthews, G. (1986) "Spread of the light response along the rod outer segment: an estimate from patch-clamp recordings", *Vision Res.* 26:535-541.

Matthews, H.R., R.L. Murphy, G.L. Fain, and T.D. Lamb (1988) "Photoreceptor light adaptation is mediated by cytoplasmic calcium concentration", *Nature* 334:67-69.

Nakamura, T. and G.H. Gold (1987) "A cyclic nucleotide-gated conductance in olfactory receptor cilia", *Nature* 325:442-444.

Nakatani, K., T. Tamura, and K.W. Yau (1991) "Light adaptation in retinal rods of the rabbit and two other nonprimate mammals", *J. Gen. Physiol.* 97:413-435.

Nakatani, K. and K.W. Yau (1988a) "Calcium and light adaptation in retinal rods and cones", *Nature* 334:69-71.

Nakatani, K. and K.W. Yau (1988b) "Guanosine 3',5'-cyclic monophosphate-activated conductance studied in a truncated rod outer segment of the toad", *J. Physiol. (Lond.)* 395:731-753.

Nakatani, K. and K.W. Yau (1988c) "Calcium and magnesium fluxes across the plasma membrane of the toad rod outer segment", *J. Physiol. (Lond.)* 395:695-729.

Nakatani, K. and K.W. Yau (1989) "Sodium-dependent calcium extrusion and sensitivity regulation in retinal cones of the salamander", *J. Physiol. (Lond.)* 409:525-548.

Palczewski, K. and J.C. Saari (1997) "Activation and inactivation steps in the visual transduction pathway", *Curr. Opin. Neurobiol.* 7:500-504.

Peng, Y.W., A.H. Sharp, S.H. Snyder, and K.W. Yau (1991) "Localization of the inositol 1,4,5-trisphosphate receptor in synaptic terminals in the vertebrate retina", *Neuron* 6:525-531.

Picones, A. and J.I. Korenbrot (1995) "Permeability and interaction of Ca^{2+} with cGMP-gated ion channels differ in retinal rod and cone photoreceptors", *Biophys. J.* 69:120-127.

Pugh, E.N.J. and T.D. Lamb (1993) "Amplification and kinetics of the activation steps in phototransduction", *Biochim. Biophys. Acta* 1141:111-149.

Ranganathan, R., D.M. Malicki, and C.S. Zuker (1995) "Signal transduction in Drosophila photoreceptors", *Annu. Rev. Neurosci.* 18:283-317.

Rattner, A., H. Sun, and J. Nathans (2000) "Molecular genetics of human retinal disease", *Annu. Rev. Genet.* In press:

Rieke, F. and D.A. Baylor (1998) "Origin of reproducibility in the responses of retinal rods to single photons", *Biophys. J.* 75:1836-1857.

Schnapf, J.L. and R.N. McBurney (1980) "Light-induced changes in membrane current in cone outer segments of tiger salamander and turtle", *Nature* 287:239-241.

Shin, J., E.A. Richard, and J.E. Lisman (1993) "Ca^{2+} is an obligatory intermediate in the excitation cascade of limulus photoreceptors", *Neuron* 11:845-855.

Stryer, L. (1986) "Cyclic GMP cascade of vision", *Annu. Rev. Neurosci.* 9:87-119.

Stryer, L. (1987) "Visual transduction: design and recurring motifs", *Chemical Scripta* 27B:161-171.(Abstract)

Tomita, T. (1965) "Electrophysiological study of the mechanisms subserving color coding in the fish retina", *Cold Spring Harb. Symp. Quant. Biol.* 30:559-566.

Tomita, T. (1970) "Electrical activity of vertebrate photoreceptors", *Q. Rev. Biophys.* 3:179-222.

Whitlock, G.G. and T.D. Lamb (1999) "Variability in the time course of single photon responses from toad rods: termination of rhodopsin's activity", *Neuron* 23:337-351.

Xiong, W., K. Nakatani, B. Ye, and K. Yau (1997) "Protein kinase C activity and light sensitivity of single amphibian rods", *J. Gen. Physiol.* 110:441-452.

Yau, K.W. (1994) "Phototransduction mechanism in retinal rods and cones. The Friedenwald Lecture", *Invest. Ophthalmol. Vis. Sci.* 35:9-32.

Yau, K.W. and D.A. Baylor (1989) "Cyclic GMP-activated conductance of retinal photoreceptor cells", *Annu. Rev. Neurosci.* 12:289-327.

Yau, K.W. and K. Nakatani (1985) "Light-induced reduction of cytoplasmic free calcium in retinal rod outer segment", *Nature* 313:579-582.

Zagotta, W.N. and S.A. Siegelbaum (1996) "Structure and function of cyclic nucleotide-gated channels", *Annu. Rev. Neurosci.* 19:235-263.

FORMATION OF "ON" AND "OFF" GANGLION CELL MOSAICS

LEO M. CHALUPA
Center for Neuroscience, Section of Neurobiology, Physiology and Behavior, and Department of Psychology, University of California, Davis, CA 95616, U.S.A.

ABSTRACT

The fact that ganglion cells are distributed in non-random arrays has been known for many years, but until recently little effort has been devoted to understanding how this fundamental feature of the retina is established. Here I describe the results of studies from my laboratory which have implicated two types of activity-mediated events in the formation of ON and OFF ganglion cell mosaics. Our work has shown that glutamate-mediated activity is responsible for the stratification of ganglion cell dendrites into ON and OFF sublaminae of the inner plexiform layer. This developmental event provides a morphological signature for the cells that process light onset or light offset. Subsequently, sodium-voltage gated activity is involved in regulating the pattern, but not the degree, of ganglion cell death across the retina. This developmental event effectively removes ganglion cells that are positioned inappropriately, thus giving rise to the mature mosaic pattern prevalent in the adult retina.

1. Introduction

In this chapter, I discuss the results of studies from my laboratory dealing with the formation of retinal ganglion cell mosaics. Readers are referred to an article I recently co-authored with Jeremy Cook which provides a broader perspective of how mosaics are formed in the developing retina (Cook and Chalupa, 2000).

In the mature retina, ON and OFF ganglion cells can be differentiated morphologically on the basis of their dendritic stratification patterns within the inner plexiform layer (IPL): ON cell dendrites branch proximal and OFF cell dendrites branch distal to their respective somas situated in the ganglion cell layer (Nelson *et al.*, 1978). These neurons are also distributed in a non-random pattern, so that ON cell and OFF cells of a given class form independent mosaics that tile the surface of the retina (Wässle *et al.*, 1978, 1981a, b). By contrast, early in development ON and OFF cells cannot be differentiated from each other because the dendrites of immature ganglion cells ramify throughout the IPL (Maslim and Stone, 1988; Ramoa *et al.*, 1988; Bodnarenko *et al.*, 1995). Thus, a necessary prerequisite for the formation of ON and OFF ganglion cell mosaics is for the dendrites of these neurons to change from a multistratified to a unistratified state.

It has also been well-established that there is a massive loss of ganglion cells during the normal development of the cat (Chalupa, 1988). Some degree of ganglion cell loss is thought to represent the correction of misprojections in developing retinofugal pathways. This could be the case in fetal cats since retinal ganglion cells projecting either to the wrong hemisphere or inappropriate loci within retinorecipient target nuclei have been documented (Williams and Chalupa, 1982).

However, by birth, the available evidence indicates that the pattern of retinal projections in carnivores is virtually identical to that found in the mature animal (Chalupa *et al.*, 1996; Chalupa and Snider, 1998). Consequently, the significance of the postnatal loss of ganglion cells presents a conundrum. The results of recent studies we will describe here provide an answer to this puzzle: ganglion cell loss in the postnatal cat retina serves to refine the early distributions of ON and OFF cells to form the regular mosaic patterns essential for the normal processing of visual information. Before considering this issue, we will provide an account of our work dealing with the stratification of dendrites in the developing cat retina.

2. Stratification of Retinal Ganglion Cell Dendrites

As indicated above, whereas in the mature cat retina, alpha and beta cells can be subdivided into ON and OFF subclasses, developing ganglion cells cannot be differentiated in this manner because of their initially multistratified dendritic branching patterns (Maslim and Stone, 1988; Ramoa *et al.*, 1988; Bodnarenko *et al.*, 1995). Examination of DiI labeled retinal cross-sections at different stages of development has revealed that by embryonic day (E) 50 virtually all beta cells are multistratified. This is two weeks before birth and the youngest age at which the three major ganglion cell classes can be distinguished in the cat retina (Ramoa *et al.*, 1988). The stratification process was found to proceed rapidly so that by the end of the second postnatal week relatively few beta cells were found to be multistratified in the central region of the retina (Bodnarenko *et al.*, 1995). The timing of this event appears to coincide with synaptogenesis in the IPL (Maslim and Stone, 1986), suggesting a role for afferent cells in regulating dendritic stratification. The unique actions of the glutamate analog, 2-amino-4-phosphonobutyrate (APB), which hyperpolarizes rod bipolar and ON-cone bipolar cells selectively (Slaughter and Miller, 1981), enabled us to investigate such a role for these afferent cells. APB blocks the release of glutamate by these interneurons thereby abolishing all visual responses in ganglion cells of dark-adapted animals (Wässle *et al.*, 1991). Intraocular injections of APB, performed during the time period of normal dendritic stratification, resulted in a virtually total arrest of this developmental process since the incidence of multistratified cells was about 40% at P2, the age at which APB treatments were initiated. The incidence of such multistratified cells was not decreased appreciably when daily APB treatment was continued as late as P13. By contrast, at P13 in the normal retina, only about 12% of the ganglion cells are still multistratified (Bodnarenko *et al.*, 1995).

These findings provided the first indication that activity plays a role in the dendritic remodeling of developing ganglion cells. More specifically, these results suggest that glutamate-mediated afferent activity regulates the dendritic stratification process. Interestingly, manipulations known to alter the formation of eye-specific domains in the developing visual system, such as intraocular injection of tetrodotoxin (TTX) and monocular deprivation, do not affect the normal restriction of RGC dendritic processes (Leventhal and Hirsch, 1983; Dubin *et al.*, 1986; Lau *et al.*, 1990; Wong *et al.*, 1991). Thus, it is not ganglion cell activity per se , but rather pre-synaptic afferent activity which appears to regulate the maintenance or elimination of RGC dendritic processes. Moreover, RGC density as well as somal and dendritic field sizes were unaffected following APB treatment, demonstrating that such afferent input has a highly selective impact on RGC dendritic development (Bodnarenko *et al.*, 1995). When short-term APB treatments (P2 to P13) were terminated, the dendritic stratification process was found to resume. At three months of age there were very few multistratified cells in the treated eyes as is the case in the normal adult cat retina (Bodnarenko *et al.*, 1995).

More recently, in a collaborative study with the laboratory of Silvia Bisti in Pisa we have found that APB treatment throughout the first postnatal month results in what appears to be a permanent arrest of dendritic stratification. This provided an opportunity to examine the visual response properties of the APB-treated eye. The obvious question we were interested in addressing was whether or not the presence of ganglion cells with multistratified dendrites resulted in receptive fields with ON-OFF discharge patterns. Extracellular recordings from the A or A1 laminae of the dorsal lateral geniculate nucleus innervated by the APB-treated eye, as well as recordings from the optic tract, revealed that this was indeed the case. Whereas virtually all of the cells driven by the normal eye responded as expected with either ON or OFF discharges, in the case of the treated eye about 40% of the units manifested ON-OFF discharge patterns (Bisti *et al.*, 1998). These observations demonstrate a clear-cut functional correlate for the morphological changes observed in ganglion cells following APB treatment of the developing retina. Moreover, they imply that at maturity the dendrites of these multistratified cells are innervated by axon terminals of ON as well as OFF bipolar cells.

3. A Role for Cell Death in Mosaic Formation

Ganglion cell death has been documented in the developing cat retina by assessment of optic nerve fiber number (Williams *et al.*, 1986) and the presence of pyknotic profiles (Wong and Hughes, 1987; Pearson *et al.*, 1993). In particular, optic nerve counts have revealed that most ganglion cells die during embryonic life, yet twice the mature number of fibers are still present in newborn cats. However, neither of these measures can provide an indication of the degree to which the different classes of cells contribute to the overall magnitude of ganglion cell loss.

By counting all alpha cells within the central region of the developing cat retina, we have recently shown that approximately 20% of these neurons are eliminated during the first postnatal month (Jeyarasasingam *et al.*, 1998). Because the central region of the retina does not expand during this developmental period (Mastronarde *et al.*, 1984; Jeyarasasingam *et al.*, 1998), it can be inferred that this loss of cells must reflect the normal death of these neurons. Moreover, the postnatal period of ganglion cell loss continues after most ganglion cells have completed their stratification process. Thus, ON and OFF subclasses can be differentiated before the final number of ganglion cells is established. This leads to an intriguing question: are regular mosaics present in the retina at a time when there is an "excess" of ganglion cells. In considering this matter, one of two possible scenarios can be envisaged: (i) regular distributions of cells might be present even though the number of cells is higher than normal; or (ii) cell regularity could be "masked" by the excess cells. In the latter case, the loss of neurons would contribute to the formation of cell mosaics.

The resolution of this matter seemed rather straightforward: label all ganglion cells in a large region of the retina so that ON and OFF cells could be distinguished, and then compare the mosaics in the developing retina with those present at maturity. For technical reasons, however, it has been problematic to label the dendrites of a large number of ganglion cells sufficiently well so as to allow classification of these neurons into ON or OFF subtypes. Consequently, it has not been feasible to directly assess mosaic patterns in the developing retina. To overcome this problem, we relied on the common observation that ON and OFF RGCs of a given class are often situated in close proximity to one another (Wässle *et al.*, 1981a, b). By means of computer simulations we first showed that the superimposition of two regular distributions consistently resulted in around 90% opposite sign pairing. By contrast, the superimposition of two random distributions repeatedly resulted in only 50% of such pairs. This relationship between the incidence of opposite sign cell pairs and the degree of regularity exhibited by two superimposed distributions was remarkably robust over a relatively broad range of cell densities, approximating those found from the central to the peripheral retina.

Using an in vitro eyecup preparation we were able to obtain Golgi-like labeling of a relatively small number of ganglion cells by making focal deposits of horseradish peroxidase (HRP) into the fiber layer. Although, not suitable for assessing mosaic patterns using conventional measures of regularity, this material permitted us to quantify the incidence of opposite sign pairs in the developing cat retina. This approach revealed that only 58% of alpha cell pairs are of opposite sign before the developmental period of cell death has ended, suggesting that at this stage the distribution of these neurons is not appreciably different from random. By contrast, in the mature retina the incidence of such opposite sign pairs was found to be around 90%, as predicted by our computer simulations.

Having demonstrated that ON and OFF alpha cell distributions become more regular during postnatal development, we next considered the possibility that this process could be regulated by sodium voltage-gated retinal activity. This would be the case if the spatial pattern of ganglion cell loss in the developing retina was dependent on activity-mediated mechanisms involving the firing of action potentials (O'Leary *et al.*, 1986a, b; Thompson and Holt, 1989). Accordingly, we treated postnatal cat retinas with TTX beginning at P9, when the density of alpha cells is greater than at maturity and before the adult complement of opposite sign alpha cell pairs is established. When these animals reached maturity, we examined both the incidence of opposite sign cell pairs and the regularity indices of the resulting distributions. Unlike in the developing retina, it is feasible to label large regions of the mature retina so as to differentiate between ON and OFF ganglion cells, permitting the calculation of regularity indices.

In control retinas the adult complement of opposite sign pairs is ~90. By contrast, in the same region of TTX treated retinas only 60% of opposite sign pairs was found, a value comparable to that seen in the developing retina. Similarly, the regularity indices for the ON and OFF cell distributions were 3.64 and 3.59, respectively, whereas the TTX treated retinas displayed regularity indices of 2.6 for each cell population illustrating a more disorderly pattern. At the same time, the density of alpha cells in the TTX-treated retinas were within normal values suggesting that sodium channel blockade altered the pattern but not the magnitude of cell loss in the developing retina. In a recent study, we have further shown that the normal magnitude of cell loss observed during RGC development is sufficient to produce a regular distribution pattern from a random one (Jeyarasasingam *et al.*, 1998). Collectively, these findings indicate that spatially selective cell death plays a key role in the formation of RGC mosaics and that this process is regulated by sodium-voltage gated activity.

4. Discussion

The results of the studies summarized above indicate that the formation of retinal ganglion cell mosaics involves two developmental events: (i) the restriction of initially multistratified dendrites to form morphologically distinct ON and OFF cells and (ii) the selective elimination of ganglion cells to change the ON and OFF distribution patterns from random to regular.

In the fetal cat, virtually all ganglion cells possess multistratified dendrites (Bodnarenko *et al.*, 1995) and, therefore, ON and OFF cells cannot be morphologically identified. Beginning in late embryonic life and continuing postnatally, RGCs begin the stratification process establishing the structural signature for ON and OFF cells. Until this process is completed, ON and OFF ganglion cell mosaics cannot be discerned. During early postnatal life, "excess" cells obscure regular mosaic patterns resulting in a low incidence of opposite sign pairs following dendritic stratification. Normal retinal activity during this postnatal period induces a spatially selective pattern of cell death, allowing for the formation of regular ON and OFF ganglion cell mosaic patterns.

These studies have also revealed that two different types of activity-based events are involved in forming ON and OFF ganglion cell mosaic patterns: (i) glutamate-mediated afferent activity and (ii) sodium-voltage gated discharges. It remains to be established, however, how these diverse activity-regulated mechanisms regulate their respective developmental changes. For example, it is clear that blockade of bipolar cell activity with APB prevents ganglion cell dendritic stratification, but how does normal afferent activity direct the retraction of diffusely branching dendrites to allow for the formation of both ON and OFF cells? Perhaps these afferent cells provide selective input to either proximal or distal dendrites of multistratified ganglion cells early in development. If this were the case, the release of glutamate by these afferents could instruct ganglion cells to maintain those processes receiving the necessary input. Alternatively, these ganglion cells may be specified intrinsically as ON or OFF, despite the multistratified dendritic state. In this case, glutamate release from afferents may simply activate a genetic program within a ganglion cell to retract the appropriate dendrites to provide the morphological signature that corresponds to its pre-determined functional state.

These possibilities can be explored by investigating the state of synaptic contacts onto multistratified ganglion cells during development. Localized unistratified synaptic input to ganglion cells would provide evidence for the instructional hypothesis whereas "diffuse" afferent input would more likely support an intrinsic specification hypothesis. Recall that the results of our recent recordings have revealed that multistratified ganglion cells in the APB-treated retina respond to light with ON-OFF discharges. As noted above, this implies that these neurons are innervated by the axonal processes of both ON and OFF bipolar cells. However, it is not known whether this reflects the maintenance of immature bipolar inputs or de novo axonal ingrowth in response to the APB treatment. For these reasons, it would be of great interest to establish the pattern of connections between bipolar cells and multistratified ganglion cell dendrites in the developing retina.

The mechanisms underlying activity-mediated selective cell death have yet to be explored. In this context, the recent findings of Rachel Wong, showing that developing ON and OFF ganglion cells in the ferret retina generate separate waves of activity, may be of relevance. Such subclass-specific waves of activity were proposed to underlie the formation of ON and OFF sublaminae in the ferret dorsal lateral geniculate nucleus. Independent waves of ON and OFF cell activity could also serve to regulate the cell loss required to form regular ON and OFF ganglion cell distributions across the retinal surface. Blocking this activity during development alters the pattern of cell death thereby disrupting the formation of ganglion cell mosaics.

The question of how sodium voltage-gated activity regulates the pattern of cell loss, however, remains to be addressed. Three possible activity-mediated mechanisms can be proposed: interactions at the level of ganglion cell afferents, terminals, and/or directly among RGCs. At the level of ganglion cell afferents, it has been shown that neuropeptide Y containing amacrine cells are arranged in mosaics in the inner nuclear layer during the embryonic development of the cat retina, much earlier than we have shown here to be the case for ganglion cells (Hutsler and Chalupa, 1995). The regular distribution pattern of these and other afferent populations could therefore act as a template directing the selective loss of cells resulting in the formation of ganglion cell mosaics. Alternatively, activity-mediated interactions at the level of ganglion cell terminals may direct the appropriate loss of neurons to form RGC mosaics. Retinal activity has been implicated in the removal of inappropriately projecting neurons in the refinement of retinocollicular topography in the rodent visual system (O'Leary *et al.*, 1986a, b; Thompson and Holt, 1989). Though there is a higher degree of topographic precision in the developing cat visual system (Chalupa *et al.*, 1996; Chalupa and Snider, 1998), a fine tuning of the topographic pattern may occur via activity-mediated interactions. If the inappropriately positioned ganglion cells (which obscure mosaic patterns during development) contribute to topographic imprecision, their removal by such activity-mediated events could refine irregular distribution patterns as well.

Electrical interactions between ganglion cells themselves may also act to regulate the pattern of cell death. For example, alpha ganglion cells are electrically coupled to one another during development (Penn *et al.*, 1994). Perhaps this communication serves to maintain cells within the coupled network at the expense of non-coupled cells. If these non-coupled cells are randomly distributed among a regular array of coupled cells, the removal of these non-coupled cells would result in the formation of mosaics during development.

In order to distinguish between these possibilities, it would be necessary to selectively block activity at each level independently. For example, if activity blockade within retinorecipient nuclei disrupted mosaic formation, this would support a target-mediated mechanism. In contrast, mosaic disruption by blockade of communication between ganglion cells with gap junction inhibitors would suggest that electrical coupling directed the selective loss of cells necessary for mosaic formation.

5. Concluding Remarks

Mature retinal mosaics are essential for spatial information processing. For this reason, it is important to understand the mechanisms underlying their formation.

The research discussed in this chapter has investigated this fundamental feature of retinal organization at a systems level by invoking such ubiquitous developmental phenomena as dendritic restructuring, cell death, and activity-mediated events. It remains for future studies to unravel the cellular and molecular mechanisms behind these events to further our understanding of the formation of retinal mosaics.

Acknowledgments

I thank my students, postdoctoral fellows and colleagues for their valuable contributions to various aspects of this work: Drs. Gaya Jeyarasasingam, Stefan R. Bodnarenko, Gimmi Ratto, Silvia Bisti and Cara Wefers. Supported by National Institutes of Health, National Science Foundation, Fogarty Institute for International Studies and NATO.

References

Bisti, S., C. Gargini and L.M. Chalupa (1998) "Blockade of glutamate-mediated activity in the developing retina perturbs functional segregation of ON and OFF pathways", *J. Neurosci.* 18:5019-5025.

Bodnarenko, S.R., G. Jeyarasasingam and L.M. Chalupa (1995) "Development and regulation of dendritic stratification in retinal ganglion cells by glutamate-mediated afferent activity", *J. Neurosci.*15(11):7037-7045.

Chalupa, L.M. (1988) "Factors underlying the loss of retinal ganglion cells", in: *Cell Interactions in Visual Development*, S.R. Hilfer and J.B. Sheffield, eds, Springer Verlag, 69-86.

Chalupa, L.M. (1995) "The nature/nuture of retinal ganglion cell development", in: *The Cognitive Neurosciences, a Handbook for the Field*, M.S. Gazzaniga ed., MIT Press, 37-50.

Chalupa, L.M. and C.J. Snider (1998) "Topographic specificity in the retinocollicular projection of the developing ferret: An anterograde tracing study", *J. Comp. Neurol.* 392:35-47.

Chalupa, L.M., C.J. Snider and M.A. Kirby (1996) "Topographic organization in the retinocollicular pathway of the fetal cat demonstrated by retrograde labeling of ganglion cells", *J. Comp Neurol*, 368:295-303.

Cook, J.E. and D.L. Becker (1991) "Regular mosaics of large displaced and non-displaced ganglion cells in the retina of the cichlid fish", *J. Comp. Neurol.* 306(4):668-684.

Cook, J.E. and L. M. Chalupa (2000) "Retinal mosaics: new insights into an old concept", *Trends Neurosci.* 23(1):26-34.

Dubin, M., L. Stark and S. Archer (1986) "A role for action potential activity in the development of neuronal connections in the kitten retinogeniculate pathway", *J. Neurosci.* 6:1021-1036.

French, A.S., A.W. Snyder and S.G. Stavenga (1977) "Image degradation by an irregular retinal mosaic", *Biological Cybernetics* 27:229-233.

Hannover, A. (1843) "Mikroskopiske undersogelser af nervesystemet", *Vid. Sel. Naturvid. Og Mathem. Afh* 10:9-112.

Hirsch, J and R. Hylton (1984) "Quality of the primate photoreceptor lattice and limits of spatial vision", *Vis. Res.* 24:347-355.

Hutsler, J.J. and L.M. Chalupa (1994) "Neuropeptide Y immunoreactivity identifies a regularly arrayed group of amacrine cells within the cat retina", *J. Comp. Neurol.* 346:481-489.

Hutsler, J.J. and L.M. Chalupa (1995) "Development of neuropeptide Y immunoreactive amacrine and ganglion cells in the pre- and postnatal cat retina", *J. Comp. Neurol.* 361:152-164.

Jeyarasasingam, G., C.J. Snider, G. Ratto and L.M. Chalupa (1998) "Activity-regulated cell death contributes to the formation of ON and OFF alpha ganglion cell mosaics", *J. Comp. Neurol.* 394:335-343.

Kirby, M.A. and L.M. Chalupa (1986) "Retinal crowding alters the morphology of alpha ganglion cells", *J. Comp. Neurol.* 251:532-541.

Kuffler, S.W. (1953) "Discharge patterns and functional organization of mammalian retina", *J. Neurophysiol.* 16:37-68.

Lau, K., K. So and D. Tay (1990) "Effects of visual or light deprivation on the morphology and the elimination of the transient features during development of type I retinal ganglion cells in hamsters", *J. Comp. Neurol.* 300:583-592.

Leventhal, A. and H. Hirsch (1983) "Effects of visual deprivation upon the morphology of retinal ganglion cells projecting to the dorsal lateral geniculate nucleus of the cat", *J. Neurosci.* 3:332-344.

Maslim, J and J. Stone (1986) "Synaptogenesis in the retina of the cat" *Brain Res.* 373:35-48.

Maslim, J. and J. Stone (1988) "Time course of stratification of the dendritic fields of ganglion cells in the retina of the cat", *Develop Brain Res.* 44:87-93.

Mastronarde, D.N., M.A. Thiebeault and M.W. Dubin (1984) "Non-uniform postnatal growth of the cat retina", *J. Comp. Neurol.* 228:598-608.

Nelson, R., E.V. Famiglietti and H. Kolb (1978) "Intracellular staining reveals different levels of stratification for on- and off-center ganglion cells in cat retina", *J. Neurophysiol.* 41:472-483.

O'Leary, D.D.M., D. Crespo, J.W. Fawcett and W.M. Cowan (1986a) "The effect of intraocular tetrodotoxin on the postnatal reduction in the numbers of optic nerve axons in the rat", *Develop. Brain. Res.* 30:96-103.

O'Leary, D.D.M, J.W. Fawcett and W.M. Cowan (1986b) "Topographic targeting errors in the retinocollicular projection and their elimination by selective ganglion cell death", *J. Neurosci.* 6:3692-3705.

Pearson, H.E., B.R. Payne and T.J. Cunningham (1993) "Microglial invasion and activation in response to naturally occurring neuronal degeneration in the ganglion cell layer of the postnatal cat retina", *Develop. Brain Res.* 76:249-255.

Peichl, L. (1991) "Alpha ganglion cells in mammalian retinae: common properties, species differences, and some comments on other ganglion cells", *Vis. Neurosci.* 7:55-169.

Penn, A.A., R.O.L. Wong and C.J. Shatz (1994) "Neuronal coupling in the developing mammalian retina", *J. Neurosci.* 14(6):3605-3615.

Ramoa, A.S., G. Campbell and C.J. Shatz (1988) "Dendritic growth and remodeling of cat retinal ganglion cells during fetal and postnatal development", *J. Neurosci* 8:4239-4261.

Slaughter, M.M. and R.F. Miller (1981) "2-amino-4-phosphonobutyric acid: A new pharmacological tool for retina research", *Science* 211:182-184.

Thompson, I. and C. Holt (1989) "Effects of intraocular tetrodotoxin on the development of the retinocollicular pathway in the syrian hamster", *J. Comp. Neurol.* 282:371-388.

Wässle, H. and B.B. Boycott (1991) "Functional architecture of the mammalian retina", *Physiol. Rev.* 71(2):447-480.

Wässle, H., B.B. Boycott and R-B. Illing (1981a) "Morphology and mosaic of on- and off-beta cells in the cat retina and some functional considerations", *Proc. Roy. Soc. Lond. B* 212:177-195.

Wässle, H., L. Peichl and B.B. Boycott (1981b) "Morphology and topography of on- and off-alpha cells in the cat retina", *Proc. Roy. Soc. London B* 212:157-175.

Wässle, H. and H.J. Riemann (1978) "The mosaic of nerve cells in the mammalian retina", *Proc. Roy. Soc. London B* 200:441-461.

Wässle, H., M. Yamashita, U. Greferath, U. Grünert and F. Müller (1991) "The rod bipolar cell of the mammalian retina", *Vis. Neurosci.* 7:99-112.

Williams, R.W. and L.M. Chalupa (1982) "Prenatal development of retinocollicular projections in the cat: an anterograde tracer transport study", *J. Neurosci.* 2:604-622.

Wong, R.O.L., K. Herrmann and C.J. Shatz (1991) "Remodeling of retinal ganglion cell dendrites in the absence of action potential activity", *J. Neurobiol.* 22:685-697.

Wong, R.O.L. and A. Hughes (1987) "Role of cell death in the topogenesis of neuronal distributions in the developing cat retinal ganglion cell layer", *J. Comp. Neurol.* 262:496-511.

Young, H.M. and D.I. Vaney (1991) "Rod-signal interneurons in the rabbit retina: I. Rod bipolar cells", *J. Comp. Neurol.* 310:139-153.

DEVELOPMENTAL SPECIFICITY OF RETINAL PROJECTIONS IN THE PRENATAL MONKEY

LEO M. CHALUPA

Center for Neuroscience, Section of Neurobiology, Physiology and Behavior, and Department of Psychology, University of California, Davis, CA 95616, U.S.A.

ABSTRACT

Retinogeniculate projections in the mature monkey are characterized by several key features that distinguish the primate visual system from that of non-primates. These include: (i) A highly precise nasotemporal decussation pattern, so that all ganglion cells in the nasal retina project to the contralateral hemisphere, while those in the temporal retina project ipsilaterally; (ii) The segregation of magnocellular (M) and parvocellular (P) functional channels to different layers of the dlgn; and (iii) The presence of 6 layers in the dorsal lateral geniculate nucleus (dlgn), each layer receiving input from either the left or the right eye. We have used timed-pregnant animals of known gestational ages to assess the sequence of events occurring from the time that the axons of ganglion cells first enter the optic stalk until the period when the mature pattern of retinal projections is established. These studies have revealed a remarkable degree of developmental specificity during the formation of M and P pathways, with respect to the establishment of eye-specific projection patterns, and also that of retinogeniculate fibers innervating the dlgn. Collectively, the results of these studies suggest that early connections in the primate visual system may differ from those that have been previously described in studies dealing with the fetal cat.

1. Introduction

In order to further our understanding of how the brain gets wired many different systems have been studied. In this daunting endeavor the connections of the eyes to retinorecipient structures in the midbrain and thalamus have long been considered favorite models. Consequently, we know a great deal about what occurs from the time that ganglion cells first innervate their target structures until the highly precise projection patterns found at maturity are established. A tacit assumption in this field has been the notion that the developmental events responsible for the formation of retinal connections are basically the same in different mammalian species. What occurs during the formation of the visual system in the rat is thought to apply, albeit at a different time scale, to the cat, monkey, and by extension to the human. At the same time, no one doubts that at maturity the salient features of the visual system are strikingly different among species. This is certainly the case when one compares the organization of retinal projections in the animals most commonly studied by developmental neurobiologists, such as the rat, cat, ferret, and monkey.

Here, I consider the results of experiments dealing with the development of the retinogeniculate projections in the monkey. The results of these studies, when compared to related work on these pathways in other species, suggest that different developmental strategies have evolved in animals occupying diverse phylogenetic and ecological niches for establishing their unique patterns of connections.

1.1. Development of Retinogeniculate Projections in the Fetal Monkey

Retinogeniculate projections in the primate are characterized by several distinguishing features. In the primate, ganglion cells are separated into nasal and temporal hemi-retinas on the basis of their decussation patterns, so that all cells in the temporal retina project ipsilaterally, while those in the nasal retina project to the contralateral side of the brain. In all other species some ganglion cells, distributed across the entire retina, project to the contralateral hemisphere, so that cells with crossed and uncrossed projections are intermingled within the temporal retina. Upon reaching the thalamus, primate ganglion cells project to the dorsal lateral geniculate nucleus in an eye-specific manner, so that the contralateral eye innervates layers 1, 4 and 6, while the ipsilateral eye projects to layers 2, 3 and 5. Eye-specific projection patterns are found in other species with highly developed binocular vision, such as the cat and ferret, but not in rodents where the projections of the two eyes innervate the same segment of the lateral geniculate nucleus. What distinguishes the primate from other species is that the major classes of ganglion cells also project in a laminar specific pattern, with the large P-alpha cells innervating magnocellular layers, 1 and 2, and the smaller P-beta neurons projecting to parvocellular layers, 3 through 6. Thus, the primate retinogeniculate pathway is characterized by laminar specificity defined in terms of ocular domains as well as cell-specific patterns. In some species, the inputs of ON and OFF subclasses of ganglion cells are further segregated within sublaminae of the dorsal lateral geniculate nucleus (Stryker and Zahs, 1983), but this is not the case in the primate where the projections of both subclasses are intermingled within a single geniculate layer.

Studies on fetal monkeys have provided information about the formation of retinal decussation patterns, eye-specific laminar projections, as well as magno- and parvo- pathways. Below we discuss the development of these fundamental features of the primate visual system and compare our findings to what has been learned from related studies on non-primate species.

1.2. Retinal Decussation

The organization of the fetal monkey's retinal decussation pattern was studied in using two different methods. First, we examined the distributions of labeled cells in both retinas in animals of known gestational ages that received large injections of HRP into the retinorecipient structures of one hemisphere (Chalupa and Lia, 1991). To gain further insight into this problem, we next assessed the organization of pioneer retinal axons in the embryonic rhesus monkey (Meissirel and Chalupa, 1994). The initial contingents of crossed and uncrossed optic axons were labeled by two different carbocyanine dyes that permitted their differentiation with confocal microscopy.

1.3. Formation of Retinogeniculate M and P Pathways

Tritiated thymidine studies have revealed a curious mismatch between the birth order of P-alpha and P-beta retinal ganglion cells and that of their target neurons in the M and P layers of the dorsal lateral geniculate nucleus. Within a given region of the retina, P-beta cells are born earlier than P-alpha cells (Rapaport *et al.*, 1992), but in the geniculate anlage neurons destined for the M segments are generated prior to those that will form the P laminae (Rakic, 1977).
At maturity, retinogeniculate arbors stemming from the P-alpha and P-beta ganglion cells can be differentiated on the basis of their distinctive morphological features (Conley and Fitzpatrick, 1989). Remarkably, such morphological differences become evident in the primate embryo as soon as terminal arbors became elaborated (by E95).

The highly specific ingrowth pattern characterizing the formation of parvocellular and magnocellular retinogeniculate pathways was unexpected because left and right eye inputs to the different laminae of this structure were shown to be initially completely intermingled (Rakic, 1976). Indeed, the retinogeniculate pathway of the fetal primate has long been considered the classic example of exuberant projections and subsequent refinement in the developing brain. Our findings reveal, however, that when functional components of this pathway are considered, namely parvocellular and magnocellular inputs, a different picture emerges. Refinements of early projections may not be required in this case because the parvo and magno subsystems seem to follow laminar-specific cues which appear to guide the axons of P-alpha and P-beta cells to the appropriate segments of the primate geniculate.

1.4. Formation of Eye-specific Projections

More than twenty years ago, Rakic (1976) discovered, by means of intraocular injections of tritiated amino acids, that the projections of the two eyes innervate the entire dorsal lateral geniculate before segregating into eye-specific laminae. The separation of intially intermingled binocular projections occurs later in development (from about E85 until E120) than the formation of parvo and magno pathways, discussed above. Analogous experiments on numerous species have shown that such binocular overlap of retinal projections is a common feature of mammalian development, although the degree of binocular overlap exhibited by different species can vary substantially (cf, Chalupa and Dreher, 1991). There is also evidence that the segregation process reflects binocular interactions. Removal of one eye at the time that projections overlap results in the maintenance of the widespread pattern from the remaining eye (Rakic, 1976; Chalupa and Williams, 1984). Such binocular interactions are thought to reflect activity-mediated events since blockade of spontaneously generated retinal activity induces marked changes in the geniculate territory innervated by the two eyes (Penn *et al.*, 1998).

A morphological analysis of single retinal fibers in the fetal cat revealed that during the prenatal binocular overlap period there are numerous axonal side-branches (Sretavan and Shatz, 1983). These often span across territories destined to become eye-specific during the course of normal development. Furthermore, the loss of such axonal processes corresponds to the time when binocular segregation occurs. Interestingly, the terminal arbors did not show any transient increase in size during the binocular overlap period. These findings provide a clear-cut account of the cellular basis of binocular overlap and segregation: the overlap reflects the presence of axonal side-branches, while the segregation is due to the loss of such processes.

In the more than two decades since Pasko Rakic showed binocular overlap of retinogeniculate projections in the fetal monkey, it seems remarkable that no one has studied retinogeniculate fibers in the prenatal primate. To some degree, this lacuna reflects the common assumption that what was found in the cat also applies to the monkey.

Recently, we undertook a study of the morphological characteristics of single retinogeniculate axons in fetal monkeys (Snider *et al.*, 1999). Such an investigation seemed a logical extension of the work described above. We were also motivated by increasing evidence of species differences in the developmental specificity exhibited by other components of the mammalian visual systems (Chalupa and Dreher, 1991). In particular, we were interested in determining whether the axonal side-branches, found to be prevalent in the cat during the binocular overlap period, would also be present in the fetal monkey.

2. Materials and Methods

2.1. Retinal Decussation

Detailed descriptions of the methodologies used for these experiments is given in Chalupa and Lia, (1991) and Meissirel and Chalupa (1994). Briefly, timed-pregnant animals were pre-anesthetized with ketamine (10mg/kg) and maintained at a surgical plane of anesthesia with 1-2% halothane in a 30% oxygen-70% nitrous oxide mixture. For one set of the experiments four fetal animals were studied at embryonic (E) days E69, E85, E115 and E129. Following uterotomy, the fetal head was exposed and injections of 3-12 microliters (depending upon age of the fetus) of 50% HRP were made through the cranial bone with a Hamilton syringe. The head was then sutured, returned to the uterus and all incisions closed. After appropriate survival times, the animals were delivered by cesarean section, given an overdose of barbiturate and perfused transcardially with 0.9% phosphate buffered saline followed by a 1% paraformaldehyde/2% gluteraldehyde mixture. In other cases, (E48, E53, E64, E74, E95, E115, E135) animals were removed by cesarean section (without having injections) given an overdose, and perfused with

4% paraformaldehyde or immersion fixed in the same solution. HRP injected brains were sectioned on a freezing microtome at 50 microns and reacted for peroxidase activity using diaminobenzidine (Adams, 1981) as the chromagen. Retinae were dissected from the eye and reacted using pyrocatecol method (Hanker *et al.*, 1977). The distribution of labeled retinal ganglion cells were plotted using DIC optics on a Zeiss microscope.

After removal of the cornea and lens, uninjected animals were used for placement of two lipophilic carbocyanine dyes DiI (1,1'diotadecyl-3,3,3',3'-tetramethylinocarbocyanine perchlorate) and DiA 4-(4-dihexadecylaminostryryl-N-methylpyridinium iodide), one into each eye. Heads were stored in fixative for 5-11 weeks, embedded in 4% agar, sectioned horizontally at 100-200 microns on a Vibratome and examined on a fluorescent microscope using appropriate filters.

2.2. Formation of Retinogeniculate M and P Pathways

A complete description of these methods can be found in Meissirel *et al.*, (1997). Fetuses at age E30, E36, E40, and E42 were obtained from timed-pregnant animals by cesarean section, as described in the previous section. Fetuses were overdosed with barbiturate and either perfused transcardially with saline followed by 4% paraformaldehyde or immersion fixed in the same solution. Blocks of tissue including the optic nerve, chiasm, tract, and dorsal lateral geniculate nucleus were dissected from the head. DiI crystals were implanted into the optic tract or nerve (depending on age), allowed transport time, embedded in agar, sectioned coronally at 100-200 microns, counterstained with bisbenzimide, and examined on a Bio Rad confocal microscope.

2.3. Formation of Eye-Specific Projections

For a detailed description see Snider *et al.*, (1999). In short, monkeys with controlled dates of impregnation (E77, E85, E95, E112) provided fetuses of known gestational ages. Pregnant monkeys were prepared for surgery under Alfatesine anesthesia. After intubation, anesthesia was continued with halothane in nitrous oxide/nitrogen (70:30) mixture. The fetuses were delivered by cesarean section, deeply anesthetized and perfused transcardially with 0.9% saline followed by 4% paraformaldehyde fixative. Additionally, two fetal cats at E50 were used to obtain comparative information. The in utero surgical procedures for harvesting fetal cat tissue have been described in detail in previous publications (Williams *et al.*, 1983; Chalupa *et al.*, 1984).

A small block of tissue including the optic tract and dlgn was isolated and embedded in 5% agar. The block was then sectioned in the coronal plane (horizontal plane for cat) on a vibratome until the optic tract close to the dlgn was visible, and two small crystals of DiI were implanted into the tract. Subsequently, the block was submerged in 4% paraformaldehyde and stored to allow for passive diffusion of the DiI to the dlgn. After 1-4 months, the dlgn was sectioned at 200 microns coronally for the monkey and horizontally for the cat, mounted on gelatinized slides, stained with bizbenzimide, to visualize the outline of the dlgn, and coverslipped.

Labeled retinogeniculate axons were examined using a Bio Rad MRC-600 confocal microscope system. Optical sections were collected in sequence as a function of tissue depth (150-200 microns) to generate a z-series. These images were then compiled, and a z-series projected to obtain a view that was in focus throughout the entire labeled area. Photographic montages of retinogeniculate axons were constructed with several z-series projections (Adobe Photoshop) and printed using a Fujix printer.

In the monkey, measurements of terminal arbors, branch points within terminal arbors, and number of side-branches along the parent axon were calculated from confocal montages of retinogeniculate axons from various loci within the dlgn. Using Imagespace software, total terminal arbor lengths were calculated by measuring and adding all axon segments belonging to the terminal arborization. In the cat, side-branches were counted in 40 retinogeniculate axons (E47-postnatal day 2) using the data provided by Sretavan and Shatz (1986, their figures 6 & 7), and four axons from an E50 animal processed in our laboratory. Using NIH Image software, the number of side branches/mm of parent axon was calculated for both species.

3. Results

3.1. Retinal Decussation

The results for this portion of the study are based on 4 fetal monkeys in which injections of HRP were placed successfully in the optic tract of one hemisphere. Such injections resulted in robust labeling of ganglion cells in the entire contralateral nasal hemiretina and in the entire ipsilateral temporal hemiretina (see Figures 2 and 3 from Chalupa and Lia, 1991). This revealed that throughout development virtually all retinal ganglion cells in the macaque monkey make a correct chiasmatic decision. Thus, even as early as E69, about 100 days before birth, less than 0.5% of all retinal ganglion cells innervate the inappropriate hemisphere.

To characterize the sequence of uncrossed and crossed retinal axon ingrowth during early development, we examined the organization of retinal projections within the optic chiasm and tract. The uncrossed axons enter the tract first. When the crossed axons begin to enter the tract they remain largely segregated from the uncrossed contingent of fibers. (Meissirel and Chalupa, 1994). This shows that crossed and uncrossed retinal projections, which initially form the primate optic tract, follow distinct temporal and spatial ingrowth patterns. Such an orderly sequential ingrowth of pioneer retinal axons would be expected if chiasmatic cues were expressed very early in development.

3.2. Formation of Retinogeniculate M and P Pathways

In the macaque embryo, the first retinal fibers (stemming from the contralateral eye) reach the geniculate anlage by E48 (Meissirel *et al.* 1997). By this age all geniculate neurons have been generated and have completed their migration, but despite this availability of target cells, the initial contingent of axons bypasses the dorsal thalamus to innervate the midbrain. The innervation of the geniculate begins several days later, when retinal fibers sprout short branches that terminate selectively within the medial segment of the nucleus. Crossed fibers innervate the medial segment first and the uncrossed fibers follow this specific ingrowth pattern several days later. During this time period the geniculate undergoes a progressive rotation, and it is only when this process is largely completed (at E74) that the ventral region of the nucleus, (formerly the lateral segment) begins to receive retinal inputs. At this stage, lamination of the geniculate has not occurred, so it is not possible to differentiate between parvo and magno layers. However, based on the outside-to-inside pattern of geniculate cell generation (Rakic, 1977), it can be inferred that the early innervated segment corresponds to what will become the parvo layers, while the later innervated lateral portion of the nucleus will differentiate into the magno laminae. Thus, the temporal sequence for "hooking-up" the retinogeniculate pathway in the primate embryo follows the order of ganglion cell generation (*i.e.*, P-beta before P-alpha rather than the temporal sequence of target cell generation, in which magno cells are generated before parvo neurons. This sequence of developmental event is illustrated in Meissirel *et al.*, (1997) figure 5.

3.3. Formation of Eye-specific Projections

We have analyzed more than 90 retinogeniculate fibers obtained from fetal animals spanning in age from E77 through E112 (Snider *et al.*, 1999). The youngest age is near the peak of the binocular overlap period and the oldest is when segregation is already well underway (Rakic, 1976). For examples of DiI labeled retinogeniculate fibers see figures 1,2 and 6 of Snider *et al.*, (1999). Several key features are clearly evident. First, axon terminals are yet to form in the youngest animal, while at older ages they become increasingly more complex and elaborate. Importantly, there is no indication that terminal arbors are greater in size at the time when binocular overlap is near its peak. Detailed measurements of a number of salient parameters, including the size and complexity of terminal arbors at 4 different ages, provided quantitative support for the impression obtained from the raw data (not shown). This was not unexpected since retinogeniculate fibers of the fetal cat also showed no sign of retrenchment during the course of development. What was distinct in the monkey embryos was the paucity of axonal side-branches. Throughout the binocular overlap period, retinal fibers in the magno and parvo segments of the geniculate were characterized by very few such processes. Moreover, the low incidence of axonal side-branches remained relatively constant throughout the development period of binocular segregation.

To convince ourselves that this lack of transient axonal branches reflected a genuine species difference rather than some spurious methodological factor, we made similar deposits of DiI into the fixed optic tract of the fetal cat at E50, near the peak of the binocular overlap period. This revealed numerous axonal side-branches, in agreement with the report of Sretavan and Shatz (1984), who used HRP deposits in an in vitro preparation to label fetal cat optic fibers. This indicates that there are genuine differences between the cat and monkey in the cellular factors responsible for the binocular overlap and subsequent segregation of retinogeniculate projections: resorption of axonal side-branches plays a role in this process in the fetal cat, but not in the monkey embryo.

4. Discussion

4.1. Retinal Decussation

We have shown that there is a remarkable degree of precision in the retinal decussation pattern of the fetal rhesus monkey. It is likely that ganglion cells generated in the temporal retina are characterized by different molecular markers than cells in the nasal retina. Such position-derived cues could guide the distinct behaviors of ganglion cell axons when they arrive at the optic chiasm.

In the embryonic mouse it has been suggested that a combination of outgrowth promoting and inhibiting molecules (termed L1/CD44 array) is expressed by neurons in the developing optic chiasm and that this acts as a template for guiding the initial decussation pattern (Sretavan *et al.*, 1994). It would seem reasonable to think that similar molecular cues are expressed by chiasmatic neurons in primate embryos, but this remains to be established. Moreover, the temporal and spatial segregation of crossed and uncrossed fibers indicates that the ingrowth of axons into the optic tract is not dependent upon interactions between fibers from the ipsilateral and contralateral eyes. Such an interaction has been inferred from studies on the developing mouse in which one eye was removed before the optic axons arrived at the chiasm (Godement *et al.*, 1990), but this idea has not been supported by time-lapse video analysis of navigational patterns of crossed and uncrossed optic fibers (Sretavan and Reichardt, 1993).

4.2. Formation of Retinogeniculate M and P Pathways

Early in development, there is precise specificity of the M and P retinogeniculate system. Our findings with regards to the formation of these pathways imply that the two main classes of primate retinal ganglion cells must express different molecular markers that permit their axons to react differentially to putative laminar-specific cues. In the adult macaque retina, antibodies generated against two different gene products (termed Brn-3a and Brn-3b) have been shown to differentially label P-alpha and P-beta ganglion cell populations (Xiang *et al.*, 1995). Application of these antibodies to the embryonic retina revealed that Brn-3a and Brn-3b positive cells could be visualized at very early stages of development.

This was observed shortly after ganglion cells had undergone their final division, and even before they had migrated from the ventricular layer to the ganglion cell layer. These findings provide evidence for the early divergence of P-alpha and P-beta ganglion cells. Furthermore, they reveal an essential link between distinct cell classes and the high specificity exhibited by these neurons when their fibers innervate the parvo and magno segments of the geniculate.

Possibly, M and P streams in the primate retinogeniculate pathway are established on the basis of the expression of molecular cues, without the involvement of activity-mediated refinements. This inference is certainly consistent with the evidence summarized above. This interpretation is also in line with what is known about the functional development of mammalian retinal ganglion cells. Patch-clamp recordings from ganglion cells isolated from the fetal cat retina indicate that very early in development these neurons are incapable of generating actions potentials to depolarizing current injections (Skaliora *et al.*, 1993). To a large degree this reflects the low density of sodium channels at early stages of development (Skaliora *et al.*, 1993), but ontogenetic fluctuations in other conductances and channel properties associated with spike generation have also been documented by means of voltage-clamp recordings (Skaliora *et al.*, 1995; Wang *et al.*, 1997; Robinson and Wang, 1998). By extrapolation from these studies on the fetal cat, it seems unlikely that ganglion cells in the embryonic primate are capable of firing action potentials at the time that the parvo and magno pathways are being established. By contrast, several weeks before the segregation of overlapping binocular projections has begun all retinal ganglion cells can discharge action potentials to depolarizing current injections (Skaliora *et al.*, 1993).

4.3. Formation of Eye-specific Projections

At a cellular level the early intermingling of left and right eye inputs could be accounted for by two non-mutually exclusive mechanisms. One possibility is that individual retinal fibers could be more extensive during development than at maturity, with axons from the two eyes innervating overlapping territories. Such exuberance at the single fiber level could reflect larger terminal arbors as well as the presence of transient axonal side-branches. An alternative hypothesis is that terminal arbors of individual fibers are not exuberant, but some innervate inappropriate territories destined to be the exclusive domain of the other eye. Thus, fibers from the ipsilateral eye might innervate layer 1 and subsequently such inappropriate projections would be eliminated during the period of developmental cell death.

During the segregation of binocular projections, there is a massive loss of optic axons in both the fetal monkey (Rakic and Riley, 1983a) and in the fetal cat (Williams *et al.*, 1986). Such loss of axons, which reflects the normal death of ganglion cells, could account entirely for the formation of eye-specific projection patterns. This idea was originally put forth by Rakic (1986), and our findings are entirely in accord with his suggestion. Moreover, this would also explain the observation that prenatal removal of one eye results in an increase of optic fibers in the remaining eye (Rakic and Riley, 1983b) which is concomitant with an expansion of the retinogeniculate projections stemming from the remaining eye. Monocular enucleation in the fetal cat during the binocular overlap period also results in an increase in the ganglion cell population (Chalupa *et al.*, 1984) and a corresponding increase in the number of fibers (Williams *et al.*, 1983) in the remaining eye and optic nerve as well as an expanded retinogeniculate projection (Chalupa and Williams, 1984). Thus, it seems reasonable to think that loss of retinogeniculate axons is involved in forming eye-specific projection patterns in both the monkey and cat. The key feature distinguishing the monkey from the cat is the presence of transient axonal side-branches in carnivores.

5. Concluding Remarks

The available evidence clearly indicates that there are marked species differences in the developmental events leading to the formation of certain key attributes of retinal projection patterns. In one respect, this complicates our efforts to obtain an understanding of how neuronal connections are formed in the developing brain. On the other hand, the ontogenetic variations exhibited by different species provide an opportunity for developmental neurobiologists to assess the problem from a fresh perspective. By considering why one class of cells (for instance, the retinocollicular projection of the rat as compared to that of the cat) behaves differently than another could expand the scope of the enquiry to a new level of analysis.

To understand how retinal ganglion cells make their precise patterns of connections, it could prove insightful to consider why ganglion cells in different animals (and even different classes of cells in the same animal) hook-up with their target neurons by means of different strategies.

What's different and what's common during the development of mammalian retinofugal pathways? Until this issue is resolved the task of developmental neurobiologists will not be completed.

Acknowledgements

Supported by grants from the National Institute of Health, the National Science Foundation and the Human Frontiers of Science Program.

References

Adams, J.C. (1981) "Heavy metal intensification of DAB-based HRP reaction product", *J. Histochem. Cytochem.* 29:775.

Chalupa, L.M. and B. Dreher (1991) "High precision systems require high precision "blueprints": A new view regarding the formation of connections in the mammalian visual system", *J. Cog. Neurol.* 3:209-219.

Chalupa, L.M. and B.L. Lia (1991) "The nasotemporal division of retinal ganglion cells with crossed and uncrossed projections in the fetal rhesus monkey", *J. Neurosci.* 11(1):191-202.

Chalupa, L.M. and R.W. Williams (1984) "Prenatal development and reorganization in the visual system of the cat", *Development of Sensory Systems in Mammals*, pp. 3-60. New York.

Chalupa, L.M., R.W. Williams and Z. Henderson (1984) "Binocular interaction in the fetal cat regulated the size of the ganglion cell population", *Neurosci.* 12:1139-1146.

Conley, M. and D. Fitzpatrick (1989) "Morphology of retinogeniculate axons in the macaque", *Visual Neurosci.* 2:287-296.

Godement, P., J. Salaün and C. Mason (1990) "Retinal axon pathfinding in the optic chiasm: Divergence of crossed and uncrossed fibers", *Neuron* 5:173-186.

Godement, P., J. Salaün and C. Métin (1987) "Fate of uncrossed retinal projections following early or late prenatal monocular enucleation in the mouse", *J. Comp. Neurol.* 225:97-109.

Hanker, J.S., P.E. Yates, C.B. Metz and A. Rustioni (1977) "A new specific, sensitive and non-carcinogenic reagent for the demonstration of horseradish peroxidase", *Histochem. J.* 9:789-792.

Meissirel, C. and L.M. Chalupa (1994) "Organization of pioneer retinal axons within the optic tract of the rhesus monkey", *Proc. Natl. Acad. Sci. USA* 91:3906-3910.

Meissirel, C., K.C. Wikler, L.M. Chalupa and P. Rakic (1997) "Early divergence of magnocellular and parvocellular functional subsystems in the embryonic primate visual system", *Proc. Natl. Acad. Sci. USA.* 94:5900-5905.

Penn, A.A., A.R. Patricio, M.B. Feller and C.J. Shatz (1998) "Competition in retinogeniculate patterning driven by spontaneous activity", *Science* 279:2108-2112.

Rakic, P. (1976) "Prenatal genesis of connections subserving ocular dominance in the rhesus monkey," *Nature* 261:467-471.

Rakic, P. (1977) "Prenatal development of the visual system in rhesus monkey", *Philos. Trans. R. Soc. Lond. (Biol.)* 278:245-260.

Rakic, P. (1986) "Mechanism of ocular dominance segregation in the lateral geniculate nucleus: competitive elimination hypothesis", *TINS* 9:11-15.

Rakic, P. and K.P. Riley (1983a) "Overproduction and elimination of retinal axons in the fetal rhesus monkey", *Science* 209:1441-1444.

Rakic, P. and K.P. Riley (1983b) "Regulation of axon numbers in the primate optic nerve by prenatal binocular competition", *Nature* 305:135-137.

Rapaport, D.H., J.T. Fletcher, M.M. LaVail and P. Rakic (1992) "Genesis of neurons in the retinal ganglion cell layer of the monkey", *J. Comp. Neurol.* 322(4):577-588.

Robinson, D.W. and G-Y. Wang (1998) "Development of intrinsic membrane properties in mammalian retinal ganglion cells", *Seminars in Cell and Dev. Bio.* 9:301-310.

Skaliora, I., D.W. Robinson, R.P. Scobey and L.M. Chalupa (1995) "Properties of K+ conductances in cat retinal ganglion cells during the period of activity – mediated refinements in retinofugal pathways", *Eur. J. Neurosci.* 7:1558-1568.

Skaliora, I., R.P., Scobey and L.M. Chalupa (1993) "Prenatal development of excitability in cat retinal ganglion cells: Action potentials and sodium currents," *J. Neurosci.* 13:313-323.

Snider, C.J., C. Dehay, M. Berland, H. Kennedy and L.M. Chalupa (1999) "Prenatal development of retinogenicuale axons in the macaque monkey during segregation of binocular inputs", *J. Neurosci.* 19(1):220-228.

Sretavan, D.W and C.J. Shatz (1984) "Prenatal development of individual retinogeniculate axons during the period of segregation", *Nature* 308:845-848.

Sretavan, D.W. and C.J. Shatz (1986) "Prenatal development of retinal ganglion cell axons: segregation into eye-specific layers within the cat's lateral geniculate nucleus", *J. Neurosci.* 6:234-251.

Sretavan, D.W., L. Feng, E. Pure and L.F. Reichardt (1994) "Embryonic neurons of the developing optic chiasm express L1 and CD44, cell surface molecules with opposing effects on retinal axon growth", *Neuron* 12:957-975.

Sretevan, D.W. and L.F. Reichardt (1993) "Time-lapse video analysis of retinal ganglion cell axon pathfinding at the mammalian optic chiasm: Growth cone guidance using intrinsic chiasm cues", *Neuron* 10:761-777.

Stryker, M.P. and K.R. Zahs (1983) "On and off sublaminae in the lateral geniculate nucleus of the ferret", *J. Neurosci.* 3(10):1943-1951.

Wang, G-Y., G-M. Ratto, S. Bisti and L.M. Chalupa (1997) "Functional development of intrinsic properties in ganglion cells of the mammalian retina", *J. Neurophysiol.* 78:2895-2903.

Williams, R.W., M.J. Bastiani and L.M. Chalupa (1983) "Loss of axons in the cat optic nerve following fetal unilateral enucleation: An electron microscope analysis", *J. Neurosci.* 3:133-144.

Williams, R.W., M.J. Bastiani, B. Lia and L.M. Chalupa (1986) "Growth cones, dying axons, and developmental fluctuations in the fiber population of the cat's optic nerve", *J. Comp. Neurol.* 246:32-69.

Xiang, M., L. Zhou, J. Macke, T. Yoshioka, S.H.C. Hendry, R. Eddy, T.B. Shows and J. Nathans (1995) "Genesis of neurons in the retinal ganglion cell layer of the monkey", *J. Neurosci.* 15:4762-4785.

HYPERPOLARIZING VS DEPOLARIZING PHOTORECEPTORS: IMPLICATIONS FOR THE LENGTH OF THE LIGHT SENSITIVE REGION AND FOR THE CONDUCTANCE OF THE PHOTOSENSITIVE CHANNELS

JEAN-PIERRE RAYNAULD

Centre de recherches en Sciences Neurologiques, Département de physiologie, Université de Montréal, BP 6128 Succ. Centre-Ville, Montréal, Qc, Canada H3C 3J7

ABSTRACT

Hyperpolarizing photoreceptors found in most vertebrate retinae have a rather short light sensitive region when compared to the depolarizing photoreceptors of cephalopods and gastropods. Another major difference between these two types of photoreceptors is the conductance of the photo-activated channels. When the conductance in hyperpolarizing photoreceptors is one of the smallest found (~ 0.1 pS), the conductance of depolarizing photoreceptors is one of the largest (~ 40 pS). The compartment model offers an explanation for the above differences.

1. Introduction

Photoreceptors can be classified in two groups when one considers the direction of voltage change produced by isomerizations. In arthropods and cephalopods light capture is signalled by a depolarization while in vertebrates an hyperpolarization is the usual response. A recent exception is the parietal eye of the lizard where depolarization seems to occur (Finn *et al.*, 1998). Clear anatomical differences, which have been known for many years also, exist between depolarizing vs hyperpolarizing photoreceptors. The light capturing region or outer segment is in general much longer in depolarizing photoreceptors when compared to the outer segment of hyperpolarizing rods and cones (Fein and Szuts, 1982). Furthermore, in vertebrates, rod outer segments are in general longer than cone outer segments (Walls, 1942). Another difference has appeared over the recent past and relates to the conductance of the photo-activated channels. In vertebrates, the conductance is extremely small, barely measurable (Detwiler *et al.*, 1986; Gray and Attwell, 1985) where in depolarizing photoreceptors, it is exactly the opposite (Bacigalupo *et al.*, 1986), the conductance of the channel is very large some 400 X larger than in hyperpolarizing photoreceptors.

Over the past ten years, I have further investigated the total occlusion model (Lamb *et al.*, 1981) for vertebrates photoreceptors. This analysis has yielded interesting dividends with regards to the prediction of the sensitivity and Weber-Fechner adaptation of this class of receptors (Raynauld, 1996, 1997). In a nutshell the «total occlusion» or «compartment» model proposes that the vertebrate photoreceptor outer segment is constructed of a number of compartments which are biochemically isolated from each other but well connected electrically such that photocurrents generated in each compartment sum algebraically. Furthermore, a single isomerization in a compartment results in the closure of all the sodium channels present in the compartment. If the response is fast rising and decays exponentially, then the system will show Weber-Fechner adaptation characteristics. Here, I would like to further this analysis and offer a reasonable explanation for the difference in length of the light capturing region of these two class of photoreceptors, and the size of the conductance of the photo-activated channels.

2. Hyperpolarizing Photoreceptors

Pipette recordings of voltage and suction electrode recordings of current have revealed that, across receptor types and across species, the resting potential in the dark is circa –40 mV and the associated current is circa 40 pA. Vertebrate photoreceptors are thus depolarized with respect to the normal resting potential of nerve cells which lies around –70 mV. An hypothesis can be made that -40 mV is the best operating point (resting potential) in the dark for the cone pedicules and rod spherules synaptic complexes. By best operating point, I mean that the transfer function, that is the change in transmitter release per mV change in membrane potential at the post synaptic site is maximal. The transfer functions of transistors and vacuum tubes have such best operating points and when one designs a circuit for maximum gain one tries to operate at this point.

In terms of total conductance, this value is in the range of 1000 picosiemens (pS) (40 pA/40 mV). This represents the amount of leakage in the outer segment required to lower the resting potential in the dark from –70 mV to –40 mV. In the vertebrate cone, the whole photo-activated biochemistry are located in the lamellae which, when looked at in 3D (Eckmiller, 1987), represents nearly isolated reaction vessel when one considers the diffusion coefficient of the molecules involved and the duration of the process.

In order to maximize the optical density of the cone outer segment one could increase the number of lamellae, but in order to maintain the same operating point of – 40 mV the number of open channels per lamellae would have to decrease. The end point of this strategy would be only one open channel per lamella. If the conductance of the photoactivated channel would be a typical 20 pS, then to obtain the required total leakage of 1000 pS only 50 channels would be required , thus limiting the number of lamellae to 50 and the probability of photon capture to 6%, a low value indeed. This number is obtained from the linear density of lamellae which is 33 per micron and the optical density of 0.016 O.D. per micron (Harosi, 1976).

It become obvious from the above analysis that reducing the channel conductance permits to increase the number of lamellae and thus the optical density while maintaining the total conductance circa 1000 pS. This is exactly what is observed. Vertebrate photoactivated channels have a conductance which is barely measurable and of the order of 0.1 pS. In a typical outer segment, which is 25 micron long, thus containing 825 lamellae, the number of open channel per lamella is 12.

Vertebrate rods can be longer than cones because the light sensitive channels are on the plasma membrane as illustrated in Fig. 1, one can therefore increase the number of disks while maintaining a constant leakage at the expense of channel surface density.

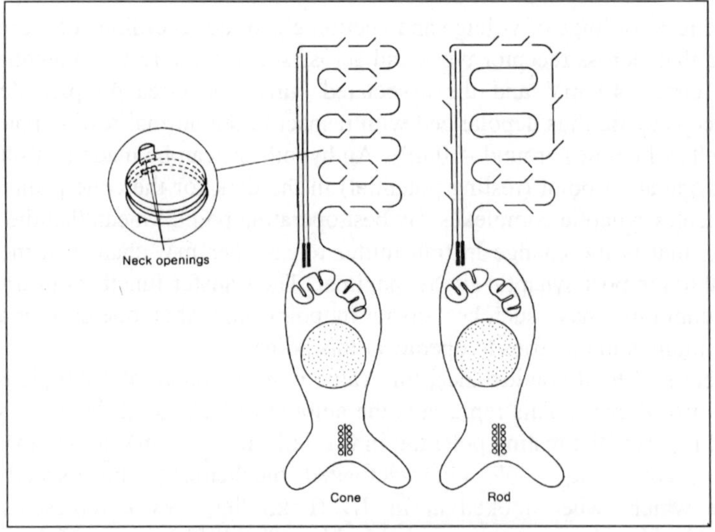

Figure 1. Cartoon showing side by side a vertebrate cone and a rod. It is postulated that the ribbon synaptic complex has a best operating point of –40 mV thus requiring that the same number of open channels be the same in both receptors. Note that for the cone, the channels are located on the lamellae, when for the rod they are on the plasma membrane.

Another reason for a low channel conductance in hyperpolarizing photoreceptors is that the channel fluctuates from the open to the closed state even in total darkness. This process generates noise and as pointed out (De Felice, 1981, Yau and Baylor, 1989) the RMS value of the noise is inversely proportional to the square root of the number of channels involved. Small conductance therefore contributes to lowering the noise of the system.

3. Depolarizing Photoreceptors

In the above, we have seen that an hyperpolarizing photoreceptor is limited in the length of its outer segment and that a very low conductance cGMP gated channel allows to maximize the length of the outer segment and the photon catching ability. Such a limitation in length does not exist in depolarizing photoreceptors such as found in cephalopods and in gastropods. In these photoreceptors the dark resting potential is -70 mV, the normal resting potential of nerve cells, the channels, which will be opened by light, are closed. One can therefore increase the number of compartments containing theses channels without changing the resting potential and this way achieve a much greater optical density. Indeed, it is not unusual for a depolarizing photoreceptor to have a light catching region of 250-300 microns in length (Fein and Szuts,1982), some 10 times the length of an hyperpolarizing photoreceptors.

Contrary to hyperpolarizing photoreceptors, it is advantageous for a depolarizing photoreceptors to have photoactivated channels of large conductance, it minimizes the number of channels that have to be opened in order to produce the required depolarization. In *Limulus*, the photochannel conductance is 40 pS (Bacigalupo *et al.*, 1986), 400 X larger that the channels in vertebrate photoreceptors (Detwiler *et al.*, 1982; Gray and Attwell, 1985). Large conductance lowers the gain requirement in term of number of G proteins excited per rhodopsin molecule and could offer an explanation why only eight G proteins are excited per rhodopsin molecule in the *Limulus* eye (Kirkwood *et al.*, 1989).

4. Conclusion

In hyperpolarizing receptors, a low channel conductance allows to maximize the length of the outer segment up to a limit. In depolarizing receptors, a large channel conductance minimizes the size of the compartment which contains the signalling complex (Montel, 1998) and the number of channels that the biochemistry has to operate on. Small size also facilitates fast transduction as illustrated by Laughlin and Weckstrom (1993) where even the so called slow receptors are as fast as primate cones.

Acknowledgements

The author would like to express his gratitude to Simon Laughlin for critical reading and suggestions.

References

Bacigalupo, J., K. Chinn, and J. Lisman (1986) "Ion channels activated by light in *Limulus* ventral photoreceptors", *J. Gen. Physiol.* 87:73-89.

De Felice, L.J. (1981) *Introduction to Membrane Noise*, New York, London: Plenum.

Detwiler, P.B., J. D. Conner, and R.D. Boboia (1982) "Gigaseal patch clamp recordings from outer segments of intact retinal rods", *Nature* 300:59-61.

Eckmiller, M (1987) "Cone outer segment morphogenesis: Taper change and distal invagination", *J. Cell Biol*. 105:2267-2277.

Fein, A. and E.Z. Szuts (1982) "Photoreceptors: their role in vision", Cambridge: Cambridge University Press.

Finn, J.T, W.H. Xiong, E.C. Solessio and K.W. Yau (1998) "A cGMP-gated cation channel and phototransduction in depolarizing photoreceptors of the lizard parietal eye", *Vision Res.* 38:1353-1357.

Gray, P. and D. Attwell (1985) "Kinetics of light-sensitive channels in vertebrate photoreceptors", *Proc. R. Soc. Lond. B Biol. Sci.* 223:379-388.

Harosi, F. (1975) "Absorption spectra and linear dichroism of some amphibian photoreceptors", *J. Gen. Physiol*. 66:3357-382.

Kirkwood, A., D. Weiner and J.E. Lisman (1989) "An estimate of the number of G regulatory proteins activated per excited rhodopsin in living *Limulus* ventral photoreceptors.", *Proc. Natl. Acad. Sci. USA* 86:3872-3876.

Laughlin, S.B. and M. Weckstrom (1993) "Fast and slow photoreceptors- A comparative study of the functional diversity of coding and conductances in Diptera.", *J. Comp. Physiol. A.* 172: 593-609.

Montell, C. (1998) "TRP trapped in fly signalling web", *Curr. Opin. Neurobiol.* 8:389-397.

Raynauld, J-P. (1996) "A compartment model for vertebrate phototransduction predicts sensitivity and adaptation" in: *Neurobiology: Ionic Channels, Neurons, and the Brain,* V. Torre and F. Conti, eds., New York, London: Plenum, pp. 201-215.

Raynauld, J-P. (1997) "The adaptation properties of a compartment system", *Proceedings of the II Workshop on Cybernetic Vision, December 9-11, Sao Carlos, Brazil*, IEEE Computer Society, pp. 27-32.

Walls, G.L. (1942) *The Vertebrate Eye and its Adaptative Radiation*, Michigan: Cranbook Press.

Yau, K-W. and D.A. Baylor (1989) "Cyclic GMP-activated conductance of retinal photoreceptor cells", *Ann. Rev. Neurosci*. 12:289-327.

SUBCELLULAR LOCALIZATION OF InsP$_3$ RECEPTOR-LIKE IMMUNOREACTIVITY IN INVERTEBRATE MICROVILLAR PHOTORECEPTORS

KYRILL UKHANOV[1], RICHARD PAYNE[2] AND BERND WALZ[1]
[1]*Institute for Zoophysiology, University of Potsdam, Potsdam D 14471, Germany*
[2]*Department of Biology, University of Maryland, College Park, MD 20742, USA*

ABSTRACT

We have attempted to localize InsP$_3$ receptor-like immunoreactivity in photoreceptors of three invertebrate species, american horseshoe crab *Limulus polyphemus*, honeybee drone *Apis mellifera* and medicinal leech *Hirudo medicinalis*. Two polyclonal antibodies raised against a defined fragment of the N-terminus of the type I InsP$_3$ receptor and against that of the C-terminus of the putative receptor from *C. elegans* were used. On Western blots these antibodies recognized protein bands with apparent molecular weights of approximately 250kD (*Apis*), 210kD (*Hirudo*) and 300kD (*Limulus*). In the retinal cryosections the subrhabdomeral region of the cytoplasm was stained most intensely using indirect immunofluorescence. This area is enriched in smooth endoplasmic reticulum known to act as a Ca store. No labeling was found in photoreceptive microvilli. Immunogold staining revealed precipitate in the cytoplasm of the photoreceptors in close proximity to the rhabdom indicating on the tight functional link between Ca stores and photoreceptive microvilli.

1. Introduction

Invertebrate microvillar photoreceptors utilize the phosphoinositide cascade to transduce light stimuli but the product of the cascade that triggers the electrical response remains elusive (O'Day *et al.*, 1997; Ranganathan *et al.*, 1995).

There is a little doubt now that the key enzyme in the cascade is phospholipase C (PLC) which breaks down phosphatidylinositol 4,5-bisphosphate (PIP$_2$) into inositol 1,4,5-trisphosphate (IP$_3$) and diacylglycerol (DAG). Remarkably that there is yet no solid evidence on how activation of PLC leads to opening of non-selective cation channels in the plasma membrane of invertebrate microvillar photoreceptors.

The demonstration that intracellular IP$_3$ injection both excites and adapts *Limulus* ventral nerve photoreceptors (Brown *et al.*, 1984; Fein *et al.*, 1984) has long suggested a role for IP$_3$ -induced Ca^{2+} release in invertebrate phototransduction. Many models of invertebrate phototransduction therefore require the existence of IP$_3$ -gated Ca^{2+} release channels in close proximity to the microvillar plasma membrane, where rhodopsin, GTP-binding protein, phospholipase C and cation channels reside (Bloomquist *et al.*, 1987; Devary *et al.*, 1988; Tsunoda *et al.*, 1997).

Recently the role of IP_3 has been challenged following the demonstration that mutant *Drosophila* photoreceptors lacking the IP_3 receptor protein (IP_3R) are fully functional (Acharaya *et al.*, 1997) and alternatively DAG signalling pathway was implicated in *Drosophila* phototransduction (Chyb *et al.*, 1999).

Therefore we have sought to examine and compare the subcellular localization of IP_3R in photoreceptors of three invertebrate species, *Limulus*, *Apis* and *Hirudo* where endoplasmic reticulum is known to form an extensive network of Ca stores beneath the photoreceptive microvilli (Baumann and Walz, 1989; Payne *et al.*, 1988; Walz, 1979).

2. Materials and Methods

2.1. SDS-PAGE and Western blotting

Retinal tissues were subjected to extraction, SDS-PAGE and Western blotting as described elsewhere (Yamamoto *et al.*, 1997; Ukhanov *et al.*, 1998).

Tissues were homogenized in a sample buffer on ice and debris were sedimented by a centrifugation at 13,000 rpm. Aliquots of the extracts were subjected to SDS-PAGE using 6% gel.

After electrotransfer of the proteins nitrocellulose filters were probed with two antibodies against fragments of IP_3R. A first antibody was raised against the N-terminus peptide of the type I and III IP_3R (anti-type I IP_3R Ab) (Cardy *et al.*, 1997) and a second one was raised against the C-terminus peptide of the putative IP_3R from the nematode *C.elegans* (anti-*C.elegans* IP_3R Ab) (Baylis *et al.*, 1999).

2.2. Immunofluoresence and immunogold labeling

Freshly excised tissues were processed according to a conventional immunohistochemical protocol. After embedding in Mowiol (Hoechst) or Vectashield (Vector Laboratories) sections were examined in a conventional epifluorescent microscope (Zeiss Axiophot) or in a laser confocal microscope (Zeiss LSM 510).

The preembedding method was used for immunogold labeling (Yamamoto *et al.*, 1997). Cryostat sections were first permeabilized with 0.1% Triton X-100 and then treated in the same manner as for the immunofluorescent labeling.

Anti-rabbit Fluoro-Nanogold (Nanoprobes) conjugate was used as a secondary antibody. Following primary inspection of the fluorescent signal, sections were postfixed with 2.5% glutaraldehyde and 0.2% tannic acid, silver enhanced with HQ Silver kit (Nanoprobes) and flat embedded in the BEEM capsules using modified Epon resin (Serva).

Finally, ultrathin sections were stained with uranyl acetate and examined in Philips CM100 electron microscope operating at 80kV.

3. Results

3.1. Western blot analysis

Both the antibodies specifically label protein bands of the same apparent molecular weight of ca.260kD on the filters prepared from the mouse cerebellum. This band corresponds to IP_3R of type I and III (Cardy *et al.*, 1997). Such a cross binding of antibodies was found only in cerebellum and not in any of the invertebrate species studied.

In every experiment an extract from the mouse cerebellum was used as an internal positive control and a molecular weight marker. In the blood and lateral eye of *Limulus* a protein band of ca.300kD was identified using anti-type I IP_3R Ab. Using anti-*C.elegans* IP_3R Ab a dominant protein band of ca.210kD was detected in the leech *Hirudo* while in the honeybee eye a protein band of ca.250kD was detected. Antibodies preadsorbed with the excessive amount of the antigen peptide or preimmune serum were used as a negative control. No labeling was found in the control experiments.

3.2. Immunocytochemistry

Indirect immunofluorescence labeling of the frozen cryostat section from the *Limulus* lateral eye, honeybee drone eye and a simple eye of the leech *Hirudo* revealed striking similarity in the staining pattern. The most intense staining was observed in the cytoplasm area situated close to the rhabdom formed by photoreceptive microvilli (Fig.1). This area is abundant in the smooth endoplasmic reticulum (ER). In leech photoreceptors, ER is distributed throughout the cytoplasm without forming distinct structures, like a palisade in *Limulus* or honeybee. Nevertheless, anti-IP_3R staining was localized mostly beneath the photoreceptive microvilli rather than spread homogenously. We have also attempted indirect immunogold labeling using the pre-embedding technique. Firstly, the photoreceptive microvilli were never labeled indicating absence of the putative

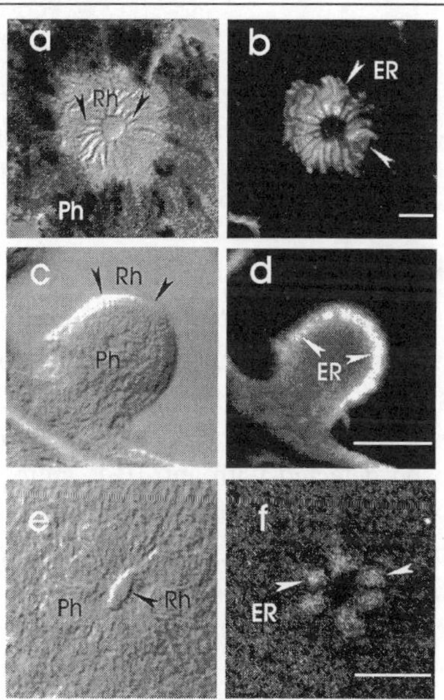

Figure 1. Immunofluorescent confocal images of anti- IP_3R labeled cryostat sections of the lateral eye of *Limulus* (a, b), the simple eye of *Hirudo* (c, d) and the compound eye of *Apis* (e, f). In each photoreceptor (Ph) microvilli form the photoreceptive rhabdom (Rh, black arrows). Endoplasmic reticulum (ER) area is most brightly stained (white arrows). Scale bar is 10 μm.

IP_3R in the photoreceptive membrane. Secondly, the immuno-gold precipitate was consis-tently localized to the subrhabdomeral cytoplasmic domain (Fig.2).

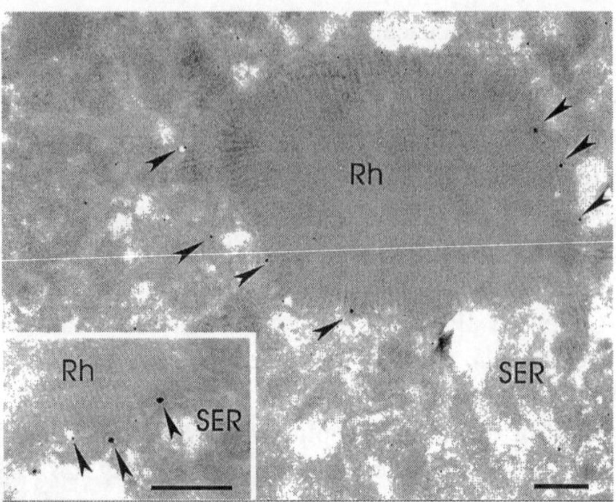

Figure 2. In the bee photoreceptors immunogold particles (arrows) are found beneath the rhabdom (Rh) in submicro-villar endoplasmic reticulum (SER) area. Scale bar is 0.5 μm.

4. Conclusions

Using two polyclonal antibodies against the defined fragments of the IP_3R we have studied its subcellular distri-bution in the visual cells of the three invertebrate species, *Limulus*, *Apis* and *Hirudo*. Previously we have shown that the putative IP_3R is localized to the ER in the *Limulus* lateral eye photoreceptors. The main protein band detected with anti-IP_3R antibody in the leech *Hirudo* refers to the molecular weight 210kD which is very similar to what has been found in the total preparation of the nematode *C.elegans* (Baylis *et al.*, 1999, personal communication). Recently a fragment of IP_3R has been cloned from the honeybee brain (Kamikouchi *et al.*, 1998). It shares some similarity with both the type I and a putative *C.elegans* IP_3Rs.

Therefore it is most likely that we have detected a putative IP_3R in the bee retina as well. The ER in invertebrate photoreceptors of most species forms an extensive network serving several functions. The most developed domain of the ER is a so called subrhabdomeral ER (SER) or palisade (Horridge and Barnard, 1965). It is situated in the close proximity to the photoreceptive microvilli and is known to actively uptake calcium (Walz, 1979; Payne *et al.*, 1988; Baumann and Walz, 1989).

Since IP_3 has been implicated in phototransduction in microvillar type of invertebrate photoreceptors it is very tempting to hypothesize that the highest density of the putative IP_3R should be found in the SER. Indeed, direct pressure injections of IP_3 in the *Limulus* ventral nerve photoreceptors produced largest Ca release and strongest excitation after being injected in the light-sensitive R-lobe (Fein *et al.*, 1984). This has been also proved recently by microphotolysis of caged IP_3 combined with simultaneous confocal Ca imaging (Ukhanov *et al.*, 1998). Our findings are in line with this data showing highest anti-IP_3R labeling in the SER. This may indicate a polarized targeting of the IP_3R to functionally determined membrane domains as has been reported recently (Fujimoto *et al.*, 1995; Wilson *et al.*, 1998).

Although our findings present additional evidence for a system that can release calcium from internal stores in microvillar photoreceptors, the role played by that calcium is far from being clear. Especially after recent reports on activation of ionic currents in photoreceptors of *Drosophila* by polyunsatured fatty acids (Chyb *et al.*, 1999) and in depolarizing photoreceptors of the mollusc *Lima* by protein kinase C activators (del Pilar Gomez and Nasi, 1998).

Acknowledgements

This project was supported in part by Alexander von Humboldt Foundation, DFG and NIH grants.

References

Acharya, J.K., K. Jalink, R.W. Hardy, V. Hartenstein and C.S. Zuker (1997) "InsP_3 receptor is essential for growth and differentiation but not for vision in *Drosophila*", *Neuron* 18:881-887.

Baumann, O. and B. Walz (1989) "Calcium- and inositol polyphosphate-sensitivity of the calcium-sequestering endoplasmic reticulum in the photoreceptor cells of the honeybee drone", *J. Comp. Physiol. A* 165:627-636.

Baylis, H.A., T. Furuichi, F. Yoshikawa, K. Mikoshiba and D.B. Sattelle (1999) "Inositol 1,4,5-trisphosphate receptors are strongly expressed in the nervous system, pharynx, intestine and excretory cell of *Caenorhabditis elegans* and are encoded by a single gene (itr-1)", *J. Mol. Biol.* 294(2):467-76.

Bloomquist, B.T., R.D. Shortridge, S. Schneuwly, M. Perdew, C. Montell, H. Steller, G. Rubin and W.L. Pak (1988) "Isolation of putative phospholipase C gene of *Drosophila*, norpA, and its role in phototransduction", *Cell* 54:723-733.

Brown, J.E., L.J. Rubin, A.J. Ghalayini, A.P. Tarver, R.F. Irvine, M.J. Berridge and R.E. Anderson (1984) "Myo-Inositol polyphosphate may be a messenger for visual excitation in *Limulus* photoreceptors", *Nature* 311:160-163.

Cardy, T.J., D. Traynor and C.W. Taylor (1997) "Differential regulation of types-1 and -3 inositol trisphosphate receptors by cytosolic Ca^{2+}", *Biochem. J.* 328:785-93.

Chyb, S., P. Raghu and R.C. Hardie (1999) "Polyunsaturated fatty acids activate the *Drosophila* light-sensitive channels TRP and TRPL", *Nature* 397:255-9

del Pilar Gomez, M. and E. Nasi (1998) "Membrane current induced by protein kinase C activators in rhabdomeric photoreceptors: implications for visual excitation", *J. Neurosci.* 18:5253-63.

Devary, O., O. Heichal, A. Blumenfeld, D. Cassel, E. Suss, S. Barash, C.T. Rubinstein, B. Minke and Z. Selinger (1987) "Coupling of photoexcited rhodopsin to inositol phospholipid hydrolysis in fly photoreceptors", *Proc. Natl. Acad. Sci. U.S.A.* 84:6939-6943.

Fein, A., R. Payne, D.W. Corson, M.J. Berridge and R.F. Irvine (1984) "Photoreceptor excitation and adaptation by inositol 1,4,5-trisphosphate", *Nature* 311:157-160.

Fujimoto, T., A. Miyawaki and K. Mikoshiba (1995) "Inositol 1,4,5-trisphosphate receptor-like protein in plasmalemmal caveolae is linked to actin filaments", *J. Cell Sci.* 108:7-15.

Horridge, G.A. and P.B. Barnard (1965) "Movement of palisade in locust retinula cells when illuminated", *Q. J. Microsc. Sci.* 106:131-135.

Kamikouchi, A., H. Takeuchi, M. Sawata, K. Ohashi, S. Natori and T. Kubo (1998) "Preferential expression of the gene for a putative inositol 1,4,5-trisphosphate receptor homologue in the mushroom bodies of the brain of the worker honeybee *Apis mellifera* L.", *Biochem. Biophys. Res. Commun.* 242:181-186.

Ranganathan, R., D.M. Malicki and C.S. Zuker (1995) "Signal transduction in *Drosophila* photoreceptors", *Ann. Rev. Neurosci.* 18:283-317.

Tsunoda, S., J. Sierralta, Y. Sun, R. Bodner, E. Suzuki, A. Becker, M. Socolich and C.S. Zuker (1997) "A multivalent PDZ-domain protein assembles signalling complexes in a G-protein-coupled cascade", *Nature* 388:243-249.

Walz, B. (1979) "Subcellular calcium localization and ATP-dependent Ca^{2+}-uptake by smooth endoplasmic reticulum in an invertebrate photoreceptor cell. An ultrastructural, cytochemical and X-ray microanalytical study", *Eur. J. Cell Biol.* 20:83-91.

Wilson, B.S., J.R. Pfeiffer, A.J. Smith, J.M. Oliver, J.A. Oberdorf and R.J.H. Wojcikiewicz (1998) "Calcium-dependent clustering of inositol 1,4,5-trisphosphate receptors", *Mol. Biol. Cell* 9:1465-1478.

Yamamoto Hino, M., A. Miyawaki, A. Segawa, E. Adachi, S. Yamashina, T. Fujimoto, T. Sugiyama, T. Furuichi, M. Hasegawa and K. Mikoshiba (1998) "Apical vesicles bearing inositol 1,4,5-trisphosphate receptors in the Ca^{2+} initiation site of ductal epithelium of submandibular gland", *J. Cell Biol.* 141:135-142.

LIGHT ADAPTIVE EFFECT OF NITRIC OXIDE ON CONE PLASTICITY IN FISH AND AMPHIBIAN RETINAE

ANNA RITA ANGOTZI[*°], JOE HIRANO[†], SILVANA VALLERGA[*†] and MUSTAFA DJAMGOZ[°]

[*]*International Marine Centre, Torregrande, 09072 Oristano, Italy*

[°]*Neurobiology Group, Department of Biology, Imperial College of Science, Technology and Medicine, London SW7 2AZ, UK*

[†]*Istituto di Cibernetica e Biofisica CNR, Sezione di Oristano, Italy*

ABSTRACT

The possible role of nitric oxide (NO) as a novel light adaptive neuromodulator of cone plasticity (photomechanical movements) in fish and amphibian retinae were studied pharmacologically using cytomorphometric techniques. Application of a NO donor [S-nitroso-N-acetyl-D, L-penicillamine] (500-700 μM) to dark-adapted retinae induced contraction of cones with an efficiency, relative to full light adaptation (CE) of around 54%. Pre-treatment with a NO scavenger [2-(4-Carboxyphenyl)-4,4,5,5-tetrametylimidazoline-1-oxil-3-oxide] (30-35 μM) produced a consistent inhibitory action on the light adaptation-induced cone contraction (CE = 15-20%) in the retinae tested. These results strongly suggest the involvement of endogenous NO in the cone contractions that occur in fish and amphibian retinae as a part of the light adaptation process.

1. Introduction

In lower vertebrates, photoreceptor inner segment length is regulated by photomechanical movements (PMMs) (Ali, 1975). The neurochemical basis of PMMs mechanism has been studied and found to include dopamine (DA) as a neuromodulator, which is released during light adaptation (Besharse and Iuvone, 1992; Kirsh and Wagner, 1989), however, DA may not be the only light adaptive neuromodulator (Baldridge and Ball, 1991; Douglas et al., 1992). In particular, applications of exogenous nitric oxide (NO) have been shown to induce light adaptive cone PMMs in the fish retina (Greenstreet and Djamgoz, 1994).

There is increasing evidence that vertebrate retinae contain an extensive system of NO (Djamgoz *et al.*, 1998). Recently, it has been shown that nicotinamide adenine dinucleotide phosphate (NADPH) diaphorase activity, a marker for neuronal NO synthase, in the retinae of rabbit and rat depends on the state of ambient illumination (Zemel *et al.*, 1996), suggesting that NO could play a role in light adaptation.

The main aim of the present study was to determine whether the apparent role of NO in cone plasticity also occurred in retinae of marine teleost fish and amphibia, using gilthead bream and the South African clawed toad as respective examples. In addition, another cyprinid species (common carp), used extensively for electrophysiological experiments, was used for further comparison.

2. Materials and methods

Gilthead bream (*Sparus auratus*), carp (*Cyprinus carpio*) and *Xenopus laevis* were maintained in darkness for 3-4 hours starting from the initial time of their normal dark period and killed by decapitation. The eye-balls were removed and cut equatorially; the eye-cups containing the retina were used for the experiments. For the *light-adapted controls*, eye-cups were exposed to bright room light for 40 min. with Ringer solution (in mM): NaCl 102, KCl 2.6, $MgCl_2$ 1, $NaHCO_3$ 28, $CaCl_2$ 1, glucose 5, pH 7.6 for fish; NaCl 100, KCl 3.3, $MgCl_2$ 2, Hepes 10, glucose 10, pH 7.6 for toad). *Dark-adapted control* data were obtained from dark-adapted eye-cups kept for a further 40 min. in darkness with Ringer solution. S-nitroso-N-acetyl-D,L-penicillamine (SNAP), a potent NO donor (final concentration: 500-700 μM) and (2) 2-(4-Carboxyphenyl)-4,4,5,5-tetramethylimidazoline-1-oxil-3-oxide (cPTIO), a NO scavenger (final concentration: 30 - 35 μM), were used. Eye-cups were treated with SNAP in the dark for 40 min. In the other set of experiments, cPTIO was applied to the eye-cups prior to light adaptation which also lasted 40 min. After the treatments, the eye-cups were embedded in Historesin. Semi-thin (2 μm) sections were cut and stained by Richardson's stain (Richardson *et al.*, 1960).

Cone index was defined as (x/b) where (x) was the distance between the distal border of the cone ellipsoid and the external limiting membrane (ELM), and (b) was the distance from the ELM to the ganglion cell layer. Cone efficiency (CE) (%) was calculated as follows: CE = (Dc - Pt) / (Dc - Lc) x 100, where Dc and Lc are the average CI values in dark- and light-adapted control retinae, respectively, and Pt is the average of the CI value for a given pharmacological treatment. Statistical analysis was performed using one-way ANOVA followed by Fisher's PLSD test.

3. Results

The dark-adapted control values of CI were as follows: 0.60 ± 0.04 (carp), 0.56 ± 0.03 (bream) and 0.24 ± 0.01 (toad). In the control light-adapted situation, the corresponding CI values became 0.25 ± 0.01, 0.32 ± 0.01 and 0.13 ± 0.02, respectively.

Application of SNAP to dark-adapted eye-cups induced contractions of the cone populations in all three species, resulting in the following values of CI: 0.41 ± 0.02 (carp), 0.43 ± 0.01 (bream) and 0.18 ± 0.02 (toad). Light adaptive effect of SNAP in all three cases was significant ($p < 0.01$ for carp and bream, $p < 0.05$ for toad). The overall effect of SNAP corresponded to a CE value of @ 54% in all three species.

Pre-treatment with cPTIO largely blocked the effect of light adaptation in the two species tested, the final values of CI being 0.53 ± 0.02 (carp) and 0.52 ± 0.01 (bream). These changes corresponded to CE values of only 20% and 17%, respectively. Statistically, the values of CI obtained in the presence of cPTIO were not significantly different from the dark-adapted control values in given species ($p < 0.05$).

Typical light microscopic radial sections are shown in Figure 1.

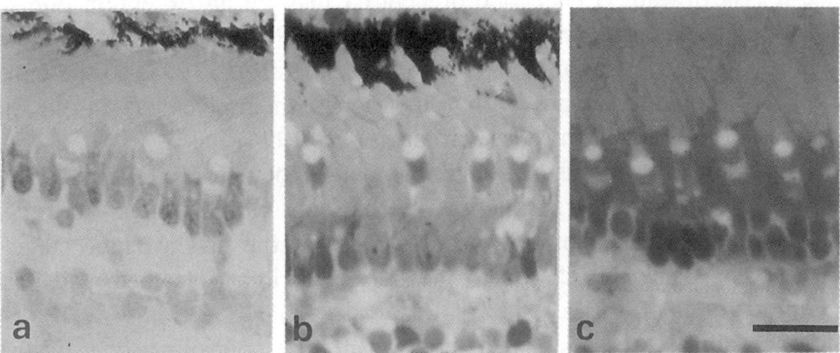

Figure 1. Light micrographs showing mainly the outer retinae of *Xenopus laevis,* in different condition: (a) light control, (b) dark control, (c) SNAP application, respectively. Scale bar 10 μm.

Table 1. Average of cone index measured under the four different adaptation conditions: Dark control. Light adapted control. cPTIO treatment. SNAP treatment. Mean data are shown with standard errors.

Cone Index values	*Xenopus*	Bream	Carp
Dark	0.24 ±0.01	0.56 ±0.03	0.60 ±0.04
light	0.13 ±0.02	0.32 ±0.01	0.25 ±0.01
SNAP	0.18 ±0.02	0.43 ±0.01	0.41 ±0.02
cPTIO	-	0.52 ±0.01	0.53 ±0.02

4. Discussion

The overall conclusion of the present study is that NO is a novel signal of light adaptation controlling cone PMMs in a variety of lower vertebrates. This extends the original observation of Greenstreet and Djamgoz (Greenstreet and Djamgoz, 1994) on a fresh water fish, roach. Thus, in all fish and amphibian species studied so far, exogenous NO application to the retinae in the dark mimicked the effect of light adaptation on cone contractions. Furthermore, the results of the experiments with cPTIO, a NO scavenger, confirmed the involvement of endogenous NO in the control of PMMs. The available evidence is consistent, therefore, with release of NO in the retina during light adaptation. Such release has recently been shown directly to occur in retinae of carp (Sekaran *et al.*, 1999) and rabbit (Neal *et al.*, 1998).

At present, the cellular origin(s) of NO inducing cone PMMs is not clear. Although it is generally thought that NO acts upon cells other than those that produce it (Garthwaite, 1991), photoreceptors may be an exception due to their functional compartmentalisation (Goldstain *et al.*, 1996). Photoreceptors themselves may synthesise NO (Djamgoz *et al.*, 1996; Kurenny *et al.*, 1995) so the contractile effect of NO could originate within the cones themselves, as in the case of modulation of ion channels in rods (Kurenny *et al.*, 1994).

The apparent involvement of multiple neuromodulators (DA and NO) with markedly different signal transduction mechanisms would appear to reflect the complexity of the light adaptation process, even at the first synaptic stage in the visual system, possibly in relation to the pattern of light (Haamedi and Djamgoz, 1996; Angotzi *et al.*, 1999).

In conclusion, NO is a signal of light adaptation controlling cone PMMs in a variety of lower vertebrates, including fresh water and marine fish, and an amphibian (*Xenopus*). Since DA is another major neuromodulator in the outer retina, it is likely that retinal light adaptation involves multiple, interactive control mechanisms.

Acknowledgements

This work has been supported by grants EC Programme STRIDE, Regione autonoma della Sardegna and P.O. Murst-CNR.

References

Ali, M.A. and H.J. Wagner (1975) in: *Distribution and Development of Retinomotor Response*, New York, London: Plenum, pp. 369-396.

Angotzi, A.R., J. Hirano, S.N. Haamedi, R. Murgia, S. Vallerga and M.B.A. Djamgoz (1999) "Comparable effects of flickering and steady patterns of light adaptation on photomechanical responses of cones in amphibian (*Xenopus laevis*) retina", *Neurosci. Lett.* 272(3):163-166.

Baldridge, W.H. and A.K. Ball (1991) "Background illumination reduces horizontal cell receptive-field size in both normal and 6-OHDA lesioned goldfish retinas", *Visual Neurosci.* 7:441-450.

Besharse, J.C. and P.M. Iuvone (1992) "Is dopamine a light-adaptive or a dark-adaptive modulator in retina?", *Neurochem. Int.* 20:193-199.

Djamgoz, M.B.A., R. Aguilo, E.H. Greenstreet, R. Reynolds and G.P. Wilkin (1996) "Histochemistry of NADPH-diaphorase - a marker for neuronal nitric oxide synthase - in the carp retina", *Neurochem. Int.* 28:283-291.

Djamgoz, M.B.A., S. Vallerga and H.J. Wagner (1998) "Functional organization of the outer retina in aquatic and terrestrial vertebrates: comparative aspects and possible significance to the ecology of vision", in: *Adaptive Mechanisms in the Ecology of Vision*, S.N. Archer, M.B.A. Djamgoz, E.R. Loew, J.C. Partridge and S. Vallerga, eds, Dodrecht, Kluwer: pp. 329-382.

Douglas, R.H., H.J. Wagner, M. Zaunreiter, U.D. Behrens and M.B.A. Djamgoz (1992) "The effect of dopamine depletion on light-evoked and circadian retinomotor movements in the teleost retina", *Visual Neurosci.* 9:335-343.

Garthwaite, J. (1991) "Glutamate, nitric oxide and cell-cell signalling in the nervous system", *Trends Neurosci.* 14:60-67.

Goldstein, I.M., P. Ostwald, and S. Roth (1996) "Nitric oxide: a review of its role in retinal function and disease", *Vision Res.* 36:2979-2994.

Greenstreet, E.H and M.B.A. Djamgoz (1994) "Nitric oxide induces light-adaptive morphological changes in retinal neurones", *NeuroReport* 6:109-112.

Haamedi, S.N. and M.B.A. Djamgoz (1996) "Effects of different patterns of light adaptation on cellular and synaptic plasticity in teleost retina: Comparison of flickering and steady lights", *Neurosci. Lett.* 206:93-96.

Kirsch, M. and H.J. Wagner (1989) "Release pattern of endogenous dopamine in teleost retinae during light adaptation and pharmacological stimulation", *Vision Res.* 29:147-154.

Kurenny, D.E., L.L. Moroz, R.W. Turner, K.A. Sharkey and S. Barnes (1994) "Modulation of ion channels in rod photoreceptors by nitric oxide", *Neuron* 13:315-324.

Kurenny, D.E., G.A. Thurlow, R.W. Turner, L.L. Moroz, K.A Sharkey and S. Barnes (1995) "Nitric oxide synthase in tiger salamander retina", *J. Comp. Neurol.* 361:525-536.

Miyachi, E., M. Murakami and T. Nakaki (1990) "Arginine blocks gap junctions between retinal horizontal cells", *NeuroReport* 1:107-110.

Neal, M., J. Cunningham and K. Matthews (1998) "Selective release of nitric oxide from retinal amacrine and bipolar cells", *Invest. Ophthalmol. Visual Sci.* 39:850-853.

Richardson, K.C., L. Jarett and E.H. Finke (1960) "Embedding in epoxy resin for ultra-thin sectioning in electron microscopy", *Stain Technol.* 35:313-323.

Sekaran, S., K.L. Mattews, J.R. Cunningham, M.J. Neal and M.B.A. Djamgoz (1999) "Nitric oxide release during light adaptation of the carp retina", *J. Physiol.* 515:102-103.

Zemel, E., O. Eyal, B. Lei and I. Perlman (1996) "NADPH diaphorase activity in mammalian retinas is modulated by the state of visual adaptation", *Visual Neurosci.* 13:863-871.

POSSIBLE RELATIONSHIPS BETWEEN THE SHAPING OF ASYMMETRICAL PROJECTIONS OF THE FRONTAL ORGAN WITH ASYMMETRICAL HABENULAR ACTIVITY DURING THE FROG BRAIN DEVELOPMENT

VITTORIO GUGLIELMOTTI
Istituto di Cibernetica del CNR, I-80072 Arco Felice, Napoli, Italy

ABSTRACT

The expression of NADPH-diaphorase (ND) activity in the frontal organ and habenular nuclei of developing and adult frog *Rana esculenta* is reported in this study. Positive cells were found in the frontal organ belonging to the embryonic period of development in which also a selective and intense neuropil staining was asymmetrically detected within the left dorsal habenula. Such ND activity was observed until the methamorphosis, while it appeared less intense in the adult frog. During metamorphosis and in adult frog, labeled fibers were stained in the frontal nerve, while only in adult specimens positive fibers were observed in the left habenular region in adult frog. Thus, the present data point out a peculiar neurochemical pattern of the habenular asymmetry in the frog, suggesting that nitric oxide may be involved in the developmental shaping, which leads to an asymmetrical configuration of the habenulae. In addition, this finding supports the postulated relationship of the habenular asymmetry with the occurrence of the frontal organ in lower vertebrates.

1. Introduction

In the frog, the frontal organ (FO) develops extracranially in the midline of the diencephalon in close association, through the frontal nerve, with the intracranic pineal organ or epyphisis. These two structures constitute the pineal complex. The FO is located in the skin between the lateral eyes and it is constituted by neurons, pineal photoreceptor cells and glial cells; its function has a photoreceptive and, probably, neurondocrine role (Guglielmotti *et al.*, 1997). Central projecting fibers of the FO display an asymmetrical organization and during their ipsilateral course cross the habenular nucleus (Eldred *et al.*, 1980; Kemali and De Santis, 1983). In the diencephalon also the habenular nuclei (HN) show a morphological asymmetry (Kemali and Braitenberg, 1969). The cell bodies of the left dorsal nucleus are distributed in two distinct medial and lateral subnuclei, whereas the right dorsal habenula is formed by a single nucleus. Conveyng information from the limbic forebrain to the midbrain, the HN represent a major relay station of the dorsal diencephalic conduction system. It has been postulated that the habenular asymmetry in lower vertebrates could be linked to the presence of the extracranial component of the pineal complex (Engbretson *et al.*, 1981). Although evidence of asymmetrical arrangements of the FO and the

HN with related circuits has been provided in the frog, very few data are available on the structural maturation of the epithalamic region in amphibians, and in particular, on the developmental events that result in differences between the left and right side in the frog.

In studies on the distribution of β-nicotinamide adenine dinucleotide phosphate (NADPH)-diaphorase (ND) enzymatic activity in the brain of amphibians, asymmetric staining of fibers has been reported in the habenular regions of frog and newt (Muñoz *et al.*, 1996; González *et al.*, 1996), and pinealocytes and nerve cells were demonstrated in the pineal organ of the frog, but the frontal organ and frontal nerve were not examined (Sato, 1990). The ND activity reveals in aldehyde-fixed nervous tissue the presence of nitric oxide synthase (NOS), the synthetic enzyme of the gaseous molecule nitric oxide (NO) (see Vincent, 1994, for a review).

NO has been suggested to play also a role in the neural development of mammals (Bredt *et al.*, 1990; Gally *et al.*, 1990), in several neuronal signaling processes of invertebrates (Jacklet, 1997) and in the retinal functions of vertebrates and invertebrates (Goldstein *et al.*, 1996; Bicker, 1998; Hirooka *et al.*, 2000). However, little is known about the involvement of NO in the developmental shaping of the nervous system in lower vertebrates.

We undertook a study on the distribution of ND activity in the diencephalon of the developing and adult frog to investigate wheter the enzymatic activity was expressed during the maturation of this region and could provide clues on the establishment of differences between the right and left sides. We have focused our study on the ND positivity of the FO, frontal nerve and HN.

2. Material and Methods

Fertilized spawns of the species *Rana esculenta* were bred in our laboratory. According to the developmental table of Manelli and Margaritora (1961) we have used five animals of the following stages: 23-26 that belong to the embryonic period, 31 and 32 to the larval period, and 33, 34, 37, 39, 44, 46, 48, and 50 to the period of metamorphosis.

For the study of ND positivity of mature frogs, 5 animals were used.

All the animals were anesthetized in a solution of tricaine methanesulfonate (MS 222, Sigma) and perfused intraperitoneally (until stage 39) or intracardially (stages 44-50 and adult). The protocol of the fixative and the histochemical demonstration of NADPH-diaforase was performed according to the study of Muñoz *et al.*,1996.

3. Results

The expression of ND activity was visualized in cells and fibers of both frontal organ and pineal organ. In the developing frog, at stage 32, the cell bodies of the frontal organ displayed a moderate ND activity (Fig. 1A). From stage 46 until complete metamorphosis, an intense ND staining was detected in cell bodies of the frontal organ (Fig. 1B), and in fibers of the frontal nerve arising from the frontal organ. Similar findings were also detected in the adult frog in which ND positivity allowed to follow the course of fibers of the frontal nerve in the skin (Fig. 1C). In its extra-encephalic course, the frontal nerve contained labeled fibers running through the choroid plexus of the third ventricle and entered the brain at the level of the habenular commissure and crossed the pineal organ to reach the pineal tract. ND-positive fibers, deriving in all likelihood from the frontal organ, were also detected in the pineal organ of the adult frog.

In the epithalamus, the left dorsal habenula displayed a marked ND positivity throughout the examined stages. These findings were consistently observed in all animals sampled at each stage.

The occurrence of ND staining in the prospective epithalamus was first detected at stage 26, when the area corresponding to the maturing HN exhibited a histochemical positivity confined to the left region, whereas the right counterpart was unstained. At stage 31, ND activity in the left habenula was even more evident than at earlier stages, and it appeared restricted to a portion of the left habenular tissue (Fig. 1D).

The asymmetry of ND activity in the HN remained confined to a neuropil compartment in the left habenula also in the subsequent stages until metamorphosis, and it was still by far more intense than elsewhere in the brain, appearing like a drop of blue ink (Fig. 1E).

The compartmental distribution of ND staining within the neuropil of the left habenula became even more evident from stage 37, when the HN had achieved a rather mature configuration. The staining revealed a defined subregion located laterally in the left medial subnucleus. The left dorsal habenula appeared thus composed by three portions: a medial subnucleus formed by two compartments, the most lateral of which exhibited an intense histochemical positivity in the neuropil, and a lateral subnucleus devoid of ND reactivity, while absence of ND reaction were observed in the right dorsal habenula (Fig. 1E). Given this configuration, we would define as "lateral neuropil" the lateral portion of the left medial subnucleus in which ND activity was consistently observed during development, and as "medial neuropil" the remaining portion of the left medial subnucleus.

In the left dorsal habenula of adult frog, ND labeling was still evident in the medial neuropil of the medial subnucleus (Fig. 1F), but this staining was rather light - certainly lighter than that observed during development. In the medial subnucleus, few labeled fibers, probably originating from the frontal organ, were also seen (Fig. 1F).

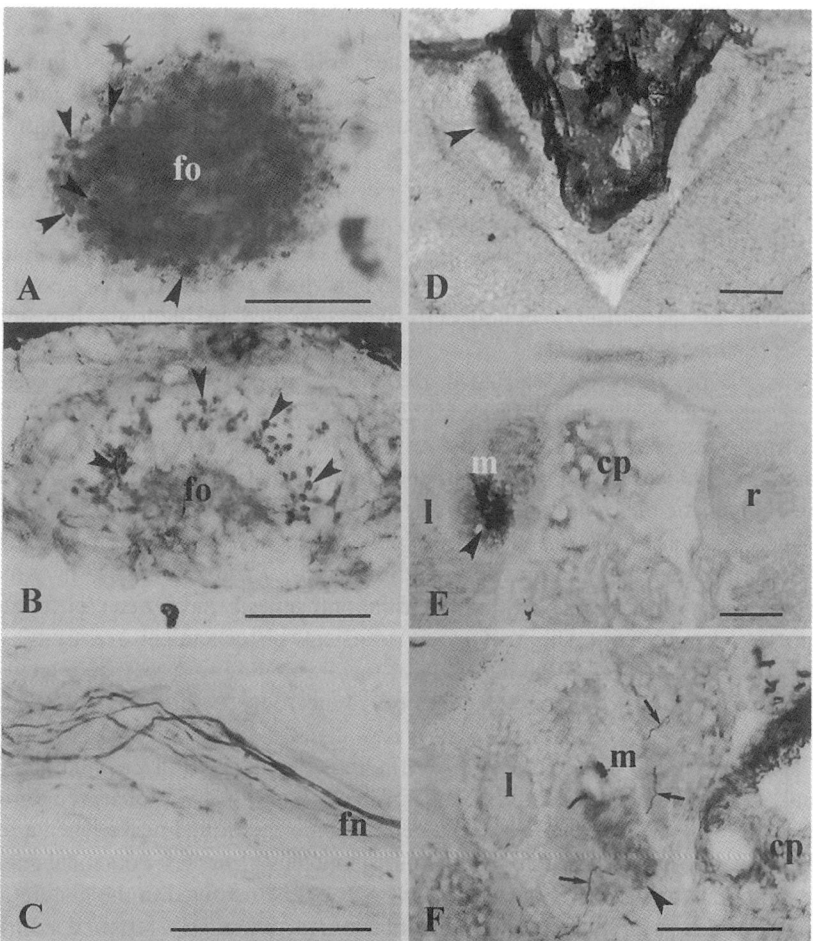

Figure 1. NADPH-diaphorase (ND) activity in the frontal organ and habenular nuclei of developing and adult frog. **A:** Horizontal view of a whole mount of the frontal organ (fo) at stage 32; note the moderate staining of cell bodies, some of which are indicated by arrowheads. **B:** Transverse section through the frontal organ (fo) at stage 50. Staining of cells is well evident (arrowheads). **C:** Sagittal section through the frontal nerve (fn) in adult frog. Note the intense staining of fibers of the frontal nerve that run in a rostro-caudal direction (from left to right, respectively) under the lower dermis. **D, E:** Transversal sections through the habenular nuclei at stages 32 and 50 respectively. In D, the ND positivity is well evident within a portion of the left habenular subnucleus (arrowhead) while in E, the intense staining (arrowhead) is evident within the medial subnucleus (m). Note the presence of the developing lateral subnucleus (l). cp, choroid plexus; r, right dorsal habenular nucleus. **F:** Transversal section through the left habenula in adult frog. Note the moderate reactivity (arrowhead) in the medial subnucleus (m) and the labeled fibers (arrows) that run towards the ventral habenula.cp, choroid plexus. Scale bars: A, B, C = 50 μm; D, E, F = 100 μm.

4. Discussion

Focusing on the distribution of histochemical ND activity in the FO and HN of the developing and adult *Rana esculenta*, the present study provides a number of novel data relevant to the developmental features of these two structures in the adult configuration in the frog.

The present findings on the habenula cannot be related with other similar studies, since no data on the ND reactivity in the brain of developing amphibians are available in the current literature. However, a massive NOS activity was found in the habenular region of the larval lamprey, in which asymmetrical ND-positive projections of the pineal ganglion cells were reported in the right habenular nucleus (Schober *et al.*, 1994).

The ND positivity we detected in the lateral neuropil of the developing left medial subnucleus of the HN strikingly decreased but was still detectable in the adult frog. These new findings points out that NOS expression in the developing habenula was at least in part transient, and suggests that NO could be implicated in the ontogenetic history of the habenular asymmetry in the frog.

The occurrence of transient NOS expression in developing brain structures has been reported also in previous studies in birds and mammals, in which different roles played by NO during brain ontogenesis have been postulated (Cramer *et al.*, 1995). The pattern of transient histochemical reactivity we observed in the neuropil of the habenula of the developing frog may be related to the establishment of a distinct pattern of connectivity within the left medial subnucleus.

This assumption is supported by the findings we obtained in the study of the frontal organ. An intense labeling of cells of the frontal organ was observed during metamorphosis, and stained fibers became evident in the frontal nerve when a few positive fibers were first seen in the medial subnucleus of the left dorsal habenula. Such temporal coincidence suggests that NOS could be involved in the maturation of the asymmetric projections of the frontal organ, previously reported both in *Rana pipiens* (Eldred *et al.*, 1980) and in *Rana esculenta* (Kemali and De Santis, 1983). The marked decrease of ND reactivity in the left habenula of the adult frog, when the course of the frontal nerve was finally traced, supports a strict relationship between the HN and the pineal complex. It should be recalled in this respect that, asymmetrical projections of the parapineal organ of the lamprey (Yáñez *et al.*,1999) and trout (Yáñez *et al.*,1996), and the parietal organ of lizard (Engbretson *et al.*, 1981) were described to reach exclusively the left habenula.

The histochemical reactivity we detected on the left side in the neuropil of the developing medial habenula, where stained cell bodies were not observed, could derive from extrinsic fibers. No data are available on the inputs to the developing habenula, but it should be considered that the mature HN in the frog are innervated by fibers deriving from regions which all contain ND-positive neurons

(Muñoz *et al.*, 1996). Thus, ND activity in the left dorsal habenula of the tadpole could derive from one of these sources and/or by a transient NOS expression in neural circuits that may be established during development, disappearing, afterwards, during the normal developmental reshaping.

Altogether our data indicate that, during frog development, NO could subserve a crucial role in the differentiation of the structural arrangement which leads to asymmetrical specializations.

References

Bicker, G. (1998) "NO in insect brains", *Trends Neurosci.* 21:349-355.

Bredt, D.S., P.M. Hwang and S.H. Snyder (1990) "Localization of nitric oxide synthase indicating a neural role for nitric oxide", *Nature* 347:768-770.

Cramer, K.S., C.I. Moore and M. Sur (1995) "Transient expression of NADPH-diaphorase in the lateral geniculate nucleus of the ferret during early postnatal development", *J. Comp. Neurol.* 353:306-316.

Eldred, W.D., T.E. Finger and J. Nolte (1980) "Central projections of the frontal organ of *Rana pipiens*, as demonstrated by the anterograde transport of horseradish peroxidase", *Cell Tissue Res.* 211:215-222.

Engbretson, G.A., A. Reiner and N. Brecha (1981) "Habenular asymmetry and the central connections of the parietal eye of the lizard", *J. Comp. Neurol.* 198:155-165.

Gally, J.A., P.R. Montague, N. Reeke and G.M. Edelmann (1990) "The NO hypothesis: possible effects of a short-lived, rapidly diffusible signal in the development and function of the nervous system", *Proc. Natl. Acad. Sci. USA* 87:3547-3551.

Goldstein, I.M., P. Ostwald and S. Roth (1996) "Nitric oxide: a review of its role in retinal function and disease", *Vision Res.* 36:2979-2994.

Gonzalez, A., A. Muñoz , M. Muñoz, O. Marín, R. Arévalo, A. Porteros and J.R. Alonso (1996) "Nitric oxide synthase in the brain of a urodele amphibian (*Pleurodeles waltl*) and its relation to catecholaminergic neuronal structures", *Brain Res.* 727:49-64.

Guglielmotti, V., U. Vota-Pinardi, L. Fiorino and E. Sada (1997) "Seasonal variations in the frontal organ of the frog: Structural evidence and physiological correlates", *Comp. Biochem. Physiol.* 116A:137-141.

Hirooka, K., D.E. Kourennyi and S. Barnes (2000) "Calcium channel activation facilitated by nitric oxide in retinal ganglion cells", *J. Neurophysiol.* 83:198-206.

Jacklet, J.W. (1997) "Nitric oxide signaling in invertebrates", *Invertebrate Neuroscience* 3:1-14.

Kemali, M. and V. Braitenberg (1969) *Atlas of the Frog's Brain*, Heidelberg: Springer-Verlag.

Kemali, M. and A. De Santis (1983) "The extracranial portion of the pineal complex of the frog (frontal organ) is connected to the pineal, the hypothalamus, the brain stem and the retina", *Exp. Brain Res.* 53:193-196.

Manelli, H. and F. Margaritora (1961) "Tavole cronologiche dello sviluppo di *Rana esculenta*", *Rendiconti Accademia Nazionale dei Lincei* 12:1-15 + 13 Tables.

Muñoz, M., A. Muñoz, O. Marín, J.R. Alonso, R. Arévalo, A. Porteros and A. González (1996) "Topographical distribution of NADPH-diaphorase activity in the central nervous system of the frog, *Rana perezi*", *J. Comp. Neurol.* 367:54-69.

Sato, T. (1990) "Histochemical demonstration of NADPH-diaphorase activity in the pineal organ of the frog (*Rana esculenta*), but not in the pineal organ of the rat", *Arch. Histol. Cytol.* 53:141-146.

Schober, A., C.R. Malz, W. Schober and D.L. Meyer (1994) "NADPH-diaphorase in the central nervous system of the larval lamprey (*Lampetra planeri*)", *J. Comp. Neurol.* 345:94-104.

Vincent, S.R. (1994) "Nitric oxide: a radical neurotransmitter in the central nervous system", *Prog. Neurobiol.* 42:129-160.

Yáñez, J., H. Meissl and R. Anadón (1996) "Central projections of the parapineal organ of the adult rainbow trout (*Oncorhynchus mykiss*)", *Cell Tissue Res.* 285:69-74.

Yáñez, J., M.A. Pombal and R. Anadón (1999) "Afferent and efferent connections of the parapineal organ in lampreys: A tract tracing and immunocytochemical study", *J. Comp. Neurol.* 403:171-189.

EFFECT OF PHOTIC STIMULATION AND PHOTODEPRIVATION IN THE TAURINE CONTENT IN DISCRETE BRAIN REGIONS AND RETINA

V.V. SUBBARAO and D. RAO*
Department of Physiology, Mamata Medical College,
507002 Khammam, Andhra Pradesh, India
**Department of Psychiatry, Indiana University, Indianapolis (IN), USA*

ABSTRACT

Intermittent light simulation causes "*grand mal*" seizures in some epileptics. Taurine is implicated in some forms of epilepsy. The elucidation of relationship between brain amino acids and alterations during conditions of photic stimulation and confinement to darkness may be useful in the better understanding of cortical reactivity to light stimulation, therefore the effect of photic simulation and photo deprivation on the alterations in taurine content of brain regions (frontal cortex and occipital cortex) and retina has been studied in the rat. Photo stimulation resulted in decreased level of taurine in the brain and retina and Photo deprivation resulted in its elevation. Earlier studies revealed changes in GABA and glycine due to photic stimulation and structural resemblance of taurine to GABA and glycine supports the view that taurine may act like neuromodulator or inhibitor in brain and retina.

1. Introduction

Cortical responses to continuous intermittent light stimuli is a method used for the investigation of the reactivity of the central nervous system. The visual system is the most widely used in this kind of study. Taurine may act as a neurotransmitter or neuromodulator in nervous tissue and retina (Barbeau *et al.*, 1975; Mandel and Passantes - Morales, 1978; Oja and Kontro, 1978) and it is also involved in maintaining the structural integrity of certain cellular layers of retina of the Cat (Hayes *et al.*, 1975; Schmidt *et al.*, 1976). Considering the importance of photic stimulation in the causation of "*grand mal*" seizures in some epileptics and involvement of taurine in epilepsy both in animal models and man (Vangelder, 1976, 1978), this study is carried out on the alteration of taurine content in the frontal cortex, occipital cortex and retina by photic stimulation and confinement to total darkness in the albino rat.

2. Material and Methods

Adult male albino rats weighing between 150 - 180g were divided into three groups of eight each. The first group that was housed for diurnal conditions served as control vehicle. The second group was kept for constant for 20 days, while the third group was phonically stimulated (12 flashes per minute) by a photo stimulator for 30 minutes. Animals of all the groups were decapitated and their brain and retina were removed and the brain tissue content was separated into frontal cortex, occipital cortex. The determination of Turin content was carried out by amino acid analysis.

3. Results

Photic stimulation has induced a significant decrease of taurine of frontal cortex (Fig. 1) occipital cortex (Fig. 2) and retina (Fig. 3) and photodeprivation has caused elevation of taurine content of frontal cortex, occipital cortex and retina (Fig. 1, 2, 3).

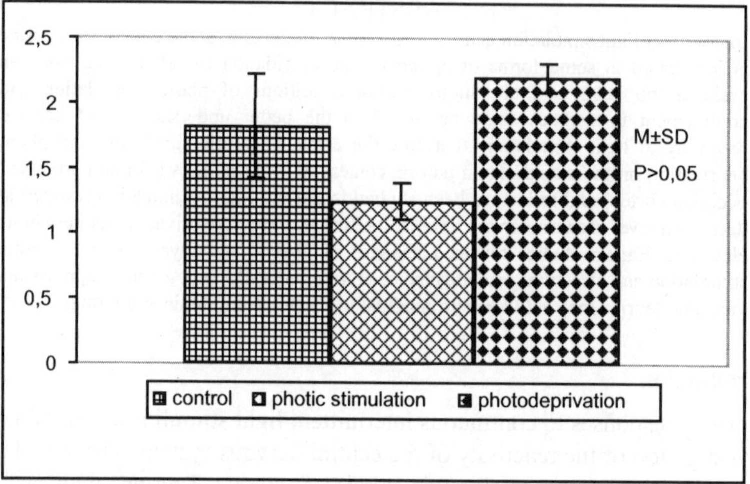

Figure 1. Effect of photic stimulation and photodeprivation on frontal cortex taurine content (μm/g).

4. Discussion

Alterations in somatic mental and behavioural disturbances including seizures could be produced by photic stimulation. Some of the factors that can increase the excitability of the "epileptogenic" circuitry enough to precipitate attacks is loud noises or flashing lights (Guyton, 1998). Taurine is implicated in epileptogenic process and the present study shows a significant depletion of taurine content of the frontal cortex, occipital cortex and retina, taurine is rapidly transported into optic axons of the gold fish and has been limited to the transport of macromolecules in the optic system. (Ingoglia *et al.*, 1978) Considerably more taurine is transported axonally in the visual system of young rats and rabbits than in mature animals and it has been suggested high concentration of taurine with in axons is maintained by this mechanism to facilitate formation of synaptic connections (Politis and Ingoglia, 1979).

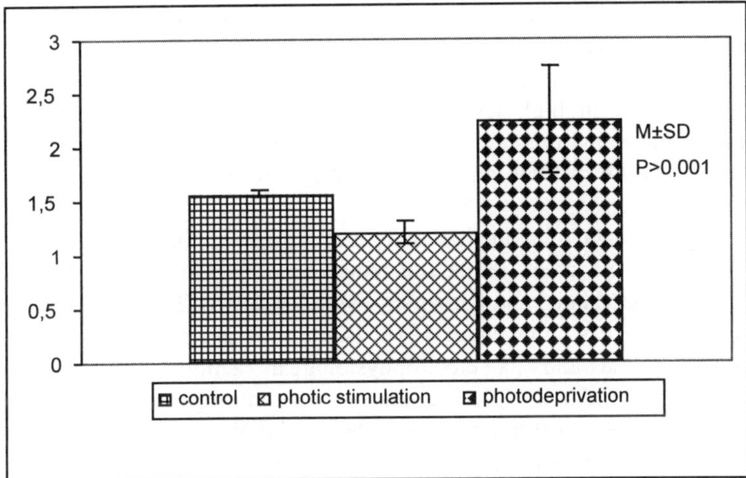

Figure 2. Effect of photic stimulation and photodeprivation on visual cortex taurine content (μm/g).

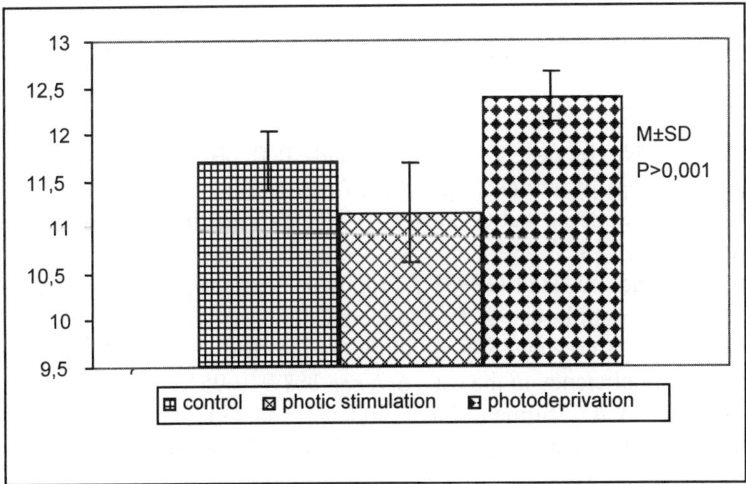

Figure 3. Effect of photic stimulation and photodeprivation on retinal taurine content (μm/g).

A reduction of brain GABA by photic stimulation (Subbarao, 1980) further supports the data presented here. The present study also shows decrease of taurine content in the visual system i.e., retina and visual cortex. In the visual system taurine is present in high levels (Guidotti *et al.*, 1972; Passantes - Morales, 1972)and in the retina this high content is maintained throughout embryonic and postnatal life while in brain and other organs there is a marked reduction of its level after birth.

The present data of reduction of retinal taurine in the rat is similar to the earlier report in the chick retina the reduction of taurine due to light and its elevation in dark (Mandel and Passantes - Morales, 1976). Taurine is implicated in epileptogenic process (Vangelder, 1978; Barbeau *et al.*, 1975). The presence of taurine and retina and its electrophysiological actions and neurochemical alterations are similar to glycine and GABA during various physiological situations. Its altered level in photic stimulation and photo deprivation support the role of taurine as a neuromodulator in brain and retina.

Acknowledgements

Grateful thanks are extended to professor. M. Ramakriasha Raju , Head of the Department of Physiology, Mamata Medical College Khammam, for his valuable suggestions and to Mr. Mohad. Mujahid, Department of Physiology, Mamata Medical Collage, Khammam for his assistance.

References

Barbeau, A., N. Inoue, Y. Tsukada and R.F. Butterworth (1975) "The neurophysiology of taurine", *Life. Sci.* 17:669-678.

Guidotti, A., G. Badsani and G. Pepeu (1972) "Taurine distribution in cat brain", *J. Neurochem.* 19:431-435.

Guyton, A.C. and J.E. Hall (1998) *Textbook of Medical Physiology*, 9th edition, Philadelphia, Saint Louis: W.B. Saunders Co.

Hayes, K.C., R.E. Carey and S.Y. Schmidt (1975) "Retinal degneration associated with taurine deficiency in the cat", *Science* 188:949-951.

Ingoglia, N.A., A.A. Sturman, T.D. Lindquat and G.E. Gaull (1976) "Axonal migration of taurine in the gold fish visual system", *Brain Res.* 115:535-539.

Mandel, P. and H. Passantes-Morales (1978) "Taurine in the nervous system" *Rev. Neurosci.* 3:157-193.

Oja, S.S. and P. Kontro (1978) "Neuro transmitter actions of taurine in the central nervous system", in: *Taurine and Neurological Disorders*, Barbeau, A. and R.J. Huxtable, eds, New York: Raven Press, pp. 181-200.

Passentes-Morales, H., J. Klethi, F. Urban and P. Mandel (1972) "The physiological role of taurine in retina. Uptake and effect on electro retinogram", *Physiol. Chem. Phys.* 4:339-345.

Passentes-Morales, H., J. Klethi, M. Ledig and P. Mandel (1972) "Free amino acids of chicken and rat retina", *Brain Research* 41:494-49.

Schmidt, S.Y., E.L. Berson and K.C. Hayes (1976) "Retinal degeneration in cats fed casien. I. Taurine Deficiency", *Invest. Opthalmol.* 15:47-52.

Subbarao, V.V.(1980) "Effect of photic stimulation and dark confinment on brain protein and GABA contents", in: *Physiology and pharmacology of Epileptogenic phenomenon*, M.R. Klee, H. Dieterlux and E.J Speckman, eds, New York: Raven Press, pp. 395-396.

Van Gelder, N.M. (1978) "Rectification of abnormal glutamic acid levels by taurine", in: *Taurine and Neurological Disorders*, A. Barbeau and R. Huxtable, eds, New York: Raven Press, pp. 293-302.

Vangelder, N.M. (1978) "Glutamic acid and epilepsy", in: *Taurine and Neurological Disorders*", A. Barbeau and R.J. Huxtable, eds, New York: Raven Press, pp. 387-402.

INTEGRATIVE LEVEL

THE ROLES OF EYE MOVEMENTS IN ANIMALS

MICHAEL F. LAND
Sussex Centre for Neuroscience, School of Biological Sciences
University of Sussex, Brighton BN1 9QG, United Kingdom

ABSTRACT

All vertebrates share a characteristic pattern of eye-movements which consists of periods of stationary fixation, separated by fast gaze-relocating saccades. The underlying reason for this strategy is the need to keep the retinal image almost stationary, to avoid blur. Primates, and a few other vertebrates, have an additional system for tracking small targets. If vision has the same basic requirements in all sighted animals, then evolutionarily unrelated creatures should share this pattern. Cuttlefish, crabs and many insects all show this pattern of fixations and saccades, with reflex compensation for body rotation. In some flying insects the same eye movements occur, but - unencumbered by contact with the ground - it is now the whole body that makes the saccades and fixations, or in some cases tracks a target. However, a few animals which employ a quite different strategy. Some sea snails, copepods, mantis shrimps and jumping spiders take in information when the eye is moving (scanning). In all these cases the scanning movements are unlike saccades in being sufficiently slow for the receptors to generate fully modulated responses.

1. Introduction

For all their apparent variety, human eye movements are controlled by a small number of well-defined mechanisms. Gaze changes are made by the fast (saccadic) system, and the eye is held almost still during the intervening fixations by two powerful reflexes - the vestibulo-ocular reflex (VOR) and optokinetic nystagmus (OKN). This "saccade and fixate" system is supplemented in primates by vergence and smooth pursuit, the former concerned with keeping the two eyes in register for objects at different distances, and the latter ensuring that small moving objects are kept on or near the fovea. Vergence and pursuit, although not confined to primates, are fairly uncommon amongst other vertebrates, as they evolved to deal with the special visual needs of front-eyed, foveate animals such as ourselves. The saccade and fixate strategy, however, seems to be nearly universal amongst vertebrates (Walls, 1962; Carpenter, 1988).

Why is this pattern so important, and how universal is it? Is it confined to the vertebrates, linked, perhaps, to our kind of camera-like eye? Or is it found in other phyletic groups with other kinds of eye? Do the cephalopod molluscs show it - animals with eyes like ours but of quite different evolutionary origins? Do insects and crustaceans with compound eyes share this strategy? The answers should tell us something important about the role of eye movements in vision. If they really *are* phylogenetically universal, then this argues very strongly that they are fundamental to the process of vision, and not just an interesting set of habits retained from our particular ancestors.

Powerful arguments have been advanced for thinking that our eye movement strategy is concerned above all with keeping retinal image velocity within a range that the retina can cope with: a few degrees per second in humans (Carpenter, 1988; 1991). The basis of the argument is that receptors with finite response times will not give fully modulated signals if structures in the image move too fast across them - in just the same way that slow shutter speeds blur photographs. If this is true then we would expect all animals with good eyesight to adopt some measures for keeping the image still, and, since animals move in the world, for shifting it from time to time as well.

I hope to show that by and large this is true, and that most animals with reasonable eyesight use a saccade and fixate strategy not very different from ours - even though, in flying insects for example, it may take unexpected forms. However, there are some remarkable exceptions, four of which I shall describe briefly later. These are animals whose eyes really do "pan" across the scene, taking in information as they do so. It would seem at first sight that their existence contradicts the "shutter time" idea just commended, but in fact it doesn't. The scanning rates seem to be nicely judged to be just below the speed at which image quality would suffer, and thus they strengthen rather than weaken the basic argument.

2. Turning a corner

During locomotion that involves turning, all animals must change the direction of their gaze from time to time. They could, of course, just let the direction of the eyes follow that of the head or body, but if avoiding image blur is important we would expect to see a saccade and fixate strategy instead. Figure 1 shows records of the eye movements made during turns by animals from three different phyla - the chordata, arthropoda and mollusca - in which eyes evolved independently (see Land and Fernald, 1992). The records all show the same features, namely that during turns the eyes make fast movements into the turn, followed by periods in which the eye counter-rotates relative to the head or body, ensuring that gaze direction (eye + head) stays more or less constant in the intervals between the fast saccadic movements. Figure 2 shows an insect example - a stalk-eyed fly -in more detail. The fly's body turns smoothly through 90°, but the eyes, built into the ends of the stalks attached rigidly to the head, make two 45° fast saccades. The head counter-rotates relative to the body in the intervening intervals, again keeping gaze direction impressively still. (Sadly, the role of the impressive eye-stalks in these animals has more to do with aggressive display than with vision). The examples given demonstrate clearly that a saccade and fixate strategy, involving a stabilising system for counter-rotating the eye, has evolved a number of times in evolution: we may tentatively assume for the same reasons.

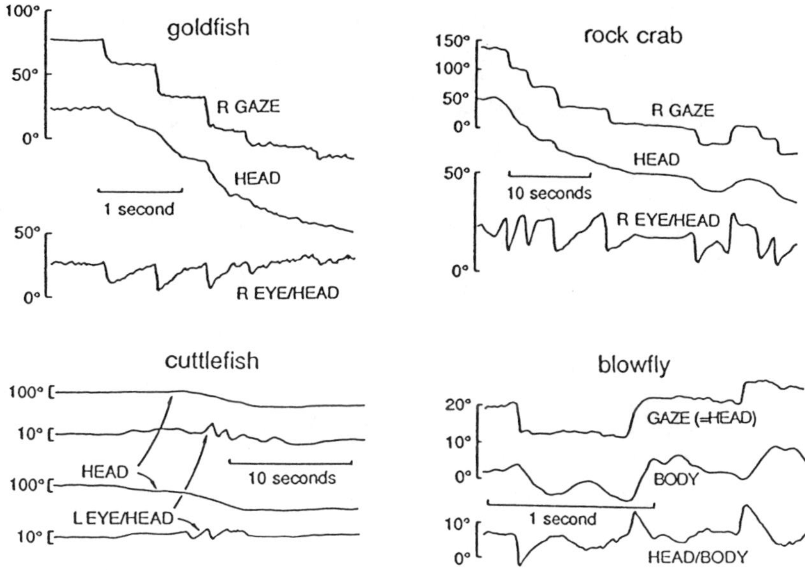

Figure 1. Four examples of eye and body movements made by animals with different phylogenetic origins, during locomotion involving rotation. In each case the eye movements (lower records) have the double function of changing gaze direction with fast saccades, and stabilizing gaze between saccades by moving the eyes in the opposite direction to the head or body movements. This results in fast gaze changes separated by almost stationary fixations (upper records, not shown for the cuttlefish). Compiled and modified from Easter *et al.*, 1974, goldfish; Paul *et al.*, 1990, rock crab; Collewijn, 1970, crayfish; Land, 1973, blowfly.

Some arthropods track small moving objects, as primates do, but as in the vertebrates this ability is uncommon. A particularly good example is the praying mantis, whose capture technique involves tracking a moving prey with the head (and hence the eyes) prior to making a lunge to catch it with the forelegs as it comes into range. Rossel (1980) found that mantids can track targets accurately and smoothly at slow speeds, but as the target movement speeds up, so the pursuit becomes increasingly saccadic in nature (as does human pursuit). Of particular interest in Rossel's study was the finding that pursuit becomes more saccadic as the contrast of the background is increased. A problem for any pursuit system is that it has to overcome the ubiquitous optokinetic response - a visual feedback loop (OKN in humans) whose function is precisely to keep the image of the background stationary on the retina. One way round the problem is to move the eyes at speeds beyond the range of the optokinetic system, and by switching to saccades, this is what the mantis seems to be doing. These issues are further discussed by Land (1992) and Collett *et al.* (1993).

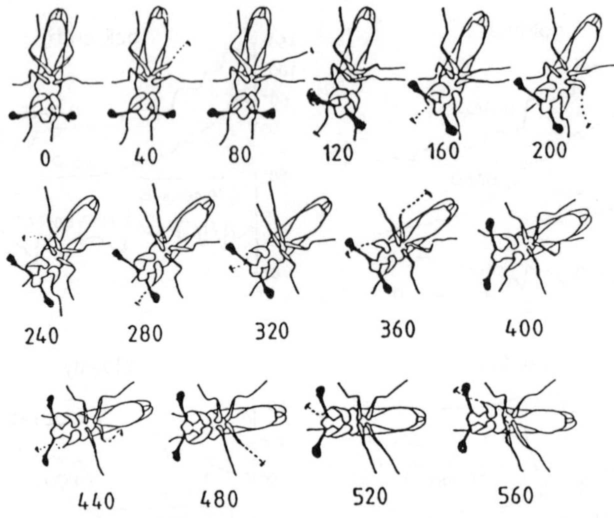

Figure 2. A stalk-eyed fly, turning through 90°, filmed from below through a glass plate. Time in ms. The head (and eyes) make two saccades (after ca 120 and 380 ms) and remain stationary in the intervals between as the body rotates continuously. Data from a film by W Wickler and U Seibt.

3. Insects as disembodied eye movements

Our eyes are attached indirectly to the ground via a body with substantial inertia, and so to shift gaze fast we must make eye movements. For a small flying insect this need not be the case. With low mass and high manoeuvrability, eye movement can be achieved by body movement. Although most insects are able to make limited head (eye) movements around all three axes, they do not always choose to do so, with the result that flight behaviour and eye movements become the same thing. An excellent example of this is the small hoverfly *Syritta pipiens*. Female flies hover around flowers, feeding on nectar, whilst the males spend much of their time in stealthy pursuit of the females (Collett and Land, 1975). The males have an advantage in that they have an "acute zone" in the front-facing part of the compound eye, where the resolution is about 3 times better than anywhere in the female eye.

Thus the males can shadow the females around until they land, whilst remaining effectively out of sight. Figure 3 shows an example of this. It is clear that the flight behaviour of the female (above) and male (below) are not the same. Although the female's flight is continuous, her turning is not. She makes rotational saccades from time to time (*e.g.* just before 3, just after 5) and between these the body does not rotate, even though translational flight may occur in any direction. The flight of non-tracking males is similar. As soon as they begin to track, however, the pattern changes dramatically. Throughout the 3.6s period shown in Figure 3 the male points directly towards the female, tracking smoothly, and keeping her within the ± 5° forward sector containing his acute zone. Notice too that he maintains a roughly constant distance of about 10cm, which is important if he is to remain undetected. Interestingly, if the female moves fast he switches to a saccadic mode of tracking, just as we do. Unlike mantids, *Syritta* is able to track smoothly against a textured background. The responses of the optokinetic and tracking control systems simply add together, with the result that a male tracking a female in a rotating environment can do so, but with a small position error (Collett, 1980). From the point of view of visuo-motor coordination, it is not far-fetched to think of male *Syritta* flight manoeuvres as analogous to primate eye movements, and those of the females to non-primate (rabbit, say) eye movements.

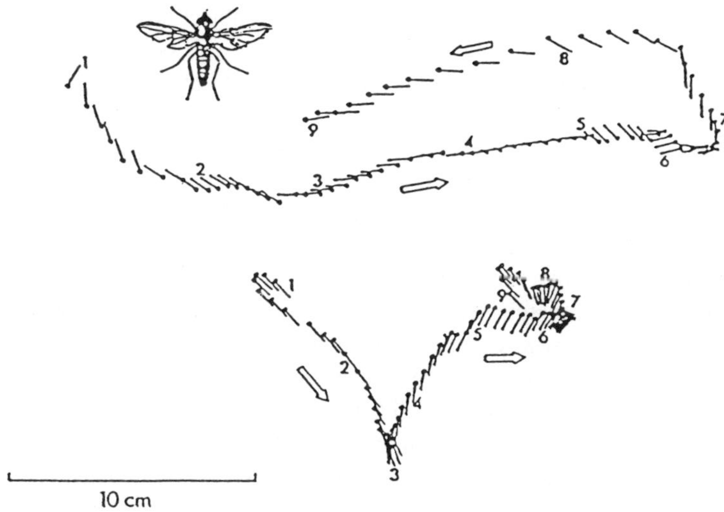

Figure 3. Insect flight manoeuvres as eye movements. The hoverfly *Syritta pipiens* filmed from above, showing the flight path of a female (above) being tracked by a male (below). Notice that the female's flight has a "saccade and fixate" pattern, with very little rotation between saccades, although there is no restriction on translation. The male, however, tracks the female smoothly, keeping her within 5° of his midline. Here there is a region of high acuity, absent in the female. Also notice that he maintains a constant distance from the female. Corresponding times are numbered every 400ms. From Collett and Land, 1975.

4. How still does the eye need to be?

Having made the case for thinking that all well-sighted animals avoid a moving retinal image where they can, and shift gaze as fast as possible when motion is unavoidable, it is useful to look in more detail at the at the factors that determine how much image slip can be tolerated. We will consider later the related question: What kinds of slip are actually desirable?

If we ask how fast the image can move before blur becomes a problem, it turns out that the answer depends not only on the response time of the receptors, but also on the the fineness with which the image is sampled by the receptor mosaic. Consider a receptor whose field of view (or acceptance angle) is, say, 1°. Images of objects 1° and larger will, if stationary, fully stimulate the receptor, but smaller ones will do so only partially. Thus *spatial* degradation begins when image detail is smaller than a receptor's acceptance angle (this assumes that the eye's optics resolve adequately). A parallel argument applies to *temporal* degradation, which will start to occur when the receptor has inadequate time to respond. Suppose the receptor takes 20ms to respond fully to a small light flash. This "flash response time" sets the minimum time required for the cell to produce a response to any type of stimulus (Howard *et al.*, 1984; Land *et al.*, 1990). Thus if an object takes less than 20ms to pass through the receptor's field of view the response will be only partial, but if it takes longer the response will be complete. Returning to the 1° object that was just fully resolved spatially, it is now clear that to elicit a full response from the receptor it must take at least 20ms to pass through its field of view, which means that it can move across the retina at a maximum velocity of $(1/0.02) = 50°.s^{-1}$. We can generalise this result to say that *the maximum tolerable velocity across the retina, without loss of usable contrast, is given by the receptor acceptance angle divided by the response time* (Srinivasan and Bernard, 1975). Interestingly, this relation predicts that the maximum acceptable velocity should increase as the spatial resolution of the eye decreases, which means that the relatively coarse (1°) mosaic of insect eyes should be more tolerant to image slip than the 0.5' foveal mosaic of humans by about two orders of magnitude, if the response times are similar. Conversely, excellent resolution like ours requires particularly good image stabilization. Applying the "one acceptance angle per response time" rule to humans would give a value of rather less than $1°.s^{-1}$ as the maximum speed that will not degrade the finest resolvable grating image. This turns out to be a little pessimistic; Westheimer and McKee (1975) estimate that measurable contrast loss begins at $2\text{-}3°.s^{-1}$.

This argument leaves no doubt about the need to stabilize the eye, if its resolving power is not to be compromised, and the better the eye the truer this is. However, it does also suggest an alternative way of acquiring visual information, particularly in eyes where the receptor mosaic is relatively coarse. Provided the speed given by the rule above is not exceeded, an eye may make movements that scan the retina across the image, without loss of spatial resolution. In conventional eyes with 2-dimensional retinae such a strategy might merely produce confusion. However, there are a few eyes with narrow, almost linear retinae, which do indeed move in a way that shows that they really are scanning the image. Four of these unusual eyes are considered in the next section.

5. Scanning vision: sea snails, water fleas, mantis shrimps and jumping spiders

The most straight-forward scanning eye I know of is in the carnivorous planktonic sea-snail *Oxygyrus* (Figure 4a). It has been known for a century that this group of gastropod molluscs, the heteropods, have very narrow retinae (Hesse, 1900), but the reason for this has only recently become apparent. *Oxygyrus* has a lens eye not unlike a fish eye, except that the retina is only 3 receptors wide by about 410 receptors long, and covers a field of about 3° by 180°. The 1-dimensional structure of this retina would make very little sense unless it moved in some way, and indeed the eyes do scan (Land, 1982). The eyes move so that the retina sweeps through a 90° arc at right angles to its long dimension. The scanning pattern is a sawtooth, and the slower upward component has a velocity of $80°.s^{-1}$. The eye scans through the dark field below the animal, and the suggestion is that it is searching for food particles glinting against the dark of the abyss.

The oceanic copepod *Labidocera* exhibits a similarly straight-forward scanning pattern (Land, 1988). The animal has a pair of eye-cups directed dorsally (Figure 4b). The combined retina has a set of 10 slab-like receptor structures - 5 per eye - arranged as a line. These are pulled to and fro by a combination of a pair of small muscles behind, and elastic ligaments in front, so that the linear retina scans the water above as shown. The muscle-powered movement is the slower one, but even this is fast, more than $200°.s^{-1}$. Interestingly, only the males have these specialized eyes, and we must assume that this scanning arrangement is part of the way they find females. These do have rather dark elongated bodies, so a scanning linear array might well be an appropriate detector.

The third example is more complicated. The mantis shrimps are quite large crustaceans, very distantly related to the more familiar decapod shrimps. Like their insect namesakes they are ambush predators, with a legendary ability to destroy their prey with smashing or spearing appendages (Caldwell and Dingle, 1976). Their eyes are basically compound eyes of the ordinary apposition type, and these provide an erect 2-dimensional image. However, stretching more or less horizontally across each eye is a band of enlarged facets, 6 rows wide (Figure 5a). This mid-band, which has a field of view only a few degrees wide, contains the animals' extraordinary colour vision system (Cronin *et al.*, 1994; Marshall *et al.*, 1991). This consists of 4 of the mid-band rows (the other two subserve polarization vision) and in each row the receptors are in three tiers. Each of these 12 tiers contains a different visual pigment, giving the animal *dodeca-chromatic* colour vision. In adopting this impressive system, however, the mantis shrimps have set their eye movement system a daunting task. The outer parts of the eye operate as normal compound eyes - and are subject to the kinds of image stability considerations discussed earlier. The mid-band, however, has to move or it will not be able to register the colour of objects in the environment outside a very narrow strip.

The result of this visual schizophrenia is a repertoire of eye movements unlike anything else in the animal kingdom (Land *et al.*, 1990). In addition to "normal" eye movements - fast saccades, tracking and optokinetic stabilizing movements - there is a special class of frequent, small (ca 10°) and relatively slow (40°.s^{-1}) movements, which give the animal a strange inquisitive appearance, perhaps because they resemble human saccades in their frequency of occurrence. They are, however, not saccades, which are much faster. These movements, illustrated in Figure 5a, are typically at right angles to the band, and the only plausible explanation is that they are the scanning movements the band uses to "colour in" the monochrome picture provided by the rest of the eye.

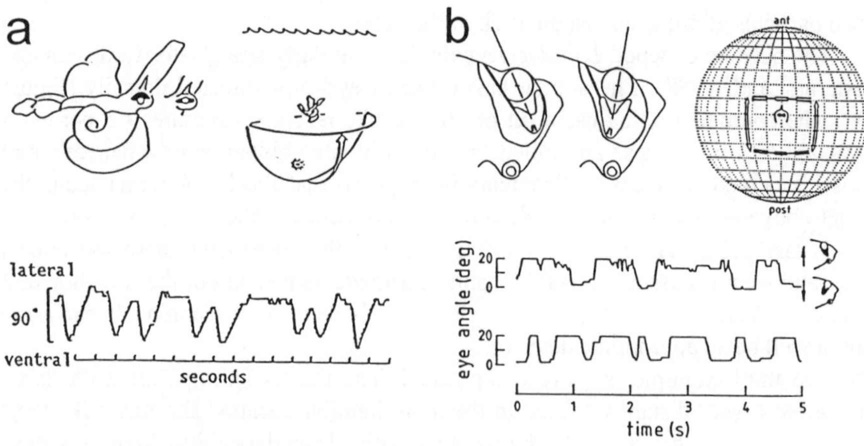

Figure 4. Simple examples of scanning eyes. **a)** The sea-snail *Oxygyrus* (left), with one eye pointing downwards. The inset shows the appearance of the eye when directed laterally. Diagram on the right shows the visual field of the eye during a scanning movement, and its probable role in detecting plankton. The time course of 8 scans is given below. Mainly from Land, 1962. **b)** Head of the copepod *Labidocera* from the side, showing the eyecup at the extreme positions of a scan (left). The plot on the right shows the upward-pointing field of view of the line of rhabdoms projected onto a hemisphere above the animal. The time course of a number of scans is shown below. Mainly from Land, 1968.

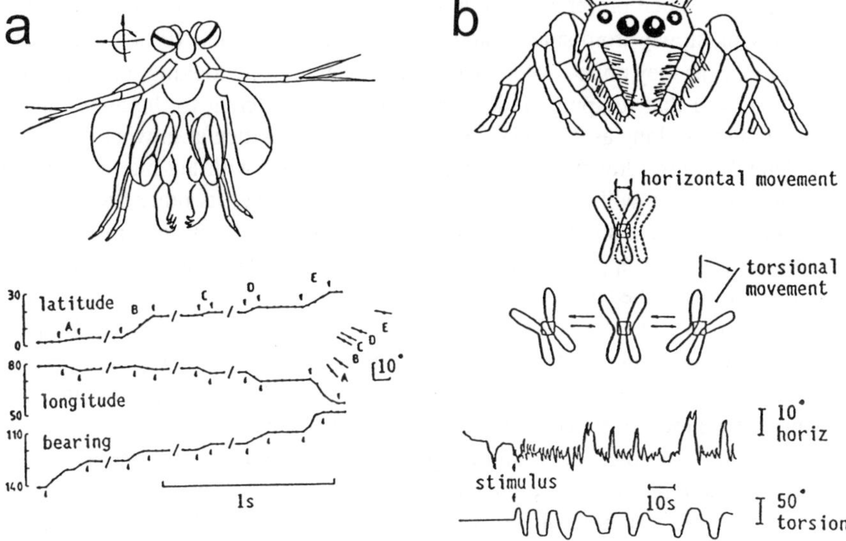

Figure 5. More complex scanning eyes. **a)** The mantis shrimp *Odontodactylus* from the front, showing the large compound eyes and the 6-row mid-bands indicated in black. Arrows indicate the three rotational axes of the eyes. Below is a record of 5 scanning movements, showing their small amplitude and low velocity, and the independence of movement around each axis. The insert (right) shows the angular trajectory the eye's centre and mid-band, projected onto a sphere. From Land *et al.*, 1990. **b)** The jumping spider *Phidippus* showing the large movable principal eyes, and smaller fixed antero-lateral eyes. Below is a diagram and record of the movements of the boomerang-shaped retinae of the two principal eyes while scanning a novel target. These movements are conjugate, and consist of a stereotyped pattern of fast horizontal oscillations and slower torsional rotations. From Land, 1969.

Jumping spiders stalk insect prey rather as cats stalk birds. They have eight simple (camera-type) eyes, although two are usually rudimentary (Figure 5b). Of the remaining six, four are fixed to the carapace and act only as motion detectors. If something moves in the surroundings these eyes initiate a turn, which results in the target being acquired by the larger, forward-facing pair of "principal" eyes (Homann, 1928). These eyes have narrow retinae shaped like boomerangs, subtending about 20° vertically by 1° horizontally in the central region, which is only about 6 receptor rows wide (Land, 1985; Blest, 1985). The resolution is very high, with receptor spacings of 10' fairly typical, and as low as 2.5' in one genus (*Portia*). The principal retinae can move, horizontally and vertically by as much as 50°, and they can also rotate about the optic axis (torsion) by a similar amount

(Land, 1969). When presented with a novel target, the eyes scan it in a stereotyped way, moving slowly from side-to-side at speeds between 3 and $10°.s^{-1}$, and rotating through ±25° as they do so. We actually know what they are looking for: legs! Drees (1952) showed that jumping spiders are relatively indifferent to the appearance of potential prey, so long as it moves, but males are quite particular in what they regard as potential mates. Drawings consisting of a central dot with leg-like markings on the sides, however, will elicit courtship displays. Whatever its other functions may be, scanning in these spiders really seems to be concerned with feature extraction, the procedure itself apparently designed to detect the presence and orientation of linear structures in the target.

The dual system of fixed and moveable eyes of jumping spiders, with one set acting as target finder and the other as analyser, does seem to have much to commend it, compared with the cumbersome time-sharing arrangement in mantis shrimps where the two functions are combined in the same eye.

The four examples of scanning given in this section represent a range of different functions, from simple detection, to colour and feature extraction. Nevertheless, they should all be expected to obey the rule given earlier, that the scanning speed should not exceed the receptor acceptance angle divided by the response time. We would not expect the speed to be much slower than this, however, as that would merely waste time. Clearly there is an optimum.

Animal	**Scan rate (s) $°s^{-1}$**	**Receptor subtense (r) °**	**"Dwell time" (t=r/s) msec**
Labidocera (Copepod)	219	3.5	16
Oxygyrus (Mollusc)	80	1.1	15
Odontodactylus (Stomatopod)	40	1.0	25
Metaphidippus (Spider)	6.2	0.15	24

Table 1. Scanning eyes: inverse relation of speed and resolution

Table I gives the scanning speeds and acceptance angles for the four animals discussed. In the Table it is clear that there is an inverse relationship between resolution and scanning speed, as indeed there should be; the high resolution jumping spider is slowest, and the copepod the fastest. The response times of the receptors, estimated as the time it takes a receptor to move through its own acceptance angle, are also shown, and they all fall nicely into the expected range of 15 to 25ms. Although we do not know the true response times for these animals, these values are well within the range of insect flash response times, for which data are available (Howard *et al.*, 1984).

6. Conclusions: good and bad retinal motion

Whilst most animals' visual systems go to considerable lengths to protect the stability of gaze from the vagaries of body movement, a few, as we have seen, actually exploit the tolerance of the receptor response to modest velocities to scan the image with systematic eye movements. The question arises: Do we all do this, in one way or another? To answer this we need to look at the sources of image motion our eyes are subject to. Gibson (1950) pointed out that whenever we move there is a pattern of image movement across the retina - the "flow field" - that is the inevitable consequence of our locomotion. Generally speaking the flow field has two components, one due to translational (linear) motion, and one to rotation (Figure 6). The translational flow field is a pattern of velocity vectors expanding from a central stationary "pole" that corresponds to the direction of motion. Rotation, by contrast, causes the image to move in the same direction everywhere, and with the same angular velocity. The only information in the rotational pattern concerns the speed of rotation itself, which is unlikely to be of great interest. On the other hand the translational flow field is rich and valuable, as it contains information about the distances of objects (nearer objects move past faster) and also the animal's current heading, as shown by the location of the pole.

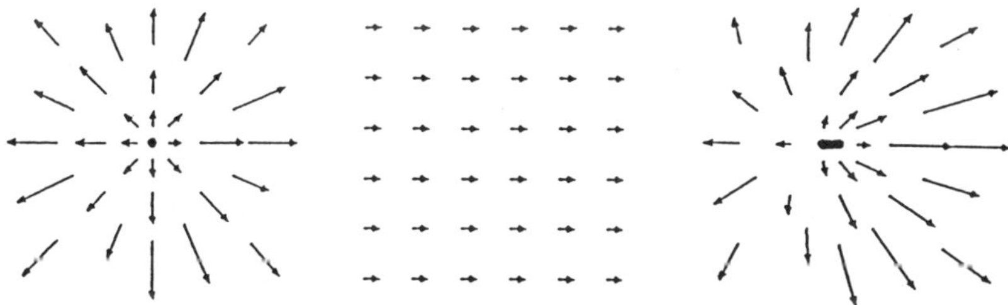

Figure 6. Diagrams illustrating the velocity flow-fields on an animal's retina resulting from pure translation (left), rotation (centre) and a combination (right). Arrows represent velocity vectors, and the variation in length of the translational vectors indicates the presence of objects at different distances. In the combined flow-field lengths and directions of vectors are distorted, and the stationary pole (dot), which gives the animal's heading direction, has become an indeterminate line.

The track of the female hoverfly in Figure 3 illustrates one way of coping with the mixture of flow-field components, and it may be typical of animals generally. Rotation is strenuously prevented, being allowed only as brief saccades during which - because of the speeds involved - the fly must have very reduced acuity. Translation, however, is not obviously impaired, and one has no sense that the fly is being held back from moving in any direction it chooses, provided this does not involve rotation. The situation in man is similar; VOR prevents eye rotation, but does not interfere with our locomotion. The inference is that in insects and in man the oculomotor system leaves the translational flow field intact, by getting rid of rotational flow before it even happens. The reason for doing this probably lies in the difficulty of extracting the useful translational flow from the combined flow-field (see for example Buchner, 1984, Fig.1). And even though there is evidence that humans can do this under some circumstances (Warren and Hannon, 1990), that capability is probably only the second line of defence.

The retinal velocities involved in translational flow are not great (Kowler, 1991), especially around the direction of motion where they fall well within the range that the receptors can deal with. Implausibly, we still lack direct proof that flow-field information usefully influences human behaviour (although driving and playing tennis seem inconceivable without it). There is, however, convincing evidence that bees can learn their distances from objects by velocity information alone (Lehrer *et al.*, 1988), and there is strong circumstantial evidence that translational flow is used by other animals (Davies and Green, 1990; Lee and Reddish, 1981). Some animals actually generate translational flow in order to measure distance. Collett (1978) showed that locusts judge their jump distance by the rate of image motion across the retina as they make stereotyped lateral "peering" movements (Figure 7). Because translational flow contains information about an animal's progress through the environment immediately ahead of it, it would be astonishing if it were not properly exploited for the control of locomotion.

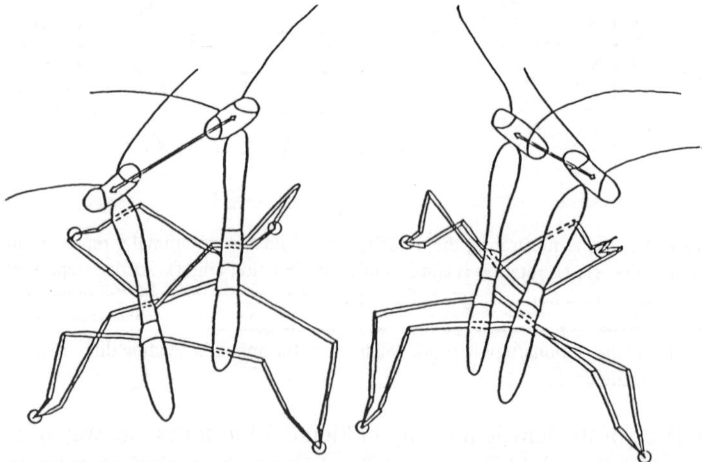

Figure 7. Side-to-side translational scanning movements of an early instar praying mantis. Note that the body moves in such a way that the head travels along a line that is almost perpendicular to the animal's forward direction of view, and that the head does not rotate relative to the surroundings during a scan. As with locust peering, these movements allow accurate range estimation from the movement of the image.

To sum up. In nearly all animals with good eyesight the main function of the oculomotor system is to prevent rotational slip of the image. The overriding reason for this is the need to prevent loss of acuity resulting from the blur caused by the finite response time of the photoreceptors. However, in a few animals rotational motion is actually used to scan the image, but when this does occur the velocities involved do not exceed a "no blur" value, given by the acceptance angle of a receptor divided by its response time (Table I). These exceptions aside, animals who have stabilised their eyes against rotation are generally free to contemplate and exploit the remaining translational image motion that results from locomotion. This is usually slow enough to avoid blur, and contains much useful information about the structure of the world ahead of the animal.

Acknowledgements

This chapter is a modified and extended version of an essay that first appeared in 1995. (Land, 1995).

References

Blest, A.D. (1985) "The fine structure of spider photoreceptors in relation to function", in: *Neurobiology of Arachnids*, F.G. Barth, ed., Berlin: Springer, pp. 79-102.

Buchner, E. (1984) "Behavioural analysis of spatial vision in insects", in: *Photoreception and Vision in Invertebrates*, M.A. Ali, ed., New York:Plenum, pp. 561-621.

Caldwell, R.L. and H. Dingle (1976) "Stomatopods", *Sci. Amer.* 234(1):80-89.

Carpenter, R.H.S. (1988) *Movements of the Eyes*, 2nd ed. London: Pion.

Carpenter, R.H.S. (1991) "The visual origins of ocular motility". in: *Vision and Visual Dysfunction, Vol 8.*, R.H.S. Carpenter, ed., Basingstoke: Macmillan, pp. 1-10.

Collewijn, H. (1970) "Oculomotor reactions in the cuttlefish, *Sepia officinalis*", *J. Exp. Biol.* 52:369-384.

Collett, T.S. (1978) "Peering - a locust behaviour pattern for obtaining motion parallax information", *J. Exp. Biol.* 76:237-241.

Collett, T.S. (1980) "Angular tracking and the optomotor response. An analysis of visual reflex interaction in a hoverfly" *J. Comp. Physiol.* 140:145-158.

Collett, T.S. and M.F. Land (1975) "Visual control of flight behaviour in the hoverfly, *Syritta pipiens* L", *J. Comp. Physiol.* 99:1-66.

Collett, T., H.-O. Nalbach and H. Wagner (1993) "Visual stabilization in arthropods", in: *Visual Motion and its Role in the Stabilization of Gaze*, F.A. Miles and J. Wallman, eds, Amsterdam: Elsevier, pp. 239-263.

Cronin, T.W. and N.J. Marshall (1994) "The unique visual system of the mantis shrimp", *Amer. Scientist* 82:356-365.

Davies, M.N.O. and P.R. Green (1990) "Optic flow-field variables trigger landing in hawk but not in pigeons", *Naturwissenschaften* 77:142-144.

Drees, O. (1952) "Untersuchungen über die angeborenen Verhaltensweisen bei Springspinnen (*Salticidae*)", *Z. Tierpsychol.* 9:169-207.

Easter, S.S., P.R. Johns and D. Heckenlively (1974) "Horizontal compensatory eye movements in goldfish (*Carrassius auratus*). I. The normal animal", *J. Comp. Physiol.* 92:23-35.

Gibson, J.J. (1950) *The Perception of the Visual World*, Boston: Houghton Mifflen.

Hesse, R. (1900) "Untersuchungen über die Organe der Lichtempfindung bei neideren Thieren. VI. Die Augen einiger Mollusken", *Z. Wiss. Zool.* 68:379-477.

Homann, H. (1928) "Beiträge zur Physiologie der Spinnenaugen. I Untersuchungsmethoden. II Das Sehvermögen der Salticiden", *Z. Vergl. Physiol.* 7:201-269.

Howard, J., A. Dubs and R. Payne (1984) "The dynamics of photo-transduction in insects. A comparative study", *J. Comp. Physiol. A* 154:707-718.

Kowler, E. (1991) "The stability of gaze and its implications for vision", in: *Vision and Visual Dysfunction. Vol 8*, R.H.S. Carpenter, ed., Basingstoke: Macmillan, pp. 71-92.

Land, M.F. (1973) "Head movements of flies during visually guided flight", *Nature* 243:299-300.

Land, M.F. (1969) "Movements of the retinae of jumping spiders (*Salticidae: Dendryphantinae*) in response to visual stimuli", *J. Exp. Biol.* 51:471-493.

Land, M.F. (1982) "Scanning eye movements in a heteropod mollusc", *J. Exp. Biol.* 96:427-430.

Land, M.F. (1985) "The morphology and optics of spider eyes", in: *Neurobiology of Arachnids*, F.G. Barth, ed., Berlin: Springer, pp 53-78.

Land, M.F. (1988) "The functions of eye and body movements in *Labidocera* and other copepods", *J. Exp. Biol.* 140:381-391.

Land, M.F. (1992) "Visual tracking and pursuit: humans and arthropods compared", *J. Insect Physiol.* 38:939-951.

Land, M.F. (1995) "The functions of eye movements in animals remote from man", in: *Eye Movement Research*, J.M. Findlay *et al.*, eds, Amsterdam: Elsevier, pp. 63-76.

Land, M.F. and R.D. Fernald (1992) "The evolution of eyes", *Annu. Rev. Neurosci.* 15:1-29.

Land, M.F., J.N. Marshall, D. Brownless and T.W. Cronin (1990) "The eye-movements of the mantis shrimp *Odontodactylus scyllarus* (Crustacea: Stomatopoda)", *J. Comp. Physiol. A* 167:155-166.

Lee, D.N. and P.E. Reddish (1981) "Plummeting gannets: A paradigm of ecological optics", *Nature* 293:293-294.

Lehrer, M., M.V. Srinivasan, S.W. Zhang and G.A. Horridge (1988) "Motion cues provide the bee's visual world with a third dimension", *Nature* 332:356-357.

Marshall, N.J., M.F. Land, C.A. King and T.W. Cronin (1991) "The compound eyes of mantis shrimps (Crustacea, Hoplocarida, Stomatopoda)", *Phil. Trans. R. Soc. B* 334:33-84.

Paul, H., H.-O. Nalbach and D. Varjú (1990) "Eye movements in the rock crab *Pachygrapsus marmoratus* walking along straight and curved paths", *J. Exp. Biol.* 154:81-97.

Rossel, S. (1980) "Foveal fixation and tracking in the praying mantis" *J. Comp. Physiol A* 139:307-331.

Srinivasan, M.V. and G. Bernard, G. (1975) "The effect of motion on the visual acuity of the compound eye: a theoretical analysis", *Vision Res* 15:515-525.

Walls, G.L. (1962) "The evolutionary history of eye movements", *Vision Res* 2:69-80.

Warren, W.H. and D.J. Hannon (1990) "Eye movements and optical flow", *J. Opt. Soc. Amer. A* 7:160-169.

Westheimer, G.A. and S. McKee (1975) "Visual acuity in the presence of retinal image motion", *J. Opt. Soc. Amer.* 65:847-850.

ENDOGENOUS NITRIC OXIDE MODULATES SIGNAL TRANSMISSION FROM PHOTORECEPTORS TO ON-CENTER BIPOLAR CELLS IN THE RABBIT RETINA

BO LEI* and IDO PERLMAN

The Bruce Rappaport Faculty of Medicine, Technion-Israel Institute of Technology and the Rappaport Institute, Haifa, Israel.

**Kellog Eye Center, University of Michigan, Ann Arbor, Michigan, USA*

ABSTRACT

The enzyme nitric oxide synthase (NOS), that synthesizes nitric oxide (NO) from L-arginine, has been demonstrated in a variety of retinal cells from different species. Therefore, NO has been implicated to play a role in visual information processing within the retina. In order to study the role of NO *in vivo*, we recorded the electroretinogram (ERG) from rabbits at different time intervals after intravitreal injection of drugs that either modulated NO formation or served as an NO donor. The receptor component of the ERG was isolated by a mixture of APB and PDA. L-arginine but not D-arginine augmented the ERG responses within the first 6 hr after injection. N^{ω}-nitro-L-arginine methyl ester (L-NAME) transiently depressed the ERG responses during the first 6 hr after injection. Neither L-arginine nor L-NAME induced any apparent effect on the receptor component of the ERG. An NO donor, Sodium Nitroprusside (SNP) induced short-term effects similar to those of L-arginine and also increased the receptor component of the ERG. These data are consistent with the notion that guanylate cyclase in photoreceptors and ON-center bipolar cells is NO-sensitive. However, NO produced endogenously primarily modulates signal transmission from photoreceptors to ON-center bipolar cells by acting upon the post-synaptic cells.

1. Introduction

Nitric oxide (NO) serves as an inter-cellular messenger molecule in a variety of physiological processes including synaptic transmission in the central and peripheral nervous systems (Bredt and Snyder, 1992; Garthwaite, 1991; Snyder, 1992). It is formed from L-arginine by the enzyme Nitric Oxide Synthase (NOS) (Bredt and Snyder, 1992) and exerts its physiological role by directly acting on cytoplasmatic guanylate cyclase (Ignarro, 1990; Knowles *et al.*, 1989).

Cyclic GMP plays a crucial role in at least three types of neurons in the distal retina. In photoreceptors, cGMP acts as the intracellular second messenger of the phototransduction process (Kaupp and Koch, 1992; Yau *et al.*, 1988). In ON-center bipolar cells, cGMP-gated cationic channels, located on the post-synaptic membrane, are involved in signal transmission from the photoreceptors (Nawy and Jahr, 1991). In horizontal cells, cGMP is one of the intracellular second messengers controlling the conductance of the gap junctions between neighboring cells (Miyachi and Murakami, 1991).

Recent morphological and physiological studies have supported a role for NO in visual information processing by the retina. Neurons and glial cells containing NOS have been demonstrated in retinae of different vertebrate species (Haverkamp *et al.*, 1999; Koistinaho and Sagar, 1995). The degree of NADPH diaphorase activity and its distribution between different retinal cells have been shown to depend upon the state of visual adaptation (Zemel *et al.*, 1996). Administration of sodium nitroprusside (SNP) or L-arginine and NADPH into isolated vertebrate rods prevented the decay of the dark current and accelerated the recovery of the photocurrent (Schmidt *et al.*, 1992; Tsuyama *et al.*, 1993). The activity of GC in ON-center bipolar cells was shown to be NO-sensitive (Koistinaho *et al.*, 1993; Shiells and Falk, 1992). Intracellular administration of L-arginine or exogenous supply of NO (Nitroprusside) reduced the conductance of gap junctions between horizontal cells in the fish retina (DeVries and Schwartz, 1989; Miyachi *et al.*, 1990; Pottek *et al.*, 1997). It has been recently reported that NO can modulate glutamate-gated currents in isolated horizontal cells from fish retina (McMahon and Ponomareva, 1996).

The above findings suggest a modulatory role for NO in the vertebrate retina. However, most of the physiological studies utilized NO donors and were conducted on isolated retinal cells *in vitro*. In one *in situ* study on isolated rabbit retina, modulation of the NO system has been shown to affect the compound action potential of the optic nerve (Maynard *et al.*, 1995). In order to test the role of NO *in vivo* and to localize its site of action, we measured the electroretinogram from rabbits after intravitreal injections of an NO donor or of drugs that either activate or inhibit endogenous NO formation.

2. Material and Methods

2.1. Animals

The experiments were performed on 30 adult albino rabbits, weighing 2-3 Kg. The rabbits were housed in separate cages with free access to water and food under 12/12 hr light/dark cycle. All experimental procedures conformed to the ARVO Statement for the Use of Animals in Ophthalmic and Vision Research and to institutional guidelines.

For intravitreal injection and for recordings of the electroretinogram, the rabbits were anesthetized by an intramuscular injection of a "cocktail" made up of ketamine hydrochloride solution (10 mg/ml), acepromazine maleate solution (10 %) and xylazine solution (2%) in the following volume proportions; 1.0: 0.2: 0.3 respectively. Dose used was 0.5 ml/kg body weight. The pupils were fully dilated with cyclopentolate hydrochloride 1% and topical anesthesia (benoxinate HCl 0.4%) was administered.

2.2. *Intravitreal injection*

Intravitreal injections were performed as previously described (Loewenstein *et al.*, 1991). Briefly, a 25-gauge needle, attached to a 1.0 ml tuberculin syringe, was inserted through the sclera 2 mm posterior to the limbus. The needle was advanced under visual control (indirect ophthalmoscope) towards the retina close to the optic disk. A volume of 0.1ml was slowly injected with the bevel of the needle pointing away from the retina in order to prevent physical damage to the retina by the injection procedure itself. A similar volume of physiological salt (0.9%) solution (saline) was injected into the vitreous of the fellow, control eye. L-arginine, D-arginine, N^{ω}-nitro-L-arginine methyl ester (L-NAME), Sodium Nitroprusside (SNP), 2-amino-4-phosphonobutyric acid (APB) and cis-2,3-piperidine dicarboxylic acid (PDA) were purchased from Sigma Chemical Co. (St Louis, USA). Spermine NONOate (SPNN) was kindly supplied by Dr. Ziad Abassi, Faculty of Medicine, Technion. Fresh solutions were prepared on the day of the experiment, filtered (0.45 μm filter) and tested for osmolarity and pH. The osmolarity of all the solutions was close to 300 mOsm. The pH of the arginine (L- or D-isomers) solutions was 11 while that of the L-NAME containing solution was 2.6. No attempt was made to titrate these solutions to a pH of 7.4 since only a small volume was injected (0.1 ml) relative to the large volume of the vitreous (about 1.5 ml).

Throughout the text, doses of drugs will be given as estimated vitreal concentrations assuming a vitreous volume of 1.5 ml (Tano *et al.*, 1980) and complete mixing of the drug in the vitreous. The latter assumption has not been tested directly and is probably incorrect since previous studies have shown that drug-induced retinal damage varies with distance from the site of injection (Loewenstein *et al.*, 1993; Zemel *et al.*, 1993).

2.3. *Electroretinogram (ERG)*

The ERG responses were recorded simultaneously from both eyes with corneal electrodes (Medical Workshop, Holland). The signals were amplified (X20,000) and filtered (0.3-300 Hz) by differential preamplifiers (Grass, USA). The ERG responses were digitized and averaged by a personal computer equipped with a LabMaster data acquisition board (Scientific Solutions, USA) at a rate of 2 KHz. Triggering the light stimuli, sampling of the ERG signals and averaging were controlled by a data acquisition program that was written locally.

Light stimuli obtained from a Ganzfeld light source (LKC Technologies, USA) with a maximum intensity of 5.76 cd-s/m^2. Light-adapted ERG responses were recorded during background illumination of 14.1 cd/m^2.

ERG analysis consisted of amplitude and latency measurements. The latency of the b-wave was measured from stimulus onset to the peak of the b-wave. The amplitude of the a-wave was measured from the baseline to the trough of the a-wave and the amplitude of the b-wave was determined from the trough of the a-wave to the peak of the b-wave. Since the ERG of rabbits varied between different recording sessions, the effects of experimental drugs were assessed from the a- and b-wave ratios (Loewenstein *et al.*, 1993). These were calculated by dividing the amplitude of the ERG wave in the experimental eye by the corresponding value of the control eye. For each ERG recording session, several ratios were calculated and averaged.

3. Results

Intravitreal injection of L-arginine (vitreal concentration of 18 mM) induced complex time dependent effects on the rabbit ERG as shown in figure 1.

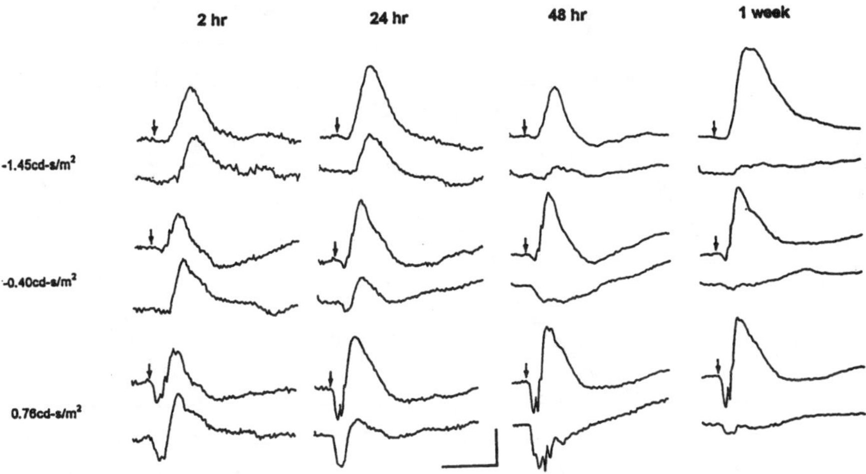

Figure 1. Representative ERG responses of one rabbit recorded at different time intervals after injection of L-arginine (vitreal concentration of 18 mM). In each pair of responses the upper and lower traces represent the ERG of the control and experimental eyes respectively. The intensities of the light stimuli are denoted in log units to the left of each row of responses. Calibration bars: vertical 100 μV; horizontal, 100 ms.

Initially (2 hr), slight augmentation of the ERG responses and slowing down of response kinetics were seen (1st column). Thereafter, the ERG b-wave started to decrease in amplitude while the a-wave increased in amplitude (2nd and 3rd columns). One week after injection, the ERG of the experimental eye was almost non-recordable (4th column). No recovery of the ERG was seen with prolonged periods of follow-up (up to 30 days). Similar long-term toxic action of L-arginine has been reported before (Loewenstein *et al.*, 1993) and here we only dealt with the short-term effects of the drug.

Within the first 5-6 hours after injection of L-arginine, the ERG responses were augmented throughout the dynamic range tested. This is shown in the intensity-response relationships for the ERG a- and b-waves of one rabbit in figure 2A. The ERG responses elicited from the experimental eye by stimuli of moderate to bright intensities were characterized by larger b-waves and no change in a-wave compared to those of the control eye. Similar effects of L-arginine were seen in a total of 14 albino rabbits. Neither of these effects could be attributed to the high pH of the L-arginine solution. Injecting a similar volume of saline that was made basic (pH = 11) with NaOH caused no effect on the rabbit ERG (not shown here).

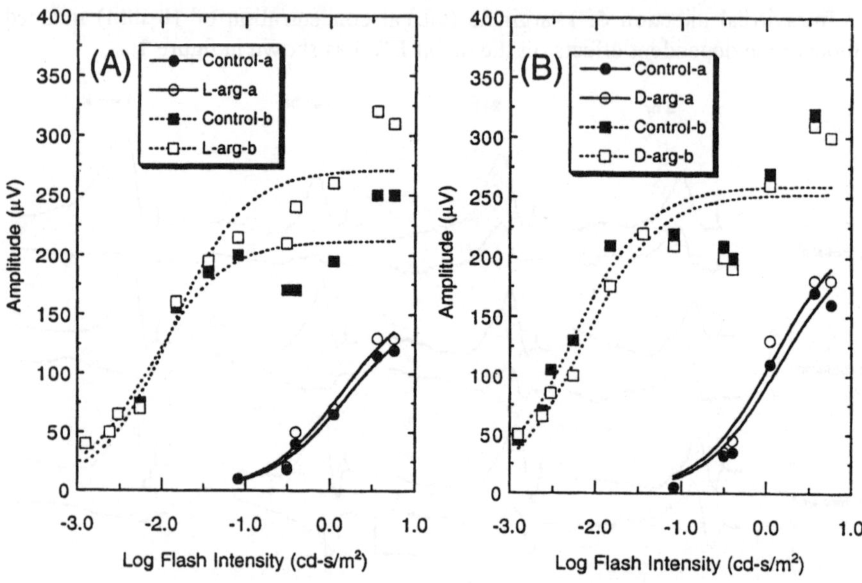

Figure 2. The short-term (5 hours) effects of L-arginine (A) and D-arginine (B) upon the dark-adapted ERG responses of rabbits. Intensity-response relationships for the ERG a- and b-waves (circles and squares respectively) of control (filled symbols) and experimental (open symbols) eyes were fitted to a Michaelis-Menten type relationship (Hood and Birch, 1992).

In order to test the specificity of the L-arginine effect, a similar dose of D-arginine (vitreal concentration of 18 mM) was injected into the vitreous of one eye in 8 rabbits. Short-term (first 6 hrs) effects of D-arginine on the ERG a- and b-wave were negligible as shown in figure 2B for one rabbit. The intensity-response curves are practically identical for the control and experimental eyes. (filled and open symbols respectively). With longer periods of follow-up (not shown here), the ERG responses of the eye injected with D-arginine deteriorated in a similar pattern and time course to that observed with L-arginine (Fig. 1).

The data shown in figures 1 and 2 suggest that the ERG augmentation measured during the first 6 hr after injection of L-arginine reflects a specific action of the drug. A specific action of L-arginine may be mediated through its *in vivo* activation of the NO system. In that case, exogenous application of NO is expected to exert similar effects. In order to test this prediction, we injected sodium nitroprusside (SNP) as an exogenous NO donor. Five rabbits were studied for SNP effects. SNP (vitreal concentration of 67 μM) caused augmentation of the ERG responses (not shown here). The SNP effects developed very fast (within 30 min) and lasted for 2 hr.

If L-arginine augmented the rabbit ERG responses by activating the NO system, then L-NAME, a competitive inhibitor of cNOS (Rees *et al.*, 1990) is expected to induce the opposite effects. Figure 3 illustrates the short-term effects L-NAME (vitreal concentration of 18 mM).

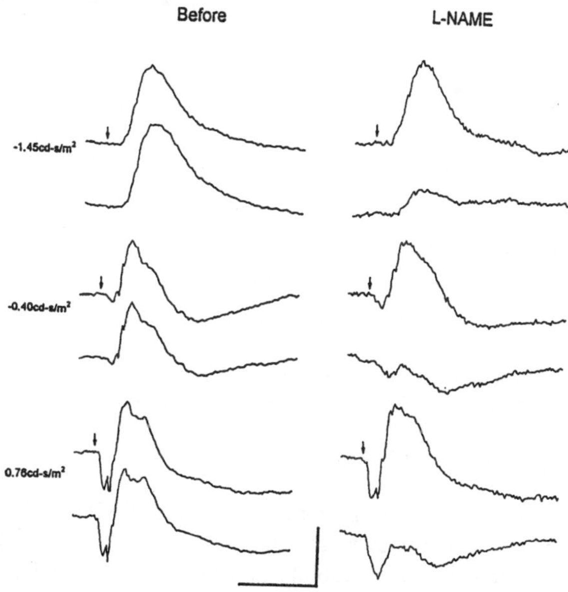

Figure 3. Dark-adapted ERG responses of one rabbit, recorded before and 4 hr after injection of L-NAME (vitreal concentration of 18 mM) into the right eye and saline into the left eye (lower and upper traces respectively). Calibration bars: vertical 100 μV, horizontal 100 ms.

The ERG responses, recorded from both eyes prior to drug injection were of similar amplitude and pattern. Four hours after injection the ERG responses from the eye injected with L-NAME were considerably reduced in amplitude with the b-wave more affected compared to the a-wave. In long-term follow-up (up to 2 weeks), the ERG responses recovered almost completely (about 80% of control). The short-term effects of L-NAME on the intensity-response relationship of the dark-adapted ERG responses are shown in figure 4. L-NAME exerted a selective reduction of the b-wave (open squares) while the a-wave (open circles) was augmented (filled circles). Similar effects of L-NAME were observed in a total of 4 rabbits.

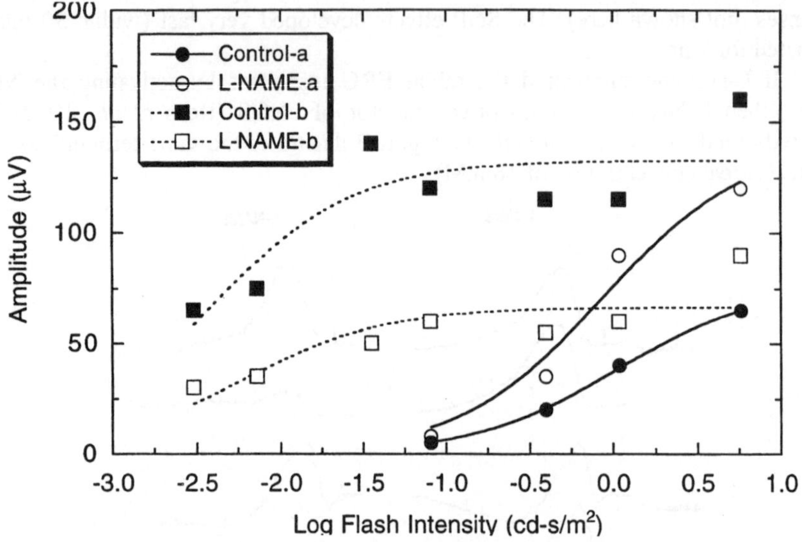

Figure 4. The short-term (4 hr) effects of L-NAME upon the dark-adapted ERG responses of a rabbit. Intensity-response relationships for the ERG a- and b-waves (circles and squares respectively) of control (open symbols) and experimental (filled symbols) eyes were fitted to a Michaelis-Menten type relationship (Hood and Birch, 1992).

The short-term effects of L-arginine, D-arginine, SNP and L-NAME on the dark-adapted ERG a- and b-waves are summarized in Table 1. The non-treated ERG data were obtained from 7 rabbits before injection and indicate the small variability between the two eyes in our ERG recording system. The ERG b-wave ratio of the L-arginine and of the SNP groups and the a-wave ratio of the SNP group were all larger than unity indicating a drug-induced augmentation. All these values differed significantly (student t-test, $p<0.005$) from those of the non-treated ERG ratios and those of the rabbits treated with D-arginine. The a-wave and b-wave ratios of the D-arginine group did not differ significantly from the non-treated group. The effect of L-arginine upon the a-wave was similar to the control. L-NAME significantly ($p<0.005$) increased the a-wave and decreased the b-wave. It should be emphasized that the values listed in Table 1, do not necessarily represent the optimum effects of the drugs since no attempt was made to construct dose-response relationships. In fact, a previous study showed that nitroprusside-induced augmentation of cGMP production in isolated slices of rat cerebellum could not be saturated (Southam and Garthwaite, 1991). Furthermore, the exact injection site could not be accurately controlled and therefore, the time needed to achieve maximum effects probably varies between experiments. All the data in Table 1 were obtained 4 hr after injection of L-arginine, D-arginine and L-NAME and 2 hr after injection of SNP.

Table 1. Short-term effects of NO related drugs on the dark-adapted ERG ratios (experimental eye/control eye). For each drug, the estimated concentration in the vitreous are given.

	Non-treated	**D-arginine** (18 mM)	**L-arginine** (18 mM)	**SNP** (67 μM)	**L-NAME** (18 mM)
b-wave (mean)	0.994	0.97	1.28*	1.17*	0.45*
b-wave (s.d.)	0.043	0.16	0.17	0.17	0.14
a-wave (mean)	0.991	1.10	1.	1.37*	1.73*
a-wave (s.d.)	0.067	0.16	0.16	0.12	0.38
N	7	8	14	5	4

*statistically ($p<0.005$) different from unity using student t-test.

If L-NAME exerted its effects on the rabbit ERG through inhibition of NO production, then, exogenous application of NO is expected to oppose the effects of L-NAME as shown in figure 5.

The ERG responses of both eyes were similar in amplitude and pattern before injection (left column). One hour after injection of L-NAME (vitreal concentration of 18 mM) into the vitreous of the right eye, the ERG responses were significantly reduced in amplitude with the b-wave more affected than the a-wave (middle column). We measured the ERG responses only 1 hr after injection of L-NAME in order to verify that the drug effects had begun to develop. At that time, SPNN was injected into the vitreous of the right eye (vitreal concentration of 670 μm) and saline into the vitreous of the left eye. Three hours after the second injection, the ERG responses of the experimental eye (lower traces) almost completely recovered in terms of b-wave amplitude and b-wave to a-wave relationship (right column). This sequence for injections and ERG recordings was chosen because in previous experiments where only L-NAME was injected, the effects of the drug on the ERG responses reached their maximal effectiveness about 3-4 hr after injection. Similar effects were seen in 2 rabbits.

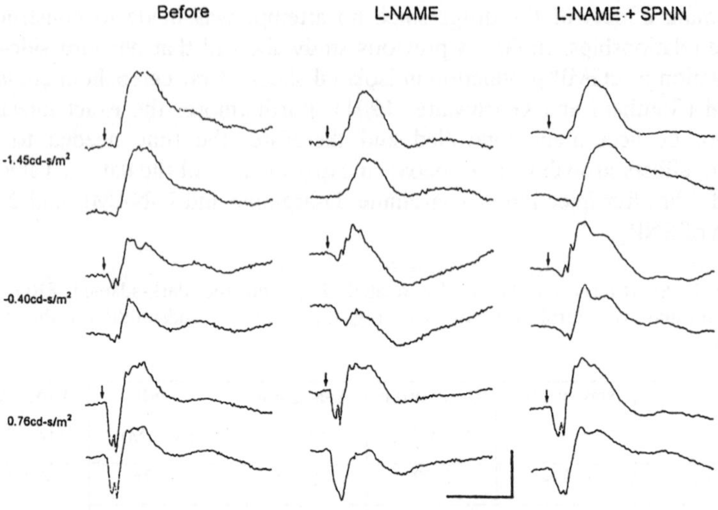

Figure 5. The effects of L-NAME and SPNN on the ERG responses of one rabbit. After recording control ERG responses from both eyes (left column), L-NAME was injected into the vitreous of the right eye and saline into the vitreous of the left eye. The effects of L-NAME were measured 1 hr after injection (middle column). Then, SPNN was injected into the right eye and saline into the left eye. The effects of SPNN were measured 3 hr after injection (right column). Calibration bars: vertical 100 μV, horizontal 100 ms.

The effects of L-arginine, L-NAME and SNP on the rabbit ERG can reflect a direct action of endogenously produced or exogenously supplied NO on any of the major contributors to the flash-evoked ERG responses; the photoreceptors, the ON-center bipolar cells and the Muller cells. In order to separate between these possibilities, all three drugs were injected after isolating the receptor component of the ERG, the P-III wave, with a mixture of APB and PDA. APB, a glutamate agonist for metabotrophic receptors, is used to block synaptic transmission from photoreceptors to ON-center bipolar cells (Slaughter and Miller, 1981; Slaughter and Miller, 1985). PDA, an antagonist of KA/AMPA type glutamate receptors, blocks synaptic transmission from photoreceptors to horizontal cells and OFF-center bipolar cells (Massey and Miller, 1987; Slaughter and Miller, 1981). Thus, the combination of APB and PDA is expected to completely block synaptic transmission from the photoreceptors to the inner retina without affecting the photoreceptors themselves (Sieving *et al.*, 1994). We used a mixture of drugs that produced an estimated vitreal concentration of 0.4 mM and 1.3 mM for APB and PDA respectively.

The effects of L-arginine (vitreal concentration 18 mM), L-NAME (vitreal concentration 18 mM) and SNP (vitreal concentration of 67 μM) on the isolated receptor component of the rabbit ERG are shown in figure 6. In each case, the mixture of APB + PDA was first injected into both eyes. One hour after injection, the ERG were measured in order to verify similar isolation of the P-III component of the ERG. Then the experimental drug was injected into the right eye and the saline into the left control eye (lower and upper traces in each pair of responses respectively). The effects of L-arginine and L-NAME were measured 4 hr after injections while those of SNP 2 hr after injection. In both eyes of each rabbit, the APB + PDA mixture effectively blocked synaptic transmission from the photoreceptors to the neurons in the inner nuclear layer (INL). The ERG responses consisted of a large negative wave that resembled the receptor component of the ERG. In the rabbits tested with either L-arginine (Fig. 6, 1st column) or L-NAME (Fig. 6, 2nd column) no apparent difference in either amplitude or temporal properties were seen between the ERG responses of the two eyes while SNP augmented the receptor component of the ERG (Fig. 6, 3rd column).

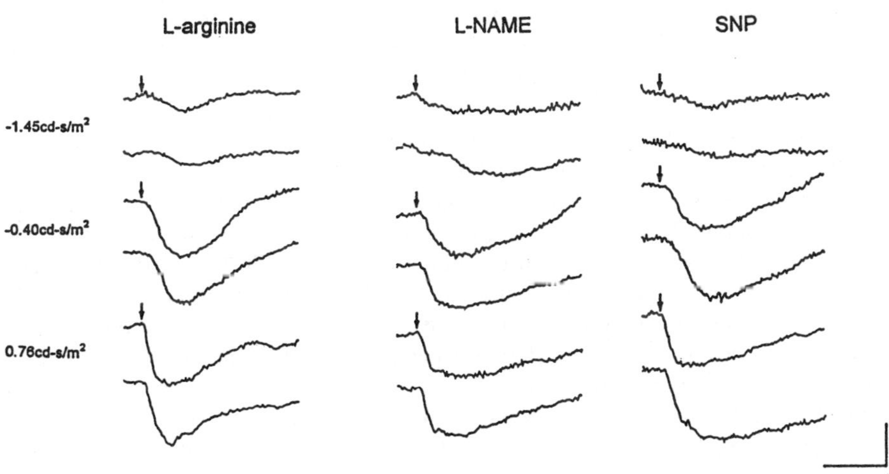

Figure 6. Short-term effects of L-arginine, L-NAME and Sodium Nitroprusside on the isolated receptor potential of rabbits. The receptor component of the ERG was isolated in both experimental and control eyes (lower and upper traces in each pair of responses respectively) with a mixture of APB + PDA that was injected 1 hr prior to drug injection. Calibration bars; vertical 100 μV, horizontal 100 ms.

4. Discussion

Comparing the effects of L-arginine and D-arginine on the rabbit ERG responses (Figs. 1-2) allows differentiating between specific and non-specific action of the amino acid. The short-term effects of L-arginine, that were characterized by an increase in the amplitude of the ERG and a slow down of the ERG kinetics (Fig. 1, first column), probably reflected specific action since this phase was not induced by the D-isomer (Fig. 2B). The slow deterioration of the ERG responses seen for longer periods of follow-up was similar for both isomers of arginine and therefore, reflected non-specific toxic action. The pattern of the ERG loss observed during the entire period of follow-up, suggested that the toxic action of the amino acids spread in accordance with the diffusion of the drug from the vitreous towards the distal retina. Retinal cells in the INL were damaged first causing a selective reduction of the b-wave (Fig1, 2nd column). Since the a-wave of the ERG reflects the summation of two opposing components (P-III and P-II), it was augmented during the initial stages of toxicity causing the ERG responses to change their wave-form into a negative pattern (Perlman, 1983). At longer time intervals when the toxic action of the drugs spread throughout the entire retinal depth, the a- and b-waves were similarly reduced.

The short-term differential effects of L- and D-arginine on the ERG responses of rabbits suggest the involvement of an enzymatic reaction that utilizes the L- and not the D-isomer of arginine. A possible candidate is the formation of nitric oxide by nitric oxide synthase (NOS) (Bredt and Snyder, 1992). The effects of Nitroprusside (SNP) on the rabbit ERG responses were qualitatively similar to those of L-arginine and thus, support the above hypothesis. This NO donor (Southam and Garthwaite, 1991) exerted a transient increase in the ERG responses that reached its maximal magnitude within 1-2 hr (Table 1). The differences in the rate of ERG augmentation seen with SNP and L-arginine probably reflect the different mode of NO supply. L-arginine is the substrate of NOS and therefore, the increased level of NO is associated with increased rate of *in vivo* NO production. This mechanism is expected to develop relatively slowly depending on L-arginine diffusion within the retina, its uptake by the NOS containing cells and activation of NOS. SNP is an NO donor that release NO at fast rate, its half life is less than 3 min (Diodati *et al.*, 1992; Maragos *et al.*, 1993). It should be noted that in a previous report, an NO donor (S-nitroso-N-acetyl-DL-penicillamine; SNAP) was found to reduce the rabbit a-wave and increase the oscillatory potential (Oku *et al.*, 1997). The reduction in the a-wave was attributed to toxic action of the drug upon the photoreceptors. The augmentation in the oscillatory potentials are basically similar to our findings and may reflect NO-dependent augmentation of post-receptoral neuronal activity.

Additional support for a physiological role for NO in the rabbit retina is offered by the experiment using L-NAME. This drug is an efficient competitive inhibitor of endothelial NOS (Rees *et al.*, 1990) and its injection into the vitreous of the rabbit eye is expected to rNO production from endogenous L-arginine. L-NAME caused a significant reduction in the ERG b-wave (Figs. 3 and 4), by about 55% (Table 1). This observation is consistent with the notion that NO is tonically produced in the rabbit retina and its effects on retinal structures are reflected in ERG augmentation. This conclusion is further supported by the ability of exogenous addition of NO to reverse the L-NAME effect on the rabbit ERG responses (Fig. 5). The idea that NO is produced at high basal rate in the intact rabbit retina also explains the relatively small augmentation induced by L-arginine (28%) and SNP (17%) (Table 1). If the basal rate of NO production is high in the normal rabbit retina, then adding more NO is not expected to induce large additional effect.

The enzymatic machinery for NO production exists in the distal retina (horizontal cells) and proximal retina (amacrine cells) of rabbits as indicated by immunostaining for NOS and by NADPH diaphorase histochemistry (Haberecht *et al.*, 1998; Koistinaho *et al.*, 1993; Osborne *et al.*, 1993; Sagar, 1986; Vaney and Young, 1988; Zemel *et al.*, 1996). Assuming that the target cells for NO modulation are those dependent on cGMP, two neurons in the outer retina that contribute to the ERG responses should be considered; the photoreceptors and the ON-center bipolar cells. In order to separate between these two putative sites of NO action, mixtures of APB + PDA were injected in order to isolate the receptor component of the ERG (Knapp and Schiller, 1984; Sieving *et al.*, 1994). Neither increasing the level of *in vivo* NO production by administration of L-arginine nor inhibiting *in vivo* NO production by L-NAME induced any measurable effect on the receptor component of the ERG (Fig. 6) indicating that endogenously produced NO do not modulate the activity of the photoreceptors. Thus, the short-term effects of L-arginine and L-NAME on the rabbit ERG probably reflect an action of NO on ON-center bipolar cells.

This conclusion does not oppose the possibility that guanylate cyclase in photoreceptors is also NO-sensitive. In fact, exogenous administration of NO by SNP augmented the isolated receptor component of the ERG (Fig. 6). SNP releases NO in a very fast rate therefore, NO can reach the photoreceptors and exerts its action on the guanylate cyclase before being neutralized by water and oxygen (Garthwaite, 1991; Stamler *et al.*, 1992). Thus, guanylate cyclase in the photoreceptors is sensitive to NO and can be activated by this messenger molecule to produce more cGMP, as suggested before based on recordings from isolated rods (Schmidt *et al.*, 1992; Tsuyama *et al.*, 1993).

A previous study reported a reduction in the ERG P-III wave when the isolated rabbit retina was exposed to SNP (Maynard *et al.*, 1995). The difference between this observation and our findings (Figs. 6) may reflect technical factors. In the study on the isolated rabbit retina, only dim light stimuli were used to elicit the ERG (Maynard *et al.*, 1995) in order to prevent bleaching of the visual pigment. When the cytosolic guanylate cyclase is activated by NO, the dark-adapted concentration of cGMP is expected to be high and the "dark" potential more depolarized. Under these conditions, a dim light stimulus may induce only minimal activation of phosphodiesterase that is insufficient to overcome the high intracellular level of cGMP and therefore, the light-induced activity is smaller than normal. Only with bright light stimuli, when the PDE is maximally activated, the photoreceptors can hyperpolarize maximally from the elevated «dark» potential and therefore, the ERG P-III is expected to be of supernormal amplitude.

How does endogenous NO affect the amplitude and kinetics of the ERG? Assuming that the a-wave amplitude depends on the opposing contributions of the P-II and P-III waves and that the P-II mainly reflects the activity of ON-center bipolar cells (Ripps and Witkovsky, 1985), we suggest that a tonic production of NO maintains a relatively high basal activity of cytoplasmatic guanylate cyclase in ON-center bipolar cells. This effect keeps the ON-center bipolar cells in a slightly more depolarized level in darkness but allows for a fast and large increase in intracellular cGMP when a light stimulus is applied and the activity of PDE in the ON-center bipolar cells is reduced. Thus, a large and fast depolarizing response is obtained. Exogenous application of NO using NO donors (SNP and SPNN) or endogenous increase in NO production by L-arginine act to augment this effect since these procedures further activate the guanylate cyclase in the ON-center bipolar cells. Under these conditions, the rate of recovery of the membrane potential at stimulus offset is expected to be slower than in the normal conditions because the PDE is activated to a normal rate but it is working against a supernormal concentration of cGMP and supernormal activity of guanylate cyclase. Therefore, the ERG responses are characterized by slow kinetics especially in the recovery phase of the b-wave (Fig. 1, first column). L-NAME reduces *in vivo* production of NO causing reduction in the activity of guanylate cyclase. The response of the ON-center bipolar cells to a light stimulus is expected to reduce in amplitude since the magnitude of light-induced increase in cGMP is smaller.

Acknowledgments

This study was supported in part by grant No. 2573 from the Chief Scientist's Office of Ministry of the Health, Israel.

References

Bredt, D.S. and S.H. Snyder (1992) “Nitric oxide, a novel neuronal messenger”, *Neuron* 8:3-11.

DeVries, S.H. and E.A. Schwartz (1989) “Modulation of an electrical synapse between solitary pairs of catfish horizontal cells by dopamine and second messengers”, *J. Physiol. (Lond.)* 414:351-375.

Diodati, J.G., A.A. Quyyumi and L.K. Keefer (1992) “Complexes of nitric oxide with nucleophiles as agents for the controlled biological release of nitric oxide: Hemodynamic effect in the rabbit", *J. Cardiovasc. Pharmaco.* 22:287-292.

Garthwaite, J. (1991) "Glutamate, nitric oxide and cell-cell signalling in the nervous system", *Trends Neurosci.* 347:768-770.

Haberecht, M.F., H.H. Schmidt, S.L. Mills, S.C.N. Massey, M. and D.A. Redburn-Johnson (1998) "Localization of nitric oxide synthase, NADPH diaphorase and soluble guanylyl cyclase in adult rabbit retina", *Vis. Neurosci.* 15:881-890.

Haverkamp, S., H. Kolb and N. Cuenca (1999) "Endothelial nitric oxide synthase (eNOS) is localized to Müller cells in all vertebrate retinas", *Vis. Res.* 39:2299-2303.

Hood, D.C. and D.G. Birch (1992) "A computational model of the amplitude and implicit time of the b-wave of the human ERG", *Vis. Neurosci.* 8:107-126.

Ignarro, L.J. (1990) "Haem-dependent activation of guanylate cyclase and cyclic GMP formation by endogenous nitric oxide: a unique transduction mechanism for trancellular signaling", *Parmcol. Toxicol.* 67:1-7.

Kaupp, U.B. and K.-W. Koch (1992) "Role of cGMP and calcium in vertebrate photoreceptor excitation and adaptation", *Ann. Rev. Physiol.* 54:153-175.

Knapp, A.G. and P.H. Schiller (1984) "The contribution of ON-bipolar cells to the electroretinogram of rabbits and monkeys. A study using 2-amino-4-phosphonobutyrate (APB)", *Vis. Res.* 24:1841-1846.

Knowles, R.G., M. Palacois, R.M. Palmer and S. Moncada (1989) "Formation of nitric oxide from L-arginine in the central nervous system: a transduction mechanism for stimulation of soluble guanylate cyclase", *Proc. Natl. Acad. Sci. USA* 86:5159-5162.

Koistinaho, J. and S.M. Sagar. 1995 "NADPH-diaphorase-reactive Neurons in the Retina", in: *Progress in Retina and Eye Research*, N.N. Osborne and G.J. Chader, eds, Oxford: Pergamon Press, pp. 69-87.

Koistinaho, J., R.A. Swanson, J. De Vente and S.M. Sagar (1993) "NADPH-diaphorase (nitric oxide synthase)-reactive amacrine cells of rabbit retina: Putative target cells and stimulation by light", *Neurosci.* 57:587-597.

Loewenstein, A., E. Zemel, M. Lazar and I. Perlman (1991) "The effects of depo-medrol preservative (MGP) on the rabbit visual system", *Invest. Ophthalmol. Vis. Sci.* 32:3053-3060.

Loewenstein, A., E. Zemel, M. Lazar and I. Perlman (1993) "Drug-induced retinal toxicity in albino rabbits: Effects of imipenem and aztreonam", *Invest. Ophthalmol. Vis. Sci.* 34:3466-3476.

Maragos, C.M., J.M. Wang, J.A. Hrabie, J.J. Oppenheim and K.L. Keefer (1993) "Nitric oxide/nucleophile complexes inhibit the in vitro proliferation of A375 melanoma cells via nitric oxide release", *Cancer Res.* 53:564-568.

Massey, S.C. and R.F. Miller (1987) "Excitatory amino acid receptors of rod- and cone-driven horizontal cells in the rabbit retina", *J. Neurophysiol.* 57:645-

659.

Maynard, K.I., P. Yanez and C.S. Ogilvy (1995) "Nitric oxide modulates light-evoked compound action potentials in the intact rabbit retina", *Neuroreport* 6:850-852.

McMahon, D.G. and L.V. Ponomareva (1996) "Nitric oxide and cGMP modulate retinal glutamate receptors", *J. Neurophysiol.* 76:2307-2315.

Miyachi, E. and M. Murakami (1991) "Synaptic inputs to turtle horizontal cells analyzed after blocking of gap junctions by intracellular injection of cyclic nucleotides", *Vis. Res.* 31:631-635.

Miyachi, E., M. Murakami and T. Nakaki (1990) "Arginine blocks gap junctions between retinal horizontal cells", *NeuroReport* 1:107-110.

Nawy, S. and C.E. Jahr (1991) "cGMP-gated conductance in retinal bipolar cells is supressed by photoreceptor transmitter", *Neuron* 7:677-683.

Oku, H., H. Yamaguchi, T. Sugiyama, S. Kojima, M. Ota and I. Azuma (1997) "Retinal toxicity of nitric oxide released by administration of a nitric oxide donor in the albino rabbit", *Invest. Ophthalmol. Vis. Sic.* 38:2540-2544.

Osborne, N.N., N.L. Barnett and A.J. Herrera (1993) "NADPH diaphorase localization and nitric oxide synthetase activity in the retina and anterior uvea of the rabbit eye", *Brain Res.* 610:194-198.

Perlman, I. (1983) "Relationship between the amplitudes of the b wave and the a wave as a useful index for evaluating the electroretinogram", *Br. J. Ophthalmol.* 67:443-448.

Pottek, M., K. Schultz and R. Weiler (1997) "Effects of nitric oxide on the horizontal cell network and dopamine release in the carp retina", *Vision Res.* 37:1091-1102.

Rees, D.D., R.M. Palmer, R. Schulz, H.F. Hodson and S. Moncada (1990) "Characterization of three inhibitors of endothelial nitric oxide synthase in vitro and in vivo", *Br. J. Pharmacol.* 101:746-752.

Ripps, H. and P. Witkovsky. (1985) "Neuron-glia interaction in the brain and retina", in: *Progress in Retinal Research*, N.N. Osborne and G.J. Chader, eds, Oxford: Pergamon Press, pp. 181-219.

Sagar, S.M. (1986) "NADPH diaphorase histochemistry in the rabbit retina", *Brain Res.* 373:153-158.

Schmidt, K.-F., G.N. Noll and Y. Yamamoto (1992) "Sodium nitroprusside alters dark voltage and light responses in isolated retinal rods during whole-cell recording", *Vis. Neurosci.* 9:205-209.

Shiells, R.A. and G. Falk (1992) "Retinal ON-bipolar cells contain a nitric oxide-sensitive guanylate cyclase", *NeuroReport* 3:845-848.

Sieving, P.A., K. Murayama and F. Naarendorp (1994) "Push-pull model of the primate photopic electroretinogram: A role for hyperpolarizing neurons in shaping the b-wave", *Vis. Neurosci.* 11:519-532.

Slaughter, M.M. and R.F. Miller (1981) "2-amino-4-phosphonobutyric acid: a new pharmachological tool for retinal research", *Science* 211:182-185.

Slaughter, M.M. and R.F. Miller (1985) "Characterization of an extended glutamate receptor of the on bipolar neuron in the vertebrate retina", *J. Neurosci.* 5:224-233.

Snyder, S.H. (1992) "Nitric oxide: first in a new class of neurotransmitters?", *Science* 257:494-496.

Southam, E. and J. Garthwaite (1991) "Comparative effects of some nitric oxide donors on cyclic GMP levels in rat cerebellar slices", *Neurosci. Lett.* 130:107-111.

Stamler, J.S., D.J. Singel and J. Loscalzo (1992) "Biochemistry of nitric oxide and its redox-activated forms", *Science* 258:1898-1901.

Tano, Y., D.B. Chandler and R. Machemer (1980) "Treatment of intraocular proliferation with intravitreal injection of triamcinolone acetonide", *Am. J. Ophthalmol.* 90:810-816.

Tsuyama, Y., G.N. Noll and K.-F. Schmidt (1993) "L-arginine and nicotinamide adenine dinucleotide phosphate alter dark voltage and accelerate light response recovery in isolated retinal rods of the frog (Rana temporaria)", *Neurosci. Lett.* 149:95-98.

Vaney, D.I. and H.M. Young (1988) "GABA-like immunoreacitivity in NADPH-diaphorase amacrine cells of the rabbit retina", *Brain Res.* 474:380-385.

Yau, K.-W., L.W. Haynes and K. Nakatani. (1988) "Study of phototransduction mechanism in rods and cones", in: *Cellular and Molecular Biology of the Retina*, D.M.K. Lam, ed., Cambridge: MIT Press, pp. 41-58.

Zemel, E., O. Eyal, B. Lei and I. Perlman (1996) "NADPH diaphorase activity in the mammalian retina is modulated by the state of visual adaptation", *Vis. Neurosci.* 13:863-871.

Zemel, E., A. Loewenstein, M. Lazar and I. Perlman (1993) "The effects of myristyl γ-picolinium chloride on the rabbit retina: Morphologic observations", *Invest. Opthalmol. Vis. Sci.* 34:2360-2366.

COLOUR MATCHING IN RED/GREEN CHROMATICITY TYPE HORIZONTAL CELLS OF THE TURTLE RETINA

HUSAM ASI, AVIRAN ITZHAKI and IDO PERLMAN
The Bruce Rappaport Faculty of Medicine, Technion-Israel Institute of Technology and the Rappaport Institute, P.O.Box 9649, Haifa 31096, Israel

ABSTRACT

The pigment theory of colour matching was examined here by conducting electrophysiological colour matching experiments in Red/Green (R/G) chromaticity-type horizontal cells of the turtle retina. In each cell, the photoresponses elicited by monochromatic orange stimuli were matched to those elicited by mixtures of red (700nm) and green (540nm) lights and compared the experimental results to the theoretical ones, calculated from the action spectra of turtle cones. The variability of colour matching between different horizontal cells could be accounted for by the filtering properties of the oil droplets located in the inner segments of the cones. For a given C-type horizontal cell, colour matching was radiance invariant within the limited range of intensities studied here (about 1.5 log units) despite nonlinear interactions between the antagonistic inputs from the different cone types to the horizontal cells.

1. Introduction

Cone photoreceptors and horizontal cells in the vertebrate retina are involved in an intricate neural network consisting of feedback and feedforward excitatory and inhibitory synapses. These interactions lead to the genesis of colour-coded photoresponses in chromaticity (C-type) horizontal cells (Naka and Rushton, 1966a; Svaetichin and MacNichol, 1958). Two types of chromaticity horizontal cells have been described physiologically in the turtle retina (Asi and Perlman, 1998; Fuortes and Simon, 1974). They differ in the wavelength at which response polarity reverses; 600-620nm for Red/Green (R/G) cells and 540-560nm for Yellow/Blue (Y/B) cells. Turtle C-type horizontal cells belonging to the same type exhibit very similar action spectra as determined from sensitivity measurements (Asi and Perlman, 1998). However, they differ significantly in the waveform of the photoresponses elicited by bright monochromatic stimuli. This may reflect variability in the relative magnitudes of the inputs from the different spectral types of cones, differences in the mode of summation of these inputs or additional inputs from other sources that are too weak to contribute to sensitivity measurements but may become apparent with bright lights. One way to test these possibilities is by colour matching.

Colour matching is one of the most fundamental psychophysical experiments in colour vision. The subject is instructed to adjust the intensities of red, green and blue lights until the mixture appears indistinguishable from a given monochromatic stimulus. The pigment theory argues that colour matching is achieved only when photon absorption by each of the visual pigments participating in colour vision is equal for the matched stimuli (Alpern, 1989; Rushton, 1972). Mathematically, this theory is described by:

$$I_\lambda \alpha_\lambda = \alpha_r I_r + \alpha_g I_g + \alpha_b I_b$$

$$I_\lambda \beta_\lambda = \beta_r I_r + \beta_g I_g + \beta_b I_b$$

$$I_\lambda \gamma_\lambda = \gamma_r I_r + \gamma_g I_g + \gamma_b I_b \quad (1)$$

Where α, β and γ are the absorption coefficients of the visual pigments in the L-, M- and S-cones respectively, The symbols λ, r, g, and b refer to wavelength λ, red, green and blue lights. I_λ, I_r, I_g and I_b are the intensities of the λ, red, green and blue lights respectively.

C-type horizontal cells in the turtle retina are ideally suited for colour matching experiments because they receive antagonistic inputs from cones of different spectral types resulting in photoresponses of complex pattern that are highly dependent on the wavelength and intensity of the stimulus. This procedure was previously applied to R/G C-type horizontal cells in the fish retina but successful matching was achieved only for responses of relative simple waveform and not for stimuli that equally excited the L- and M-cones (Naka and Rushton, 1966b). In this study, we performed colour matching experiments in turtle R/G C-type horizontal cells using as monochromatic stimuli, wavelengths in the range of transition between depolarizing and hyperpolarizing responses, in order to test the basic assumptions of the pigment theory of colour matching.

2. Materials and Methods

The experiments were performed on the eyecup preparation of the fresh water turtle *Mauremys caspica*. The retina of this turtle is very similar to that of the *Pseudemys scripta elegans* except that in the *Mauremys* retina only one type of oil droplet (red ones) has been assigned to single L-cones (Kolb *et al.*, 1988).

The eyecup was prepared according to procedures that have been previously described in detail (Baylor *et al.*, 1971; Itzhaki and Perlman, 1984). Following decapitation, one eye was enucleated and the anterior part was hemisected away with a razor blade and the vitreous humor was cleared. The eyecup was placed in a holding chamber, surrounded by moist cotton to minimize retinal dehydration. A continuous flow of 95%O2/5%CO2 mixture, saturated with water vapor, was directed above the preparation.

Microelectrodes were pulled (Sutter Instruments P80/PC, USA) from capillary tubing and filled with 3M potassium acetate solution. Microelectrode resistance was typically in the range of 100-300MΩ. The amplified signal (Almost Perfect Electronics, Switzerland) of the microelectrode was displayed on an oscilloscope screen, monitored by a pen recorder (Gould 2200, Cleveland, OH, USA) and digitized by a personal computer equipped with a LabMaster DMA (Scientific Solutions, USA) data acquisition board.

The photostimulation system consisted of two light beams originating from a single light source (250-watt tungsten halogen). One beam was used for delivering the monochromatic light stimuli while both beams were used simultaneously to obtain mixtures of red and green lights. The intensity and spectral content of each beam was independently controlled by neutral density and narrow-band (half bandwidth 5-10nm) interference filters. The duration of the light stimuli and the intervals between successive stimuli were controlled by electronic shutters (Vincent Associates, USA). The intensities of the light beams were calibrated with a PIN 10 photodiode (United Detector Technology, USA).

R/G C-type horizontal cells were identified according to physiological criteria which have been previously established by extensive retinal research on a variety of fresh water turtles (Asi and Perlman, 1998; Fuortes *et al.*, 1973; Fuortes and Simon, 1974). These cells were characterized by large receptive fields (diameter>2mm) and exhibited colour opponency. They responded with hyperpolarization to short- and middle-wavelengths and with depolarization to long-wavelengths. Response polarity reversed around 600-620nm.

After impaling an R/G C-type horizontal cell, the photoresponse was elicited by a monochromatic stimulus of 600 or 620nm. Then, the eyecup was stimulated with a mixture of red (700nm) and green (540nm) lights. The intensity of each channel was independently changed until the evoked photoresponse was most similar to that elicited previously by the monochromatic stimulus. After obtaining a match, the monochromatic stimulus was applied again in order to ensure the stability of the recording. All light stimuli, monochromatic and mixtures were of 500ms duration. The stimuli covered retinal areas of large diameter (2.9mm) in order to completely illuminate the receptive fields of the cells (Itzhaki and Perlman, 1984; Lamb, 1976; Simon, 1973). Signal/noise ratio was improved by on-line averaging of 10 responses elicited at 3sec intervals.

3. Results

3.1. Colour matching functions of the turtle retina

The pigment theory assumes that colour matching depends only on the rate of photon absorption by the visual pigments. Therefore, colour matching functions can be derived from the absorption spectra of the cone visual pigments. The action spectra of cone photoreceptors, measured in the eyecup preparation deviate from the aspectra of the visual pigments because of the coloured oil droplets that filter the impinging light. Therefore, we used the action spectra of turtle cones that were measured in the inverted isolated retina preparation where light was shown from the outer segments' side (Schneeweis and Green, 1995). Average data, kindly supplied to us by Dr. D. Schneeweis, are shown as continuous curves in figure 1 for L-cones (circles) and M-cones (crosses). The action spectra of S-cones were not measured in the isolated retina. However, these cones contain clear oil droplets (Kolb *et al.*, 1988) which do not change the spectral content of the incident light (Lipetz, 1985). Thus, the absorption spectrum of the S-cone visual pigment (Fig. 1, stars) can be described by the action spectrum of an S-cone measured in the eyecup (Itzhaki *et al.*, 1992).

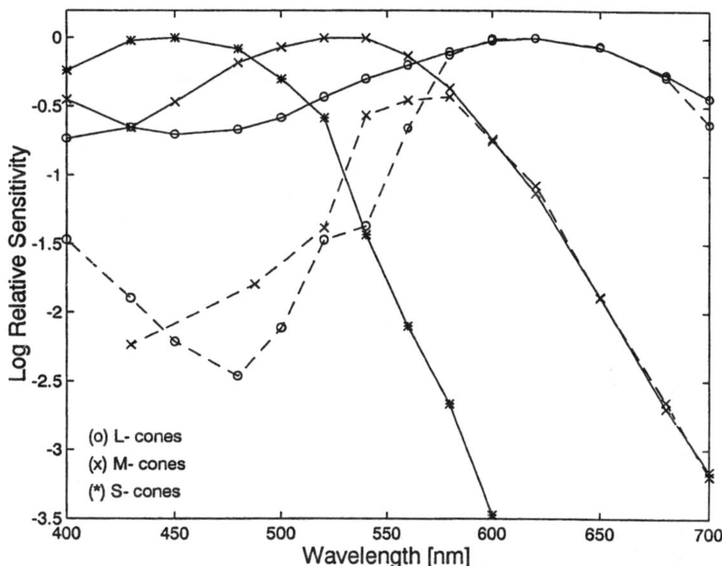

Figure 1. Mean action spectra of L- and M-cones (circles and crosses respectively) measured in the isolated turtle retina (Schneeweis, personal communication) and the action spectrum of an S-cone (stars) recorded in the turtle eyecup (continuous curves). Action spectra of an L- cone (circles) and an M-cone (crosses) recorded in the turtle eyecup (dashed curves).

The set of absorption spectra (continuous curves in Fig. 1) was used to calculate the rate of photon absorption by the three visual pigments in turtle cones for any given wavelength within the visible spectrum. Then, the intensities of the red (700nm), green (540nm) and blue (450nm) lights in the mixture needed to match any monochromatic light were derived by solving the set of equations (Eq. 1). According to the pigment theory, colour matching is radiant-invariant (Grassmann's Law of scalar multiplication). Therefore, the colour matching functions were calculated for 1 photon/sec/μm^2 of each wavelength in the visible range. Figure 2A shows the intensities of the 700nm (circles), 540nm (crosses) and 450nm (stars) needed to achieve the above matches.

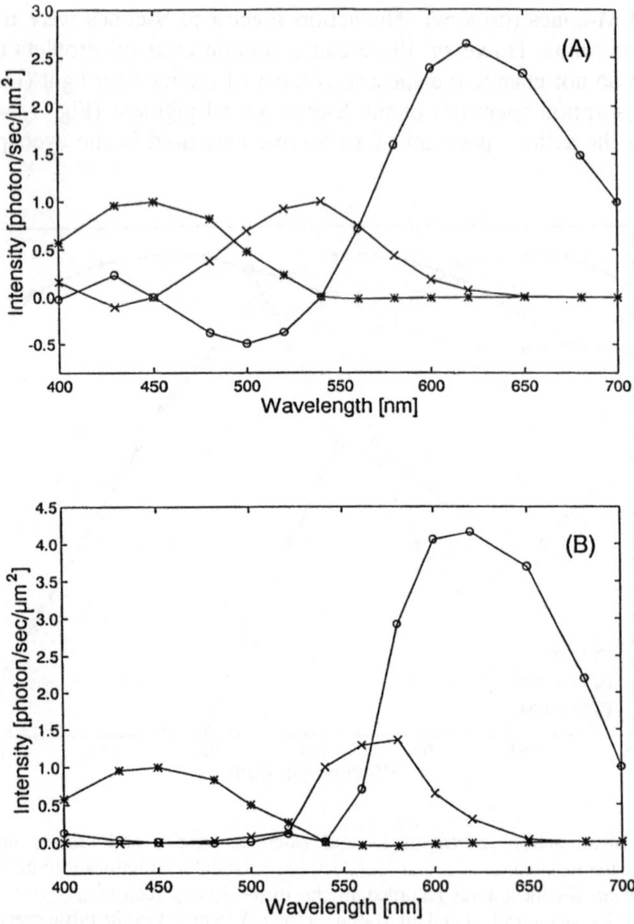

Figure 2. Colour matching functions for the turtle retina calculated by Eq. (1) from the cone action spectra measured in the isolated retina (A) and from the action spectra of the cones in the eyecup (B). The data points describe the intensity of each of three primary lights in the mixtures; 700nm (circles), 540nm (crosses) and 450nm (stars), required to match any wavelength within the visible spectrum of intensity 1 photon/sec/μm^2.

The colour matching functions shown in figure 2A are very similar to those calculated previously for monkey cones though different wavelengths were used for the light components of the mixture (Baylor *et al.*, 1987). As in human colour matching, for certain test wavelength one of the primaries is used as a desaturant. In the turtle, the red light needs to be added to the monochromatic stimulus in the range 450 - 540nm in order to achieve a match to a mixture of green and blue.

The above analysis is a quantitative description of the pigment theory of colour matching for the turtle retina. However, the cone photoreceptors in the turtle retina, unlike the cones in the human retina, contain coloured oil droplets that filter the light before it reaches the outer segment. Therefore, colour matching in the eyecup depends on the action spectra of the cones and not on the absorption spectra of the visual pigments. The effects of the oil droplets on the action spectra of turtle cones depend on the fraction of light reaching the outer segment after bypassing the oil droplets due to scatter and reflection and on the density of the pigment within the oil droplets (Lipetz, 1984). Therefore, the action spectra of cones belonging to the same spectral type vary between different eyecup preparations and even between different recording sites in a given preparation (Baylor and Hodgkin, 1973; Perlman *et al.*, 1994; Schneeweis and Green, 1995). We used as an extreme case action spectra of an L-cone and an M-cone, measured in the eyecup preparation, that exhibited the largest deviation from the visual pigment absorption spectra in the short-wavelength range of the visible spectrum. These are shown as dashed curves in figure 1 (circles and crosses respectively). The action spectrum of the S-cone is not affected by its clear oil droplet. The set of action spectra, measured in the eyecup of the turtle *Mauremys caspica*, are very similar to those measured previously in the eyecup of the turtle *Pseudemys scripta elegans* (Baylor and Hodgkin, 1973).

The colour matching functions were calculated also using the eyecup set of cone action spectra. The intensities of the red (circles), green (crosses) and blue (stars) lights in the mixture needed to match a single photon of any wavelength in the visible spectrum are shown in figure 2B. The two sets of colour matching functions (Figs 2A, B) are similar in shape but vary in a manner that affect colour matching prediction. Both curves indicate that for wavelengths longer than 560nm, a match can be achieved by a mixture of red and green lights. This is the dichromatic range of the turtle where the short-wavelength sensitive visual pigment, located in the S-cones, does not absorb light and therefore does not participate in colour vision. Since the colour matching experiments were performed on R/G C-type horizontal cells with orange light stimuli (600-620nm), the contribution of the S-cones was ignored and colour matching was performed with a mixture of red (700nm) and green (540nm) lights. The prominent difference between the two sets of curves is seen for wavelengths between 560nm and 630nm. In this range, the sensitivities of L- and M- cones are significantly affected by the oil droplets. As a result, a higher intensity of green light relative to the red light is need for the mixture to match the monochromatic light and the ratio I_{540}/I_{700} becomes larger. The theoretical combinations of 700nm and 540nm lights which are needed to match monochromatic light stimuli of unit intensity at wavelengths, 600 and 620nm are given in Table 1 as the range between the two extreme cases discussed above.

3.2. Electrophysiological colour matching

Chromaticity-type horizontal cells are highly sensitive to wavelength and intensity of the light stimuli for wavelength in the transition zone between depolarizing and hyperpolarizing photoresponses as shown n figure 3. The photoresponses shown in this figure were recorded from one dark-adapted R/G C-type horizontal cell in response to light stimuli (50 ms duration) of different wavelength but similar quantal content (350-750 photons/flash/μm^2).

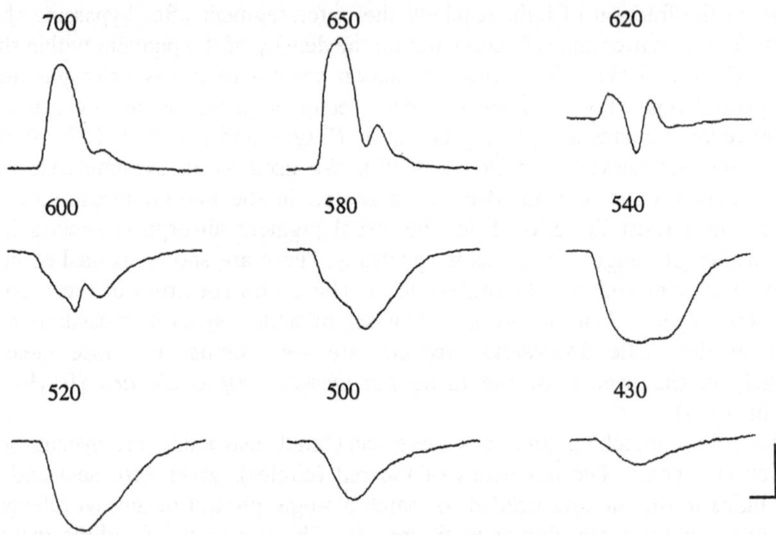

Figure 3. The photoresponses of one R/G C-type horizontal cell elicited in the dark-adapted state by light stimuli of 50ms duration. The wavelength of the stimulus is denoted above each photoresponse. All stimuli were of similar intensity (350-700 photons/flash/μm^2). Calibration bars; vertical 10mV, horizontal 100ms.

The photoresponses, elicited by long wavelength stimuli (650-700nm), are of similar depolarizing waveform while those elicited by short wavelength (430-560nm) stimuli are of similar hyperpolarizing shape. Therefore, colour matching in these ranges is insensitive to small changes in the stimulus parameters and in many cases matching can be achieved between two monochromatic lights *e.g.* 650nm to 700nm or 500 to 540nm. However, when the wavelength of the stimulus is chosen within the transition zone between depolarizing and hyperpolarizing responses (580-620nm), the photoresponse is characterized by a complex waveform and a match can be achieved only with a unique combination of the three light primaries.

Achieving a good match between a monochromatic stimulus and a mixture of red and green lights required stable recordings for at least 15min. Such stable recordings were successful only in 6 R/G C-type horizontal cells. Color matching results for 3 out of 6 different cells studied in different eyecups shown in figure 4 and 5 using for monochromatic light stimuli 600 and 620.

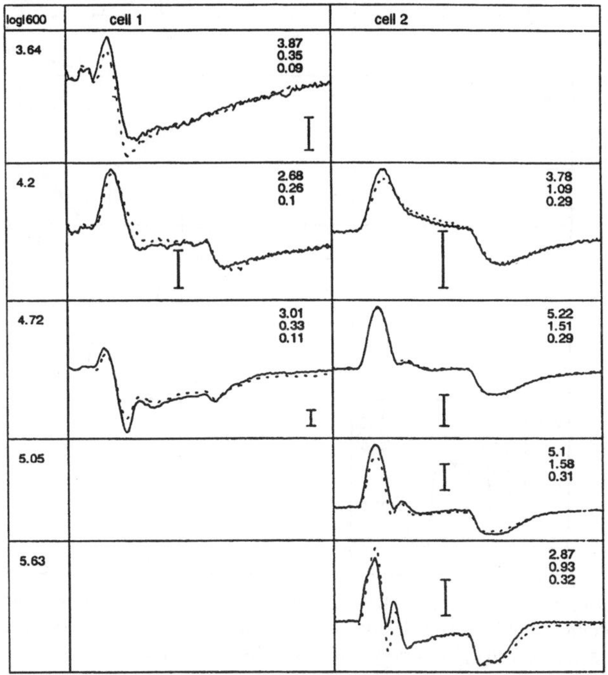

Figure 4. Colour matching in 2 R/G C-type horizontal cells using 600nm light as the monochromatic stimulus. In each square, the photoresponse to the 600nm stimulus is compared to that elicited by a mixture of red (700nm) and green (540nm) lights (continuous and dotted traces respectively). Matching were performed for different intensities of the 600nm stimulus as denoted in logarithmic units (photons/sec/μm^2) at the left of each row. The normalized intensities of the 700nm and 540nm lights and their ratio (I_{540}/I_{700}) needed for each match are indicated from top to bottom at the upper right corner of each square. All records are of 1200ms duration. Vertical bars denote 1 and 5mV for cells # 1 and 2 respectively.

Colour matching to 600nm light stimuli is shown in figure 4 for 2 R/G horizontal cells. In cell #1 (first column) the I_{540}/I_{700} ratios needed to match stimuli of 600nm were about 0.10 for three intensities of the 600nm light which covered a range of 1.08 log units. This ratio is within the range calculated to match 600nm light (Table 1). In cell #2 (second column) the I_{540}/I_{700} ratios needed to match the same test wavelength varied slightly between 0.29 and 0.32 for 4 intensities of the monochromatic light covering a range of 1.43 log units. These ratios are about 3 folds higher than the upper limit of (0.16) of the calculated range (Table 1).

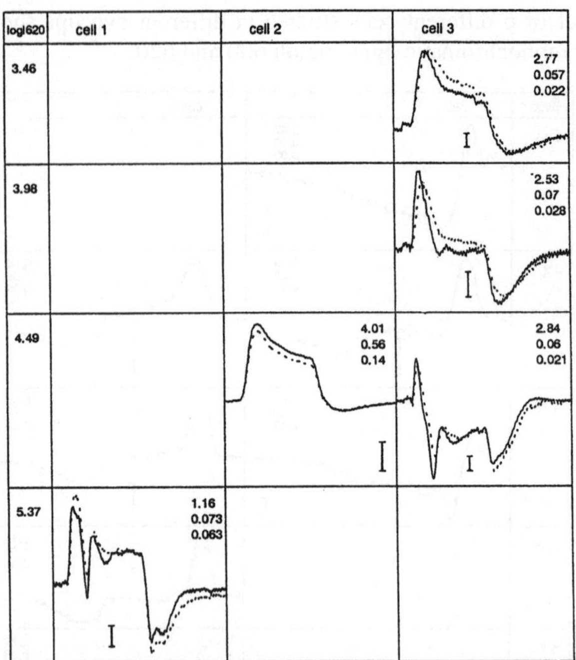

Figure 5. Colour matching in 3 R/G C-type horizontal cells using 620nm light for the monochromatic stimulus. The details of the figure are identical to those of Fig. 4. The vertical bars denote 1 mV, 5mV and 1 mV for cells # 1, 2 and 3 respectively.

Colour matching to 620nm light stimuli is shown for 3 R/G C-type horizontal cells in figure 5. In cell #1 and #2 only one intensity of the monochromatic light was used for colour matching. The I_{540}/I_{700} ratios needed for these cells were 0.063 and 0.14 respectively. The first ratio is within the calculated range and the second slightly above it (Table 1). Cell #3 (third column) was tested with 3 intensities of 620nm light covering a range of 1.03 log units. The I_{540}/I_{700} ratios needed to obtain the matches varied from 0.022 for the lowest intensity to 0.028 for the intermediate intensity to 0.021 for the brightest one. These values are close to the lower limit of the calculated range (Table 1).

The data from all 6 R/G C-type horizontal cells that were studied with 600 and/or 620nm stimuli are summarized in Table 1. In this table, the data labeled "expected" denote the calculated ranges of intensity of the 700nm and 540nm components of the mixture that are needed to match 1 photon of 600nm or 620nm stimuli. These ranges were calculated from the two extreme cases; light incident bypasses the oil droplets (Fig. 2A) or goes through them (Fig. 2B). For each of the cells, the experimental data are given either as single values or as a range. A single value means that only one monochromatic stimulus was tested. A range of values means that several intensities of the same monochromatic test were matched to the green/red mixture. The number of intensities used, are denoted in parenthesis in the columns labeled I_{540}/I_{700}.

Table 1. Normalized light intensities of the 700nm and 540nm components of the mixture and their ratios (I_{540}/I_{700}) needed to match monochromatic light stimuli of wavelength 600nm or 620nm for 6 R/G C-type horizontal cells studied. The Range row gives the expected values that were calculated from theoretical considerations. The values in parenthesis in the columns labeled I_{540}/I_{700} denote the number of intensities of monochromatic stimulus that were tested.

		600nm			**620nm**	
	I_{700}	I_{540}	I_{540}/I_{700}	I_{700}	I_{540}	I_{540}/I_{700}
Range	2.4-4.1	.19-.65	.08-.16	2.6-4.2	.07-.30	.03-.07
Cell #1	2.7-3.9	.26-.35	.09-.11 (3)	1.16	.073	.063 (1)
Cell #2	2.9-5.2	.93-1.51	.29-.32 (4)	4.01	.56	.14 (1)
Cell #3				2.5-2.8	.06-.07	.02-.03 (3)
Cell #4	3.14	.099	.031 (1)			
Cell #5	1.1-3.2	.15-.35	.11-.13 (4)	3.4-6.2	.10-.18	.030 (2)
Cell #6				3.6	.044	.012 (1)

The data in Table 1 clearly indicate the difficulties of the experiments. The green/red combination needed for matching vary between cells and within a given cell between different intensities. Moreover, only few cells were characterized by data that were well within the expected range.

3.3. Linearity of the responses to light mixtures

The pigment theory assumes that colour matching is independent of the type of neural interactions beyond the visual pigments. This assumption was examined by testing for linearity the interactions between the depolarizing and hyperpolarizing inputs in the light mixtures needed for colour matching in R/G C-type horizontal cells. After obtaining a match between a given monochromatic stimulus and the red/green mixture, the photoresponses to the red (700nm) and green (540nm) lights alone were recorded. The photoresponse elicited by the mixture of lights was compared to the algebraic summation of the individual photoresponses. Such a linearity test is shown in figure 6 for cell #1 using the mixtures of 700nm and 540nm lights that were found to match the photoresponses elicited by three intensities of 600nm light (Fig. 4, first column).

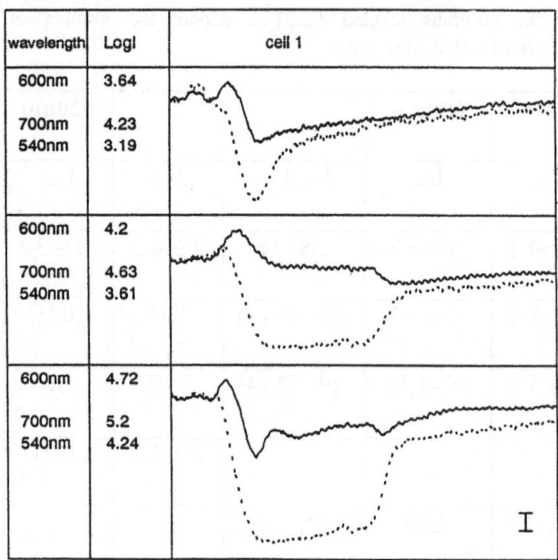

Figure 6. Testing the interactions between the depolarizing and hyperpolarizing inputs in one R/G C-type horizontal cell. The responses elicited by the mixtures of red (700nm) and green (540nm) lights needed to match 600nm stimuli are compared to the sums of the individual responses elicited by each light component alone (continuous and dotted traces respectively). The absolute intensities (photons/sec??m^2) of the 600nm stimulus and of the 700nm and 540nm components of the matched mixtures are indicated in logarithmic units in the left column. All records are of 1200ms length. Vertical calibration: 1mV.

The photoresponses elicited by the mixtures are shown in figure 6 as continuous traces while the algebraic summations of the individual responses are shown as dotted traces. It is clear that the summation of the depolarizing and hyperpolarizing inputs in this horizontal cell is not linear. The deviation from linearity increases as the lights are made brighter.

4. Discussion

The photoresponses of R/G C-type horizontal cells are characterized by a complex intensity-dependent waveform especially for wavelengths in the transition zone (600-620nm) between depolarizing and hyperpolarizing responses (Fig. 3). These photoresponses can be matched to those elicited by mixtures of red and green lights of a unique combination and were used here to test the assumptions of the pigment theory of colour matching.

The pigment theory of colour matching predicts that the combination of green and red lights needed to elicit a photoresponse that matches the one evoked by a given monochromatic stimulus will be similar for turtle R/G C-type horizontal cells studied in different experiments. This should hold even though the photoresponses recorded from different cells in different experiments may differ significantly in waveform and amplitude. The findings described here do not support this prediction (Figs 4, 5 and Table 1). The intensities of the two components and their ratios varied considerably between cells.

However, the turtle retina differs from that of the human by the coloured oil droplets that occupy the inner segments of the cone photoreceptors. The oil droplets do not change the absorption spectra of the cone visual pigments but alter the spectral properties of the light reaching the visual pigment compared to those determined for the light incident upon the eyecup. Since the effects of the oil droplets depend on variety of factors such as the angle of the incident light relative to the long axes of the cone, the size of the oil droplet and the density of its pigment, the measured action spectra of the cones vary considerably between experiments (Baylor and Hodgkin, 1973; Schneeweis and Green, 1995). Therefore, the green/red combination needed to match a given monochromatic stimulus is also expected to vary between diffeexperiments.

We calculated the theoretical range of colour matching from Eq. (1), using two extreme cases. In the first case, the entire incident light bypassed the oil droplets (Fig. 2A) while in the second; the oil droplets' effects were the maximum measured in our laboratory (Fig. 2B). For three of the 6 cells studied with 600 and/or 620nm light (Table 1) colour matching was within the expected range. The I_{540}/I_{700} ratios for cell #2 were higher than the upper limit of the calculated range while cells #4 and 6 exhibited ratios smaller than the lower limit of the calculated range. The higher ratios of cell #2 probably reflect a case where the effects of the oil droplets were stronger than those considered. The low ratios of cell #4 and 6 can not be explained in a simple manner. It should be stressed that in each of these two cells only one colour matching was done compared to a total of 18 matching in the other 4 cells.

These findings indicate that small changes in the contribution of the coloured oil droplets to the action spectra of turtle cones can lead to significant variability in colour matching between different R/G chromaticity type horizontal cells. Similar variability of colour matching between human observers was attributed to small inter-individual variability in the absorption spectra of the visual pigments (Alpern, 1979; Neitz *et al.*, 1993). Neither of these contradicted the pigment theory of colour matching.

The pigment theory also predicts that matches must be radiant-invariant and additive (Grassmann's Law of scalar multiplication and additivity, Krantz, 1975). The experiments performed here support this notion, though to a limited extent. The largest range of intensities tested was 1.43 log units for 600nm in cell #2 (Fig. 4, 2nd column). The reason for this limited range was that dim stimuli elicited monophasic responses of simple waveform that did not require a unique combination of the green/red mixture to obtain a match. Bright stimuli were saturating one or both inputs to the C-type horizontal cells and therefore, colour matches were insensitive and could be obtained for different combinations of the green/red mixture. In general, for each horizontal cell studied with several intensities, the ratio I_{540}/I_{700} needed to match a specific wavelength was preserved at different intensities even though the pattern of the photoresponse changed dramatically (Figs. 4, 5). The small variability seen in the I_{540}/I_{700} ratios could be accounted for by several factors. (1) The optical system did not allow continuous changes in intensities but discrete changes with the smallest step being of 0.08 log units. (2) Colour matching exhibited different degrees of accuracy for different intensities (Fig. 4, 2nd column). (3) With bright light stimuli, saturation of one or both inputs to the R/G C-type horizontal cell might have prevented obtaining a unique match.

The pigment theory assumes that colour matching is independent of the neural interactions proximal to photon absorption by the visual pigments. This assumption is supported by the observation that the I_{540}/I_{700} ratio needed to match a given stimulus was independent of the pattern of the photoresponse. Cell #2 was dominated by the depolarizing (red) input compared to cell #1 when 600nm stimuli were used (Fig. 4), but the ratio I_{540}/I_{700} was almost three fold larger in this cell compared to cell #1 (0.29 compared to 0.11). Similarly, cell # 3 was dominated by the green, hyperpolarizing input compared to cell # 2 when a 620nm stimulus was used (Fig. 5), yet the green/red ratio needed for a match was almost 7 fold smaller in this cell (0.021 compared to 0.14). Furthermore, the algebraic summation of the photoresponses elicited by the red and green lights alone always differed from the photoresponse elicited by their mixture indicating a non-linear summation of the two antagonistic inputs (Fig. 6). Thus, the green/red combination needed to match a given monochromatic stimulus does not depend on the relative magnitude of the two opposing inputs or on their mode of summation but only on the rate of photon absorption by the visual pigments and the effects of the oil droplets.

The data presented here support the notion that colour matching experiments can be used to reveal the absorption spectra of the visual pigments and the filtering effects of the oil droplets circumventing all the neural interactions proximal to the site of photon absorption. It has been previously shown that the voltage responses of turtle cone photoreceptors do not obey the principle of univariance (Itzhaki *et al.*,

1992; Normann *et al.*, 1984; Perlman *et al.*, 1994). These interactions should not affect colour matching because they occur at a stage proximal to photon absorption. Furthermore, these deviations from univariance were only apparent for small amplitude photoresponses where small contributions from one spectral type of cones to another could be significant. The colour matching experiments were performed at intensities well above those needed to reveal colour mixing in the cone photoreceptors. Thus, the colour matching experiments, described here, indicate that univariance is obeyed by the visual pigments in turtle cones even when bright light stimuli (below saturation) are used.

Acknowledgments

This research was supported by grants from the Israel Science Foundation, Israel Academy of Sciences and Humanities (to I.P.) and from the Niedersachsen-Technion Research Foundation (to I.P.).

References

Alpern, M. (1979) "Lack of uniformity in colour matching", *J. Physiol. (Lond.)* 288:85-105.

Alpern, M. (1989) "The Charles Prentice Award Lecture 1988: The directionality of color matches and its relation to secondary protanomalous trichromacy", *Opt. Vis. Sci.* 66:339-354.

Asi, H. and I. Perlman (1998) "Neural interactions between cone photoreceptors and horizontal cells in the turtle (*Mauremys caspica*) retina", *Vis. Neurosci.* 15:1-13.

Baylor, D.A., M.G.F. Fuortes and P.M. O'Bryan (1971) "Receptive fields of cones in the retina of the turtle", *J. Physiol. (Lond.)* 214:265-294.

Baylor, D.A. and A.L. Hodgkin (1973) "Detection and resolution of visual stimuli by turtle photoreceptors", *J. Physiol. (Lond.)* 234:163-198.

Baylor, D.A., B.J. Nunn and J.L. Schnapf (1987) "Spectral sensitivity of cones of the monkey macaca fascicularis", *J. Physiol. (Lond)* 390:145-160.

Fuortes, M.G.F., E.A. Schwartz and E.J. Simon (1973) "Colour dependence of cone responses in the turtle retina", *J. Physiol. (Lond)* 234:199-216.

Fuortes, M.G.F. and E.J. Simon (1974) "Interactions leading to horizontal cell responses in the turtle retina", *J. Physiol. (Lond.)* 240:177-198.

Itzhaki, A., S. Malik and I. Perlman (1992) "The spectral properties of short wavelength (blue) cones in the turtle retina", *Vis. Neurosci.* 9:235-241.

Itzhaki, A. and I. Perlman (1984) "Light adaptation in luminosity horizontal cells in the turtle retina: role of cellular coupling", *Vision Res.* 24:1119-1126.

Kolb, H., I. Perlman and R.A. Normann (1988) "Neural organization of the retina of the turtle *Mauremys caspica*: A light microscope and Golgi study", *Vis. Neurosci.* 1:47-72.

Krantz, D.H. (1975) "Colour measurement and colour theory. I. Representation of theorem for Grassmann structure", *J. Math. Psychol.* 12:283-303.

Lamb, T.D. (1976) "Spatial properties of horizontal cell responses in the turtle retina", *J. Physiol. (Lond)* 263:239-255.

Lipetz, L.E. (1984) "Pigment types, densities and concentrations in cone oil droplets of *Emydoidea blandingle*", *Vision Res.* 24:605-612.

Lipetz, L.E. (1985). "Some neuronal circuits of the turtle retina", in: *The Visual System*, A. Fein and J.S. Levine, eds, New York: Alan R. Liss Inc., pp. 107-132.

Naka, K.I. and W.A.H. Rushton (1966a) "S-potentials from colour units in the retina of fish (*Cyprinidae*)", *J. Physiol. (Lond)* 185:536-555.

Naka, K.I. and W.A.H. Rushton (1966b) "An attempt to analyze colour perception by electrophysiology", *J. Physiol. (Lond)* 185:556-586.

Neitz, J., M. Neitz and G.H. Jacobs (1993) "More than three different cone pigments among people with normal colour vision", *Vision Res.* 33:117-122.

Normann, R.A., I. Perlman, H. Kolb, J. Jones and S.J. Daly (1984) "Direct excitatory interactions between cones of different spectral types in the turtle retina", *Science* 224:625-627.

Perlman, I., A. Itzhaki, S. Malik and M. Alpern (1994) "The action spectra of cone photoreceptors in the turtle (*Mauremys caspica*) retina", *Vis. Neurosci.* 11:243-252.

Rushton, W.A.H. (1972) "Pigments and signals in colour vision", *J. Physiol. (Lond.)* 220:1-31P.

Schneeweis, D.M. and D.G. Green (1995) "Spectral properties of turtle cones", *Vis. Neurosci.* 12:333-344.

Simon, E.J. (1973) "Two types of luminosity horizontal cells in the retina of the turtle", *J. Physiol. (Lond.)* 230:199-211.

Svaetichin, G. and E.F. MacNichol (1958) "Retinal mechanisms for chromatic and achromatic vision", *Ann. N.Y. Acad. Sci.* 74:385-404.

NOW YOU SEE IT, NOW YOU DON'T: SHUNTING INHIBITION IN EARLY VISION

LYLE BORG-GRAHAM, CYRIL MONIER and YVES FREGNAC
Unité de Neurosciences Intégratives et Computationnelles
Institut Féderatif de Neurobiologie Alfred Fessard, CNRS
91198 Gif-sur-Yvette, France

ABSTRACT

Synaptic inhibition, in particular that mediated by $GABA_A$, has been implicated at various points in early visual pathways by electrophysiological and pharmacological experiments. Here we focus on the role of this input in the generation of directional selectivity (DS) in retinal ganglion cells in the turtle, and in spatial, orientation, and of directional selectivity in neurons of primary visual cortex in the cat. In the first case, previous intracellular recordings had suggested a critical role for $GABA_A$ at the level of the ganglion cell proper. In contrast, intracellular recordings have argued against a functional role of $GABA_A$ in visual cortex. However, by applying a new technique for the quantitative measurement of synaptic dynamics we show that, on one hand, $GABA_A$ *does not* provide a critical computational input at retinal ganglion cells but, on the other hand, *does* provide a functionally significant input to cortical cells. Furthermore, our result in the retina suggests models for the DS circuit which in turn may have important implications for mechanisms underlying the biophysics of computation in cortex in general. The results from visual cortex suggest that canonical models for the generation of stimulus selectivity may oversimplify what may be a more baroque wiring scheme.

1. Introduction

Inhibitory neural responses are observed with functional stimuli throughout the various sensory systems of the brain. In the early visual pathways, which in this chapter we take to include the retina and primary visual cortex, inhibition is involved in the extraction of basic features of the visual scene, such as spatial organization (Hubel and Wiesel, 1962), movement (Barlow and Levick, 1965), and orientation (Sillito, 1975). In both the retina and cortex, pharmacological studies have implicated $GABA_A$ receptor-mediated inhibition as being crucial for the genesis of stimulus selectivity. The importance of this type of inhibition is of interest from a computational perspective, since its interaction with synaptic excitation is more non-linear as compared to inhibition mediated by $GABA_B$ receptors. This distinction arises since the action of $GABA_A$-controlled chloride channels, whose reversal potential is near the operating point of the neuron, is primarily by shunting the membrane. In contrast, since the reversal potential of the $GABA_B$-controlled potassium channels is significantly more hyperpolarized, their interaction with synaptic excitation is subtractive rather than divisive, that is in a more linear manner.

1.1. Searching for the Electrophysiological Signature of Shunting

Depending on the electrotonic architecture of the neuron, and the locations of its inputs relative to the site for the generation of the action potential, functional shunting by $GABA_A$-mediated inhibition of excitatory post-synaptic potentials (EPSPs) should increase the neuron input conductance, G(t), on the order of 50-100% or greater (Koch *et al*, 1990), when compared to the resting condition, G_0. To measure this modulation, we have applied a technique (Figure 1) using whole-cell patch recordings that estimates both G(t) and the apparent reversal potential driving the composite synaptic input, $E_{rev}(t)$, as continuous functions of time (Borg-Graham *et al.*, 1998) (to be more precise, $E_{rev}(t)$ is derived whenever G(t) is significantly larger than the resting input conductance).

As will be shown, we apply this method not only to search for functionally relevant shunting, but also to distinguish between various models of synaptic interaction underlying stimulus selectivity.

This method avoids several limitations of previous methods, such as measuring the amplitude modulation of responses applied in current clamp either to injected current pulses (*e.g.*, Douglas *et al.*, 1988) or of electrically-evoked EPSPs (Ferster and Jagadeesh, 1992) during the sensory (visually) evoked response. For example, the continuous estimation of the conductance modulation has a greater bandwidth than that of the methods based on repetitive current injection or electrical stimuli (e.g. as set by the repetition rate). In addition, for the second technique the amplitude of independently evoked fast EPSPs is dominated by the inverse of the capacitance of the neuron, which thus minimizes the modulation due to synaptic conductances activated by the sensory response. The use of low access resistance (R_a) whole-cell patch recordings avoids the non-specific shunt introduced by conventional microelectrodes. Since the size of this shunt can be on the order of the neuron's resting input conductance, it will reduce the relative modulation of G(t) by the evoked input by about half. Low values of R_a (typically between 15 and 50 Mohms) allow accurate estimates of electrode artifacts. Finally, when G(t) is estimated using voltage clamp protocols, the contribution of transient voltage-dependent channels and membrane capacitance local to the electrode is minimized.

2. Directional Selectivity in the Retina

2.1. Background and Motivation

A hierarchy of computational models for neurons starts with the basic integrate and fire response, where the algebraic sum of the inputs, represented as linear synaptic currents, is applied to a fixed action potential threshold.

However, a more accurate description of excitatory and inhibitory synapses considers a non-linear interaction (Torre and Poggio, 1978), primarily depending on whether inhibition acts mainly to hyperpolarize the membrane or to shunt the membrane conductance.

Neurons are also extended in space via elaborate dendritic trees, and therefore it is natural to consider the functional dependence of the inputs and outputs with respect to their dendritic location (Koch *et al.*, 1982).

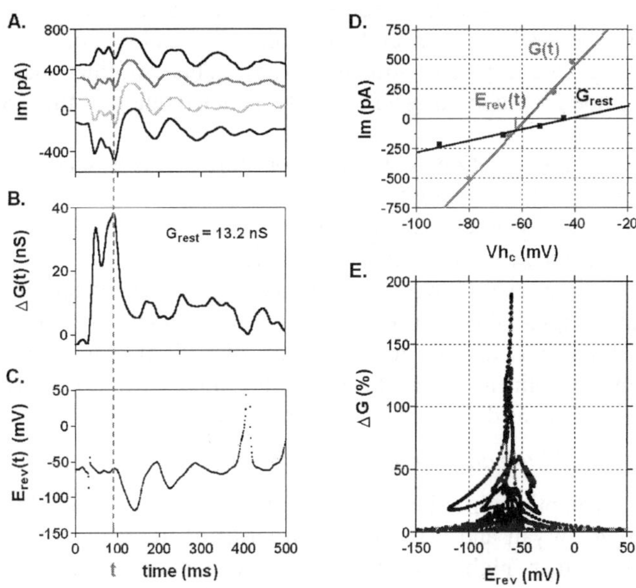

Figure 1. The continuous measurement of the modulation of the neuron input conductance, G(t), is done by repeating a given stimulus while holding the cell at two or more holding potentials (under voltage clamp) or holding currents (under current clamp). The resulting waveforms, and the associated holding potentials or currents (A), are combined according to Ohm's law to give an estimate of the somatic input conductance at each instant in time (B). In addition, the voltage of the intersection of the IV characteristics of the evoked response and the resting characteristic G_{rest} (= G_0) (D) gives the apparent reversal potential $E_{rev}(t)$ of the total synaptic input (C). In the example shown here, taken from an *in vivo* recording from cat visual cortex, four voltage clamp protocols are used to extract the synaptic dynamics in response to a light bar flashing ON at time = 0ms. Plotting G(t) and $E_{rev}(t)$ against each other in a phase plot (E) reveals clearly the tendency for the maximum evoked conductance change to occur around the reversal potential for $GABA_A$ input, estimated at about -65mV for the recording conditions.

In the visual stream these distinctions have functional consequences already at the retinal level. The retina is a good candidate for exploring the relationship between neural computation and circuit, given its physically *peripheral* location and its physiologically *central* status. A classic example of a non-linear spatio-temporal computation is that of retinal directional selectivity (review in Vaney *et al.,* 2000). We then may ask what biophysical mechanisms implement this computation, and what is the anatomical substrate for these mechanisms.

2.2. The Necessary and Sufficient Conditions for Direction Selectivity

We can define a directionally selective system as one whose overall response to a stimulus moving in one direction (the PREF response) is greater than for the opposite direction (the NULL response). For any system that accomplishes this computation - be it silicon or neural - the necessary and sufficient conditions for the underlying circuitry are straightforward. This succinct list provides a convenient guide for elaborating an explicit characterization of the DS circuitry in the retina:

1. Lateral interaction between two or more points in space, with memory, in order to detect motion.
2. Lateral asymmetry, in order to detect direction.
3. Non-linearity, for stimulus selectivity.

The last requirement (Poggio and Reichardt, 1973) reflects the fact that a purely linear single-valued measure - *e.g.*, the integral - cannot distinguish between a signal S(t) and its time-reversed twin S(-t). An obvious choice of discriminating non-linearity would be a threshold operation, such as action potential generation; as discussed below, a non-linear shunt can also be effective.

2.3. Shunting Inhibition and Direction Selectivity

$GABA_A$ synaptic inhibition has been shown to be crucial for the normal retinal DS (*e.g.*, Wyatt and Daw, 1976). In terms of the requirements above this implies at least that inhibition interacts with excitation to suppress the NULL response. In addition, the shunting action of $GABA_A$ input has been proposed as the critical non-linearity (Torre and Poggio, 1978). What has not been determined is the cellular location of this putative crucial input. Thus, given a DS retinal ganglion cell, we can consider three basic classes of underlying synaptic dynamics that will yield the selectivity of the spike response.

In the first case we assume that there is no inherent time-averaged directionality to the inputs; the DS computation results from the interaction between synaptic inputs within the dendritic tree and/or soma of the ganglion cell. This type of model has been termed *post-synaptic* since the DS computation occurs downstream from the inputs to the ganglion cell. This situation requires that there be an asymmetric interaction between a pair of spatially displaced inputs, one of which has a temporal delay (this last requirement satisfies the neccesity for memory mentioned above). If both inputs were excitatory, then a temporal coincidence in the PREF direction could produce a suprathreshold output. This alternative, however, can be ruled out by the experimental evidence that inhibition is necessary for DS. On the other hand if one input was excitatory and the other inhibitory, then a NULL direction coincidence would lead to a cancellation. Note that in both cases a sufficient non-linearity could be the ganglion cell spike threshold. In support of the second scenario Marchiafava (1979) showed explicit evidence for $GABA_A$ input to DS ganglion cells in the turtle retina from intracellular recordings; most experimental and theoretical studies of retinal DS in the last 20 years have focused on this model (Koch *et al.*, 1982; He and Masland, 1997).

In contrast, the two basic classes of *pre-synaptic* models of DS assert that the computation of a functional DS signal is prior to the ganglion cell. This directionality would then be reflected in an inherent directionality in one or both of the ganglion cell's inputs. In one case the cell could receive a non-directional excitatory input that is cancelled by directionally selective inhibitory input (larger during the NULL response). Finally, the cell could receive equivocal inhibitory input with excitatory input that is larger for the PREF response.

2.4. Possible Biophysical Signatures of DS Ganglion Cells

In order to quantitatively distinguish these three models (excitatory-inhibitory post-synaptic, inhibitory-NULL pre-synaptic and excitatory-PREF pre-synaptic), we can consider biophysical predictions at the level of the recorded cell membrane (more precisely, for practical reasons, at the cell soma). Specifically, these predictions focus on whether and how inhibition acts at a given DS cell in order to suppress the NULL response.

In the simplest form of the post-synaptic model, the crucial distinction for the individual excitatory and inhibitory inputs during the PREF and NULL responses is in their relative timing. If the cell is modelled as a single compartment, then this interaction would predict that the total input conductance of the cell as a function of time, G(t), would have an equal area for the PREF and NULL responses.

However, the model predicts that the peak of G(t) would be greater for the NULL response, reflecting the temporal correlation between the two pathways. Because of the on-the-path shunting interaction between synaptic inputs in dendritic cables (Koch *et al.*, 1982), this prediction becomes less distinct if we consider the integration of inputs onto a true dendritic tree. Nevertheless, the peak of the NULL G(t) waveform must always be greater than or equal to that for the PREF G(t) (equality holds when a perfectly shunting input - that is infinite synaptic conductance - is proximal to any excitatory input).

A second prediction of the post-synaptic model considers the absolute size of the expected synaptic input, in particular the inhibitory input. Assuming that $GABA_A$ is responsible for the crucial inhibition, we can consider the degree of shunting necessary to effectively suppress an excitatory input which otherwise would result in the PREF response.

From our voltage clamp experiments, we can estimate the expected size of the PREF versus NULL EPSP, 15 and 10mV, respectively, with only small distortions from voltage-dependent channels. Assuming a resting potential of -

70mV and an excitatory reversal potential of 0mV, the simple voltage divider cicuit gives a lower bound of the excitatory conductance at about 30% of G_0.

If we now require that this EPSP is reduced by at least 30% for the NULL response, and assuming an inhibitory reversal potential of -70mV (pure shunting), then the minimum conductance increase due to inhibition is about 75% of the resting G_0. Adding up the two inputs yields a lower bound of the total modulation of G(t), about 100% of G_0.

If we consider synaptic inputs remote from the soma, then these estimates are reduced since the transfer impedance of the dendrites serves to "hide" a fraction of the synaptic conductance (Koch *et al.*, 1990). To estimate this effect, we have made detailed simulations of retinal ganglion cells of realistic dendritic trees. These simulations show that the error is on the order of 10%.

The predictions for the inhibitory-NULL pre-synaptic model – an asymmetric inhibitory pathway onto the ganglion cell – are similar to the post-synaptic case. That is, in order for the NULL inhibition to be effective at attenuating the (non-directional) excitation, the combined inputs would be expected to increase the cell's input conductance by at least 100%. With respect to the relative dynamics of G(t) for the PREF and NULL direction, since the inhibition is selective for the NULL response, the prediction of a greater NULL amplitude of G(t) would be stronger than the post-synaptic case. In addition, for the inhibitory DS pre-synaptic model, the area of G(t) in the NULL response would also be greater than the PREF response.

Finally, for the excitatory-PREF pre-synaptic model, in which excitation is greater for the PREF response and inhibition is equivocal, the predictions for G(t) are reversed. Specifically, G(t) would be expected to have both a greater area and amplitude for the PREF versus NULL response. The difference between the two responses for both measures, though, would be less than the asymmetric inhibition case since the larger driving force underlying synaptic excitation requires a smaller conductance change to be effective.

For this model the relative strength of the synaptic input can be significantly less than the previous cases. Following the calculations earlier, a lower bound of modulation of G(t) in the PREF response, due to the minimum excitatory input, would be about 30% of G_0.

2.5. The Excitatory Input to Retinal DS Ganglion Cells is Already DS

To test these predictions, whole-cell patch recordings were made from DS ganglion cells in the intact isolated turtle retina (Figures 2 through 5; Borg-Graham, 1991). Motion stimuli included both gratings and single bars. For these cells the apparent reversal potential, Erev(t), rarely went below –90mV (for values of Grel(t)>110% the average minimum Erev(t) was -49mV; compare with the more hyperpolarized extrema for the *in-vivo* cortical records in Figure 7). If the cell is modeled as a single compartment, this implies that somatic voltage clamp at -90mV or less will eliminate or reverse inhibitory synaptic currents. Therefore the total clamp current will give a lower bound on the relative amount of excitatory current during the PREF versus NULL response. If we consider the non-linear interaction of synapses distributed along dendrites, then the imperfectly clamped post-synpatic membrane will mean that true elimination or reversal of inhibitory currents will require more polarized (that is, more negative) holding potentials (Spruston and Johnston, 1993).

Simulations on this point with realistic ganglion cell morphologies suggest that a holding potential of -90mV is adequate to suppress the majority of inhibitory outward current. Nevertheless, a more quantitative assessement of the synaptic input may be obtained by the G(t) and Erev(t) measurement described earlier (see below).

To quantify the directionality of the various measures, we use the following "PN" index:

$$PN(x) = (x_P - x_N) / (x_P + x_N)$$

where x_P and x_N refer to the value of the measured variable x (assumed non-negative) for the PREF and NULL stimuli, respectively (PREF and NULL being defined by statistically significant DS spike responses). This index ranges from 1 to -1; the extreme values mean that x is only non-zero for the PREF or NULL stimuli, respectively.

In our recordings we find that PN indices of both the integral and the peak of the voltage clamp current are in general positive, and thus consistent with the original spike directionality. This result suggests that the excitatory input is greater for the PREF response.

The more direct measure of synaptic input given by G(t) and $E_{rev}(t)$ give the same result: PN indices of both the integral and the peak of G(t) are in general positive (and thus also consistent with the original spike directionality).

This suggests that the total synaptic input is greater for the PREF response and, by implication, that the excitatory input must be larger for the PREF response.

Phase plots of composite synaptic reversal potential versus total input conductance show that the strongest peaks tend to converge around the reversal potential for $GABA_A$ receptor-mediated channels (not shown, but similar to the result for cortical cells described in Figure 7). However, the values of these peaks for both PREF and NULL responses are mostly below 100% modulation of the cells' input conductance, and on average are about 50% of G_0.

Thus, any shunting input for either the PREF or NULL response is relatively small, compared to that required for a strong functional role in mediating DS.

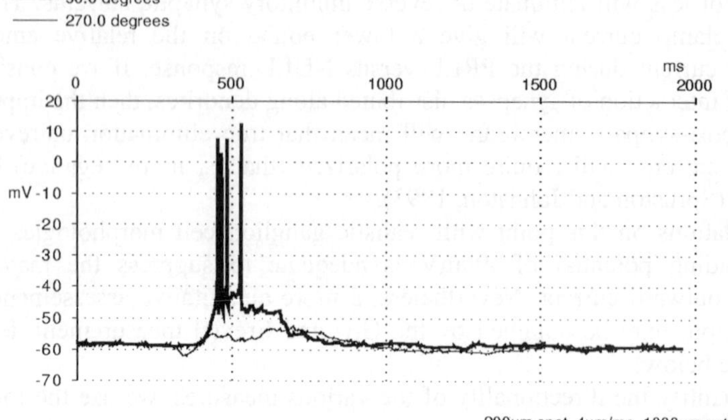

Figure 2. Whole-cell patch current clamp recording of a directionally selective ganglion cell in an intact isolated turtle retina. Stimulus is a 200 micron spot moving at 4 microns/millisecond over a distance of 1000 microns. The stimulus moving in the PREF direction of 90 degrees elicits several action potentials, while the opposite NULL direction of 270 degrees elicits a much smaller, subthreshold EPSP and a small IPSP.

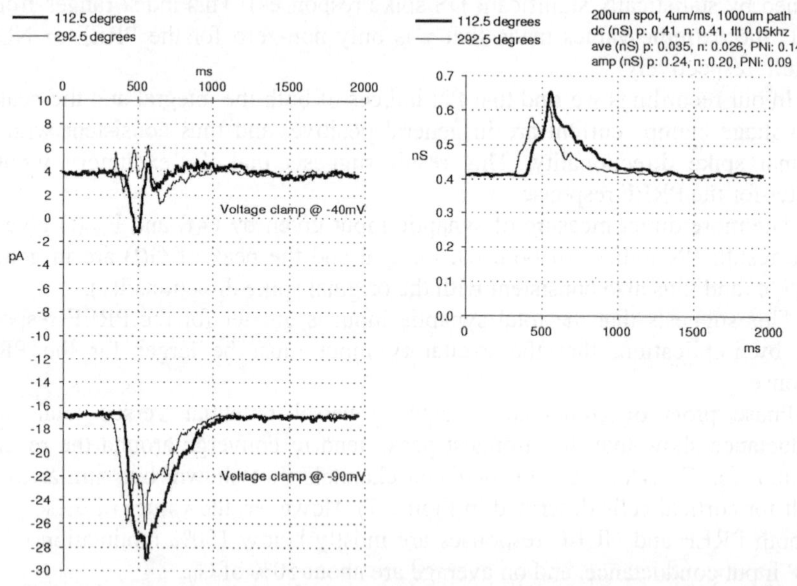

Figure 3. Left: Voltage clamp recordings of responses to same PREF (112.5 degrees) and NULL (292.5 degrees) stimuli as shown in Figure 2, at two holding potentials. Right: The evoked G(t) waveform for the PREF stimulus has larger area (PN = 0.14) and peak (PN = 0.09) than the NULL response, indicating that for this cell the excitatory input is greater for the PREF versus NULL response.

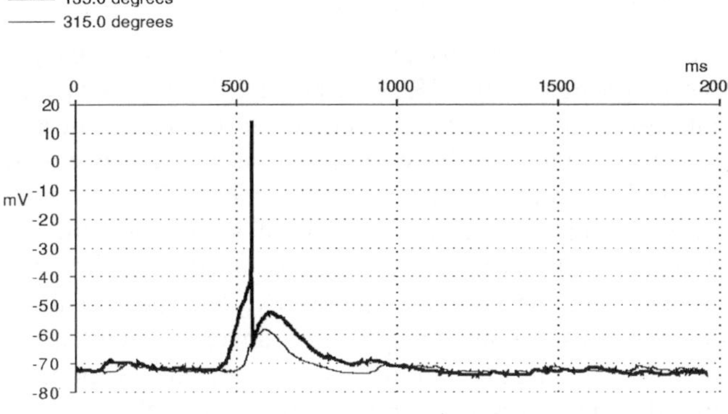

Figure 4. Current clamp recording of a DS ganglion cell in the turtle retina. Stimulus is a 200 micron spot moving at 2 microns/millisecond over a distance of 1000 microns. The stimulus moving in the PREF direction of 135 degrees elicits an action potential, while the opposite NULL direction of 315 degrees elicits a much smaller, subthreshold EPSP.

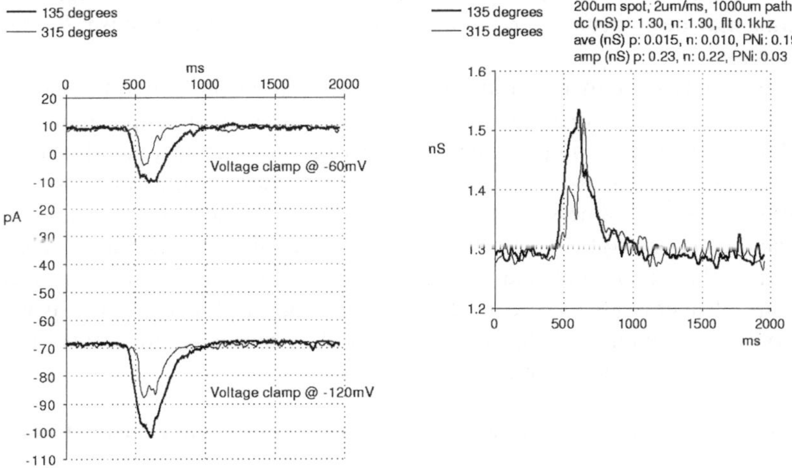

Figure 5. Left: Voltage clamp recordings of responses to same PREF (135 degrees) and NULL (315 degrees) stimuli as shown in Figure 4, at two holding potentials. Right: The evoked G(t) waveform for the PREF stimulus has larger area (PN = 0.19) and slighter larger peak (PN = 0.03) than the NULL response, indicating that for this cell the excitatory input is greater for the PREF versus NULL response.

2.6. A Model for Retinal Directional Selectivity

Since the data show that the excitatory input to DS retinal ganglion cells is already DS (that is, greater for the PREF response), the necessary and sufficient conditions mentioned above for this computation in the retina must be satisfied prior to the ganglion cell. We have proposed a model (Borg-Graham and Grzywacz, 1992) in which excitatory and (shunting) inhibitory synaptic interactions on individual amacrine cell dendrites provides the fundamental substrate for DS (Figure 6) - these cells are then assumed to provide the excitatory DS input to the DS ganglion cell. This model exploits the fact that in general output synapses (and also input synapses) of amacrine cells may be found anywhere in their *dendritic* tree (recall that amacrine means "without axon") - this arrangement provides the necessary lateral asymmetry for DS.

The basic elements of this non-linear spatio-temporal filter would be expected to be present on more conventional neurons in the central nervous system. In this case we can consider an array of inputs spread along an unbranched section of the dendritic tree. The filter output at the proximal end of the segment, consisting of synaptically-generated current for final integration at the soma, would be larger for sequential or "directional" activation of the inputs in the distal to proximal order. In summary, we suggest that the mechanism for directional selectivity in the retina may be an instantiation of a canonical biophysical mechanism for neural computation.

The activity of retinal neurons other than ganglion cells are on the most part mediated by graded potentials, not spikes.This provides an implication of the model from a neural computational perspective: the classical non-linearity of spike threshold is not *necessary* in order to implement a functional non-linear operation in the central nervous system. Here, the crucial non-linearity is provided by ligand-gated synaptic channels (as opposed to the voltage-gated channels underlying the action potential). This mechanism therefore allows a greater repertoire of operations to be considered at the single cell level than is available with only an output non-linearity (threshold), especially for more formal descriptions of network computation.

3. Spatial, Orientation and Directional Selectivity in Primary Visual Cortex

We shall now discuss some of our recent findings regarding functional inhibition involved in receptive field properties of neurons in primary visual cortex. The synaptic basis for the receptive field in visual cortex has been the subject of extensive research since the pioneering work of Hubel and Wiesel in the early 1960's (Hubel and Wiesel, 1962). As in the retina, the role of inhibition has been examined by electrophysiological protocols augmented by pharmacological manipulations.

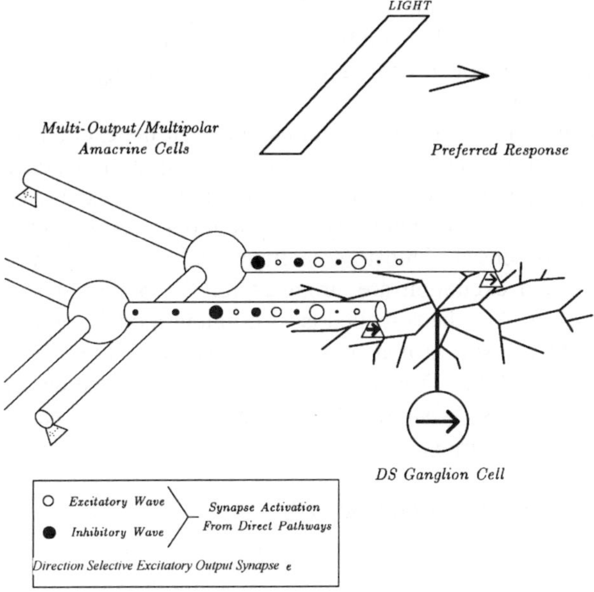

Figure 6. Schematic model of the DS circuit in vertebrate retina. In the general form of this model excitatory and ($GABA_A$) inhibitory input is homogeneously distributed along amacrine cell dendrites, with the kinetics of excitation being faster than that of the inhibition. Excitatory synaptic outputs (shown here at the dendritic tips), that are off center with respect to the amacrine cell dendritic tree, provide DS input to the DS ganglion cell, consistent with the data presented in this chapter. The asymmetry between the inputs along a given amacrine cell dendrite and the DS output provides the asymmetric lateral interaction necessary for DS. The memory in the system is provided by the slower kinetics of inhibition. For the PREF stimulus, as shown here, the wave of excitation along each amacrine cell dendrite reaches the output's presynaptic site at the dendrite tip during the entire time the stimulus sweeps from left to right, trailed by the slower wave of activated inhibition. For light moving from right to left, that is the NULL direction, for the majority of time the activated inhibition is interposed between the activated excitatory input zone and the presynaptic output site, thus shunting the excitatory synaptic current before it can activate the output. In this model the crucial non-linearity is the shunting effect of inhibition, suggesting that the classical neuronal non-linearity of the spike threshold is not *necessary* for non-linear computations in the vertebrate central nervous system. Furthermore, the dendritic substrate for this non-linear spatio-temporal filter is likely to exist in more conventional neurons in the CNS.

3.1. Background and Motivation

The results to date, however, have been somewhat contradictory. It has been demonstrated that $GABA_A$ antagonists can eliminate both spatial (Sillito, 1975) and orientation and directional selectivity (Sillito, 1980) in cat primary visual cortex. However, various intracellular protocols have failed to show shunting in visual cortex (*e.g.*, Douglas *et al.*, 1988; Ferster and Jagadeesh, 1992; review in Berman *et al.*, 1992). In addition, attempts to block $GABA_A$ receptors intracellularly appear not to change the qualitative tuning properties of the cell (Nelson *et al.*, 1994).

A related question is how the stimulus tuning of the excitatory and inhibitory input to a given cell is related to the final spiking output. In the case of orientation selectivity, for example, one may consider a variety of possible combinations of the tuning of excitatory and inhibitory synaptic input, relative to the output tuning (Figure 7). The consensus of most intracellular studies to date have concluded that both inhibition and excitation are strongest for the preferred stimulus (but see Volgushev *et al.*, 1993), the so-called iso-iso tuning model (review in Ferster and Miller, 2000). One result is that most theoretical work (*e.g.*, Ben-Yishai *et al.*, 1995; Somers *et al.*, 1995; Troyer *et al.*, 1998; but see Wörgötter and Koch, 1991) have focused on this synaptic tuning paradigm as the canonical synaptic arrangement in visual cortex.

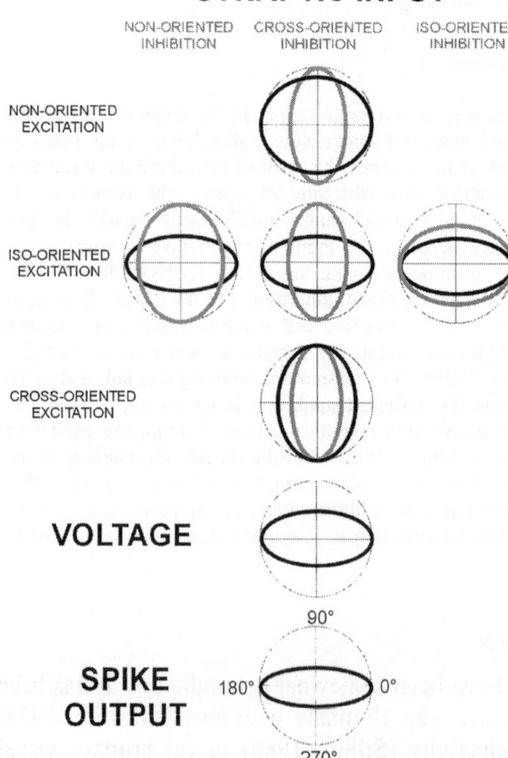

Figure 7. Scheme of the successive transformation, from top to bottom, of the different possible combinations of excitatory (in black) and inhibitory (in gray) synaptic tunings to tuned membrane voltage and, finally, tuned spike output. Here, tuning can be most directly interpreted as that of selectivity to stimuli moving towards 0 or 180 degrees (*i.e.*, pure orientation tuning). Previous experimental and theoretical studies of orientation selectivity in visual cortex have emphasized the combination of iso-oriented excitation with iso-oriented inhibition (the iso-iso model).

3.2. *Transient and Strong Shunting Inhibition in Visual Cortical Neurons in Response to Visual Stimuli*

We have studied these issues by recording visual responses from neurons in cat primary visual cortex, using the continuous estimate of G(t) described earlier. In these recordings we find that various visual stimuli - including static flashing bars as well as moving bars - can evoke $GABA_A$ input that shunts the cell by well over 100% (Borg-Graham *et al.*, 1998). This result answers in the affirmative the basic issue of whether or not functional shunting inhibition exists in cortex at all. The size of this shunt can be seen in the peaks of the phase plot shown on the left in Figure 8, where the response to both static and moving stimuli of several cells recorded *in-vivo* are superimposed. Interestingly, or static flashing stimuli this input is inriably transient, occurring at the initial phase of the response, and are *not* correlated with the total spike response (*e.g.*, dominant spiking to ON or OFF transitions of the stimulus for Simple cells). Thus, although this shunting input is strong enough to be functionally relevant, its precise functional role is not immediately obvious, e.g. by playing a direct inhibitory role to suppress the spiking response. Building on the pharmacological evidence cited earlier that this input is nevertheless *necessary* for the distinction between the ON and OFF stimuli in Simple cells, we suggest that this early and powerful shunt acts to influence the dynamic evolution of the entire response as activity percolates through the intracortical network.

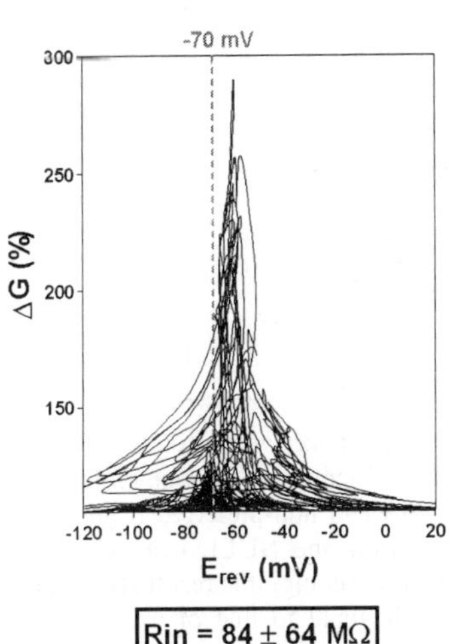

Figure 8. Phase plots of the modulation of G(t) relative to the resting input conductance G_0, versus the apparent reversal potential of the composite visually-evoked synaptic input, $E_{rev}(t)$, for neurons recorded *in-vivo* in cat primary visual. *In vivo* visual stimuli included both flashing static bars and moving bars. Note that the phase plot trajectories (which are composed from several different cells) are constrained such that they peak around the reversal potential for $GABA_A$, falling off for reversal potentials associated with glutamatergic excitatory synapses and with $GABA_B$ inhibitory synapses. The size of the peaks associated with $GABA_A$ indicate modulation of the input conductance by well over 200%, which indicates that functional stimuli can elicit a functionally relevant shunting inhibitory input.

3.3. *The Tuning of Excitation and Inhibition Underlying Spike Tuning in Primary Visual Cortex*

In our *in vivo* recordings in visual cortex we have also examined the specific tuning of excitation and inhibition underlying orientation and directional selectivity. We have found that, contrary to previous reports, non-preferred stimuli can evoke much stronger shunting input than the preferred stimuli (Figure 9).

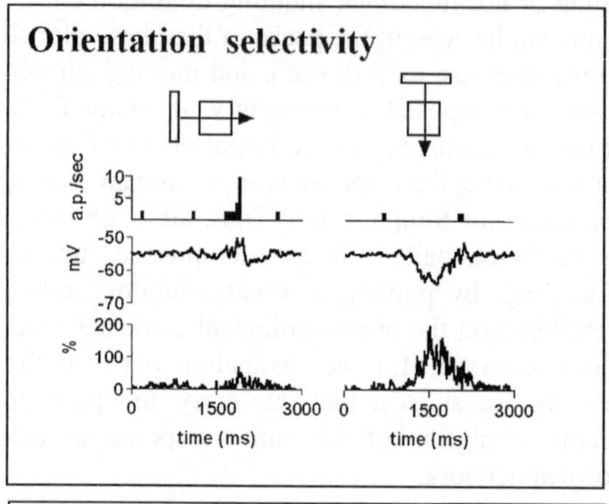

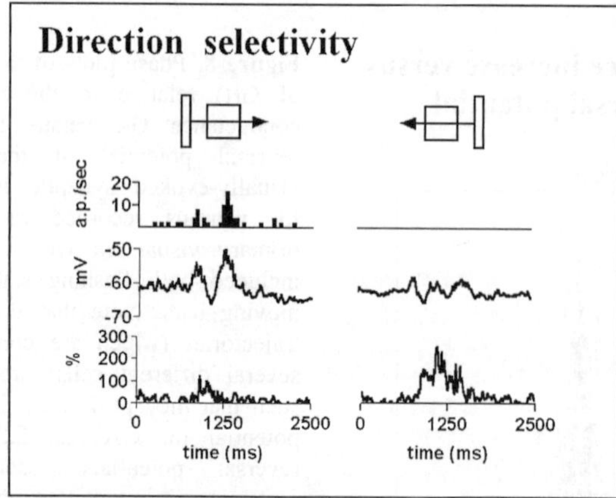

Figure 9. Example responses to oriented moving bars in primary visual cortex. Responses consist of post-stimulus time histograms (top), averaged membrane voltage (middle), and relative conductance changes (bottom) for preferred (left) and non-preferred (right) stimuli. Here strong shunting conductances can be observed during the NULL response for both orientation tuning and directional selectivity. For orientation tuning, this result shows that for some cells at least the tuning of inhibition may be orthogonal to that of excitation. With respect to directional selectivity, the example above is in contradistinction to the case for the retina, where the relatively small conductance changes are seen to be larger for the PREF response.

These findings are made more explicit when first-order estimates of excitatory and inhibitory synaptic inputs are derived from the G(t) and $E_{rev}(t)$ measurements. This is done by assuming a simple lumped circuit neuron model comprised of a parallel combination of a leak conductance, an excitatory synapse and an inhibitory synapse, each associated with known, fixed reversal potentials. The tuning of the two types of input are then quantified by taking the integral of the derived synaptic waveforms over the entire response.

Our results indicate that there is no single recipe for the synaptic input generating these functional selectivities: different cells show different combinations of the relative tuning of excitation and inhibition with respect to the spiking response (Figure 9). It will be interesting to see how these relationships hold when the dynamics of the inputs are taken into accout, for example, will the integral relationships between excitation and inhibition predict their correlational relationships.

It is important to note that these measurements are of the *effective* synaptic input as a function of the stimulus properties. In particular, given the contribution of recurrent cortical circuitry, these activities do not necessarily directly reflect the anatomic synaptology (Douglas *et al.*, 1999). Nevertheless, the fact that we observe a variety of tuning combinations argues against a single fixed canonical circuitry underlying neuronal interactions within primary visual cortex (*e.g.*, Douglas *et al.*, 1999; Ferster and Miller, 2000).

4. Conclusions

While the functional effect of synaptic inhibition is seen essentially everywhere in the visual system, the precise locus for this action, at least of non-linear inhibition, appears to be only at specific stages. To quantitatively investigate this input, we have described a sensitive method for the measurement of synaptic dynamics using whole-cell voltage clamp recordings. We have applied this technique to characterize functional responses in the visual system both *in vitro* in the retina and *in vivo* in visual cortex: the sensitivity of this method may account for the fact that in both cases the results are somewhat contrary to earlier findings.

We have described cellular responses that mark two steps of early visual processing - the extraction of motion direction at retinal ganglion cells, and the extraction of spatial, orientation and direction information in neurons of the primary visual cortex. In the first case we find that crucial inhibitory processes *do not* take place at the ganglion cell. Given the particular properties of the neuronal circuitry afferent to the ganglion cell, where the computation must therefore take place, this finding has ramifications for specific biophysical mechanisms underlying non-linear spatio-temporal filtering in neural computation. While this particular case concerns the detection visual motion direction, a similar problem is invoked when considering what biophysical mechanisms can allow the detection of temporal patterns in afferent spike trains. In contrast to the retinal ganglion cell, in the visual cortex we find inhibitory input in a large percentage of neurons that is in principle sufficient for functional selectivity. We also find, also in contrast to the retina, that there exist a variety of basic relationships in the tunings of excitation and inhibition to these cells with respect to their final spike output. These findings provide constraints on the network architecture responsible for feature extraction in the visual system and, perhaps, in other sensory areas.

Acknowledgements

The cortical experiments described here were funded by grants from Progres 99-012, AFIRST 970-MAEN 11, and HFSP RG-10398 to Y.F.

References

Barlow, H.B. and W.R. Levick (1965) "The mechanism of directionally selective units in rabbit's retina", *J. Physiology* 178:477-504.

Ben-Yishai, R., R.L. Bar-Or and H. Sompolinksy (1995) "Theory of orientation tuning in visual cortex", *Proc. Natl. Acad. Sci. USA* 92:3844-3848.

Berman, N., R. Douglas and K. Martin (1992) "GABA-mediated inhibition in the neural networks of visual cotex", *Prog. Brain Res.* 90:443-476.

Borg-Graham, L. (1991) "On directional selectivity in the vertebrate retina: An experimental and computational study", *PhD thesis*, Harvard-MIT Division of Health Sciences and Technology, Boston, USA.

Borg-Graham, L. and N.M. Grzywacz (1992) "A model of the directional selectivity in retina: Transformations by neurons singly and in concert", in: *Single Neuron Computation,* T. McKenna, J. Davis and S.F. Zornetzer, eds, New York: Academic Press, pp. 347-376.

Borg-Graham, L., C. Monier and Y. Frègnac (1998) "Visual input evokes transient and strong shunting inhibition in visual cortical neurons", *Nature* 393:369-373.

Douglas, R., K. Martin and D. Whitteridge (1988) "Selective responses of visual cortical cells do not depend on shunting inhibition", *Nature* 332:642-644.

Douglas, R., C. Koch, M. Mahowald and K. Martin (1999) "The role of recurrent excitation in neocortical circuits", in: *Cerbral Cortex, Vol. 13: Models of Cortical Circuits*, P. Ulinski, E.G. Jones and A. Peters, eds, New York: Kluwer Academic/Plenum Publishers, pp. 251-282.

Ferster, D. and B. Jagadeesh (1992) "EPSP-IPSP interactions in cat visual cortex studied with *in vivo* whole-cell patch recording", *J. Neurosci.* 12:1262-1274.

Ferster, D. and K.D. Miller. (2000) "Neural mechanisms of orientation selectivity in the visual cortex", *Annu. Rev. Neurosci.* 23:441-471.

He, S. and R.H. Masland (1997) "Retinal direction selectivity after targeted laser ablation of starburst amacrine cells", *Nature* 389:378-382.

Hubel, D. and T. Wiesel (1962) "Receptive fields, binocular interaction and functional architecture in the cat's visual cortex", *J. Physiol.* 160:106-154.

Koch, C., R. Douglas and U. Wehmeier (1990) "Visibility of synaptically induced conductance changes: Theory and simulations of anatomically characterized cortical pyramidal cells", *J. Neurosci.* 10:1728-1744.

Koch, C., T. Poggio and V. Torre (1982) "Non-linear interactions in a dendritic tree: Localization, timing, and role of information processing", *Proc. Natl. Acad. Sci. USA* 80:2799-2802.

Marchiafava, P.L. (1979) "The responses of retinal ganglion cells to stationary and moving visual stimuli", *Vision Res.* 19:1203-1235.

Nelson, S., L. Toth, B. Sheth and M. Sur (1994) "Orientation selectivity of cortical neurons during intracellular blockade of inhibition", *Science* 265:774-777.

Poggio, T. and W.E. Reichardt (1973) "Considerations on models of movement detection", *Kybernetics* 13:223-227.

Sillito, A. (1975) "The effectiveness of bicuculline as an antagonist of GABA and visually evoked inhibition in the cat's striate cortex", *J. Physiol.* 250:287-304.

Sillito, A. (1977) "Inhibitory processes underlying direction specificity of simple, complex, and hypercomplex cells in cat's striate cortex", *J. Physiol.* 271:699-720.

Somers, D., S. Nelson and M. Sur (1995) "An emergent model of orientation selectivity in cat visual cortical simple cells", *J. Neurosci.* 15:5448-5465.

Torre, V. and T. Poggio (1978) "A synaptic mechanism possibly underlying directional selectivity to motion", *Proc. R. Soc. Lond. B* 202:409-416.

Troyer, T.W., A. Krukowski, N.J. Priebe. and K.D. Miller. (1998) "Constrast-invariant orientation tuning in cat visual cortex: Feedforward tuning and correlation-based intracortical connectivity", *J. Neurosci.* 18:5908-5927.

Vaney, D.I., S. He, W.R. Taylor and W.R. Levick (2000), "Direction-selective ganglion cells in the retina", in: *Computational, Neural and Ecological Constraints of Visual Motion Processing*, J. Zanker and J. Zeil, eds, Berlin: Springer Verlag, pp. 2-44.

Volgushev, M., X. Pei, T.R. Vidyasagar and O.D. Creutzfeldt (1993) "Excitation and inhibition in orientation selectivity of cat visual cortex neurons revealed by whole-cell recordings *in vivo*", *Visual Neurosci.* 10:1151-1155.

Wörgötter, F. and C. Koch (1991) "A detailed model of the primary visual pathway in the cat: Comparison of afferent excitatory and intracortical inhibitory connectoin schemes for orientation selectivity", *J. Neurosci.* 11:1959-1979.

Wyatt, H.J. and N.W. Daw (1976) "Specific effects of neurotransmitter antagonists on ganglion cells in rabbit retina", *Science* 191:204-205.

PERCEPTUAL LEARNING AS A SIGN OF ADULT CORTICAL PLASTICITY

NICOLETTA BERARDI and ADRIANA FIORENTINI
Istituto di Neurofisiologia del C.N.R., Via S. Zeno 51, 56127 Pisa, Italy

ABSTRACT

Procedural learning ("learning how") is a form of implicit (non-declarative) learning which involves the acquisition of a new skill through practice. Motor learning is very familiar since most of us have learned to ride a bike or to skate or to play an instrument. Improvement in perceptual skills is perhaps more elusive; however, the ability of musicians to discriminate tones and our ability to discriminate european faces (as opposed to, for instance, chinese) is probably the result of perceptual learning. Perceptual learning is a sign of neural plasticity in adults specific for the sensory modality and has offered clear evidence that changes in cortical areas induced by sensory experience are not limited to a restricted postnatal developmental period. In humans and non-human primates laboratory investigations have been conducted mostly on visual, somatosensory and acoustic perceptual learning. Common to most results obtained is a striking selectivity of the learning process for the characteristics of the sensory stimuli used for training. For instance the effects of practice with visual stimuli of limited spatial extent are restricted to the trained portion of the visual field and are not transferred to a different, untrained area. The selectivity of the process has suggested that the plastic modifications underlying perceptual learning could take place also at a relatively early stage in cortical sensory processing, possibly even in primary sensory areas, where neurons are still narrowly tuned for the characteristics of the stimuli, such as the location in the visual field or the frequency of an acoustic tone. Visual perceptual learning will be dealt with separately. Here studies on perceptual learning in other sensory modalities will be reviewed with particular attention to electrophysiological and imaging studies in order to draw hypotheses on the cortical areas involved in, and on the cellular mechanisms possibly mediating, the neural plasticity for perceptual learning.

1. Introduction

The classical view of neural plasticity held that neural connections in the brain were no longer susceptible to modifications after the end of the critical period. It was accepted that changes in synaptic efficacy could be induced by learning also in adults, but these changes were thought to occur only in association cortices or in the hippocampus. As a corollary, experience dependent plasticity in sensory and motor cortices was considered impossible in the adult, also because stability of connections in these areas appeared to be *conditio sine qua non* to have the precise topography and receptive field organization necessary to give stability to our motor and perceptual functions.

On the contrary, the data we shall review here demonstrate that learning effects may be apparent also at very early levels of sensory and motor pathways.

Perceptual and motor learnings are examples of implicit learning and in particular are the acquisition of perceptual or motor skills through practice. Motor learning is quite familiar, since most of us have at least learnt to ride a bike or to play an instrument. Perceptual learning is more elusive, however, the ability of musicians to discriminate tones and our ability to discriminate european faces (as opposed to, for instance, chinese) is probably the result of perceptual learning.

The ability to discriminate simple attributes of a sensory stimulus improves with practice. Improvements in performance have been found for discrimination between pairs of visual stimuli differing for the spatial pattern or for the direction of movement, between pairs of acoustic stimuli differing in tone or between different sequences of tactile stimulations.

Performance improvements can require a long practice period (days or weeks) but can also be apparent within a single experimental session of a few hundred trials. The final result of practice is, in all cases, the ability to discriminate easily between stimuli that appeared as utterly indistinguishable to begin with, and this ability is retained over long periods of time. It is conceivable to assume that these long lasting performance improvements reflect changes in neural circuits within the brain: but where, and how do these changes take place is still a matter of investigation. We shall first discuss evidences on the lack of transfer of learning across stimulus conditions which suggest that these changes may take place at early stages of the sensory information processes. Then we shall review psychophysical and electrophysiological studies which have addressed the problem of which cellular mechanisms may underlie perceptual learning.

2. Perceptual learning characteristics: what do they tell about the site of plasticity

In most studies on perceptual learning, the effects of practice are specific for the characteristics of the stimuli used for training. For instance, if a visual discrimination task between gratings of different waveforms is practiced with vertically oriented stimuli the effects of practice are lost when the stimuli are rotated by 90 degrees (Fiorentini and Berardi, 1980, see also Berardi and Fiorentini in this volume). Similar results have been found in tactile discrimination tasks. Sathian and Zangaladze (1997) trained subjects to discriminate differences between gratings engraved on a surface by scanning it with a fingerpad; roughness was varied either increasing the width of the groove or the width of the ridge. Once the subject had practiced to discriminate between groove width, and performance had improved and reached a stable level, discrimination based on ridge width was tested; there was no transfer of groove difference discrimination to the situation of ridge width discrimination. Similarly, the effects of practicing acoustic tone discrimination around, say, 5 kHz are not transferred to discrimination of tones around 8 kHz.

The strong selectivity of learning for the parameters of the sensory input implies that only a subset of the sensory neurons are affected by experience. From what is known of the information processing in sensory systems, it is evident that many physical parameters of the stimuli are selectively represented only at low-level processing stages, while neurons in higher order sensory areas respond in an invariant manner to these parameters. This, theoretically, poses an "upper limit" to the level of where experience driven neural plasticity occurs. Another consideration to be made is that selectivity for a given physical parameter in neural responses can be absent or very broad at the very first stage of sensory processing and occur (or strongly increase) after a few processing stages. A first example is the selectivity for the orientation of a visual stimulus: it is absent in the retina and becomes apparent only in the primary visual cortex. Another example is the selectivity for the temporal frequency of an acoustic stimulus: the tuning curves of acoustic neurones get narrower and narrower as one moves from the ganglion cells of Corti to the cochlear nuclei to the primary acoustic cortex. Also properties such as binocularity or binaurality emerge after a few processing stages, namely in the primary visual cortex and in the superior olivary complex. Thus, the existence and the degree of selectivity of learning for stimulus parameters or the presence of interocular/internaural transfer of learning effects poses also a "lower limit" to the site of neural plasticity. For instance, lack of transfer of learning effects in a visual discrimination task for changes in the orientation of the gratings accompanied by interocular transfer would suggest a site at least at the level of the primary visual cortex. Lack of transfer of learning effects in an acoustic tone discrimination task for changes in temporal frequency of the acoustic tones of the order of 1 octaves would suggest a site at least at the level of the primary auditory cortex.

Another striking characteristic of perceptual learning is the specificity of learning effects for the portion of sensory space used for training: the effects of practice with visual stimuli of limited spatial extent are restricted to the trained portion of the visual field and are not transferred to a different, untrained area, even if the latter is adjacent to the former (Fiorentini and Berardi, 1997; Ahissar and Hochstein, 1997). There is an exception to this rule: transfer of learning effects has been found between portions of the visual field symmetric with respect to the vertical meridian; these regions however, are known to be connected by callosal fibers, and stimuli presented on one side of the meridian will activate neurons with receptive fields on both sides of it. This again reinforces the concept that only neurons activated by the stimulus during the training period may subserve plastic changes and that these changes may occur in cortical areas where the receptive fields are small ("upper limit").

It has been proposed that learning dependent changes will occur at the lowest level of processing at which differential neural responses (selectivity) are present for those stimulus parameters (orientation, tone, roughness type) that are critical for the performance of a given task is available (minimum level hypothesis) (Gilbert, 1994; see for a review Karni and Bertini 1997).

Recently, this proposal has been challenged (Mollon and Danilova, 1996; Ahissar and Hochstein 1997). The argument is dealt with extensively in the accompanying paper, since the supporting evidences come from visual perceptual learning experiments. The main point is that selectivity of the learning process for the stimulus parameters cannot be taken as a proof that the underlying plastic changes take place only or start at low processing levels (bottom-up effects). Learning may be triggered by changes at cortical levels higher in the hierarchy than the sensory areas and which generalize with respect to the stimulus parameters; these changes would then direct changes at the level of lower cortical areas (top-down effects), which exhibit specificity for the stimulus parameters, possibly even the primary sensory areas.

An indication of the involvement of higher order cortical areas comes also from a series of experiments aiming to find whether repeated exposure is sufficient to trigger learning. The answer to this question is no, the mere repetition of stimulus presentation is not sufficient to improve performance. Discriminative ability increase only if the subjects direct their attention to the sensory stimuli, and more precisely, to a given parameter of the sensory stimuli. For instance, training on the evaluation of the brightness of line elements (attended parameter) did not improve the ability to discriminate their orientation (unattended parameter, Shiu and Pashler, 1992). Training on a tactile discrimination while listening to tones of different frequency produces improvements only on tactile discrimination (attended stimuli) but not on tone discrimination (irrelevant stimuli), and *viceversa* (Recanzone *et al.*, 1993).

This points out to a role in triggering perceptual learning for higher order cortical areas and for those non sensory mechanisms, such as ascending reticular inputs or inputs from the basal forebrain, which are known to modulate cortical neuron responses as a function of the behavioural state and to gate neural plasticity during development (Singer 1995 for review) and in adults (Weinberger 1995 for review, Kilgard and Merzenich 1998). The "highlighting" effects of attention would not extend from modulating plasticity in a neural circuits elaborating one stimulus parameter to circuits elaborating different stimulus aspects (Gilbert, 1994; Ahissar and Shapley, 1997; Bertini and Karni, 1997).

3. Neural correlates of perceptual learning

What happens at cellular level while our discriminative ability improves with practice?

Two neural correlates of perceptual learning have been proposed:

1. changes in neural selectivity;
2. recruitment of cortical territories, with the consequent modifications of cortical maps.

Examples supporting the hypothesis of changes in selectivity are relatively scanty. Kobatake *et al.*, (1998) trained adult monkeys to discriminate 28 moderately complex shapes and then examined the effects of training on stimulus selectivity of cells in area TE of the inferotemporal cortex, which is thought to mediate visual object discrimination and recognition. Recording from trained and untrained monkeys have shown that the proportion of TE cells responsive to some member of the training set was higher in the trained than in the control monkeys. The subset of training stimuli to which individual cells responded differed from cell to cell with only partial overlaps, suggesting that the cell responded (increased selectivity) to features common to different stimuli.

In a series of experiments designed to study the correlates between MT cell responses and behavioural performance in visual motion discrimination tasks, Newsome and collaborators (Salzmann and Newsome, 1994; Britten *et al.*, 1996) have found that there is a clear correlation between the strength of directionally selective MT neuron discharge and the behavioural choice of the monkey in favour of the preferred direction of the cell. Neurons that were more sensitive to weak motion signals had a stronger relationship to behaviour than those that were less sensitive. However, the authors conclude that signals from many neurons are pooled to inform psychophysical decisions. Therefore, even if improvement in sensitivity were to take place with training, which they do not mention, this would be anyhow accompanied by recruitment of a larger population of cells. The changes in MT cell response strength was dependent upon the direction of the monkey's spatial attention (Seidemann and Newsome, 1999), supporting the notion that changes in cell responses in sensory areas are subjected to attentional modulation (top-down effects). The interplay between MT and higher, "attentional" areas in a motion discrimination task in humans (Vaina *et al.*, 1998).is discussed in the accompanying paper.

Another evidence against increase in sensitivity being the principal neural correlate of perceptual learning comes from data of Recanzone *et al.*, (1992) who find that training in acoustic tone discrimination in monkeys increases both the selectivity of cortical neurons and the number of neurons responsive to the trained frequencies in the primary acoustic cortex, but only this latter phenomenon does correlate with improved performance due to training, since the former is present also in monkeys exposed to the acoustic tones but performing a tactile discrimination and who do not exhibit any improvement in acoustic discrimination performance.

Much more abundant are the evidences for the recruitment of cortical neurons as a correlate of perceptual learning.

Sugita (1996) fitted adult monkeys with Dove prisms goggles which produce left-right reversed vision and studied both the visually guided behaviour and the neuronal activity of cells in V1. For the first two weeks after goggle fitting, the monkeys are unable not only to move but also to feed themselves; however, as it is the case for humans, they slowly begin to lead a normal life and their visually guided behaviour normalizes within 1 month and a half. At this point, recordings were performed in V1 and it was found that 30% of neurons had developed a new receptive field in the ipsilateral visual hemifield, symmetrically positioned with respect to the normal, contralateral receptive field. These new receptive fields quickly disappeared after removal of the goggles, which was followed by a very fast (one day) return to normal behaviour. This result suggests that learning to cope with field reversal is mediated, at least in part, by a large scale functional reorganization at an early stage in the visual processing pathway which seems to rely on plasticity of interhemispheric connections.

Physiological correlates of learning in acoustic tone discrimination have been investigated by Recanzone *et al.* (1993). Adult monkeys were trained for several weeks to discriminate small differences in the frequencies of single acoustic tones presented sequentially. Animals showed a clear perceptual learning, in that they became able to discriminate tone differences progressively smaller. At the end of training, the tonotopic organization of the primary auditory cortex was found to be altered in the trained animals: the cortical area where single cells responded preferentially to the trained frequencies was enlarged at the expenses of other frequencies, with respect to the extent of this area in control inexperienced monkeys. This change in the tonotopic cortical map was proved to be correlated with the learning process, rather than being due to the mere exposure to the tones. Monkeys that had listened to the same tones and for the same time as the experimental animals, but were involved in a tactile discrimination task, did not show either an improvement in auditory frequency discrimination or a change in the corresponding cortical map.

The increase in the area of representation could result from a reinforcement, mediated by a Hebbian mechanism, of the synaptic inputs for the trained frequencies, and the corresponding weakening of the inputs for the frequencies close to the trained ones. Ahissar *et al.* (1992) provided evidence showing that such a mechanism is really at work. In their experiment either a positive or a negative correlation was established (with a paradigm similar to classical conditioning) between the spike discharge of two auditory cortex neurons. This caused a rapid increase (within 15 min) of the strength of the functional connections between the two neurons (and a related increase of the receptive field), if the correlation was positive, and a decrease, if the correlation was negative. For these changes to take place, however, it was necessary that the auditory stimuli used to correlate the cell activities were behaviourally relevant for the monkey, the animal being engaged in a task that required to pay attention to the stimulus.

These studies have suggested the possibility that a training procedure similar to that applied in the Recanzone's experiment on monkeys (1992) could modify the neural responses of children impaired in language understanding because of a perceptual impairment for the discrimination of brief sensory stimuli (less than 100 ms duration) presented rapidly in succession (Merzenich *et al.*, 1996). For instance, when two different fingers are touched rapidly one after the other, these children cannot say which fingers have been touched. This perceptual difficulty is such that these children cannot discriminate between two syllables, as *ba* and *da*, that differ only for their consonant sound which is very brief, and consequently they have great difficulty in understanding language.

A group of these children was trained for 4 weeks to discriminate sounds (not necessarily phonemes) that were initially of long duration and well separated in time, but then progressively decreased both in duration and temporal separation (Merzenich *et al.*, 1996). The children improved remarkably with training and learned to discriminate stimuli of shorter and shorter duration and time interval. At the end of the four weeks of training the children had gained an amount of linguistic competence equivalent to one or two years of age, and this improvement was still present 6 weeks later. Thus, a perceptual improvement that presumably modified the properties of cells of the primary auditory cortex, relieved these children from a serious state of discomfort.

In the somatosensory system examples have been reported of expansions during training of the cortical representation of a skin area. In the primary somatosensory cortex of blind subjects who learn to read in Braille there is an expansion of the cortical representation of the cutaneous area corresponding to the finger used to read. A similar phenomenon has been obtained in monkeys trained to discriminate two different frequencies of tactile stimulation, applied to a small area of one forefinger (Recanzone, 1992). In these animals there is a progressive improvement of the discrimination performance, restricted to the stimulated area. In parallel, there is a remarkable increase of the cortical representation of this skin area, and an increase of the correlation of the spike activity of different cells in this cortical region. These effects of training observed in the cortex are originated by plastic changes occurring at a cortical level, since no modification is present in the somatosensory thalamic nuclei. The fact that in the expanded area the neural activity tends to be correlated suggests that the repeated simultaneous activation is the crucial phenomenon that promotes the map expansion. This hypothesis has been confirmed by experiments showing that, by a repeated simultaneous stimulation of the distal portions of the fore-finger, middle-finger and fourth-finger, the cortical map was modified and there where receptive fields covering the whole range from forefinger to fourth-finger in the whole stimulated area (Wang *et al.* 1995). Normally these receptive fields are not present, since the maps of the various fingers are clearly separated from each other. And the reason for this separation, that probably emerges during development and is then maintained during adulthood, is that the stimulation of the skin of adjacent fingers is normally asynchronous. A similar effect is obtained in the somatosensory cortex of the rat following the repeated simultaneous stimulation of two vibrissae: this causes an increase of the correlated activity and a map collapse.

The synchronous stimulation would tend to adjoin the receptive fields of the stimulated cells, so that the borders between the representations of adjacent cutaneous areas would disappear, while the asynchronous stimulation would do the reverse. And indeed, in people who have two or more fingers fused together, with consequent simultaneous stimulation of the skin, the cortical representations of the fingers are not separated. After surgical separation of the fingers, with consequent asynchronous skin stimulation, within a few weeks a clear separation between the cortical representations of the fingers appears. This indicates that the cortical topography is dynamic, and it can vary under the control of the afferent neural activity (see Buonomano and Merzenich, 1998, for review).

4. Conclusions

The electrophysiological and imaging data suggest that experience dependent plasticity subserving perceptual learning can take place at early levels of cortical hierarchy, even at primary cortices level. It is likely that higher cortical areas play also a role in guiding these low level changes. The underlying neural mechanisms of plasticity seem to be very similar for visual, auditory or tactile learning, and show strong similarities to the mechanisms governing experience dependent plasticity during development. However, during development plasticity is a way to select between neural circuits while fine tuning the brain wiring, and the presence of critical periods makes the selection irreversible. In the adult, plasticity is a way to enlarge our behavioural repertoire, in order to respond to the different challenges of life, and, jumping to a more cognitive level, to change our mind when necessary. Literally.

References

Ahissar, E, E. Vaadia, M. Ahissar, H. Bergman, A. Arieli and M. Abeles (1992) "Dependence of cortical plasticity on correlated activity of single neurons and on behavioural context", *Science* 257:1412-1415.

Ahissar, M. and S. Hochstein (1996) "Learning pop-out detection: specificities to stimulus characteristics", *Vision Res.* 36:3487-3500.

Ahissar, M. and S. Hochstein (1997) "Task difficulty and the specificity of perceptual learning", *Nature* 387:401-406.

Buonomano, D.V. and M.M. Merzenich (1998) "Cortical plasticity: from synapses to maps", *Ann. Rev. Neurosci.* 21:149-186.

Fiorentini, A. and N. Berardi (1980) "Perceptual learning specific for orientation and spatial frequency", *Nature* 287:43-44.

Fiorentini, A. and N. Berardi (1997) "Visual perceptual learning: a sign of neural plasticity at early stages of visual processing", *Arch. Ital. Biol.* 135:157-167.

Gilbert, C.D. (1994) "Early perceptual learning", *Proc. Natl. Acad. Sci.* USA 91:1195-1197.

Karni, A. and G. Bertini (1997) "Learning perceptual skills: behavioural probes into adult cortical plasticity", *Curr. Op. Neurobiol.* 7:530-535.

Kilgard, M.P. and M.M. Merzenich (1998) "Cortical map reorganization enabled by nucleus basalis activity", *Science* 279:1714-1718.

Kobatake, E., G. Wang and K. Tanaka (1998) "Effects of shape-discrimination training on the selectivity of inferotemporal cells in adult monkeys", *J. Neurophysiol*. 80:324-330.

Merzenich, M.M, W.M. Jenkins, P. Johnston, C. Schreiner, S.L. Miller and P. Tallal (1996) "Temporal processing deficits of language learning impaired children ameliorated by training" *Science* 271:77-84.

Mollon, J.D. and M.V. Danilova (1996) "Three remarks on perceptual learning", *Spatial Vision* 10:51-58.

Recanzone, G.H., M.M. Merzenich and C.E. Schreiner (1993) "Changes in the distributed temporal response properties of S1 cortical neurons reflect improvement in performance on a temporally based tactile discrimination task", *J. Neurosci*. 13:87-103.

Recanzone, G.H., C.E. Schreiner and M.M. Merzenich (1992) "Plasticity in the frequency representation of primary auditory cortex following discrimination training in adult owl monkeys", *J. Neurophysiol*. 67:1031-1056.

Sathian, K. and J. Zangaladze (1997) "Tactile leaning is task specific but transfers between fingers", *Percept. Psychophys*. 59:119-128.

Seidemann, E. and W.T. Newsome (1999) "Effect of spatial attention on the responses of area MT neurons", *J. Neurophysiol*. 81:1783-1794.

Shiu, L. and H. Pashler (1992) "Improvement in line orientation is retinally local but dependent on cognitive set", *Percept. Psychophys*. 52:582-588.

Singer, W. (1995) "Development and plasticity of corticl processing architectures", *Science* 270:758-764.

Sugita, Y. (1996) "Global plasticity in adult visual cortex following reversal of visual input", *Nature* 380:523-526.

Vaina, L., J.W. Belliveau, E.B. des Roziers and T.A. Zeffiro (1998) "Neural systems underlying learning and representation of global motion", *Proc. Natl. Acad. Sci. USA* 95:12657-12662.

Wang, X, M.M. Merzenich, K. Sameshima and W.M. Jenkins (1995) "Remodelling of hand representation in adult cortex determined by timing of tactile stimulation", *Nature* 378:71-74.

Weinberger, N.M. (1995) "Dynamic regulation of receptive fields and maps in the adult sensory cortex", *Ann. Rev. Neurosci*. 18:129-158.

PIGEONS' VISUAL FIELD WHEN BINOCULARITY IS KEPT OUT AT DIFFERENT LIFE STAGES

DANIELA MUSUMECI, GIOVANNI CESARETTI and CLAUDIA KUSMIC*
Department of Physiology and Biochemistry, University of Pisa, Via S. Zeno 31, 56127 Pisa, Italy
**Institute of Clinical Physiology, CNR, Via Savi 8, 56126 Pisa, Italy*

ABSTRACT

We investigated the role of visual or both visual and proprioceptive inputs from the eyes during the development and in adult stage of pigeon life comparing the effects of retinal ablation or eye-ball enucleation on the frontal visual field extension by means of free moving perimetry method. Five groups were used: monocular reversible control group; two early groups (retinal ablated and eye-ball enucleated respectively) operated in two days post-hatching and tested one year later; and two late groups (one year old) tested two weeks after the surgery. The main effect of the lack of binocular inputs in the operated birds was the reduction of the contralateral visual field extension compared to the control group, except for the early eye-ball enucleated pigeons in which visuomotor and head postural strategies kept the gaze on the frontal field.

1. Introduction

Studies on the visual system pointed out that the eyes interact one each other to succeed in gaining specific brain areas in consequence of processes of competition during the development. One of the approaches in studying binocular interaction is to rule out the interaction itself early in the life by removing visual inputs from one eye (Shaz and Srovatan, 1986).

We were interested in investigating the behavioural effects on visual field extension in pigeon when the binocular competition is kept out during the development of the visual system.

The wideness of binocular field depends on various factors: the eyes position on the head, the optic axes projection and the ocular movements. In the pigeon the eyes are laterally placed with the optic axis pointing to the lateral field determining a wide and mainly monocular panoramic vision (about 316°) except for the frontal field in which the two monocular areas overlap around the eye-bill axis allowing the binocular vision (Walls, 1963). Coordinate eye movements are present and the vergence movements are especially effective during the pecking. The measured amplitude of the binocular window on the horizontal axis ranges from 18° to 41° (Hayes, 1987; Jahnke, 1984; Martin and Katzir, 1994; Martinoya *et al.*, 1981; Martinoya *et al.*, 1984; Nalbach *et al.*, 1990). McFadden (1989), however, estimated a value of about 71° during the last phase of the pecking.

The retinal fibres are totally crossed at chiasmatic level, nevertheless a few retino-recipient brain structures involved in visual perception as well as in oculo-motor control make available binocular interactions (Bulkharter and Cuènod, 1978; Donaldson and Knox, 1991; Gamlin and Cohen, 1988; Hayman and Donaldson, 1995; Hayman *et al.*, 1995; Karten, 1979; Meier, 1971). Since in pigeons the visual system is immature at hatching (Bagnoli *et al.*, 1987; Bagnoli *et al.*, 1989), retinal removal early in the life allows to grow-up animals having no binocular experience at all.

We investigated the effects of the loss of visual or both visual and proprioceptive inputs from one eye on the frontal field extension by means of a free moving perimetry method. At the beginning of the study three experimental groups were planned: control and retinal ablated both early and late in the life. However, anatomical results revealed that the operated eyes were significantly shrunk in diameter compared to the intact ones. Thus, two other animal groups were added to the study: eye-ball enucleated pigeons both early and late in the life.

2. Materials and Methods

2.1. Animals and surgery

23 animals were divided in five groups: monocular reversible control (*Control*, n=5); *early* retinal ablated (ERA, n=4), *late* retinal ablated (LRA, n=4), *early* eye-ball enucleated (EEE, n=6) and *late* eye-ball enucleated (LEE, n=4). Retina or eye-ball were gently drawn away by means of a vacuum pump under deep ethyl-ether anaesthesia[1]. The *early* groups were operated in two days post-hatching and tested one year later. The *late* groups (one year old pigeons) were tested two weeks after surgery. To obtain the monocular reversible performance *Control* group wore thin rings of Velcro permanently fastened around their eyes, and a light metal cup (0.3 g) was alternatively coupled on the ring of one eye during the training.

2.2. Apparatus and measures

Pigeons, trained in a Skinner-box, according to a Go NoGo procedure, had to discriminate and peck a black spot (positive stimulus) randomly displayed on a VGA monitor (sloping 45° out from the frontal panel of the box) and to refrain from pecking a white background (negative stimulus). The spot (3.8 mm) could appear in one out of 33 positions laying along two Cartesian axes and the diagonals, and it lasted 600 ms on the screen. In order to get the stimulus presentation pigeons had to peck a black square (starter) occurring at the origin of the axes. A single pellet reward was delivered after a peck to the spot (positive

---

[1] Experiments were performed according to the Animal Experimentation Legislation of the National Committee (law on animal care n° 116/92).

stimulus). A Personal Computer drove the entire set-up and collected all data. The experiments were videorecorded and off-line frame by frame analysed.

Visual perimetry were measured using the *detection limit* and the *distance of fixation*. The former is defined as the endmost detected spot position, *i.e.*, the position at which the performance level dropped down 25% of correct Go responses. The latter represents the length between the centre of eye and the origin of the Cartesian axes on the monitor, along the eye-bill axis.

In addition, the *reaction times* (the time elapsing between a peck to the starter and a peck to the spot) were collected to compute the visual area of the fastest reaction times (< 300 ms) without any head saccadic movement (*coding area*).

2.3. Computation of the visual field extension and data analysis

The *extension of the visual field* (perimetry) was computed by interpolating the visual angles derived from each axis. The visual angle is expressed by the following formula: $tang\alpha = l/d$, where l is the distance between the origin of axes and the detection limit; d is the *distance of fixation*.

A two ways analysis of variance (ANOVA) and post hoc Fischer test were used to compare all collected parameters within subjects and between groups. A comparison between the eyes within the *Control* group showed no difference in all parameters measured, thus, monocular data were averaged and referred as ipsi- and contra-lateral with respect to the viewing eye.

3. Results

3.1. Behaviour, Performance level, Distance of fixation, Reaction times

The training procedure assured a stereotype sequence of movements: the starter peck, the *withdrawal* after the pecking to the starter, the *head fixation* and the ballistic *motor output*. The head fixation phase was the critical point making a difference between EEE and all the other groups. In fact, *Control*, *LRA*, *ERA* and *LEE* groups always showed an alignment of the eye-bill axis with the starter position and the sagittal plane of head perpendicular to the monitor surface, independently of the kind of stimulus displayed on the screen (Cesaretti *et al.*,1997; Goodale, 1982). On the contrary, *EEE* pigeons showed a fixation phase in moving (making little steps backward) and inclined the head to put their seeing eye in front of the monitor.

In Figure 1 are shown the performances and the reaction times of all groups. Operated animals (with the exception of *EEE* birds) showed a marked and steeper drop in the performance level on the contralateral field than monocular performances of the Control group ($p<0.001$) along the horizontal axis. Reaction times did not differ among groups in the range of corresponding performances, except for *ERA* group along ipsilateral horizontal axis whose reaction times were faster ($p<0.01$) than the other groups.

The distance of fixation did not differ among groups (*Control* = 59±5 mm; *ERA* = 60.3±3 mm; *LRA* = 61.9±5 mm; *LEE* = 59±2 mm; *EEE* = 59.9±7 mm). As the *EEE* group regards, we used the distance measured at the median head position.

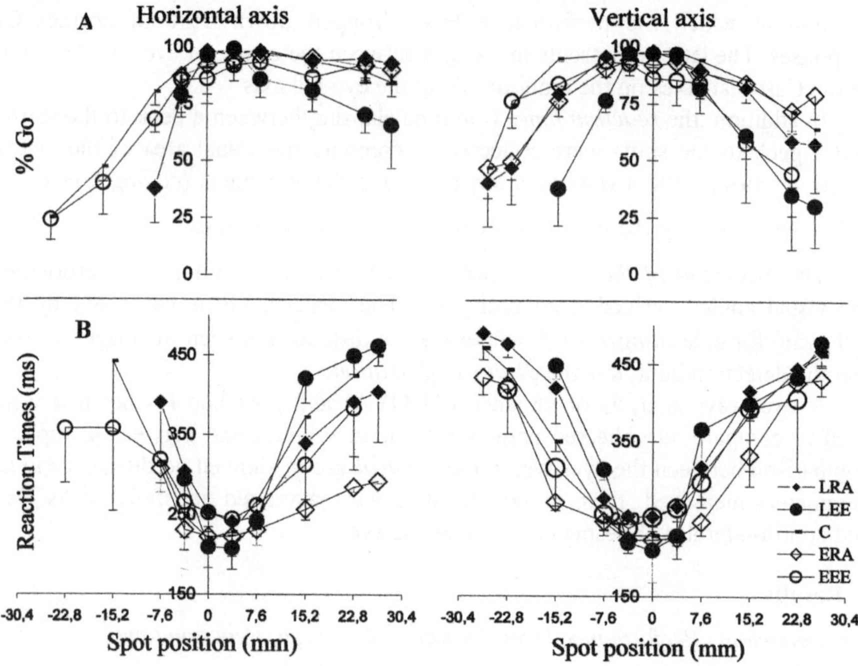

Figure 1. Mean value (S.E.) of performance (A) and reaction time (B) on different spot positions for each experimental group. Negative values indicate contralateral and down direction on the horizontal and vertical axes, respectively.

3.2. Visual field computation

Figure 2 shows the visual field computation for all groups: the pole represents the direction of the eye-bill axis, lying about 28° below the horizon and coincident with the starter position. Two main results were found: i) the notable reduction of the contralateral detection limit for *LRA*, *LEE*, and *ERA* groups (9.08°±1.1; 5.51°±0.9; 5.3°±0.1) but not of *EEE* group (15°±3) compared to the *Control* one (18.9°±1.5) and, ii) a remarkable increase in the extension of the coding area in *ERA* group.

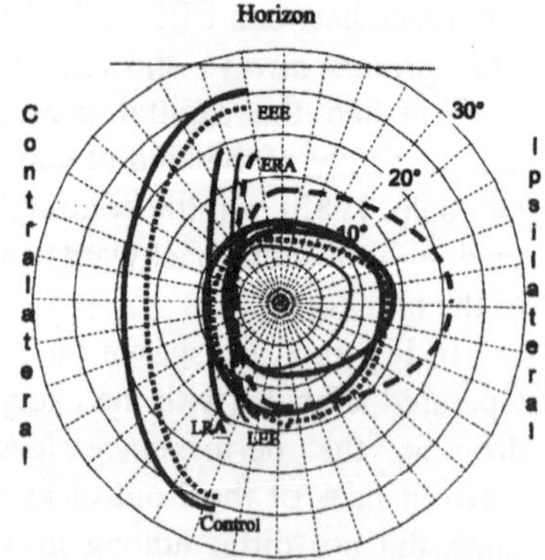

Figure 2. *Detection limit:*

4. Discussion and Conclusions

Our experimental paradigm leads the pigeons to work at their near point of accommodation (Macko and Hodos, 1985) with the eyes at their maximum convergence status (Martinoya *et al.*,1984; McFadden,1985).

How should the unbalanced visual inputs from the eyes affect the extension of the contralateral field in operated pigeons? The simplest interpretation is that the unbalanced visual inputs decrease the convergence angle of the eye during fixation phase with worse effects when also extraretinal proprioceptive inputs are involved. In this case the reduction of the field width (*i.e.*, the estimated binocular window) should correspond to a convergence angle of about 13.6° (Control value – ERA value), which is a measure consistent with data in literature.

If the vegence is reduced, the detection limit will correspond to the most eccentric portion of retina, close to the *ora terminalis*. Pigeons possess a retinal region, the red area, with foveal-like functions (Galifret, 1968) which is assumed to project in the frontal binocular field (Hayes *et al.*, 1987; Martinoya *et al.*, 1981, Nalbach, 1990); it is very wide and it extends just near to the *ora terminalis*. This fact could explain the existence of a coding area in the frontal field of operated animals despite of the convergence impairment and the detection limit reduction.

The unbalance of sensorial inputs from one eye can trigger an enhanced remodelling of retinal and visual structures during development rather than in the adult stage. In this contest, the enlargement of the coding area of ERA group could be the consequence of a retinal remodelling involving the red area, which is the most important retinal area during the fixation phase. In the case that other than retinal inputs also extraretinal proprioceptive inputs are unbalanced (as in the early enucleated pigeons) the animals have to develop compensating oculo-motor mechanisms and/or head postural adjustments to keep their red area projecting in the frontal field.

Acknowledgements

The Authors thank Dr. Angelo Gazzano for the surgical assistance and Mrs. Rita Andreotti for her helpful hand during pigeon training.

References

Bagnoli, P., V. Porciatti, G. Fontanesi and L. Sebastiani (1987) "Morphological and functional changes in the retinotectal system of the pigeon during the early posthatching period", *J. Comp. Neurol.* 256:400-411.

Bagnoli, P., V. Porciatti, G. Fontanesi and L. Sebastiani (1989) "Reorganization of visual pathways following posthatching removal of one retina in pigeons", *J. Comp. Neurol.* 288:512-527.

Bulkharter, A. and M. Cuénod (1978) "Changes in pattern discrimination learning induced by visual deprivation in normal and commissurotomized pigeons", *Exp. Brain Res.* 31:369-385.

Cesaretti, G., C. Kusmic and D. Musumeci (1997) "Binocular field in pigeons: behavioral measures of stimulus detection and coding", *Arch. Ital. Biol.* 135:131-145.

Donaldson, I.M.L. and P.C. Knox (1991) "Afferent signals from pigeon extraocular muscles modify the vestibular responses of units in the abducens nucleus", *Proc. R. Soc. London B* 244:233-239.

Galifret, Y. (1968) "Les diverses aires fonctionelles de la rètine du pigeon", *Z. Zellforsch. Mikrosk. Anat.* 86:535-545.

Gamlin, P.D.R. and D.H. Cohen (1988) "Retinal projections to the pretectum in the pigeon (Columbia livia)", *J. Comp. Neurol.* 269:1-17.

Goodale, M.A. (1982) "Visually guided pecking in the pigeon (Columba livia)", *Brain Behav. Ev.* 22:22-41.

Hayes, B.P., W. Hodos, A.L. Holden and J.C. Low (1987) "The projection of the visual field upon the retina of the pigeon", *Vision Res.* 27:31-40.

Hayman, M.R. and I.M.L. Donaldson (1995) "Deafferentation of pigeons extraocular muscles disrupts eye movements", *Proc. R. Soc. Lond. B* 261:105-110.

Hayman, M.R., J.P. Donaldson and I.M.L. Donaldson (1995) "The primary afferent pathway of extraocular muscle proprioception in the pigeon", *Neurosc.* 69:671-683.

Janke, H.J. (1984) "Binocular visual field differences among various breeds of pigeon", *Bird Behav.* 5:96-102.

Karten, H.J. (1979) "Visual lemniscal pathways in birds", in: *Neural mechanisms of behavior in the pigeon*, A.M. Granda and J.H. Maxwell, eds, Plenum Press, New York, pp. 409-430.

Macko, K.A. and W. Hodos (1985) "Near point of accommodation in pigeons", *Vision Res.* 25:1529-1530.

Martin, G.R. and G. Katzir (1994) "Visual field and eye movements in herons (Ardeidae)", *Brain Behav. Evol.* 44:74-85.

Martinoya, C., J. Le Houezec and S. Bloch (1984) "Pigeon's eye convergence during feeding: evidence for frontal binocular fixation in a lateral-eyed bird", *Neurosci. Let.* 45:335-339.

Martinoya, C., J. Rey and S. Bloch (1981) "Limit of pigeon's binocular field and direction for best binocular viewing", *Vision Res.* 23:911-915.

McFadden, S.A. (1989) "Eye design for depth and distance perception in the pigeon: an observer orientated perspective", *Int. J. Comp.* 3:101-130.

Meier, R.E. (1971) "Interhemisphärischer Transfer visueller Zweifachwahlen bei kommissurotomierten Tauben", *Psychol. Forschung* 34:220-245.

Nalbach, H-O., F. Wolf-Oberhollenzer and K. Kirschfeld (1990) "The pigeon's eye viewed through an opthalmoscopic microscope: orientation of retinal landmarks and significance of eye movements", *Vision Res.* 30:529-540.

Shatz, C.J. and D.W. Srevatan (1986) "Interactions between retinal ganglion cells during the development of the mammalian visual system", *An. Review Neurosci.* 9:171-207.

Walls, G.L. (1963) *The Vertebrate Eye and its Adaptive Radiation*, Hafner, New York.

DECISION TIME FOR CORRECT AND INCORRECT RESPONSES IN SIZE DISCRIMINATION

S. V. CHUKOVA*, A. J. AHUMADA, JR.° and E. A. VERSHININA*
*Pavlov Institute of Physiology, Emb. Makarova 6, St. Petersburg, 199034, Russia
°NASA Ames Research Center, Mail Stop 262-2, Moffett Field, CA 94035-1000, USA

ABSTRACT

A strong "temporal crowding" effect in both discrimination accuracy and decision time (DT) can not be explained by time-accuracy trade-off - the observers show the poorest performance at the longest decision time in the long inter-stimulus interval, short inter-trial interval condition. To understand context and sequential effects in discrimination accuracy, we compare the DT's for correct and incorrect judgements in threshold discrimination at different inter-stimulus intervals (ISI's of 0.05, 0.2 and 2.0 sec) and inter-trial intervals (ITI's of 0.5 and 2.5 sec). For all observers, for all ISI's, for all ITI's, the DT's for incorrect responses are much longer than the DT's for correct responses. Statistical decision theory brought signal detection theory to perception and then it was applied to memory. We are using sequential statistical decision theory adapted to perceptual detection to explain results that we attribute to the accumulation over time of information from memory.

1. Introduction

1.1. Role of memory in perceptual performance

The psychophysical measurement of sensory thresholds has usually been regarded as a window into the information processing capabilities of sensory systems. Sensory thresholds are also regarded as a measurement of the intrinsic noise of the sensory system (Ahumada, 1987). Signal detectability theory developed in the 1950's (Green and Swets, 1966; Peterson, Birdsall, and Fox, 1954; Tanner and Swets, 1954) popularized the separation of perceptual (*d'*) from decisional effects (bias). The underlying theory also includes a memory component in the form of the observer's template; however, this component has usually been ignored, assumed to be ideal or, at least, constant. Recently, Beard and Ahumada (1998, 1999) have been developing methods for examining the memory template and models for its dynamic.

There are two points of view on the role of memory in discrimination performance. First, the visual system is able to store a neural representation of the parameters of the reference stimulus over long intervals. Magnussen and Dyrnes (1994) and Regan (1995) reported that varying the interval between the stimuli to be compared from 0.4 to 30 sec does not affect the discrimination threshold. Consequently, the precision of perceptual short-term memory is determined by the *neural code*. Second, discrimination judgments are based on the memory of the stimuli presented during experiment, rather than on the fixed standard (Helson,

1947; Sandusky, 1974; Westheimer and Mckee, 1977; Lages and Treisman, 1998). Thus, the accuracy in discrimination performance is determined by the *memory recall effects*.

Our prior research (Chukova and Ahumada, 1998) has demonstrated strong context and sequential effects in spatial interval discrimination that are consistent with memory-based adaptation level-theory (Helson, 1947; Sandusky, 1974). Contrary to reports showing "perfect" visual memory for long periods, we find that threshold do increase significantly ($p<0.001$) as the interval between the stimuli to be compared increases from 0.2 to 2 sec. The increase is much greater for the 0.5 sec interval between the trials than for the 2.5 sec interval. This "temporal crowding effect" suggested that we look at how the previous trial stimulus affects performance on the next trial. The analysis showed that the "temporal crowding effect" could be explained as a result of the effects of the prior stimuli, and that the higher thresholds associated with intermixing standards can be regarded as memory effects.

1.2. Accuracy and decision time for size discrimination

Some data have been reported on the relation between accuracy for spatial-frequency discrimination and decision time (Vassilev and Mitov, 1976; Harwerth and Levi, 1978; Magnussen, Idas, and Myhre, 1998). However, judgment times or reaction times have usually been studied with *suprathreshold* stimuli, incorrect responses being excluded from consideration (Greenlee and Breitmeyer, 1989; Magnussen et al., 1998).

1.3. Aim of the work

The analysis of the decision time for correct and incorrect responses was designed to find more clues to the mechanism for the dramatic changes we have previously found in discrimination performance at different temporal intervals. To explain the results that we attribute to the accumulation over time of information from memory, we used sequential statistical decision theory adapted to perceptual detection.

2. Methods

2.1. Observers

Three observers with normal or corrected-to-normal visual acuity participated in the experiments.

2.2. Apparatus

The experiments were programmed in C and run on PC (Dell 75 MHz Pentium) computer. Stimuli were generated on active-matrix liquid crystal display (AMLCD) monitor (MultiSync LCD200) with a pixel spacing of 0.24 mm both horizontally and vertically. Viewing was binocular with natural pupils at the distance of 315 cm from the screen, at which the angular pixel's size was 0.26 arc min.

2.3. *Stimuli*

Stimuli were two vertical *black* lines (2 x 82 pixels) on a *bright* background presented in the center of the screen:

||

The reference separation (R) between lines varied from 9.6 to 16.6 arc min, in increments of 4 or 5 pixels. There were seven references in the range (9.6, 10.7, 12.0, 13.0, 14.3, 15.6 and 16.6 arc min). Six test (T) spatial intervals were distributed symmetrically about R. To minimize "introducing an internal subjective criterion in deciding whether each faint ambiguous percept deserves a 'yes' or a 'no'" (Farell and Pelli, 1999), we did not employ a pair of the same stimuli (R - R).

2.4. *Procedur*

The observer started the experiment by pressing the appropriate key. Each trial was structured as follows: (i) inter-trial interval (0.5 or 2.5 sec), (ii) reference (0.5 sec), (iii) inter-stimulus interval (0.05, 0.20 or 2.00 sec), (iv) test stimulus (0.5 sec), (v) the observer's response indicated by pressing appropriate key, which initiated the next trial.

The *observer's task* was to decide, whether the second stimulus is narrower or wider than the first, and to indicate the judgements by pressing "1" key ("narrower") or "2" key ("wider") on the keyboard.

Computer registered the time between two consecutive pressings of the key (T). As to decision time, the observers were not given any instructions.

2.5. *Size discrimination*

To measure the discrimination threshold and size estimation error, we used a classical method of constant stimuli. Psychometric curves were obtained by plotting the percentage of "wider" responses as a function of difference between the reference and the test stimulus. To calculate the threshold (Δ**s**) and size estimation error (**m**), we used Probit analysis (Finney, 1971). For each reference, the threshold value was calculated based on 500 presentations.

2.6. *Decision time*

For all conditions, the duration of the reference (T_R) and the test stimulus (T_T) held constant at 0.5 sec. The *ISI* is the time from the end of the reference stimulus to the beginning of the test stimulus. The *ITI* is the time from the observer's response to the beginning of the reference stimulus. Given the time between two consecutive pressings of the key indicating the observer's response is the trial duration T, the decision time is calculated as following:

$$DT = T - ITI - T_R - ISI - T_T \quad (1)$$

We performed the decision time analysis in terms of median characterizing the distribution central tendency, because the distributions were positively skewed with a few extremely long values.

3. Results

As we can see from Figure 1, the observer shows the poorest performance at the longest DT's in the long inter-stimulus interval, short inter-trial interval condition.

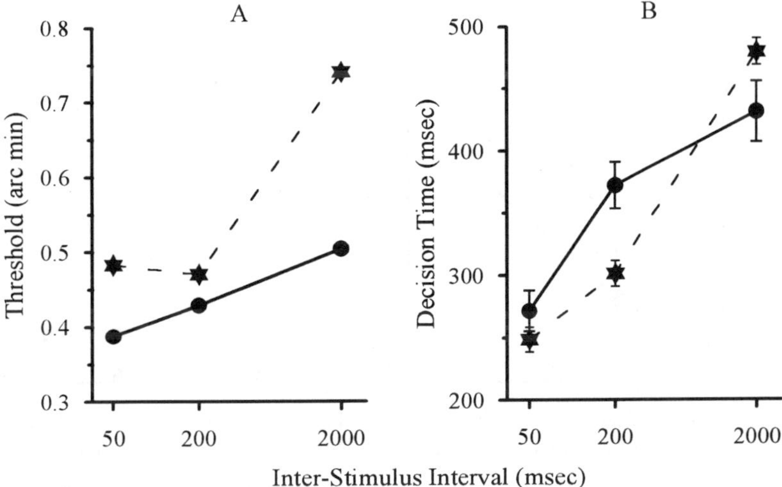

Figure 1. Threshold (*A*) and decision time (*B*) as a function of inter-stimulus interval, with inter-trial interval as the parameter (observer SV). *Circles, solid line* – 2,5 sec ITI; *stars, dashed line* – 0.5 sec ITI. The values were obtained by averaging over 7 references. Each point represents 3,500 presentations.

For all ISI's and all ITI's, incorrect dicrimination has much longer latency than correct discrimination does (Mann-Whitney rank test, $p<0.001$; Figure 2a). At the same time, the difference in the latencies is much more pronounced for the biggest size differences between the reference and test stimulus, than for the smallest ones (Figure 2b), showing that the guessing has approximately the same latencies for both correct and incorrect responses.

4. Discussion

4.1. Perceptual value of the stimuli

The poorest performance was observed at the longest decision times; so, the observers perform poorly not because they hurry their responses. If there is no a trade-off between the threshold and the DT, what might induce the decrease both the values at the short ISI's? We could use for explanation of this fact a widespread belief that the response time to a stimulus component is a function of the perceptual value of that component, which makes discrimination fairly easy, with the result that DT is short (Ashby and Townsend, 1986), *if* the stimulus component (size) would not be the only one in our experiments. Given this circumstance, we think that the stimulus perceptual value itself is a function of sequential variables, and in particularly, at the short ISI's, the perceptual value of size might be increased due to the motion component being involved.

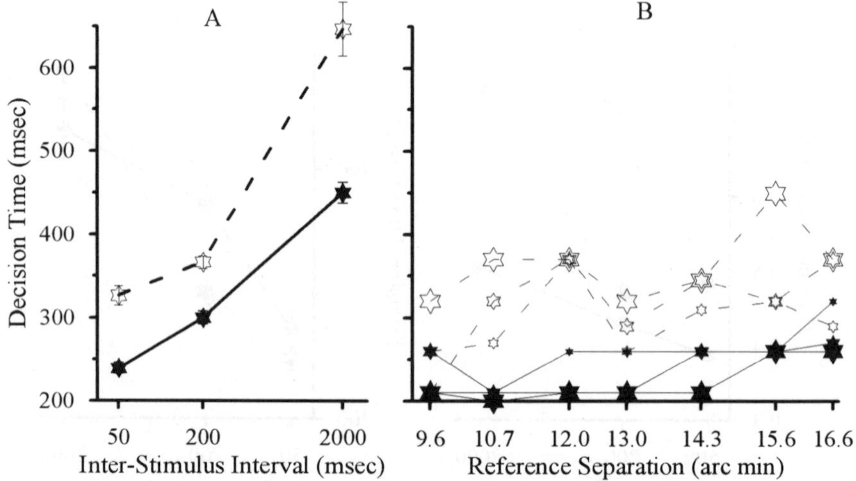

Figure 2. *A*: Decision time as a function of inter-stimulus interval for correct and incorrect responses (0.5 sec ITI). *Filled stars, solid lines* – correct responses; *empty stars, dashed lines* – incorrect responses. The values were obtained by averaging over 7 references. *B*: Decision time as a function of reference separation, with the size difference between the reference and test stimulus as the parameter (0.05 sec ISI, 0.5 sec ITI). Star size on *B* corresponds to the size difference between the reference and test stimulus. All the data for the observer SV.

Actually, Magnussen, Idas, and Myhre (1998) did perform the analysis for incorrect responses as well. (personal communication by Dr. Svein Magnussen, ECVP '99). However, no difference has been found between the choice reaction times for correct and incorrect responses; that is why they presented the data in terms of correct responses. The observers are likely to use different criteria for threshold and suprathreshold. discrimination.

4.2. Time-dependent version of signal detection theory

To explain the low accuracy on long latency trials in choice reaction time experiments with weak, continuous signals, we have developed a time-dependent version of signal detection theory.

In the usual signal detectability theory, the observer is assumed to have an internal variable x and a single criterion c and to make response 1 if $x > c$ and response 2 otherwise. The variability in x determines the threshold or detectability and c determines the bias.

In the latency situation, information about the signal is accumulating over time t in an internal variable $x(t)$ and there are two criteria $c_1(t)$ and $c_2(t)$ that change with time.

If $x(t)$ ever exceeds $c_1(t)$ the observer initiates response 1 at time t;

if $x(t)$ falls below $c_2(t)$, the observer initiates response 2 at time t;

and while $x(t)$ is between the two criteria, the observer allows more information to accumulate.

The observer adjusts the criteria functions according to instructions and payoffs for speed and accuracy. The criteria may cross at some time, the longest time the observer is willing to collect information. For these longest observations, the observer did not get any clear evidence and is essentially guessing.

We interpret the similarity of the inverse correlation of accuracy and latency in our experiment and the continuous detection experiments to mean that the memory trace information also becomes available with a time course and that the observer waits until the evidence exceeds criteria functions before responding.

References

Ahumada, Jr., A.J. (1987) "Putting the noise of the visual system picture", *Journal of the Optical Society of America A* 4:2372-2378.

Ashby, F.G. and J.T. Townsend (1986) "Varieties of Perceptual Independence back", *Psychological Review* 93:154-179.

Beard, B.L. and A.J. Ahumada, Jr. (1998) "Technique to extract relevant image features for visual tasks", in: *Human Vision and Electronic Imaging III. Proc. SPIE,* Vol. 3299, B.E. Rogowitz and T.N. Pappas, eds, San Jose, CA (USA), pp. 79-85.

Beard, B.L. and A.J. Ahumada, Jr. (1999) "Detection in fixed and random noise in foveal and parafoveal vision explained by template learning", *Journal of the Optical Society of America A* 16:755-763.

Chukova, S.V. and A.J. Ahumada, Jr. (1998) "Temporal "crowding" effect in size discrimination", *Investigative Ophthalmology and Visual Science* 39(4)S410.

Farell, B. and D.G. Pelli (1999) "Psychophysical methods, or how to measure a threshold, and why", in: *Vision Research. A practical Guide to Laboratory Methods*, R.H.S. Carpenter and J.G. Robson, eds, Oxford, New York, Tokio: Oxford University Press, pp. 129-136.

Finney, D.T. (1971) *Probit analysis* (3rd edn), Cambridge: Cambridge University Press.

Green, D.M. and J.A. Swets (1966) *Signal Detection Theory and Psychophysics*, New York: Wiley and Sons.

Harwerth, R.S. and D.M Levi. (1978) "Reaction time as a measure of suprathreshold grating detection", *Vision Research* 18:1579-1586.

Helson, H. (1947) "Adaptation-level as a frame of reference for prediction of psychophysical data", *American Journal of Psychology* 60:1-29.

Lages, M. and M.Treisman (1998) "Spatial-frequency discrimination: Visual long-term memory or criterion setting?", *Vision Research* 38:557-572.

Magnussen, S. and S. Dyrnes (1994) "High-fidelity perceptual long term memory", *Psychological Science* 5:99-102.

Magnussen S., E. Idas and S.H. Myhre (1998) "Representation of orientation and spatial frequency in perception and memory: a choice reaction-time analysis", *Journal of Experimental Psychology: Human Perception and Performance* 24:707-718.

Peterson, W.W., T.G. Birdsall and W.C Fox. (1954) "The theory of signal detectability", *Transactions of the IRE Professional Group on Information Theory* 4:171-212.

Regan, D. (1995) "Storage of spatial-frequency information and spatial-frequency discrimination", *Journal of Optical Society of America A* 2:619-621.

Sandusky, A. (1974) "Memory Processes and Judgement", in: *Handbook of Perception, Vol. II. Psychophysical Judgement and Measurement*, New York, San Francisco, London: Academic Press, Inc.

Tanner, W.P. and J.A. Swets (1954) "A decision-making theory of visual detection", *Psychological Review* 61:401-409.

Vassilev, A. and D. Mitov (1976) "Perception time and spatial frequency", *Vision Research* 16:82-89.

Westheimer, G., and S. McKee (1977) "Spatial configurations for visual hyperacuity", *Vision Research* 17:941-947.

LEARNING OF COMBINED-FEATURES SEARCH: SPECIFICITY OF STIMULUS CHARACTERISTICS

GIANLUCA CAMPANA and CLARA CASCO
Dipartimento di Psicologia Generale and Centro Interdipartimentale di Scienze Cognitive, Università di Padova, via Venezia 8, 35100 Padova, Italy

ABSTRACT

We investigated the specificity of learning in visual search for within-object conjunction (that is: combined) features, with homogeneous or heterogeneous distractors. When collinearity was perturbed learning did not persisted in homogeneous display only, suggesting high spatial resolution for a group of elements. Learning never transferred to new stimuli in which either the display was rotated or target and distractors were exchanged, suggesting that learning is specific for combined-features orientation.

1. Introduction

In perceptual learning of several visual tasks, a substantial improvement is obtained after training which is enduring after months, suggesting a long-term structural modification of basic perceptual modules (Karni and Sagi, 1991).

By analysing whether learning transfers across stimulus changes, it has been found that learning effects are specific for the stimulus, indicating that it produces a modification of the selectivity tuning of neurons activated by the task in the appropriate cortical areas (Fiorentini and Berardi, 1981; Karni and Sagi, 1991). In a simple search task, Ahissar and Hochstein (1996) found a strong selectivity for position, size and orientation, and suggested that learning with these stimuli occurred within an early cortical computing site.

In the present study we investigated the site of perceptual learning for search tasks in which target and background consisted, instead of simple features, of conjunction of features belonging to the same neural map (orientation). Time course and specificity of learning should be similar to that found in simple features learning if, as several authors claim (Humphreys *et al.*, 1989; Casco and Campana, 1999), both stimuli activate a parallel search mechanism.

2. Materials and methods

We compared the time course of learning in search for combined-features consisting of rotated Ls (51' wide and high) in either homogeneous or heterogeneous field of collinear distractors. The homogeneous display consisted of a 180° rotated L in a 90° counterclockwise-rotated L. The heterogeneous display consisted of a 90° clockwise-rotated L in three differently oriented Ls. Subjects were 24 naive adults with normal or corrected-to-normal vision. Each block of 120 trials consisted of 15 repetitions of two stimuli (target present or absent) and 4 distractor numerosities (4, 16, 36, 64) randomly presented on an IBM PC colour monitor. The learning session (16 blocks) was followed by the test session (3 blocks), in which the stimuli were somehow changed.

3. Results

Figure 1 shows mean RTs, averaged across distractors numerosities, as a function of block number separately for present and absent responses. The effect of block was significant for both homogeneous ($p<.000$) and heterogeneous ($p<.000$) conditions, indicating that learning significantly reduces RTs in both conditions. To study the receptive field tuning of the detectors involved in our task we asked whether learning transfers to other stimuli. Mean RTs of block 16 (learning session) was compared with that obtained with 3 different stimuli (test session).

3.1. Position specificity

The difference in mean RT between block 16 (learning session) and that obtained in a stimulus in which positional collinearity of the elements was perturbed (up to 30') was not significant with heterogeneous but it was with homogeneous distractors (present: $p<.001$; absent: $p<.004$), indicating that, in this search task, learning does not transfer to a non-collinear display. This suggests that improvement was related to collinearity in the homogeneous condition only. The collinearity effect is indicative of high spatial resolution, resulting from relatively early computing site. The difference between homogeneous and heterogeneous display indicates, as previously suggested (Humphreys *et al.*, 1989), that, with homogeneous distractors only, the underlying mechanism is selective for the position of a group of elements rather than a single element.

3.2. Orientation specificity

The difference between mean RT in block 16 and that obtained with a in 45° rotated display was significant for both homogeneous (present: $p<.001$; absent: $p<.001$) and heterogeneous conditions (present: $p<.042$; absent: $p<.039$) suggesting that the underlying mechanism is selective for orientation. Since the display rotation changes both single and combined-features orientation, the question is whether the learning process increases the orientation selectivity of neurons responding to simple or to combined-features. To answer this question we compared reaction times obtained in block 16 (learning session) with that obtained in a block in which target and non-targets were exchanged. The difference is significant in both homogeneous (present: $p<.039$; absent: $p<.032$) and heterogeneous (present: $p<.003$; absent: $p<.001$) distractors. Since in these conditions orientation of combined-features change but that of single features remains the same, these results suggest that the underlying mechanism is selective for the orientation of combined-features.

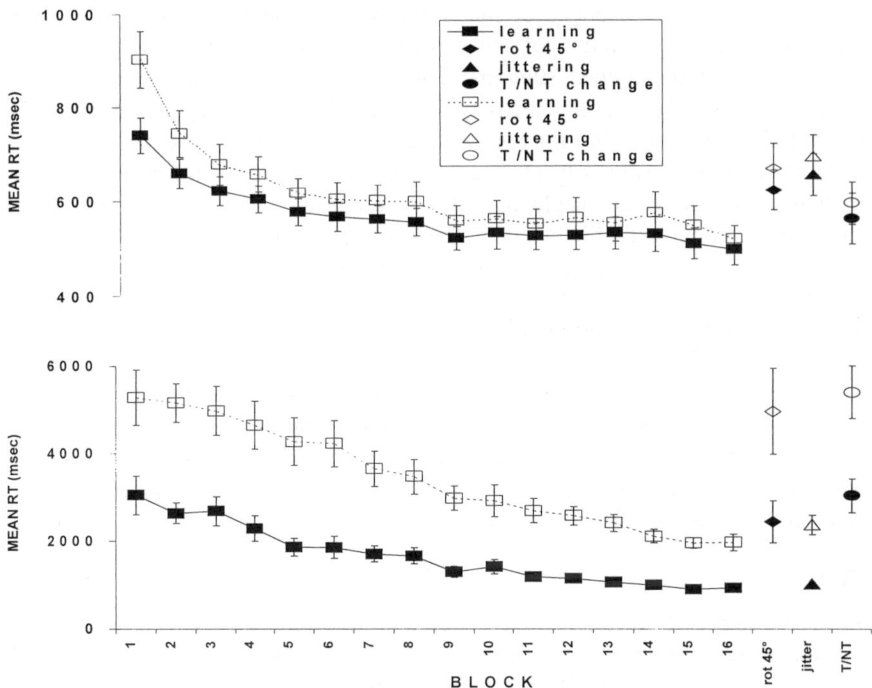

Figure 1. Mean RTs separately for present (filled symbols) and absent (unfilled) target conditions, and for homogeneous (upper graph) and heterogeneous (lower graph) target conditions are shown.

4. Discussion

In agreement with previous findings (Ahisser and Hochstein, 1996), we found that learning in visual search is specific for spatial position, suggesting that this results from restrictions of spatial extent of receptive field in visual areas higher than V1 but lower than IT. Indeed neurons in high-level visual areas (*i.e.*, IT) do not present high specificity for spatial position (Maunsell, 1995; Gross *et al.*, 1972). With our complex stimuli, the position specificity of learning it is also likely to take place in a visual area higher than V1, maybe in V2 or V4. The finding that the position specificity of learning only occurs with homogeneous display where distractor grouping may take place suggests that our homogeneous stimulus activates positionally specific neurons which present similar characteristics of collator units (Moulden, 1994) that respond to collinear disconnected elements.

Our second result is that learning is specific for the orientation of the combined-features elements. Indeed, learning does not transfer to a stimulus in which target and distractors were exchanged or the whole display was 45° rotated, suggesting that the change of orientation of the combined-features stimulus was behind the receptive field of the trained stimulus. The orientation specificity in our case is different from the orientation selectivity for single features (Karni and Sagi, 1991; Ahisser and Hochstein, 1996) for two reasons. First, one of our orientation changes is only at combined-features level. Second, our conjunction search task cannot be executed on the bases of single bar orientation discrimination, since target and distractors are composed of bars having the same orientation.

To summarise, our findings suggest that learning may reflect of phenomenon of plasticity in the mature visual system, which consists in within-maps facilitatory links. We claim that the position effect and the orientation specificity are due to two different kinds of interactions. The first kind of interaction could be excitatory and short- range, allowing simple combination of form features (as rotated Ls) to be coded in "specific" junction maps, perhaps linked hierarchically to single- features maps and depending on orientation coding. The second kind of interaction is long-range, allows grouping of distractors (Moulden, 1994; Casco and Campana, 1999) and depends on coding of position over space.

References

Ahissar, M. and S. Hochstein (1996) "Learning pop-out detection: specificities to stimulus characteristics", *Vision Res.* 36(21):3487-3500.

Casco, C. and G. Campana (1999) "Spatial interactions in simple and combined-feature visual search", *Spatial Vis.* 12(4):467-483.

Fiorentini, A. and N. Berardi (1981) "Learning in grating waveform discrimination: specificity for orientation and spatial frequency", *Vision Res.* 21:1149-1158.

Gross, C.G., C.E. Rocha-Miranda and D.B. Bender (1972) "Visual properties of neurons in inferotemporal cortex of the macaque", *J. Neurophysiol.* 35:96-111.

Humphreys, G.W., P.T. Quinlan and M.J. Riddoch (1989) "Grouping processes in visual search: effects with single and combined-features targets", *J. Exp. Psychol.: General* 118:258-279.

Karni, A. and D. Sagi (1991) "Where practice makes perfect in texture discrimination: evidence for primary visual cortex plasticity", *Proc. Nat. Acad. Sci. USA* 88:4966-4970.

Maunsell, J.H.R. (1995) "The brain's visual world: Representation of visual targets in cerebral cortex", *Science* 270(5273):764-768.

Moulden, B. (1994) "Collator units: second-stage orientational filters", in: *Higher-order Processing in the Visual System*, G.R. Bock, J.A. Goode *et al.*, eds, Chichester, UK: Ciba Foundation, pp. 170-192.

PARAFOVEAL PREVIEW FACILITATION IN A LEXICAL DECISION TASK IS VISUALLY BASED

MAURO ORIOLI and CLARA CASCO
Dipartimento di Psicologia Generale, Università degli Studi di Padova, via Venezia 8, 35100 Padova, Italy

ABSTRACT

We investigated whether the facilitatory effect in a lexical decision task carried out on a letter string (either word or non-word) due to parafoveal preview of the same letter string is tied to the visual information. Lexical decision time was shorter when the preview was a letter string (presented at 2.5° or 5° eccentricity) than when it was either absent or constituted of a string of X at 2.5° eccentricity. Moreover, it always increased linearly with string length except in two word preview conditions: a string of letters (at 5°) or of Xs (at 2.5°). We suggest that here global processing occurs because previewing enhance word length, shape, boundaries or texture.

1. Introduction

Parafoveal information available during reading facilitates word identification in the next fixation (Pollasteck *et al.*, 1982; Rayner and Bertera, 1979). High-level explanations relate the facilitatory effect to semantic (Underwood *et al.*, 1988), contextual (Balota *et al.*, 1985) and orthographic (Evett and Humphreys, 1981) parafoveal information. A more low-level explanation suggests that the facilitatory effect of parafoveal preview is visually based: either local and derived from word sub-units (Rayner *et al.*, 1982) or letters (Morris *et al.*, 1981) or global (Pollatsek and Rayner, 1982; Findlay *et al.*, 1993; Menz and Groner, 1987), consisting in information on word length, size or shape information (the boundaries or the texture layout of the word). Parafoveal preview could provide global visual information by means of grouping of elements, texture segmentation (Findlay *et al.*, 1993; Menz and Groner, 1987) or parallel processing (Fiorentini, 1989). We investigated whether during peripheral preview global aspects are made available and whether this reduces the need, in the successive lexical decision on the same patterns foveally presented, to use local information as letters or features.

2. Materials and Methods

Stimuli were letter strings viewed on a computer monitor. Each block of trials consisted of 64 strings (words and non-words with equal probability) randomly presented. String length varied according to 4 levels: 2-3, 4-5, 6-7 and 8-9 letters. For each length level, 8 different words were used in which frequency was balanced. Legal and non-legal non-words were balanced and obtained by randomly substituting 1 letter (randomly chosen within the alphabet) of the word.

In each trial, after 400 ms from fixation mark onset, the parafoveal stimulus was presented (except in Condition 1) for 100 ms and was either the successive letter string for the lexical decision, presented at 2.5° (Condition 2: string/2.5°) and 5° eccentricity (Condition 3: string/5°), or a string of Xs of the same length as the successive foveal string (Condition 4: nX/2.5°). The parafoveal stimulus was followed with no interval by the central letter string, which lasted on the screen until the subject's response. Subjects were 7 naive adults with normal (non-corrected) vision. They were required to maintain fixation and to decide whether the fixated letter string was a word or non-word by pressing the appropriate key.

3. Results

The effect of errors was not significant. An ANOVA, conducted on lexical decision time averaged across word length (LTD), showed that the two main factors but none of the interactions were significant. Results are shown in Figure 1.

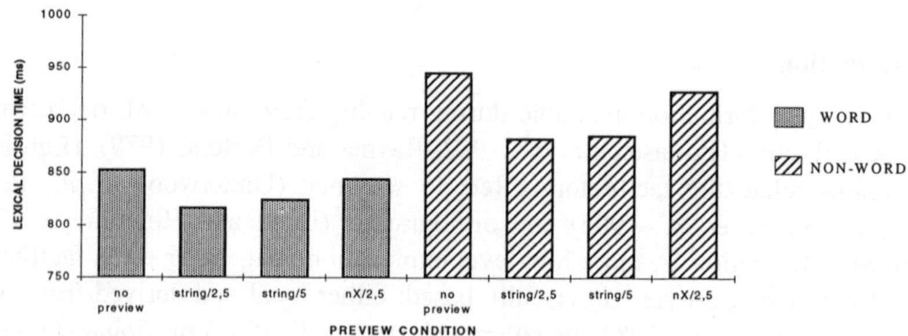

Figure 1. Lexical decision time (ms) is shown independently for each preview condition (no preview, string/2.5, string/5, nX/2.5) and each stimulus condition (words and non-words).

3.1. The effect of stimulus type on lexical decision time

LTD is shorter for words than for non-words (838.25 vs. 916.27; $F[5,1]= 5.58$, $p<.05$).

3.2. The effect of peripheral preview on lexical decision time

The effect of peripheral preview is significant (no preview: 891.7; string/2.5°:850.2; string/5°:855.5; nX/2.5°:886.5; $F[15.3]= 5.69$ $p<.01$). *Post hoc* comparison (Newman Kewls) shows that LTD is longer ($p<.05$) in Condition 1 (no preview) than in Conditions 2 (string/2.5°) and 3 (string/5°). On the other hand, it was not significantly different in Condition 1 with respect to Condition 4 (nX/2.5°), suggesting that word length information alone does not produce any facilitatory effect. Again, the difference between Condition 2 (string/2.5°) and Condition 3 (string/5°) was not significant suggesting that the facilitatory effect is not reduced when the single letter information is reduced. Finally, an important result is that LTD in Condition 4 (nX/2.5°) is shorter than both in Condition 2 (string/2.5°) and 3 (string/5°), suggesting that when string layout or texture is present the facilitatory effect is larger with respect to when only word length is present.

3.3. The effect of peripheral preview on visual processing

To establish whether parafoveal preview reduces local analysis during the lexical decision on the foveal string, we investigated whether LTD is dependent on word length more without than with preview. Regression line (Figure 2) was fitted (in all conditions) to the data points relating LTD to word length. For non-words, LTD linearly increases with word length in all conditions. The regression line slope (indicating time-per-letter) were: 29 ms (no-preview), 25 ms (string/2.5°), 26 ms (string/5°) and 36 ms (nX/2.5°). Clearly, these values indicate that local elements are serially scanned with and without preview. For words, similar results were obtained in the absence of preview (slope: 47 ms) and in the string/2.5° condition (slope: 23 ms). On the other hand LTD did not increase linearly with word length in the other two conditions (slopes were equal to 4.5 ms in string/5° and to 5 ms in nX/2.5). Time-per letter equal or smaller than 5 ms are indicative that a global process is involved in lexical decision. A result, in contradiction with this interpretation, is that despite the fact that no facilitatory effect is present in condition nX/2.5°, LTD is not affected by word length.

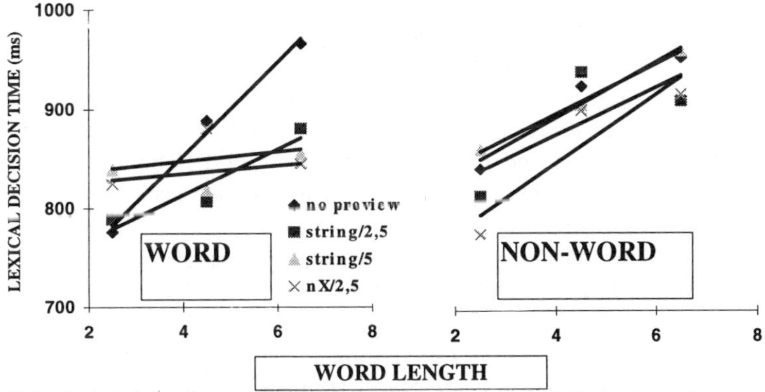

Figure 2. Lexical decision time is shown as a function of word length for both words and non-words independently for each preview condition (no preview, string/2.5°, string/5°, nX/2.5°) and each stimulus condition (words and non-words). The equations of the regression line fitted to the data are also reported for each preview condition.

Discussion

We found a facilitatory effect when the parafoveal stimulus is a string of letters (at both 2.5° and 5° eccentricity) but not when it is a string of Xs. These results rule out any single factor explanation (word length or initial letter) as responsible of the facilitatory effect. At 2.5° eccentricity, word length information is the same whether the string is made up of Xs or letters but the facilitation is only present in string/2.5°. On the other hand, initial letter is much more detectable in string/2.5°, but the facilitation is not reduced in string/5° with respect to string/2.5°. Global factors present at both 2.5° and 5° eccentricity could account for the facilitatory effect. Indeed, when this information is absent (with string of X) the facilitatory effect disappears. These results suggest that parafoveal preview increases salience of word length, border information or texture layout. The second finding of our study is that, when a parafoveal preview is present, lexical decision time becomes relatively independent on word length (as word length increases form 2 to 7 letters).

This suggests that global processing in the parafovea reduces the need of local analysis during the foveal lexical decision task. These findings can be explained if we assume that word recognition involves access to a stored visual pattern of that word, and that this knowledge can have itself a hierarchical spatial structure. Thus a common word like LOVE may have a full definition (in terms of primitive features like corners and line ends and their relations) and a partial definition in terms of texture layout or borders

References

Balota, D.A., A. Pollatsek and K. Rayner (1985) "The interaction of contextual constraints and parafoveal visual information in reading", *Cognitive Psychology* 17:364-390.

Evett, L.J. and G.W. Humphreys (1981) "The use of abstract graphemic information in lexical access", *Quart. J. Exp. Psychol.* 33A:325-350.

Findlay, J.M., D. Brogan and G. Wenban-Smith (1993) "The spatial signal for saccadic eye movements emphasise visual boundaries", *Perception and Psychophysics* 53(6):633-641.

Fiorentini, A. (1989) "Differences between fovea and parafovea in visual search process", *Vision Research* 29(9):1153-1164.

Menz, C. and R. Groner (1987) "Saccadic programming with multiple targets under different task conditions", in: *Eye movements: from physiology to cognition,* J.K. O'Regan & Levy-Schoen, eds, North-Holland, Amsterdam, pp. 5-103.

Morris, R.K., K. Rayner and A. Pollatsek (1990) "Eye movement guidance in Reading: The role of parafoveal letter and space information", *J. Experimental Psychology: Human Perception and Performance* 16(2):268-281.

Pollatsek, A. and K. Ryner (1982) "Eye movement control in reading: the role of word boundaries", *J. Experimental Psychology: Human Perception and Performance* 8:817-833.

Rayner, K. and J.H. Bertera (1979) "Reading without the fovea", *Science* 206:468-469.

Rayner, K., A.W. Inhoff, R.E. Morrison, M.L. Slowiaczek and J.H. Bertera (1981) "Madking of foveal and parafoveal vision during eye fixations in reading", *J. Experimental Psychology: Human Perception and Performance* 7:167- 179.

Rayner, K., A.D. Well, A. Pollatsek and J.H. Bertera (1982) "The availability of useful information on the right of fixation in reading", *Perception and Psychophysics* 31:537-550.

Underwood, G., R. Blomfield and S. Clews (1988) "Information influences the patter of eye fixations during sentence comprehension", *Perception* 17:267-278.

MASKING EFFECT IN ORIENTING OF ATTENTION

ALEC VESTRI
Dipartimento di Psicologia dello Sviluppo e della Socializzazione, Università di Padova, via Venezia 8, Padova, 35100, Italy

ABSTRACT

Several authors show that the orienting of attention may be acted at least by two kinds of strategies: space-based coordinate strategy or object-based coordinate strategy. They show that subject uses one or the other strategy depending on the task. In this work the two strategies are compared by modulating the instructions in order to verify the possibility to use them voluntarily. The task administered to 48 subjects consisted in following (not with the eyes but with the attentional focus) a moving circle in a squared grid. A cue-target paradigm was used and the subject was induced to fix the spatial position in the grid (instructions 1) or the moving circle (instructions 2). Cue validity was referred to the instructions and the dependent variables were the reaction times. Results show that there is a validity main effect. Although the reaction times related to the target in the spatial position of the cue were always shorter, relative reaction times were conditioned by the instructions in a coherent way. When the grid was visible, there was a masking effect on space-based strategies. Other analyses showed that the grid crosses slowed down the reaction times. Two explanations may give account of these visual effects on the orienting of attention: a purely visual component or an attentional mechanism. In both cases the effect is due to the stimulus shape. In conclusion, it is showed that the visual component can modulate the attentional strategies and that this effect is so strong as to condition the top-down processes.

1. Introduction

Attention is selection of information (Allport, 1989). The reason why our mind has to select information is that our brain has limits in terms of mental resources: there are many tasks that are impossible to be performed without a straight control of the steps of cognitive processes. Voluntary control of a mental process requires a high charge of mental resources (ibid.; Heslenfeld *et al.*, 1997; see also Treisman, 1988). In contrast, there are many tasks for which control is not so necessary. In order to decide whether to use straight control or not, neuropsychologists theorized the concept of Supervisor Attentional System (Shallice, 1988). The SAS should modulate the aware control of mental elaboration. One great problem about the SAS is something very similar to the homunculus dilemma: how does the SAS decide what is automatic and what is voluntary? This problem is very close to the one that was the object of the studies by Broadbent and other authors: they were disputing whether the attentional filter acts early or late during the information processing (Duncan, 1984; Broadbent, 1982). Early and late selection were studied in order to show if attention operates since the very first steps of the information processing or only later at a higher level of that processing. The question was about whether it is possible to select information not yet elaborated and whether it is possible to elaborate information not yet selected. Now it is at least clear that both voluntary and automatic processes have charge limits and that early and late selection are different mechanisms dependent on sensorial or semantic steps of the information processing (Heslenfeld *et al.*, 1997). Studying attention over the years, authors are becoming less drastic in their theoretical positions about attention in general, because it has been shown that there are a lot of mechanisms that interact in special conditions. Inhibition of return, dimensioning of attentional focus, spatial distribution of attentional resources, orienting of attention, attention on moving objects, multimodal compatibility are some of the mechanisms involved in visual spatial orienting of attention (Gibson and Egeth, 1994; Boucart and Humphreys, 1994; La Berge and Brown, 1989; Castiello and Umiltà, 1992; Umiltà *et al.*, 1995). Some of them are automatic, others voluntary, other may act sometimes one way sometimes the other. Some mechanisms are very close to sensorial processes (inhibition of return operates automatically and very early in visual elaboration), whereas others need a semantic elaboration (some task of orienting of attention). Out of the experimental conditions these mechanisms often act in combined way. In the experimental conditions all the mechanisms that can generate disturbance are often avoided by manipulating the stimuli. That is correct in order to study the single mechanisms, but it gives no chance to see how attention mechanisms interact with the complex reality.

The studies on the orienting of attention showed that there are at least two strategies: one based on space coordinates, the other based on object coordinates (see Cheal *et al.*, 1994). In the first case, subjects fix attention depending on the position that the target has in the space delimited by the visual field: every thing that is present in the visual field occupies a position marked by space coordinates (Posner, 1980). In the second case, coordinates are dependent on the object that is the target at that moment: all the things that are present in the visual field occupy a position that is dependent on the target of the moment (Baylis and Driver, 1993). One of the consequences of these different ways of orienting attention occurs when we have to move the attentional focus from one target to a new one. When using space coordinates, the objects present in the visual field are not influential. In contrast, using object coordinates may be crucial in determining which way the new target is related to the reference object. Several experiments show that subjects can use one or the other strategy depending on the stimulus conditions. Anyway it is still not clear how much voluntarily subjects can choose between these strategies (Egly *et al.*, 1994; Vecera and Farah, 1994).

The experiments of this work were built in order to verify to what extent instructions could modulate the strategies of the orienting of attention in a visual task. Such visual task consisted in detecting stimuli preceded by a cue. The cue-target paradigm is a typical technique of the study of attention: reaction times are dependent on the good use of a valid cue, while an invalid cue slows them down. In other experiments the cue indicates the direction in which the target will appear, whereas in our case the cue appears in a spatial position and at the same time on an object (a moving object). The target is the stimulus that has to be detected and it appears after the cue. Between the appearance of the cue and that of the target the object moves. The target can appear in two different positions: in the same spatial position as the cue or on the moving object.

Subjects could theoretically use the cue by following two different strategies. According to a space-based strategy, the cue indicates the spatial position where the target will probably appear. If the target really appears in the same spatial position of the cue, the cue is considered valid; if the target appears on the moving object, the cue is considered invalid. According to an object-based strategy, the cue indicates the object where the target will probably appear. In this case, the cue is considered valid if the target appears on the moving object and invalid if it appears in the same spatial position of the cue. The instructions were written in accordance to these two strategies and divided into two blocks, both of which were administered to each subject.

The experimental design was drawn to show how voluntary choice acts in an attentional task. Some visual aspects of the stimuli were carefully controlled to see if the independent visual variables condition the use of the two strategies.

2. Materials and Methods

Four experiments were administered to a total of 48 subjects (16 males, ranging from 21 to 27 years of age, all volunteers, all with normal or correct to normal sight). The stimulus condition was always the same in all the trials of the 4 different experiments: a red circle moved along the ways of a green squared grid (see Figure 1).

In details, the sequence of events in each trial was the following: the grid and the moving circle appear. After a variable period of time (from 1400 ms to 2600 ms), a cue appeared, that is to say the red circle became yellow for 100 ms. When the circle became red again, there was a variable (from 800 to 1600 ms) Inter-Stimulus Interval (ISI). After the ISI a target appeared: in 50% of the cases the red circle became gray, in the other 50% a gray circle appeared in the same position of the cue. The target (either moving with the moving circle or static in the cue position) lasted for 1000 ms.

The general instruction given to all the subjects was to fixate a little white cross in the middle of the screen and to press the space bar of the computer keyboard whenever the gray circle appeared. The 340 trials were divided in two equivalent blocks. Each block was preceded by one of two different instructions. The specific-for-block instructions suggested in one case to use the space-based strategy, in the other to use the object-based strategy. Subjects were induced to use the first strategy by telling them that most of the times the target would appear in the space position of the cue. They were induced to use the second strategy by telling that the target would appear on the moving circle. Since the real probability of appearance was always 50% for each position, it is clear that the subjects were on purpose driven to follow a certain strategy. The block order was balanced between subjects.

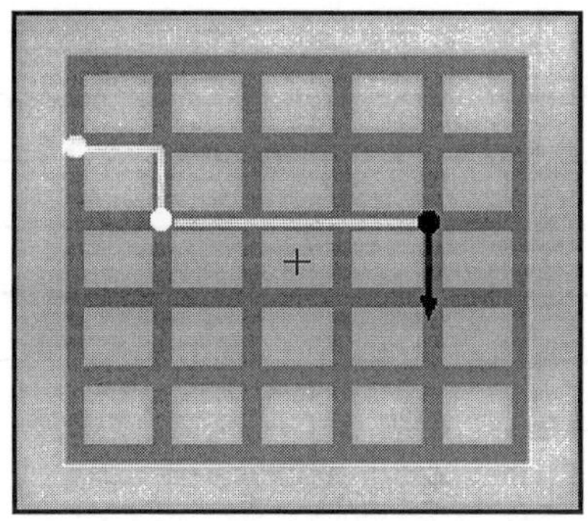

Figure 1. Stimuli used in the experiment: a circle running along the squared grid. (Measures: grid = 23 x 16 degrees of visual angle, line width = 42' degrees of v.a., circle diameter = 42' of v.a., central cross = 1 x 1 degree of v.a.; the subject distance from the screen was 22.44 inches).

There were two main between-subject variables: grid visibility and Reaction Time (RT) feedback. In two experiments the grid was visible and in two it was not visible. The grid made visible should create a better condition by giving a support to the orienting of attention. When the target appeared, the cue position and the moving circle were always at the same distance from the central fixation cross. There were three possible different distances from that cross. In two experiments the subjects could see their reaction times after each trial, in the other two they could not see them. The presence of the feedback should motivate the subjects to press the bar more quickly. Different lapses of time before and after the cue were given to avoid habituation; for the same reason, the circle followed 10 different paths. All these visual and temporal conditions were balanced for each subject and each block of trials. In the different paths the target might appear at the crosses or along the sides of the squares. When it appeared at the crosses, there should be a visual masking effect due to the presence of more lines and colors than along the sides. I expect this masking effect to slow down the RT when attention is engaged in the cue position and the grid is visible, and not to slow them down when it is engaged on the moving circle. The reason for this hypothesis is that when the attention is engaged on the cue position the grid should be directly involved, whereas when the attention is engaged on the moving circle the grid should be a non-influential background.

3. Results and Discussion

As in all experimental studies on attention, giving more time between cue and target made subjects respond more quickly (time significant effect: $F\ (1,40) = 77.750$, $p<0.001$; see Table 1 for all the RT means).

Time between cue and target							
T1 = 382				T2 = 350			
Target position							
P1 = 351				P2 = 381			
Feedback							
F1= 345				F2 = 387			
Interaction between Target Position and Instructions							
P1-I1 = 360		P1-I2 = 342		P2-I1 = 373		P2-I2 = 388	
Interaction between Distance and Instructions							
D1-I1 = 362	D1-I2 = 342	D2I1 = 348	D2-I2 = 357	D3-I1 =383	D3-I2 = 387		
Interaction between Distance and Target Position							
D1-P1 = 342	D1-P2 = 378	D2-P1 = 342	D2-P2 = 363	D3-P1 = 369	D3-P2 = 401		
Interaction among Grid visibility, Target Position and Instructions							
G1-P1-I1 = 360	G1-P1-I2 = 361	G1-P2-I1 = 376	G1-P2-I2 = 394	G2-P1-I1 = 361	G2-P1-I2 = 324	G2-P2-I1 = 370	G2-P2-I2 = 382

Table 1. Means (all the measures are expressed in ms). T1: 800 ms; T2: 1600ms; P1: in the cue position; P2: on the moving circle; F1: feedback presented; F2: feedback not presented; I1: instructions for the object; I2: instructions for the space; D1: target near to the fixation point, always in the crosses; D2: target in a middle position, out of the crosses; D3: target far from the fixation point, always in the crosses; G1: grid visible; G2: grid not visible.

Subjects were quicker if the target was on the cue position ($F\ (1,40)=57.704$, $p<0.001$). This was true both when the cue was valid (with space-based instructions) and when the cue was invalid (with object-based instructions) - (see Figure 2). The presence of the feedback gave shorter RT ($F\ (1.40)=17.63$, $p<0.0001$), thus showing that motivational aspects induce an attentional improvement.

The interaction between the position of the target and the instructions was significant (F (1.40)=16.729, p<0.0001): the direction of the preferences in the different position is coherent with the given instructions, showing that the subjects are able to follow them (see Figure 2).

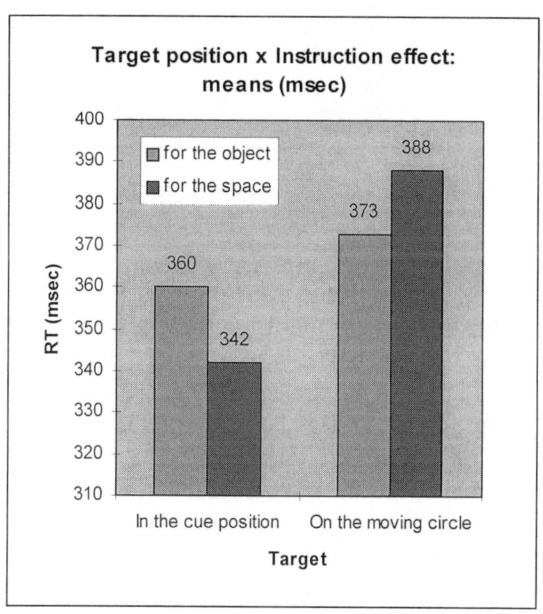

Figure 2.

This means that while the space based strategy is probably more efficient in that stimulus condition, subjects are still able to change their strategies voluntarily. There were three levels of distance from the fixation cross and only on the second level the target could appear out of the grid crosses. The significant interactions between distance and instructions (F (2,40)=12.455, p<0.001), between distance and target position (F (2,80)=3.718, p<0.05) and among grid visibility, instructions and target position (F (1,80)=4.069, p<0.05) suggested that there are two visual masking effects. Firstly, the grid crosses interfered only when instructions suggested object-based and when the target appeared in the cue position. Secondly, the grid slowed down the responses in the cue position with space-based instructions.

The masking effect of the crosses may be explained by an attentional hypothesis: only in the crosses the circle could modify its direction, so these points in the grid need more attention (a greater charge of attentional resources). Another explanation of that is purely visual (see Figure 3);); the changing colour of the circle pops out more clearly against a simpler surround (the line) than against a more complex shape (the cross).

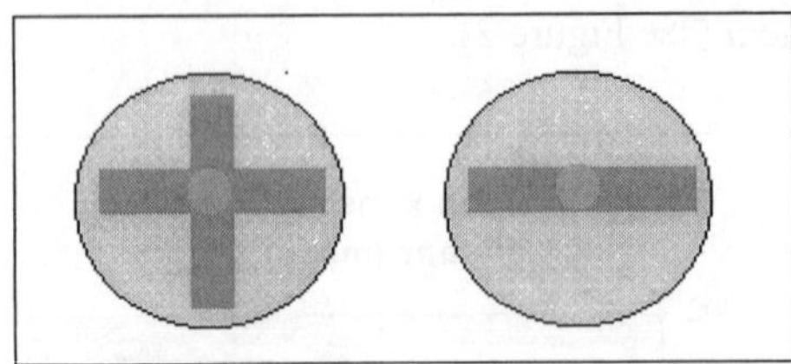

Figure 3. The visual masking given by the cross is greater than that given along the lines.

4. Conclusions

In conclusion, it is showed that in these stimulus conditions subjects are more able to orient the attentional focus following the space-based coordinate strategy, but they are able to use the object-based strategy, too. Independently of the voluntary and automatic use of the strategies, there are some visual conditions that interfere with the space-based strategy but not with the object-based strategy. The interference is given by a masking effect: the masking effect interferes only when the attention is allocated on the masked part of the stimuli, but it does not interfere when the attention, at the same stimulus conditions, make no use of that part of the stimuli. This gives quite strong evidence to the fact that the use of early (more sensorial) or late (more semantic) selection depends on task demands and visual components. It also depends on the strategies chosen or automatically used by the subjects.

References

Allport, D.A. (1989) "Visual attention", in: M.I. Posner, ed., *Foundation of Cognitive Science* Cambridge: MIT Press.

Baylis, G.C. and J. Driver (1993) "Visual attention and objects: Evidence for hierarchical coding of location", *J. Exp. Psychol.: H.P.P.* 19(3):451-470.

Boucart, M. and G.W. Humphreys (1994) "Attention to orientation, size, luminance, and color: attentional failure within the form domain", *J. Exp. Psychol.: H.P.P.* 20(1):61-80.

Broadbent, D.E. (1982). "Task combination and selective intake of information", *Acta Psychologica* 50:253-290.

Castiello, U. and C. Umiltà (1990) "Size of attentional focus and efficiency of processing", *Acta Psychologica* 73:195-209.

Cheal, M.L., D.R. Lyon and L.R. Gottlob (1994) "A framework for understanding the allocation of attention in location-precued discrimination", *Quart. J. Exp. Psychol.* 47A(3):699-739.

Duncan, J. (1984) "Selective attention and the organization of visual information", *J. Exp. Psychol.: G.* 113(4):501-517.

Egly, R., J. Driver, R.D. Rafal (1994) "Shifting visual attention between objects and locations: Evidence from normal and parietal lesion subjects", *J. Exp. Psychol.: G.* 123(2):161-177.

Gibson, B.S. and H. Egeth (1994) "Inhibition of return to object-based and environment-based locations", *Percept. Psychophys.* 55(3):323-339.

Heslenfeld, D.J., J.L. Henemans, A. Kok and P.C.M. Molenaar (1997) "Feature processing and attention in the human visual system: an overview", *Biol. Psychol.* 45:183-215.

LaBerge, D. and V. Brown (1989) "Theory of attentional operations in shape identification", *Psychol. Rev.* 96:101-124.

Posner, M.I (1980) "Orienting of attention", *Quart. J. Exp. Psychol.* 32:3-25.

Shallice, T. (1988) *From Neuropsychology to Mental Structure*, Cambridge: Cambridge University Press.

Treisman, A. (1988) "Features and objects", *Quart. J. Exp. Psychol.* 40A:201-237.

Umiltà, C., U. Castiello, M. Fontana and A. Vestri (1995) "Object-centered orienting of attention", *Visual Cognition* 2(2/3):165-181.

Vecera, S.P. and M.J. Farah (1994) "Does visual attention select objects or locations?", *J. Exp. Psychol.: G.* 123(2):146-160.

THE MODULATION OF MULTISTABLE VISUAL PERCEPTION AND THE INTENTIONAL PENETRABILITY OF VISUAL PROCESSING

C. TADDEI-FERRETTI, C. MUSIO, S. SANTILLO and A. COTUGNO
Istituto di Cibernetica, CNR, I-80072 Arco Felice (NA), Italy

ABSTRACT

With our experiments we aimed to analyze the reciprocal contribution of visual stimulation and high level mental activities to the modulation of the perception of a visually ambiguous pattern such as a (modified) Necker cube. The results suggest that an effort of will may have access to the decision-making stage, at which the choice between the possible interpretations of the pattern is made, as well as to the stability stage, at which the chosen interpretation is maintained until the mechanism is reset.

1. Introduction

A great part of the current debate on visual perception is concentrating on the discrimination between levels of information processing that are not related to consciousness and levels at which the access to the available information is conscious. Such distinction may be usefully considered in the general frame of the specific differences between implicit and explicit knowledge (the knowledge being defined as the functional use of internal representations), procedural and declarative memory, unconscious and conscious state, nonverbalizable and verbalizable content, automatic and voluntary control and so on, as well as on unconscious and conscious accessibility to different types of representations (*e.g.*, Dienes & Perner, 1999).

Among these topics, each considered in its unconscious and conscious condition, our interest deals both with visual perception and voluntary control, and in particular with the possible influence of voluntary control on visual perception.

For that which concerns the visual perception, it is known that several building blocks at increasingly higher levels of information processing contribute to human vision: related to unconscious vision there are the levels of neuroendocrine responses, reflexive responses and blindsight responses; related to conscious vision there are the further levels of phenomenal vision, object vision and object recognition (Farah, 1990; Stoerig, 1996). A rough distinction between processing stages is that of early (low) vision, in which the raw intensity data are organized into representations, and of late (high) vision, in which representations are used for object recognition. The recognition mechanism seems to be based on view interpolation of multiple-views representation (Bulthoff *et al.*, 1995).

For that which concerns the voluntary control, it has been ascertained that a previous unconscious initiative contributes with the subsequent conscious will to build up a voluntary action (Libet, 1985; 1998).

We considered that the perception multistability of ambiguous visual patterns, due to the alternance of different interpretations of each pattern, could be a suitable model to test the possible effect of the conscious component of volition on the unconscious component of perception.

Some experimental data have given evidence of the will effect on visual-perception reversal of ambiguous patterns (Liebert & Burk, 1985; Pelton & Solley, 1968; Peterson, 1986; Peterson & Hochberg, 1983; Phillipson & Harris, 1984).

While previous attempts (from 1968 to 1986), confirmed by Goddard (1998), showed that the temporal sequence of the perception alternance of ambiguous visual patterns may be modified by will, we intended to go beyond these results and investigate the interactions between upward factors (subliminal stimulation and blurring of non-fixated zones of a pattern) and downward factors (will and other factors, such as attention and imagery), as well as to clarify at which level of the information processing the effect of will is exherted in the perception alternance mechanism.

2. Methods

We used as an ambiguous pattern a modified Necker cube. In order to avoid the possibility that the automatic perception-reversal be driven by unattended shifts of fixation between the two internal vertices of the Necker cube [that constitue two focal areas, the occasional fixation of which implies the blurring of the other zones and facilitates selective interpretations of the pattern through the modification of the transfer function of the pattern information (García Pérez, 1992)], these two internal vertices of the cube coincide at the centre.

A special purpose, specifically designed and implemented programming language was used for pattern presentation, monitoring of perception changes, and data organization (Colucci *et al.*, 1994).

The pattern was presented on a personal computer screen, with the line between the two more distant vertices aligned horizontally. The pattern may be interpreted as a cube with the face in front of the observer oriented either to the right, or to the left. A possible cultural bias due to the familiarity of seeing real cubes from above is thus avoided.

The centre of the pattern was continuously fixated during each experiment, except in some cases in which another focal area was continuously fixated, *i.e.* either the left more distant vertex of the pattern, or the right one, with consequent blurring of the non-fixated areas of the pattern. Also the attention was continuously allocated to the fixated point in all cases, in order to avoid covert (Leopold & Logothetis, 1999) attention to blurred areas.

Subliminal visual stimuli could be administered to the subjects, *i.e.* biased versions of the pattern, with a duration of 14-42 ms, that replaced the pattern itself at given phases of the perceptive alternance cycle, either seconding or contrasting a previous perception event.

In order to assure that the duration of the subliminal stimuli to be administered to the subjects is as short as 14 ms, a Genoa Phantom 64 graphic accelerator board, with a vertical frequency of 70 Hz and a resolution of 320 x 200 pixels (16 colors), was used.

A continuous will effort to obtain and/or maintain a given perception could be previously requested of the subjects. From the operational point of view, the will effort, that could have to implement a shift to a given perception and anyway has to implement a hold of such perception, is a mental event different from either fixation/attention (in fact, it may be directed toward another pattern-zone with respect to the object of fixation/attention) or mental imagery.

The pattern was viewed by the subjects either i) in undisturbed conditions (control), or ii) under the administration of subliminal stimuli, 14 ms duration, either single ones that seconded or contrasted each previous perception event, or periodically alternating seconding and contrasting ones, or iii) with the continuous application (by the subject) of a will effort to obtain and/or maintain a given perception, while in some cases also a visual- and kinestetic-imaging activity was continuously performed, or iv) with subliminal stimuli and will effort.

In case iv), the experimental protocol included: a) administration of single subliminal stimuli, 14 ms duration, that either seconded or contrasted both each wanted and each not wanted perception, b) administration of single subliminal stimuli, 14 ms duration, that either seconded each wanted perception and contrasted each not wanted one or *viceversa*, c) administration of single subliminal stimuli, either seconding or contrasting, of different durations (14 ms, 28 ms, 42 ms), d) repeated administrations (every 1,500 ms) of a subliminal stimulus, 14 ms duration, either seconding or contrasting, after each perception event.

3. Results

Previous results (Taddei-Ferretti *et al.*, 1993; 1994; 1995; 1996), performed under several experimental protocols, ascertained that:

1) the fixation with attention allocation to a focal area different from the pattern centre influences the above time patterning according to the side of the fixated vertex;
2) the subliminal (seconding or contrasting) stimuli influence the time patterning of the perceptive alternance according to the administered suggestion;
3) the will application influences the above time patterning according to the side of the cube face that is intended by the subject to be in front of him/her;
4) the will effect is enhanced by a concomitant visual- and kinestetic-imagery activity applied in the same direction;
5) the will application balances and even overbalances in the opposite direction the effects of concomitant fixation with attention allocation to a focal area different from the pattern centre;
6) the will application balances and even overbalances in the opposite direction the effects also of concomitant subliminal stimulation.

In particular,

6.1) the effect of will is higher when single subliminal stimuli contrast both the wanted and the not wanted perception than when they second both types of perception.

Our more recent results (Taddei-Ferretti *et al.*, 1999; 2000) ascertained that:

6.2) the effect of will, when single or repeated subliminal stimuli contrast the wanted perception and second the not wanted one, is not different from when they second the wanted perception and contrast the not wanted one;

6.3) the effect of will is not different when the duration of subliminal stimuli is increased;

6.4) the effect of will, when the subliminal stimuli, of whichever polarity and duration, are applied repeatedly, is sensibly higher than when they are applied only once.

4. Discussion

According to the modular model of input systems (Fodor, 1983), each input system is special purpose, domain specific, autonomous, stimulus driven, informationally encapsulated (*i.e.*, cognitively impenetrable) and not influenced by top-down processes that may affect only the outputs of the modules. This view does not exclude some few interactions among submodules (Vallortigara, 1999).

A lot of data are in favour of downward influences in the representation mechanisms of all perception modalities (*e.g.*, Taddei-Ferretti & Musio, 1998). A problem arises which is related to the level at which such influences are exerted in each case: either inside a module (which, being not encapsulated from the informational point of view, should thus be no longer considered a module), or upon its output alone.

Pylyshyn (1999) considers that visual processes, at the stages of early vision, are cognitively impenetrable, encapsulated and not affected by expectations, beliefs, knowledge, utilities, values and goals, although vision as a whole is cognitively penetrable in two ways, *i.e.* it may be affected by attentional selection before early vision processing (*e.g.*, Julesz, 1990) or by unconscious evaluation and inference at the stage of decisions involved in recognizing and identifying patterns after early vision processing. Thus, early vision should be considered an impenetrable component of the cognitively penetrable visual process.

The fact that perceptual illusions (such as the Müller-Lyer one) are resistant to rational cognitive influences, and do not disappear even when they are known to be illusions, is well explained by the cognitive impenetrability of early vision, at which stage the information processing [such as the filtering of the input image (Di Maio, 1997)], which causes the altered estimations characteristic of the illusions, occurs.

Pylyshyn (1984, 130-140; 1999) distinguishes sharply between top-down influences in early vision and cognitive penetration. The first case implies a within-vision effect, for which visual interpretations emerging in early vision are affected by other visual interpretations [*e.g.*, the top-down processing involved in some cases of "filling in" (Pessoa *et al.*, 1998), in which the interpretation of parts of a stimulus depends on the interpretation of other parts of the stimulus]. In the second case, the influence originates outside the visual system and affects the content of visual perception in a meaning-dependent, logically coherent way (*e.g.*, similarly to the cognitive skill of attributing a cause to the noise outside the window). However, in the normal course of visual perception we fail to distinguish whether our subjective experience arises from the visual system alone or also from our beliefs (Pylyshyn, 1999).

Pylyshyn (1999) acknowledges also the existence of an empirical question related to the top boundary of the cognitively encapsulated early vision. In fact, while he suggests that early vision, defined functionally, involves computation of stereo, motion, size and lightness constancies, leading to 3-D descriptions, he admits that it is an object of speculation, but still undecided, if early-vision output is constitued by an explicit representation of a 3-D layout of visible surfaces (*e.g.*, Nakayama *et al.*, 1995), *i.e.* by a presemantic canonical shape category. If such a type of categorization involves inferences, it should no longer be considered a stage of early vision; thus, one should better hypothesize that early vision provides in each case not a unique canonical 3-D shape descriptor, but a set of descriptors to be selected at a further stage by knowledge-based processes of a cognitive system. This assumption is indirectly suggested also by neuropsychological (clinical and neuroimaging) studies giving evidence of a dissociation between perceptual categorization and object recognition (Warrington 1982; Warrington & Taylor, 1978). In any case, according to Moore (1999), perception should be defined as the stage at which the interpretation of early-vision output begins, leading to object recognition.

At this point we may speculate on the level of visual processing at which the will effect, evidenced by our experiments, is exerted. Taking into account the fact that different cortical areas are engaged in the perceptual reversal activation and in the perceptual stability maintenance (Kleinschmidt *et al.*, 1998), we are lead by our results to hypothezise that the conscious will effort has an effect both at the decision-making stage (when the usually unconscious choice between the two possible interpretations of the pattern is exerted at continuously successive time intervals) and at the stability stage (when the chosen interpretation is usually unconsciously maintained until the decision-making mechanism is again automatically reset).

In fact, a will effort has an increased possibility of exerting its effect on the decision-making mechanism when the mechanism is reset more often. This may happen when a single subliminal, contrasting, stimulus is administered after each (wanted or not wanted) perception (Result 6.1), and especially when a repetitive subliminal, either contrasting or seconding, stimulus is administered after a (wanted or not wanted) perception (Result 6.4). On the contrary, if a stimulus is alternately contrasting a perception (either wanted, or not wanted) and seconding the other perception, the alternance rate does not change sensibly (Result 6.2), and in this case the increase of the stimulus duration does not produce effects (Result 6.3).

It has to be noted that: not only the will effort (a downward, extra-visual factor) influences the visual perception of the ambiguous pattern (Result 3), as also imagery (a downward, extra-visual factor) does (Result 4); but especially that the will effort exerts its influence in spite of the previously evidenced (Results 1 and 2), contrasting influences either of blurring of some parts of the pattern (an upward, whithin-vision factor) due to fixation accompanied by attention allocation (two downward, extra-visual factors) (Result 5), or subliminal stimulation (an upward, within-vision factor) (Result 6).

Two assumptions may be derived from these results, one on late vision and the other on early vision.

The results mean that – as may be observed in the case of persistent ambiguity of the information provided by the sensory system that does not allow a definite interpretation of such information – the mechanisms related to the late stages of visual processing are programmable from above, intentionally penetrable (which means cognitively penetrable, as the goals are included among the cognitive factors), *i.e.* the will may have access to such mechanisms.

The results also mean that the output of the cognitively impenetrable early vision should not be a single final canonical 3-D shape description, but a set of such descriptions to be further unconsciously selected. Otherwise one should admit that the semantic category of ambiguity of the pattern – which is due to the fact that, the anterior face of the perceived cube being always equal to the posterior one, both the possible interpretations of the cube are false – is cognitively evaluated still at the stage of the cognitively impenetrable early vision.

These views are in agreement with the hypothesis that vision perception emerges always from the continuous impact of (possibly iterative) unconscious activity of high brain centres, that lie outside the visual cortex and organize the input data to obtain a stable and reliable interpretation of such data; and that, in the case in which, due to persistent ambiguity, the perception is not reliable, it is not even constant but subject to a continuous reorganization driven by the same high non-visual, frontal (Blundo & Ricci, 1998; Cohen, 1959; Meenan & Miller, 1994; Ricci & Blundo, 1990) centres, while a conscious volitional control of the organizational mechanism may be exerted (Leopold & Logothetis, 1999; Taddei-Ferretti *et al.*, 1996).

Acknowledgment

We wish to thank Mike Briggs Smith for the correction of the English manuscript.

References

Blundo, C. and C. Ricci (1998) "The frontal lobe role in perception of ambiguous figures", *Int. J. Psychophysiol.* 20(1-2):81-82.

Bulthoff, H.H., S.Y. Edelman and M.Y. Tarr (1995) "How are three-dimensional objects represented in the brain?", *Cereb. Cortex* 5(3):247-260.

Cohen, L. (1959) "Perception of reversible figures after brain injury", *Arch. Neurol. Psychiatry* (Chicago) 81:765-775.

Colucci, R.F., C. Musio and C. Taddei-Ferretti (1994) "EXPLAN - A programming language for complex visual stimuli presentation", *Int. J. Bio-Med. Comput.* 37:29-39.

Dienes, Z. and J. Perner (1999) "A theory of implicit and explicit knowledge", *Behavioral and Brain Sciences* 22:735-808.

Di Maio, V. (1997) "Filtering of the input image and visual perception of geometrical figures", in: *Biocybernetics of Vision. Integrative Mechanisms and Cognitive Processes*, C. Taddei-Ferretti, ed., Singapore, New Jersey, London, Hong Kong: World Scientific, pp. 94-103.

Farah, M.J. (1990) *Visual Agnosia. Disorders of Object Recognition and What They Tell Us about Normal Vision*, Cambridge, MA, London, UK: MIT Press.

Fodor, J.A. (1983) *The Modularity of Mind*, Cambridge, MA, London, UK: MIT Press.

García-Pérez, M.A. (1992) "Eye movement and perceptual multistability", in: *The Role of Eye Movements in Perceptual Processes*, C. Chekaluk and R.K. Lleuellyn, eds, Amsterdam, London, New York: Elsevier, pp. 73-109.

Goddard, P. (1998) "Conscious control of perceptual reversals: Eye movements or volitions?", Abstract *Toward a Science of Consciousness 1998 "Tucson III"*, p. 102.

Julesz, B. (1990) "Early vision is bottom up, except for focal attention", *Cold Spring Harbor Symp. on Quantitative Biology* 55:973-978.

Kleinschmidt, A., C. Buchel, S. Zeki and R.S. Frackowiak (1998) "Human brain activity during spontaneous reversing perception of ambiguous figures", *Proc. R. Soc. Lond. B Biol. Sci.* 265(1413):2427-2433.

Leopold, D.A. and N.K. Logothetis (1999) "Multistable phenomena: Changing views in perception", *Trends Cogn. Sci.* 3(7): 254-264.

Liebert, R.M. and B. Burk (1985) "Voluntary control of reversible figures", *Percept. & Mot. Skills* 61:1307-1310.

Libet, B. (1985) "Unconscious cerebral initiative and the role of conscious will in voluntary action", *Behav. Brain Sci.* 8:529-566.

Libet, B. (1998) "Unconscious cerebral initiative and the role of conscious will in voluntary action", in: *Downward Processes in the Perception Representation Mechanisms*, C. Taddei-Ferretti and C. Musio, eds, Singapore, New Jersey, London, Hong Kong: World Scientific, pp. 449-462.

Meenan, J.P. and L.A. Miller (1994) "Perceptual flexibility after frontal or temporal lobectomy", *Neuropsychologia* 32:1145-1149.

Moore, C.M. (1999) "Cognitive impenetrability of early vision does not imply cognitive impenetrability of perception", *Behavioral and Brain Sciences* 22:385-386.

Nakayama, K., Z.J. He and S. Shimojo (1995) "Visual surface representation: A critical link between lower-level and higher-level vision", in: *Visual Cognition*, S.M. Kosslyn and D.N. Osherson, eds, *An Invitation to Cognitive Science*. Vol. 2, Cambridge, MA, London, UK: MIT Press, pp. 1-70.

Pelton, L.H. and C.M. Solley (1968) "Acceleration of reversals of a Necker cube", *Amer. J. Psychol.* 81:585-588.

Pessoa, L., E. Thompson and A. Noë (1998) "Finding out about filling in: A guide to perceptual completion for visual science and philosophy of perception", *Behavioral and Brain Sciences* 21(6):723-780.

Peterson, M.A. (1986) "Illusory concomitant motion in ambiguous stereograms: Evidence for nonsensory components in perceptual organizations", *J. Exp. Psychol.: Human Perception and Performance* 12:50-60.

Peterson, M.A. and I. Hochberg (1983) "Opposed-set measurement procedure. A quantitative analysis of the role of local cues and intention in form perception", *J. Exp. Psychol: Human Percept. & Performonce* 9:183-193.

Phillipson, O.T. and J.P. Harris (1984) "Effects of chlorpromazine and promazine on the perception of some multi-stable visual figures", *Quart. J. Exp. Psychol.* 36A:291-308.

Pylyshyn, Z. (1994) *Computation and Cognition. Toward a Foundation for Cognitive Science*, Cambridge, MA, London, UK: MIT Press.

Pylyshyn, Z. (1999) "Is vision continuous with cognition? The case for cognitive impenetrability of visual perception", *Behavioral and Brain Sciences* 22:341-423.

Ricci, C. and C. Blundo (1990) "Perception of ambiguous figures after focal brain lesions", *Neuropsychologia* 28:1163-1173.

Stoerig, P. (1996) "Varieties of vision: From blind responses to conscious recognition", *Trends in Neurosciences* 9(9):401-406.

Taddei-Ferretti, C. and C. Musio, eds (1998) *Downward Processes in the Perception Representation Mechanisms*, Singapore, New Jersey, London, Hong Kong: World Scientific.

Taddei-Ferretti, C., C. Musio and R.F. Colucci (1994) "Top-down intrerference in visual perception", in: *ICANN '94*, M. Marinaro and P.G. Morasso, eds, Berlin, Heidelberg, New York: Springer, pp. 30-33.

Taddei-Ferretti, C., C. Musio and S. Santillo (1995) "Non linearity of the interactions between bottom-up and top-down signals in multistable visual perception", *Il Nuovo Cimento* 17D:941-947.

Taddei-Ferretti, C., C. Musio, R.F. Colucci and M. Martino (1993) "Multistable ambiguous pattern perception", *WCNN-INNS '93*. Vol. II, Hillsdale, NJ: Erlbaum, pp. 155-158.

Taddei-Ferretti, C., C. Musio, S. Santillo and A. Cotugno (1999) "Conscious and intentional access to unconscious decision-making module in ambiguous visual perception", in: *Foundations and Tools for Neural Modeling*, J. Mira and J.W. Sánchez-Andrés, eds, *Lecture Notes in Computer Sciences*. Vol. 1606, Berlin: Springer, pp. 766-775.

Taddei-Ferretti, C., C. Musio, S. Santillo and A. Cotugno (2000) "Will: A vague idea or a testable event?", in: *No Matter Never Mind*, K. Yasue, M. Jibu and T. Della Senta, eds, Amsterdam: John Benjamins, in press.

Taddei-Ferretti, C., C. Musio, S. Santillo, R. Colucci and A. Cotugno (1996), "The top-down contribution of will to the modulation of bottom-up inputs in the reversible perception phenomenon", in: R. Moreno Díaz and J. Mira-Mira, eds, *Brain Processes, Theories and Models*, Cambridge, MA, London, UK: MIT Press, pp. 446-455.

Vallortigara, G. (1999) "Segregation and integration of information among visual modules", *Behavioral and Brain Sciences* 22:398-399.

Warrington, E.K. (1982) "Neurological studies of object recognition", *Phil. Trans. Roy. Soc. London B: Biological Science* 298(1089):15-33.

Warrington, E.K. and A.M. Taylor (1978) "Two categorical stages of object perception", *Perception* 7(6):695-705.

INFLUENCE OF DOT NUMBER AND ANGLE AMPLITUDE ON MÜLLER-LYER ILLUSION

VITO DI MAIO
Istituto di Cibernetica del CNR
Via Toiano 6, I-80072 Arco Felice (NA), Italy
{vdm@biocib.cib.na.cnr.it}

ABSTRACT

Müller-Lyer illusion, as well as other visual illusions, can be partially explained by the filtering operated by the visual system on the input image. In the present paper will be presented two groups of experiments that, by using as stimuli interpolated forms (dot forms) of Müller-Lyer patterns will stress this basic assumption. Two experimental setups have been changed to consider several different variables that can influence the magnitude of the illusion.

1. Introduction

Müller-Lyer illusion is one of the most studied visual illusions. Basically it consists of two lines, each one ending with an arrow at each extremity. In one line the arrows point toward the center of the line (apex-in pattern) while in the other one they point far from the center (apex-out pattern). The illusion is given by the overestimation of the length of the former pattern when compared to the length of the latter one. This illusion has been defined for a long time as a pure cognitive illusion and several different theories have been proposed to explain its basic mechanisms (Coren, 1970; De Lucia, 1993). Nevertheless, several different evidences have shown that a large part of the illusion is due to the filtering of the input image. Ginsburg (1986), by using the presentation of filtered patterns on the screen of a digital computer, concluded that filtering of the input image contributes to the 80% of the magnitude of this illusion. In some recent papers (Di Maio *et al.*, 1992; Di Maio and Lánský, 1998) we have addressed the problem by using a different approach. To avoid the problem of oblique lines that produces the aliasing effect on the screen of a computer, we used not complete and filtered figures (as made in the quoted Ginsburg) but figures made only of dots (interpolated figures). The principle is that, if part of the figure is lacking, the filtering effect is reduced and as a consequence the error committed in visual estimation is reduced (Di Maio and Lánský, 1994; Di Maio, 1997; Di Maio and Lánský, 1998; Di Maio, 1998). The results of our experiments confirmed the findings by Ginsburg (1986), giving an overestimation of the apex-in figure due to the filtering of the input image close to 90%. In the present paper, the results of a new series of experiments performed by using interpolated Müller-Lyer figures, made of patterns with variable angles forming the arrows, are presented in order to further verify these findings.

2. Methods

In the present experiments nine subjects, 5 males and 4 females (age ranging 24-51) took part. All of them had normal or corrected to normal visual ability. The subject sat at a distance of 60 cm (head position) from a 14 inches computer monitor (640x480 pixels resolution). Each subject was presented with a set (see below) of Müller-Lyer figures made of two patterns one fixed and one that could be adjusted in size pushing the PLUS or MINUS key on the computer keyboard respectively to increase or decrease its length. When the subject judged the two figures as having the central line of the same length he/she pushed the ENTER key and a new task was automatically presented. Patterns were randomly chosen by the computer in each trial so that to have, at the end, the same number of trials where the fixed pattern was the apex-in and the apex-out. The magnitude of the error (E) estimated as (apex-in length – apex-out length) was computed on line for each of the trials and saved on the computer disk. The positive value of E indicated both the presence and the magnitude of the illusion. Statistical analysis was carried off line by using the Es committed in a sequence of 72 stimuli one half of which had the shaft (central line) composed by 2 dots only and one half had a shaft made of a complete line. For each of these two configurations of the shaft, 3 configurations for the arrows and 3 different angles for each arrow configuration were used as stimuli (respectively arrows made of 3, 5 and 9 dots and angles made of 30, 50 and 70 degrees). The factorial design was then 2 shaft-configurations x 3 arrow configurations x 3 angle configurations x 4 trials.

3. Results

The results of ANOVA factorial analysis are shown in Figure 1A and 1B, Table 1 and Table 2.

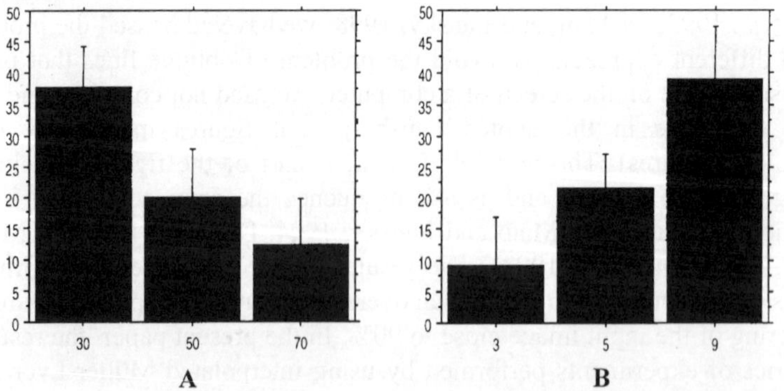

Figure 1. Error and SD expressed in pixels as function of A, amplitude in degrees of the angle forming the arrows, and B, number of dots forming the arrows.

Angle Comp.	Mean Diff.	Crit. Diff.	P-Value	Remark
30-50	17.702	10.317	0.0008	S
30-70	25.548	10.317	<0.0001	S
50-70	7.8451	0.317	0.1358	

Table 1. Fisher's PLSD for Error. Effect of amplitude of arrow angle. Significance level: 5 %.

Dots Comp.	Mean Diff.	Crit. Diff.	P-Value	Remark
3-5	-12.476	10.231	0.0169	S
3-9	-30.095	10.231	<0.0001	S
5-9	-17.619	10.231	0.0008	S

Table 2. Fisher's PLSD for Error. Effect of number of arrow dots. Significance level: 5 %.

It is evident, from both Figure 1 and Tables 1-2, that: a) interpolated figures still present the illusion effect; b) the magnitude of the error depends on the number of dots forming the arrows; c) the magnitude of the error depends on the amplitude of the angles forming the arrows. The comparisons show that errors between different amplitude of the angles (Table 1) and between different number of dots (Table 2) are always significant with the only exception of the comparison between error committed with angles of 50 and 70 degrees. However, also in this case the trend is that increasing the angle, the error decreases (Figure 1A). The trend for the dot number forming the arrows goes in the opposite direction since increasing the number of dots forming the arrows the error increases (Figure 1B). The factor shaft (two dots or complete line) was not significant as well as no significance was found for the factors sex and age (data not shown).

4. Discussion

The results of the present experiment showed once again that filtering of the input image is an important factor to be considered when visual illusions are analyzed. When the angle forming the arrows increased, the magnitude of the illusion decreased since the filtering effect in the proximity of the tips of the arrows is decreased. By reducing the effect of dots (reduction of dot number) in the same proximity of the tips the magnitude of the illusion still decrease significantly. This effect was already observed in experiments with fixed angles (Di Maio and Lánský, 1998). Also when only three dots form the pattern, the illusion is still present since the error is still significantly positive as already shown in Di Maio and Lánský (1998). All these findings are in agreement with the model for visual perception based on the Image Function theory, that we have already proposed in previous papers to explain the visual perception of geometrical figures and orientation in random dot patterns (Di Maio and Lánský, 1990; Lánský *et al.*, 1991; Mates *et al.*, 1992; Di Maio, 1997).

Acknowledgements

Many thanks to Mr. Luigi Serio for his valuable help in preparing and conducting the experiments.

References

Coren, S. (1970) "Lateral inhibition and geometrical illusions", *Quart. J. Psychol.* 22:274-278.

De Lucia, P. (1993) "A quantitative analysis of illusion magnitude predicted by several averaging theories of the Müller-Lyer illusion", *Percept. Psychophys.* 53:498-504.

Di Maio, V. (1997). "Filtering of the input image and visual perception of geometrical figures". In: *Biocybernetics of Vision: Integrative Mechanisms and Cognitive Processes,* C. Taddei-Ferretti, ed., Singapore, New Jersey, London, Hong Kong: Word Scientific, pp. 94-103.

Di Maio, V. (1998) "Threshold effect on visual perception of geometrical figures", *Percept. Motor Skills*, 87:340-342.

Di Maio, V., M. Frucci. and P. Lánský (1992). "The image function and some visual illusions" In: *Cybernetics and Systems Research '92*, R. Trappl, ed., Singapore, New Jersey, London, Hong Kong: World Scientific, pp. 805-812.

Di Maio, V. and P. Lánský. (1990). "Area perception in simple geometrical figures", *Percept. Motor Skills* 71:459-466.

Di Maio, V. and P. Lánský. (1994). "Perception of area of interpolated figures", *Cybernetics and Systems* 25:567-579.

Di Maio, V. and P. Lánský (1998). "The Müller-Lyer illusion in interpolated figures", *Percept. Motor Skills* 87:499-504.

Ginsburg, A.P. (1986). "Spatial filtering and visual form perception", in: *Handbook of Perception and Human Performance,* K.R. Boff, L. Kaufman, and J. P. Thomas, eds, New York: Wiley, pp. 34.1-34.41.

Lánský, P., N. Yakimoff, J. Males and V. Di Maio (1991) "Matching the orientation of random dot patterns with real, interpolated and extrapolated lines", *Spatial Vision* 5:219-230.

Mates, J., V. Di Maio and P. Lánský (1992) "A model of perception of area", *Spatial Vision* 6:101-116.

COMPUTATIONAL AND COGNITIVE LEVEL

VISUAL PERCEPTUAL LEARNING

NICOLETTA BERARDI and ADRIANA FIORENTINI
Istituto di Neurofisiologia del C.N.R., Via S. Zeno 51, 56127 Pisa, Italy

ABSTRACT

Several examples of visual perceptual learning have been reported. Some of them show an improvement in the discrimination of simple visual stimuli (*e.g.*, orientation of lines) but the great majority deals with tasks that require identification or discrimination of more complex stimuli, and show an effect of practice that is reflected either in the decrease of the discrimination threshold or in the shortening of the time required for identification. These include global stereopsis, discrimination of complex gratings with different luminance profiles, texture segregation, discrimination of direction of motion. Common to visual learning in the various types of visual tasks is the retention of the effects of practice for a very long time, typically for months or years. Enduring practice effects can be acquired within a single experimental session and/or progressively from one session to the next one, often continuing to improve until thousands of trials have been performed. Improvement in performance does not require that the subject is informed on the correctness of his/her response, but needs attention to the task: learning does not take place for the stimulus attributes that are not attended to. All this suggests that visual perceptual learning involves plastic changes at early neural processing levels which are dependent for their induction on the general behavioural state of the subjects such as attentiveness and motivation. The possible visual areas and neural mechanisms involved in visual perceptual learning are discussed also in the light of recent imaging studies.

1. Learning in various visual tasks

The most widely investigated type of perceptual learning in humans is visual perceptual learning. Among the various examples reported recently and in the past, some consist of an improvement in stimulus detection, *e.g.*, contrast threshold for detecting oblique gratings (Mayer, 1983), or pop-out detection (Ahissar and Hochstein, 1996), or in the discrimination of visual stimuli (*e.g.*, the orientation of lines) (Schoups *et al.*, 1995; Shiu and Pashler, 1992; Vogels and Orban, 1985). Some types of hyperacuities, like steroacuity (Fendick and Westheimer, 1983), vernier acuity (Fahle and Edelman, 1993; Fahle, 1997; McKee and Westheimer, 1978; Poggio *et al.*, 1992) and peripheral bisection acuity (Crist *et al.*, 1997) have also been shown to improve with practice. In all these visual tasks the effects of practice are retained for a very long time. Interestingly, however, some basic functions like foveal grating resolution (Bennett and Westheimer, 1991), contrast detection threshold for a sinusoidal (Gabor) pattern (Dorais and Sagi, 1997) and the discrimination of gratings differing for their spatial frequency (Fiorentini and Berardi, 1981) do not show any improvement with practice.

On the contrary, a number of other tasks that require identification or discrimination of more complex stimuli show a substantial effect of practice, that is reflected either in the decrease of the discrimination threshold or in a shortening of the time required for identification. These include discrimination of complex gratings with different luminance profiles (Fiorentini and Berardi, 1980; 1981), contrast masking (Dorais and Sagi, 1997), discrimination of directions of motion (Ball and Sekuler, 1982, 1987; Zanker, 1999), motion direction in noise (Vaina *et al.*, 1995), global stereopsis (O'Toole and Kersten, 1992; Ramachandran and Braddick, 1973; Sowden *et al.*, 1996), texture segregation (Karni and Sagi, 1991), global texture identification (Ahissar and Hochstein, 1993), spatio-temporal integration in depth (De Luca and Fahle, 1999).

2. Selectivity of learning for stimulus attributes and retinal location

Despite the differences in the stimuli and tasks, most visual learning processes display selectivity for the stimulus attributes, such as the orientation of lines or gratings (Ahissar and Hochstein, 1993, 1996; Crist et al., 1997; Fahle and Edelman, 1993; Fiorentini and Berardi, 1980, 1981; Karni and Sagi, 1991; Schoups *et al.*, 1995) or the direction of motion (Ball and Sekuler, 1987; Vaina *et al.*, 1995). In some cases, the learning process is strictly selective for a certain stimulus attribute: when this is changed, a comparable number of trials are necessary for the task to be relearned as in the previous conditions. In other cases there is a partial transfer of the effects of practice between two classes of stimuli differing for a stimulus attribute (*e.g.,* Ahissar and Levi, 1995; Liu and Vaina, 1998). When training is monocular, it may be restricted to the trained eye (*e.g.*, Fahle, 1994, but see Fahle *et al.*, 1995) but in many other cases learning transfers completely or partially to the untrained eye (Ahissar and Hochstein, 1996; Ball and Sekuler, 1982; Beard *et al.*, 1995; Dorais and Sagi, 1997; Fiorentini and Berardi, 1981; Schoups *et al.*, 1995) indicating that the learning process occurs at or centrally to the site where the inputs from the two eyes converge. For a texture segregation task the fast learning that occurs in the first experimental session transfers interocularly, while the slow process requiring several dayly sessions is strictly monocular (Karni and Sagi, 1993). Learning can also be specific for the size of the stimulus (Ahissar and Hochstein, 1996) and, with gratings, for their spatial frequency (Fiorentini and Berardi, 1980, 1981).

In some cases the effects of learning are limited very precisely to the area of the visual field where the stimuli were presented during training. For instance, for the discrimination of complex gratings with different luminance profiles, viewed at a retinal eccentricity of 1 deg, learning is restricted to the trained area of 1 deg width (Berardi and Fiorentini, 1987) and in the case of feature-detection, to an area of only 0.7 deg (Ahissar and Hochstein, 1996). For the discrimination of the direction of motion of random dot patterns there is some degree of transfer if the test stimuli overlap at least partially the trained area (Ball and Sekuler, 1987). Vernier discrimination (Fahle *et al.*, 1995) and orientation discrimination (Rivest *et al.*, 1997; Schoups *et al.*, 1995; Shiu and Pashler, 1992) are also specific for stimulus position. In a texture discrimination task, where the location of the training stimuli varies within a retinal quadrant from one trial to the next, learning is restricted to the trained quadrant (Karni and Sagi, 1991). In this as in other tasks there is no transfer of learning for stimuli presented on opposite sides of the vertical meridian, in the left and right visual hemifield. An exception has been reported: there is transfer of learning if the trained and tested areas are mirror-symmetric with respect to the vertical meridian and at a short distance from it. This occurs for complex grating discrimination (Berardi and Fiorentini, 1987) as well as for pop-out detection (Ahissar and Hochstein, 1996). It is possible that in this case transfer of learning is mediated by callosal connections. Some interhemispheric transfer for stimuli at a greater eccentricity (5 deg) has been reported for a vernier discrimination and a resolution task, but learning in this case has been tentatively interpreted as being due to a cognitive process (Beard *et al.*, 1995).

Information about different stimulus attributes (*e.g.*, luminance and colour; form and motion) is carried by parallel neural channels and processed at partially segregated cortical sites within the primary and secondary visual cortex. It is therefore not surprising that perceptual learning of complex grating discrimination is selective for the chromatic attributes of the stimuli: learning to discriminate gratings having a pure luminance contrast does not transfer to equiluminant gratings having a pure chromatic contrast, and viceversa (Fiorentini and Berardi, 1997), nor does learning transfer from gratings with a certain chromatic contrast (*e.g.*, equiluminant red-green gratings) to gratings with a different chromatic contrast (equiluminant blue-yellow). Somewhat unexpected however, is the fact that learning to discriminate equiluminant chromatic gratings is not selective for the orientation of the gratings, contrary to what occurs with gratings defined by pure luminance contrast.

No such selectivity for the type of contrast has been found in an experiment where the subjects had to discriminate the orientation of bars that differed from their background either in luminance- or in colour-contrast: in this case the improvement was not restricted to the type of contrast seen during training (Rivest *et al.*, 1997).

Perceptual learning has been shown to occur in further two visual processes. One is the perception of form from global motion (Vidyasagar and Stuart, 1993), the other is visual search (Sireteanu and Rettenbach, 1995; Steinman, 1987). Global motion was obtained by kinematograms consisting of tilted line elements;

a rectangular portion of the pattern was shifted to and fro horizontally or vertically, alternating at short time intervals (300 ms). These pattern generate a percept of a rectangle moving horizontally with respect to the background, but for inexperienced subjects it takes many seconds (20 - 25 s or more) to perceive the rectangle. By repetition of trials, the time required to perceive form-from-motion decreases considerably, similarly to what occurs for perceiving global stereopsis. However, the effects of practice do not transfer from the perception of form-from-motion to global stereopsis (Vidyasagar and Stuart, 1993). Differently from what occurs with global stereopsis, the perception of form-from-motion, once learned, transfers to patterns with elements of different orientation. This is reminiscent of what was found by Ahissar and Hochstein (1993) for global identification of texture.

Visual search, *i.e.* search for a target item in a pattern consisting of a number of different items (distractors), may imply either a parallel, preattentive process or a serial, attentive process. In the first case the time required to detect the target is practically independent of the number of distractors, while in the second case, the time increases with the number of distractors. Steinman (1987) and Sireteanu and Rettenbach (1995) have provided evidence that a visual search task that is initially "serial" can become "parallel" with practice. In other words, a search that initially requires more time if the number of distractors increases, becomes independent of the number of distractors. Learning is not specific for the item to be searched (for instance a broken circle among full circles) since it transfers to search of completely different items and distractors (for instance a pair of convergent bars among pairs of parallel bars). Thus it seems to imply an improvement in search strategy, rather that an improved perception of the items (see also Ito et al., 1998).

Both these recently reported examples of visual learning differ from many of the examples reported above, in that are not selective for the basic stimulus features present in the patterns. They seem to imply plasticity of the visual system at a level where stimulus generalization is present.

3. Time course of visual perceptual learning, consolidation and retention

For some visual tasks a rapid improvement of performance is observed with repetition of trials during a single experimental session and learning effects are partially retained in the second session and totally retained thereafter. In these cases learning is practically complete after a few hundreds of trials (*e.g.*, Fiorentini and Berardi, 1980, 1981). For other tasks, there is also a rapid improvement during the first session, but performance continues to improve from one daily session to the next, until a stable optimal level is reached (Ahissar and Hochstein, 1996; Karni and Sagi, 1993; Poggio *et al.*, 1992; Zanker, 1997). On the whole, thousands of trials may be required to complete the learning process (Ball and Sekuler, 1987; Crist *et al.*, 1997; Fahle and Edelman, 1993; McKee and

Westheimer, 1978). In any case, retention is very prolonged: the effects of visual perceptual learning may last for months (Fiorentini and Berardi, 1981) and even for years (Karni and Sagi, 1993). Karni and Sagi (1993) found that, for a texture segmentation task, following the first learning session, performance improves after a period of rest of at least 6-8 ours, spent either in normal life activity or sleeping. This time is therefore required for the effects of practice to consolidate. The phases of REM sleep seem to be crucial to assure consolidation when this occurs in sleeping time.

An apparently different type of visual learning, that also seems to lasts for ever, but that occurs abruptly rather than gradually, has something in common with cognitive forms of learning, like "insight" in problem solving. For instance, figures that at first appear as a random array of black and white blobs may suddenly perceptually organize into a figure, as occurs in the well known picture of the camouflaged Dalmatian dog. And still, there is evidence that the abrupt learning process is selective for some basic characteristics of the stimuli, like stimulus size (Rubin *et al.*, 1997). A recent attempt to localize by PET imaging, the cortical regions activated during rapid learning of the perception of objects or faces indicates that abrupt learning enhanced the activity in inferior temporal regions involved in object and face processing, as well as in lateral parietal regions that have been implicated in attention and visual imagery (Dolan *et al.*, 1997).

A phenomenon possibly related to abrupt learning, is the so-called "eureka" effect, that is a facilitation in the performance of a difficult feature-detection task induced by a preliminary presentation of an easy version of the task (Ahissar and Hochstein, 1997). The difficult task may not otherwise show any improvement with practice. This "eureka" effect that enables learning in a feature detection task is not specific for stimulus orientation.

4. The role of feedback

In most of the experiments reviewed above, the subject was informed after each trial of the correctness of his/her response. To better qualify the role of feedback for perceptual learning it is useful to compare the effects of practice in the presence or absence of feedback. McKee and Westheimer (1978) obtained a practice effect in vernier acuity whether or not they gave their subject feedback. Ball and Sekuler (1987) also found that for cardinal directions of movement subjects' accuracy in direction discrimination improved at the same rate whether they were provided of trial by trial feedback or not. For oblique movement directions learning was slower in absence of feedback.

Shiu and Pashler (1992) confirmed these observations using a different task, which involved the discrimination of a pair of straight lines differing by 3 deg. They found that subjects learned at the same pace with trial by trial feedback and with block feedbak (percentage of correct responses at the end of a block of 44 trials). This was a "within session" learning, completed in 8-9 blocks. If no feedback at all was provided subjects learned within 3 to 4 days. A further indication that feedback is not necessary for learning comes from the results of Karni and Sagi (1991). In their experiments, subjects performed letter discrimination followed by texture discrimination in the same complex stimulus. Their results show significant learning in the texture task, even though feedback was given only for the letter discrimination.

More recently, Fahle and Edelman (1993) and Fahle, Edelman and Poggio (1995) using a vernier-displacement discrimination, found that lack of feedback did not prevent learning, although the slope of the learning curves for observers who received auditory feedback was significantly steeper than for those who did not. In a subsequent experiment, only a part of the subjects were able to improve their performance in the absence of feedback (Herzog and Fahle, 1997).

On the whole these results suggest that perceptual learning can proceed in an unsupervised manner, albeit at a slower pace. It is possible, however, that a sort of “internal” feed-back is useful, although not strictly required (Fahle, Edelman and Poggio, 1995), for learning to occur. This would be consistent with the fact that presenting the subject with an easy task greatly facilitates the subsequent improvement in a more difficult discrimination (Ahissar and Hochstein, 1997).

5. The role of attention

If feedback is not a necessary prerequisite for improving visual performance, is the mere repetition of stimuli sufficient to promote perceptual learning? A number of experiments have concluded that this is not the case. Practice effects are not determined solely by activity in stimulus driven mechanisms but also by high level attentional mechanisms, which probably control, with a top-down effect, changes taking place at early visual processing levels and which may themselves be subject to training (Ito *et al.*, 1998).

Ahissar and Hochstein (1993) designed an elegant experiment, in which two independent tasks could be performed on the same set of visual stimuli, with very similar degree of difficulty. The stimulus was a briefly presented array of short line segments. Segments were arranged in a vertical or horizontal array and one of the segments had a different orientation from the others. Observers either identified the orientation of the array (global task) or detected the presence of the odd man out (local task). As a consequence, after practice on one task, the experimenters were able to test the presence of any improvement in the other task, while keeping the retinal stimuli constant. The result was very clear: learning did not take place for the stimulus attributes that were not attended to.

Shiu and Pashler (1992) obtained very similar results with observers practicing in discriminating the brightness of two lines which also differed for their orientation: orientation discrimination was not improved at the end of the brightness task.

For certain hyperacuity tasks, in particular for vernier tasks, there seems to be a great intersubject variability as regards the improvement with practice. The performance of a substantial proportion of subjects seems to be largely unaffected by practice (Beard *et al.*, 1995; Fahle and Henke-Fahle, 1996; Fahle, 1997; Kumar and Glazer, 1993; Saarinen and Levi, 1995). One has to say that for these tasks, also the optimal performance is very different in different subjects, but the large variability in the rate and amount of improvement with practice may be due to the role of attention in perceptual learning. Subjects who can concentrate and pay attention to the task better than others might be more proficient in learning. This might explain certain discrepant results, obtained in different laboratories, as for instance those relative to the hyperacuity threshold for three points alignment, and whether this can (Poggio et al., 1992; Fahle and Morgan, 1996) or cannot (Bennett and Westheimer, 1991) be improved by practice.

The suggestion which comes out is that attention controls performance of neural networks which operate at low, stimulus related processing levels. Attributes which are processed by different networks do not share the benefits of being highlighted by attention.

Evidence that attention plays a crucial role in visual perceptual learning has been provided recently by an fMRI study on learning in a task of direction discrimination of global motion (Vaina *et al.*, 1998). This task can be learned with only a few hundred trials and is therefore suitable for fMRI experiments. The results show substantial changes of activity during learning in the anterior lateral cerebellum, in the superior colliculus and the anterior cingulate, *i.e.* in structures that are likely to have a role in attention. The activation of these structures is remarkable at the beginning of the experiment, but vanishes during the repetition of trials, indicating that it is related to the learning process. In contrast, the activity in the middle-temporal visual complex known to be responsive to global-motion stimuli, expands considerably from the beginning to the end of the experiment, while the subjects' performance increases from chance to almost perfect discrimination.

The debate about the level at which plastic changes subserving visual perceptual learning occur, whether at an early level, perhaps even in the primary visual cortex (Crist et al., 1997; Karni and Sagi, 1991), or at higher levels, and whether a possible plastic modification of neural properties at early levels is subject to a top-down control, has recently received an important contribution by Ahissar and Hochstein (1997). Having proved that in learning a pop-out detection task, the selectivity of the learning process depends on the degree of difficulty of the task, easy tasks leading to generalization, but difficult tasks to specificity of learning for the stimulus properties, the authors propose a reverse hierarchy model, where the improvement begins at high generalizing hierarchical levels and then guides learning under more difficult stimulus conditions, at the lower levels which provide their input. It is indeed at relatively early cortical lelvels that the single neurons present a high degree of selectivity for stimulus properties (such as orientation, receptive field size, ecc.). According to this model, "the selection of learning site follows a basic rule of centripetal, attention guided search, along the hierarchical levels of visual representation". This model permits to account for the "eureka" effect as well as for the apparently discrepant findings of different studies about the different degrees of selectivity of visual learning processes. Further evidence favouring this model has been presented in a subsequent work (Ahissar *et al.*, 1998).

6. Concluding remarks

The various types of visual tasks that underly improvement with practice and can be classified as examples of perceptual learning share a selectivity for stimulus attributes that is known to be a property of cortical neurons at the earlier stages of visual processing, but is not present, or at least is much less definite, at cortical areas farther removed from the primary visual cortex. However, the lack of such selectivity for some visual tasks, for instance visual search (Sireteanu and Rettenbach, 1995), as well as for "easy" tasks leading to a facilitation in the learning of "difficult" task (Ahissar and Hochstein, 1997), indicates that the plastic changes induced by practice at early visual areas are likely to be driven by changes occurring at more central sites. While feedback is not strictly necessary, thus excluding models based on supervised learning, attention appears to play a crucial role in gating the improvement in performance, it is a sort of *conditio-sine-qua-non* for visual perception to improve with practice. And this is consistent with the PET and fMRI findings so far available. One could speculate that attention is necessary to build up an internal "template" of the stimuli to which the successive visual inputs can be referred.

References

Ahissar, M. and S. Hochstein (1993) "Attentional control of early perceptual learning", *Proc. Natl. Acad. Sci.* 90:5718-5722.

Ahissar, M. and S. Hochstein (1996) "Learning pop-out detection: specificities to stimulus characteristics", *Vision Res.* 36:3487-3500.

Ahissar, M. and S. Hochstein (1997) "Task difficulty and the specificity of perceptual learning", *Nature* 387:401-406.

Ahissar, M., R. Laiwand, G. Kozminsky and S. Hochstein (1998) "Learning pop-out detection: building representations for conflicting target-distractor relationships", *Vision Res.* 38:3095-3107.

Ball, K. and R. Sekuler (1982) "A specific and enduring improvement in visual motion discrimination", *Science* 218:697-698.

Ball, K. and R. Sekuler (1987) "Direction-specific improvement in motion discrimination", *Vision Res.* 27:953-965.

Beard, B.L., D.M. Levi and L.N. Reich (1995) "Perceptual learning in parafoveal vision", *Vision Res*. 35:1679-1690.

Bennett, R. and G. Westheimer (1991) "The effect of training on visual alignment discrimination and grating resolution", *Percept. Psychophys.* 49:541-546.

Berardi, N. and A. Fiorentini (1987) "Interhemispheric transfer of visual information in humans: spatial characteristics", *J. Physiol.* 384:633-647.

Crist, R.E., M.K. Kapadia, G. Westheimer and C.D. Gilbert (1997) "Perceptual learning of spatial localization: specificity for orientation, position and contest", *J. Neurophysiol*. 278:2898-2894.

De Luca, E. and M. Fahle (1999) "Learning of interpolation in 2 and 3 dimensions", *Vision Res*. 39:2051-2062.

Dolan, R.J., G.R. Fimk, E. Rolls, M. Booth, A. Holmes, R.S.J. Frackowiack and K.J. Friston (1997) "How the brain learns to see objects and faces in an impoverished context", *Nature* 389:596-599.

Dorais, A. and D. Sagi (1997) "Contrast masking effects change with practice", *Vision Res*. 37:1725-1733.

Fahle, M. (1994) "Human pattern recognition: parallel processing and perceptual learning", *Perception* 23:411-427.

Fahle, M. (1997) "Specificity of learning curvature, orientation, and vernier discriminations", *Vision Res*. 37:1885-1895.

Fahle, M. and S. Edelman (1993) "Long-term learning in vernier acuity: effects of stimulus orientation, range and of feedback", *Vision Res.* 33:397-412.

Fahle, M., S. Edelman and T. Poggio (1995) "Fast perceptual learning in hyperacuity", *Vision Res*. 35:3003-3013.

Fahle, M. and S. Henke-Fahle (1996) "Interobserver variance in perceptual perfomance and learning", *Invest. Ophthal. Vis. Sci*. 37:869-877.

Fahle, M. and M. Morgan (1996) "No transfer of perceptual learning between similar stimuli in the same retinal position", *Curr. Biol*. 6:292-297.

Fendick, M. and G. Westheimer (1983) "Effects of practice and the separation of test targets on foveal and peripheral stereoacuity", *Vision Res.* 23:145-150.

Fiorentini, A. and N. Berardi (1980) "Perceptual learning specific for orientation and spatial frequency", *Nature* 287:43-44.

Fiorentini, A. and N. Berardi (1981) "Learning in grating waveform discrimination: specificity for orientation and spatial frequency", *Vision Res.* 21:1149-1158.

Fiorentini, A. and N. Berardi (1997) "Visual perceptual learning: a sign of neural plasticity at early stages of visual processing", *Arch. Ital. Biol*. 135:157-167.

Herzog, M.H. and M. Fahle (1997) "The role of feedback in learning a vernier discrimination task", *Vision Res.* 37:2133-2141.

Ito, M., G. Westheimer and C.D. Gilbert (1998) "Attention and perceptual learning modulate contextual influences on visual perception", *Neuron* 20:1191-1197.

Karni, A. and D. Sagi (1991) "Where practice makes perfect in texture discrimination: evidence for primary visual cortex plasticity", *Proc. Natl. Acad. Sci. USA* 88:4966-4970.

Karni, A. and D. Sagi (1993) "The time course of learning a visual skill", *Nature* 365:250-252.

Kumar, T. and D.A. Glaser (1993) "Initial performance, learning and observer variability for hyperacuity tasks", *Vision Res.* 33:2287-2300.

Liu, Z. and L. Vaina (1998) "Simultaneous learning of motion discrimination in two directions", *Cognit. Brain Res.* 6:347-349.

Mayer, M.J. (1983) "Practice improves adults' sensitivity to diagonals", *Vision Res.* 23:547-550.

McKee, S.P. and G. Westheimer (1978) "Improvement in vernier acuity with practice", *Percept. Psychophys.* 24:258-262.

O'Toole, A.J. and D.J. Kersten (1992) "Learning to see random-dot stereograms", *Perception* 21:227-243.

Poggio, T., M. Fahle and S. Edelman (1992) "Fast perceptual learning in visual hyperacuity", *Science* 256:1018-1021.

Ramachandran, V.S. and O. Braddick (1973) "Orientation-specific learning in stereopsis", *Perception* 2:371-376.

Rivest, J., I. Boutet and J. Intriligator (1997) "Perceptual learning of orientation discrimination by more than one attribute", *Vision Res.* 37:273-281.

Rubin, N., K. Nakayama and R. Shapley (1997) "Abrupt learning and retinal size specificity in illusory-contour perception", *Curr. Biol.* 7:461-467.

Saarinen, J. and D.M. Levi (1995) "Perceptual learning in vernier acuity: what is learned?", *Vision Res.* 35:519-527.

Schoups, A.A., R. Vogels and G.A. Orban (1995) "Human perceptual learning in identifying the oblique orientation: retinotopy, orientation specificity and monocularity", *J. Physiol.* 483:797-810.

Shiu, L. and H. Pashler (1992) "Improvement in line orientation is retinally local but dependent on cognitive set", *Percept. Psychophys.* 52:582-588.

Sireteanu, R. and R. Rettenbach (1995) "Perceptual learning in visual search: fast, enduring, but not specific", *Vision Res.* 35:2037-2043.

Sowden, P.I.D. and D. Rose (1996) "Perceptual learning in stereoacuity", *Perception* 25:1043-1052.

Steinman, S.B. (1987) "Serial and parallel search in pattern vision?", *Perception* 16:389-398.

Vaina, L., V. Sundareswaran and J.G. Harris (1995) "Learning to ignore: psychophysics and computational modeling of fast learning of direction in noisy motion stimuli", *Cognit. Brain Res.* 2:155-163.

Vaina, L., J.W. Belliveau, E.B. des Roziers and T.A. Zeffiro (1998) "Neural systems underlying learning and representation of global motion", *Proc. Natl. Acad. Sci. USA* 95:12657-12662.

Vidyasagar, T.R. and G.W. Stuart (1993) "Perceptual learning in seeing form from motion", *Proc. R. Soc. Lond. B* 254:241-244.

Vogels, R. and G.A. Orban (1985) "The effect of practice on the oblique effect in line orientation judgements", *Vision Res.* 25:1679-1687.

Zanker, J.M. (1999) "Perceptual learning in primary and secondary motion vision", *Vision Res.* 39:1993-1304.

FUNCTIONS OF THE PRIMATE TEMPORAL LOBE CORTICAL VISUAL AREAS IN INVARIANT VISUAL OBJECT AND FACE RECOGNITION

EDMUND T. ROLLS
Department of Experimental Psychology, University of Oxford, South Parks Road Oxford, OX1 3UD, England

ABSTRACT

Neurophysiological evidence is described showing that some neurons in the macaque temporal cortical visual areas have responses that are invariant with respect to the position, size and view of faces and objects, and that these neurons show rapid processing and rapid learning. Which face or object is present is encoded using a distributed representation in which each neuron conveys independent information in its firing rate, with little information evident in the relative time of firing of neurons. This ensemble encoding has the advantages of maximising the information in the representation useful for discrimination between stimuli using a simple weighted sum of the neuronal firing by the receiving neurons, generalisation, and graceful degradation. In a clinical application of these findings, it is shown that humans with ventral frontal lobe damage have in some cases impairments in face and voice expression identification. These impairments are correlated with and may contribute to the problems some of these patients have in emotional and social behaviour.

1. Introduction

This paper draws together evidence on how information about visual stimuli is represented in the temporal cortical visual areas and the brain areas to which these are connected; on how these representations are formed; and on how learning about these representations occurs. The evidence comes from neurophysiological studies of single neuron activity in primates. The recordings described are made mainly in non-human primates, firstly because the temporal lobe, in which this processing occurs, is much more developed than in non-primates, and secondly because the findings are relevant to understanding the effects of brain damage in patients, as will be shown. In this paper, particular attention will be paid to neural systems involved in processing information about faces, because with the large number of neurons devoted to this class of stimuli, this system has proved amenable to experimental analysis; because of the importance of face recognition and expression identification in primate social behaviour; and because of the application of understanding this neural system to understanding the effects of damage to this system in humans.

2. Neuronal Responses Found in Different Temporal Lobe Cortex Visual Areas

Visual pathways project by a number of cortico-cortical stages from the primary visual cortex until they reach the temporal lobe visual cortical areas (Seltzer and Pandya, 1978; Maunsell and Newsome, 1987; Baizer *et al.*, 1991) in which some neurons which respond selectively to faces are found (Desimone and Gross,

1979; Bruce *et al.*, 1981; Desimone *et al.*, 1984; Gross *et al.*, 1985; Rolls, 1981, 1984, 1991, 1992b, c; Perrett, Rolls and Caan, 1982; Desimone, 1991). The inferior temporal visual cortex, area TE, is divided on the basis of cytoarchitecture, myeloarchitecture, and afferent input into areas TEa, TEm, TE3, TE2 and TE1. In addition there is a set of different areas in the cortex in the superior temporal sulcus (Seltzer and Pandya, 1978; Baylis, Rolls and Leonard, 1987) (see Fig. 1). Of these latter areas, TPO receives inputs from temporal, parietal and occipital cortex; PGa and IPa from parietal and temporal cortex; and TS and TAa primarily from auditory areas (Seltzer and Pandya, 1978).

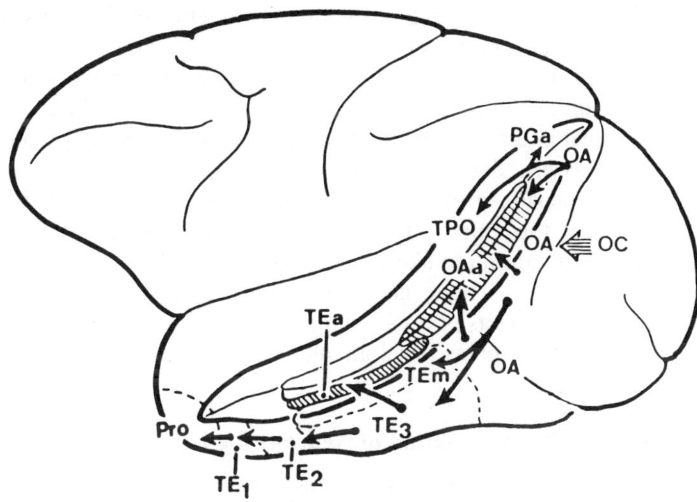

Figure 1. Lateral view of the macaque brain (left) and coronal section (right) showing the different architectonic areas (*e.g.* TEm, TPO) in and bordering the anterior part of the superior temporal sulcus (STS) of the macaque (see text). The coronal section is through the temporal lobe 133 mm P (posterior) to the sphenoid reference (shown on the lateral view). HIP - hippocampus; RS - rhinal sulcus.

In order to investigate the information processing being performed by these parts of the temporal lobe cortex, the activity of single neurons was analysed in each of these areas in a sample of more than 2600 neurons in the rhesus macaque monkey during the presentation of simple and complex visual stimuli such as sine wave gratings, three-dimensional objects, and faces; and auditory and somatosensory stimuli (Baylis, Rolls and Leonard, 1987). Considerable specialization of function was found. For example, areas TPO, PGa and IPa are multimodal, with neurons which respond to visual, auditory and/or somatosensory inputs; the inferior temporal gyrus and adjacent areas (TE3,TE2,TE1,TEa and TEm) are primarily unimodal visual areas; areas in the cortex in the anterior and dorsal part of the superior temporal sulcus (*e.g.* TPO, IPa and IPg) have neurons specialized for the analysis of moving visual stimuli; and neurons responsive primarily to faces are found more frequently in areas TPO, TEa and TEm (Baylis *et al.*, 1987), where they comprise approximately 20% of the visual neurons responsive to stationary stimuli, in contrast to the other temporal cortical areas in which they comprise 4-10%. The stimuli which activate other cells in these TE regions include simple visual patterns such as gratings, and combinations of simple stimulus features (Gross *et al.*, 1985; Tanaka *et al.*, 1990). Although face-selective neurons are thus found in the highest proportion in areas TPO within the superior temporal sulcus and TEa and TEm on the ventral lip of the sulcus, their extent is great in the anteroposterior direction (they are found in corresponding regions within the anterior half of the sulcus), and they are present in smaller proportions in many other temporal cortical areas (*e.g.* TE3, TE2 and TE1) (Baylis, Rolls and Leonard, 1987). Due to the fact that face-selective neurons have a wide distribution, it might be expected that only large lesions, or lesions that interrupt outputs of these visual areas, would produce readily apparent face-processing deficits. Further, as described below, neurons with responses related to facial expression, movement, and gesture are more likely to be found in the cortex in the superior temporal sulcus, whereas neurons with activity related to facial identity are more likely to be found in the TE areas (see Hasselmo, Rolls and Baylis, 1989). These neurophysiological findings suggest that the appropriate tests for the effects of STS lesions will include tests of facial expression, movement, and gesture; whereas facial identity is more likely to be affected by TE lesions.

3. The Selectivity of One Population of Neurons for Faces

The neurons described in our studies as having responses selective for faces are selective in that they respond 2-20 times more (and statistically significantly more) to faces than to a wide range of gratings, simple geometrical stimuli, or complex 3-D objects (see Rolls, 1984; Baylis *et al.*, 1985, 1987; Rolls, 1992b). (In fact, the majority of the neurons in the cortex in the superior temporal sulcus classified as showing responses selective for faces responded much more specifically than this. For half of these neurons, their response to the most effective face was more than five times as large as to the most effective non-face stimulus, and for 25 % of these neurons, the ratio was greater than 10:1. The degree of selectivity shown by different neurons studied is illustrated in Fig. 6 of Rolls, 1992c and by Baylis, Rolls and Leonard, 1985, and the criteria for classification as face-selective are elaborated further by Rolls, 1992c.) The responses to faces are excitatory, sustained and are time-locked to the stimulus presentation with a latency of between 80 and 160 ms. The cells are typically unresponsive to auditory or tactile stimuli and to the sight of arousing or aversive stimuli. The magnitude of the responses of the cells is relatively constant despite transformations such as rotation so that the face is inverted or horizontal, and alterations of color, size, distance and contrast (see below). These findings indicate that explanations in terms of arousal, emotional or motor reactions, and simple visual feature sensitivity or receptive fields, are insufficient to account for the selective responses to faces and face features observed in this population of neurons (Perrett *et al.*, 1982; Baylis *et al.*, 1985; Rolls and Baylis, 1986). Observations consistent with these findings have been published by Desimone *et al.*, (1984), who described a similar population of neurons located primarily in the cortex in the superior temporal sulcus which responded to faces but not to simpler stimuli such as edges and bars or to complex non-face stimuli (see also Gross *et al.*, 1985).

Further evidence has been obtained that these neurons are tuned to provide information about which face has been seen, but not about which non-face has been seen (Rolls and Tovee, 1995a). In this study a wide range of different faces (23) and non-face images (45) of real-world scenes was used. This enabled the function of this brain region to be analysed when it was processing natural scenes. The information available about which stimulus had been shown was measured quantitatively using information theory. This analysis showed that the responses of these neurons contained much more information about which (of 20) face stimuli had been seen (on average 0.4 bits) than about which (of 20) non-face stimuli had been seen (on average 0.07 bits). Multidimensional scaling to produce a stimulus space represented by this population of neurons showed that the different faces were well separated in the space created, whereas the different non-face stimuli were grouped together. The information analyses and multidimensional scaling thus provided evidence that what was made explicit in the responses of these neurons was information about which face had been seen (see further Rolls *et al.*, 1997). Information about which non-face stimulus had been seen was not made explicit in these neuronal responses. These procedures provide an objective and quantitative way to show what is "represented" by a particular population of neurons.

4. The Selectivity of These Neurons for Whole Faces Or for Parts of Faces

Masking out or presenting parts of the face (*e.g.* eyes, mouth, or hair) in isolation reveal that different cells respond to different features or subsets of features. For some cells, responses to the normal organization of cut-out or line-drawn facial features are significantly larger than to images in which the same facial features are jumbled (Perrett *et al.*, 1982; Rolls *et al.*, 1994). These findings are consistent with the hypotheses developed below that by competitive self-organisation some neurons in these regions respond to parts of faces by responding to combinations of simpler visual properties received from earlier stages of visual processing, and that other neurons respond to combinations of parts of faces and thus respond only to whole faces. Moreover, the finding that for some of these latter neurons the parts must be in the correct spatial configuration shows that the combinations formed can reflect not just the features present, but also their spatial arrangement. This provides a way in which binding can be implemented in neural networks (see further Elliffe *et al.*, 2000b).

5. Distributed Encoding of Face and Object Identity

An important question for understanding brain function is whether a particular object (or face) is represented in the brain by the firing of one or a few gnostic (or "grandmother") cells (Barlow, 1972), or whether instead the firing of a group or ensemble of cells each with somewhat different responsiveness provides the representation. We have investigated whether the face-selective neurons encode information which could be used to distinguish between faces and, if so, whether gnostic or ensemble encoding is used. First, it has been shown that the representation of which particular object (face) is present is rather distributed. Baylis, Rolls and Leonard (1985) showed this with the responses of temporal cortical neurons that typically responded to several members of a set of 5 faces, with each neuron having a different profile of responses to each face. At the same time, the neurons discriminated between the faces reliably, as shown by the values of d', taken in the case of the neurons to be the number of standard deviations of the neuronal responses which separated the response to the best face in the set from that to the least effective face in the set. The values of d' were typically in the range 1-3. A measure of the breadth of tuning which takes the value 0 for a local representation and 1 if all the neurons are equally active for every stimulus (Smith and Travers, 1979), had values that were for the majority of neurons in the range 0.7-0.95.

In a more recent study, the responses of another set of temporal cortical neurons to 23 faces and 45 non-face natural images was measured, and again a distributed representation was found (Rolls and Tovee, 1995a). The tuning was typically graded. The measure used of the tuning of the neurons was one useful in analyzing the quantitative implications of sparse representations in neuronal networks, namely

$$a = (\Sigma_{s=1,S}\, r_s/S)^2 \,/\, \Sigma_{s=1,S}(r_s^2/S)$$

where r_s is the mean firing rate to stimulus s in the set of S stimuli (see Rolls and Treves, 1998). If the neurons were binary (either firing or not to a given stimulus), then a would be 0.5 if the neuron responded to 50% of the stimuli, and 0.1 if a neuron responded to 10% of the stimuli. It was found that the sparseness of the representation of the 68 stimuli by each neuron had an average across all neurons of 0.65. This indicates a rather distributed representation. It is of interest to note that if neurons had a continuum of responses equally distributed between zero and maximum rate, a would be 0.75; while if the probability of each response decreased linearly, to reach zero at the maximum rate, a would be 0.67; and if the probability distribution had an exponentially decreasing probability of high rates, a would be 0.5. If the spontaneous firing rate was subtracted from the firing rate of the neuron to each stimulus, so that the changes of firing rate, *i.e.* the *active responses* of the neurons, were used in the sparseness calculation, then the 'response sparseness' had a lower value, with a mean of 0.33 for the population of neurons, or 0.60 if calculated over the set of faces rather than over all the face and non-face stimuli.

Thus the representation was rather distributed.

The distributed nature of the representation can be further understood by the finding that the firing rate distribution of single neurons when a wide range of natural visual stimuli are being view is approximately exponentially distributed, with rather few stimuli producing high firing rates, and increasingly large numbers of stimuli producing lower and lower firing rates (Rolls and Tovee, 1995a; Baddeley *et al.*, 1996; Treves *et al.*, 1999) (see Fig. 2). The sparseness of such an exponential distribution of firing rates is 0.5. The distribution may arise from the threshold non-linearity of neurons combined with short term variability in the responses of neurons (Treves *et al.*, 1999).

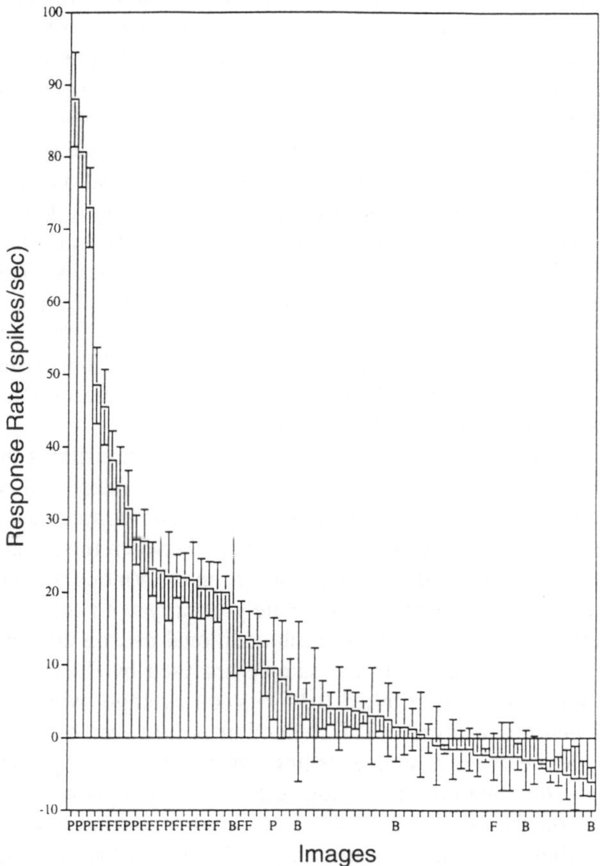

Figure 2. Firing rate distribution of a single neuron in the temporal visual cortex to a set of 23 face (F) and 45 non-face images of natural scenes. The firing rate to each of the 68 stimuli is shown. (After Rolls and Tovee, 1995a)

Complementary evidence comes from applying information theory to analyse how information is represented by a population of these neurons. The information required to identify which of S equiprobable events occurred (or stimuli were shown) is $\log_2 S$ bits. (Thus 1 bit is required to specify which of two stimuli was shown, 2 bits to specify which of 4 stimuli was shown, 3 bits to specify which of 8 stimuli was shown, etc. The important point for the present purposes is that if the encoding was local, the number of stimuli encoded by a population of neurons would be expected to rise approximately linearly with the number of neurons in the population. In contrast, with distributed encoding, provided that the neuronal responses are sufficiently independent, and are sufficiently reliable (not too noisy), the number of stimuli encodable by the population of neurons might be expected to rise exponentially as the number of neurons in the sample of the population was increased. The information available about which of 20 equiprobable faces had been shown that was available from the responses of different numbers of these neurons is shown in Fig. 3. First, it is clear that some information is available from the responses of just one neuron - on average approximately 0.34 bits. Thus knowing the activity of just one neuron in the population does provide some evidence about which stimulus was present. This evidence that information is available in the responses of individual neurons in this way, without having to know the state of all the other neurons in the population, indicates that information is made explicit in the firing of individual neurons in a way that will allow neurally plausible decoding, involving computing a sum of input activities each weighted by synaptic strength, to work (see below). Second, it is clear (Fig. 3) that the information rises approximately linearly, and the number of stimuli encoded thus rises approximately exponentially, as the number of cells in the sample increases (Rolls, Treves and Tovee, 1996; Abbott, Rolls and Tovee, 1997).

This direct neurophysiological evidence thus demonstrates that the encoding is distributed, and the responses are sufficiently independent and reliable, that the representational capacity increases exponentially. The consequence of this is that large numbers of stimuli, and fine discriminations between them, can be represented without having to measure the activity of an enormous number of neurons. Although the information rises approximately linearly with the number of neurons when this number is small, gradually each additional neuron does not contribute as much as the first (see Fig. 3). In the sample analysed by Rolls, Treves and Tovee (1997), the first neuron contributed 0.34 bits, on average, with 3.23 bits available from the 14 neurons analyzed. This reduction is however exactly what could be expected to derive from a simple ceiling effect, in which the ceiling is just the information in the stimulus set, or $\log_2 20 = 4.32$ bits, as shown in Fig. 3. This indicates that, on the one hand, each neuron does not contribute independently to the sum, and there is some overlap or redundancy in what is contributed by each neuron; and that, on the other hand, the degree of redundancy is not a property of the neuronal representation, but just a contingent feature dependent on the particular set of stimuli used in probing that representation. The data available is consistent with the hypothesis, explored by Abbott, Rolls and Tovee (1996) through

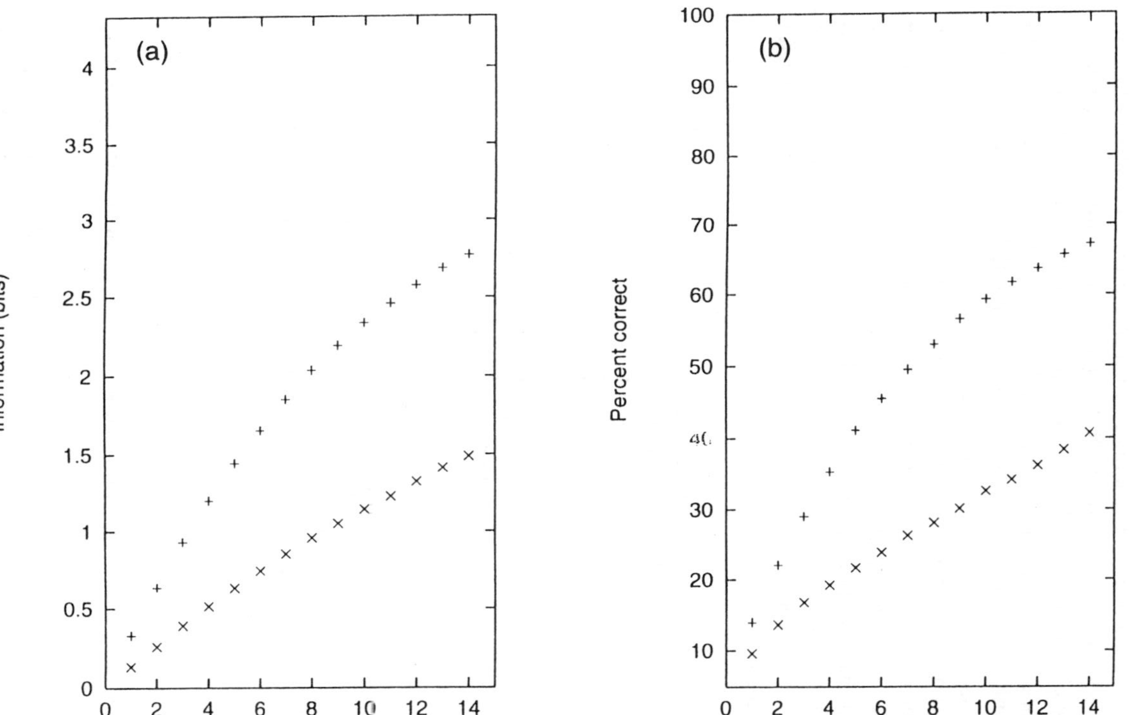

Figure 3. **(A)** The values for the average information available in the responses of different numbers of these neurons on each trial, about which of a set of 20 face stimuli has been shown. The decoding method was Dot Product (DP, diamonds) or Probability Estimation (PE, crosses), and the effects obtained with cross validation procedures utilising 50% of the trials as test trials are shown. The remainder of the trials in the cross-validation procedure were used as training trials. The full line indicates the amount of information expected from populations of increasing size, when assuming random correlations within the constraint given by the ceiling (the information in the stimulus set, I=4.32 bits). **(B)** The percent correct for the corresponding data to those shown in Fig. 3A. (After Rolls, Treves and Tovee, 1997)

simulations, that if the ceiling provided by the limited number of stimuli that could be presented were at much higher levels, each neuron would continue to contribute as much as the first few, up to much larger neuronal populations, so that the number of stimuli that can be encoded still continues to increase exponentially even with larger numbers of neurons (Fig. 4; Abbott, Rolls and Tovee, 1996). The redundancy observed could be characterised as flexible, in that it is the task that determines the degree to which large neuronal populations need to be sampled. If the task requires discriminations with very fine resolution between many different stimuli (*i.e.* in a high-dimensional space), then the responses of many neurons must be taken into account. If very simple discriminations are required (requiring little information), small subsets of neurons or even single neurons may be sufficient. The importance of this type of flexible redundancy in the representation is discussed below. The important point is that the information increases linearly with the number of cells used in the encoding, subject to a ceiling due to the fact that cells cannot add much more information as the ceiling imposed by the amount of information needed for the task is approached.

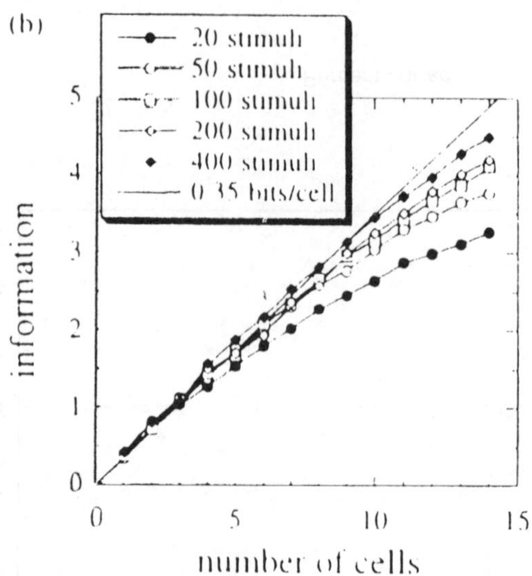

Figure 4. The information available about which stimulus was seen when the responses of many neurons to many stimuli are simulated (after Abbott, Rolls and Tovee, 1996).

It has recently been shown that there are neurons in the inferior temporal visual cortex that encode view invariant representations of objects, and for these neurons the same type of representation is found, namely distributed encoding with independent information conveyed by different neurons (Booth and Rolls, 1998).

The analyses just described were obtained with neurons that were not simultaneously recorded, but we have recently shown that with simultaneously recorded neurons similar results are obtained, that is the information about which stimulus was shown increases approximately linearly with the number of neurons, that is the neurons convey information that is nearly independent (Panzeri *et al.*, 1999b). (Consistently, Gawne and Richmond 1993 showed that even adjacent pairs of neurons recorded simultaneously from the same electrode carried information that was approximately 80% independent. In the research described by Panzeri *et al.* (1999b) we developed a method for measuring the information in the relative time of firing of simultaneously recorded neurons, which might be significant if the neurons became synchronised to some but not other stimuli in a set, as postulated by Engel *et al.* (1992). We found that for the set of cells currently available, almost all the information was available in the firing rates of the cells, and almost no information was available about these static images in the relative time of firing of neurons (Panzeri *et al.*, 1999b; Rolls *et al.*, 1999b). Thus the evidence is that for representations of faces and objects in the inferior temporal visual cortex (and of space in the primate hippocampus and of odours in the orbitofrontal cortex, see Rolls *et al.*, 1998, 1996), most of the information is available in the firing rates of the neurons.

It may be noted that it is unlikely that there are further processing areas beyond those described where ensemble coding changes into grandmother cell (local) encoding. Anatomically, there does not appear to be a whole further set of visual processing areas present in the brain; and outputs from the temporal lobe visual areas such as those described, are taken to limbic and related regions such as the amygdala and via the entorhinal cortex the hippocampus. Indeed, tracing this pathway onwards, we have found a population of neurons with face-selective responses in the amygdala, and in the majority of these neurons, different responses occur to different faces, with ensemble (not local) coding still being present (Leonard *et al.*, 1985; Rolls, 1992a). The amygdala in turn projects to another structure which may be important in other behavioural responses to faces, the ventral striatum, and comparable neurons have also been found in the ventral striatum (Williams *et al.*, 1993).

6. Advantages of the Distributed Representation Found of Objects for Brain Processing

Three key types of evidence that the visual representation provided by neurons in the temporal cortical areas, and the olfactory and taste representations in the orbitofrontal cortex, are distributed have been provided, and reviewed above. One is that the coding is not sparse (Baylis, Rolls and Leonard, 1985; Rolls and Tovee, 1995a). The second is that different neurons have different response profiles to a set of stimuli, and thus have at least partly independent responses (Baylis, Rolls and Leonard, 1985; Rolls and Tovee, 1995a). The third is that the capacity of the representations rises exponentially with the number of neurons (Rolls, Treves and Tovee, 1997; Abbott, Rolls and Tovee, 1996). The advantages of such distributed encoding are now considered, and apply to both fully distributed and to sparse distributed (but not to local) encoding schemes.

6.1. Exponentially High Coding Capacity

This property arises from a combination of the encoding being sufficiently close to independent by the different neurons (*i.e.* factorial), and sufficiently distributed. Part of the biological significance of such exponential encoding capacity is that a receiving neuron or neurons can obtain information about which one of a very large number of stimuli is present by receiving the activity of relatively small numbers of inputs from each of the neuronal populations from which it receives. For example, if neurons received in the order of 100 inputs from the population described here, they would have a great deal of information about which stimulus was in the environment. In particular, the characteristics of the actual visual cells described here indicate that the activity of 15 would be able to encode 192 face stimuli (at 50% accuracy), of 20 neurons 768 stimuli, of 25 neurons 3,072 stimuli, of 30 neurons 12,288 stimuli, and of 35 neurons 49,152 stimuli (Abbott, Rolls and Tovee, 1996; the values are for the optimal decoding case). Given that most neurons receive a limited number of synaptic contacts, in the order of several thousand, this type of encoding is ideal. It would enable for example neurons in the amygdala and orbitofrontal cortex to form pattern associations of visual stimuli with reinforcers such as the taste of food when each neuron received a reasonable number, perhaps in the order of hundreds, of inputs from the visually responsive neurons in the temporal cortical visual areas which specify which visual stimulus or object is being seen (see Rolls, 1990, 1992a,b; Rolls and Treves, 1998). Such a representation would also be appropriate for interfacing to the hippocampus, to allow an episodic memory to be formed, that for example a particular visual object was seen in a particular place in the environment (Rolls, 1989a-c; Treves and Rolls, 1994; Rolls and Treves, 1998). Here we should emphasize that although the sensory representation may have exponential encoding capacity, this does not mean that the associative networks that receive the information can store such large numbers of different patterns. Indeed, there are strict limitations on the number of memories that associative networks can store (Rolls and Treves, 1990; Treves and Rolls, 1991; Rolls and Treves, 1998). The particular value of the exponential encoding capacity of sensory representations is that very fine discriminations can be made as there is much information in the representation, and that the representation can be decoded if the activity of even a limited number of neurons in the representation is known.

One of the underlying themes here is the neural representation of objects. How would one know that one has found a neuronal representation of objects in the brain? The criterion we suggest that arises from this research (Rolls and Treves, 1998) is that when one can identify the object or stimulus that is present (from a large set of stimuli, that might be thousands or more) with a realistic number of neurons, say in the order of 100, then one has a representation of the object. This criterion appears to imply exponential encoding, for only then could such a large number of stimuli be represented with a relatively small number of units, at least for units with the response characteristics of actual neurons. Equivalently, we can say that there is a representation of the object when the information required to specify which of many stimuli or objects is present can be decoded from the responses of a limited number of neurons.

The properties of the representation of faces, and of olfactory and taste stimuli, have been evident when the readout of the information was by measuring the firing rate of the neurons, typically over a 500 ms period. Thus, at least where objects are represented in the visual, olfactory, and taste systems (*e.g.* individual faces, odours, and tastes), information can be read out without taking into account any aspects of the possible temporal synchronization between neurons (Engel *et al.*, 1992), or temporal encoding within a spike train (Tovee *et al.*, 1993; Rolls, Treves and Tovee, 1997; Panzeri *et al.*, 1999b). Further, as shown in section 11, the information is available so rapidly in the responses of these neurons that temporal encoding is unlikely to be a fundamental aspect of neuronal spike trains in this part of the brain.

6.2. Ease With Which The Code Can Be Read By Receiving Neurons

For brain plausibility, it would also be a requirement that the decoding process should itself not demand more than neurons are likely to be able to perform. This is why when we have estimated the information from populations of neurons, we have used in addition to a probability estimating measure (PE, optimal, in the Bayesian sense), also a dot product measure, which is a way of specifying that all that is required of decoding neurons would be the property of adding up postsynaptic potentials produced through each synapse as a result of the activity of each incoming axon (Rolls, Treves and Tovee, 1997; Abbott, Rolls and Tovee, 1996). It was found that with such a neurally plausible algorithm (the Dot Product, DP, algorithm), which calculates which average response vector the neuronal response vector on a single test trial was closest to by performing a normalised dot product (equivalent to measuring the angle between the test and the average vector), the same generic results were obtained, with only a 40% reduction of information compared to the more efficient (PE) algorithm. This is an indication that the brain could utilise the exponentially increasing capacity for encoding stimuli as the number of neurons in the population increases. For example, by using the representation provided by the neurons described here as the input to an associative or autoassociative memory, which computes effectively the dot product on each neuron between the input vector and the synaptic weight vector, most of the information available would in fact be extracted (see Rolls and Treves, 1990, 1998; Treves and Rolls, 1991).

6.3. Higher Resistance to Noise

This, like the next few properties, is in general an advantage of distributed over local representations, which applies to artificial systems as well, but is presumably of particular value in biological systems in which some of the elements have an intrinsic variability in their operation. Because the decoding of a distributed representation involves assessing the activity of a whole population of neurons, and computing a dot product or correlation, a distributed representation provides more resistance to variation in individual components than does a local encoding scheme (cf Panzeri *et al.*, 1996).

6.4. Generalization

Generalization to similar stimuli is again a property that arises in neuronal networks if distributed but not if local encoding is used. The generalization arises as a result of the fact that a neuron can be thought of as computing the inner or dot product of the stimulus representation with its weight vector. If the weight vector leads to the neuron having a response to one visual stimulus, then the neuron will have a similar response to a similar visual stimulus. This computation of correlations between stimuli operates only with distributed representations. If an output is based on a single X,Y (input firing, synaptic weight) pair, then if the X or the Y is lost, the correlation drops to zero (see further Rolls and Treves, 1998).

6.5. Completion

Completion occurs in associative memory networks by a similar process. Completion is the property of recall of the whole of a pattern in response to any part of the pattern. Completion arises because any part of the stimulus representation, or pattern, is effectively correlated with the whole pattern during memory storage. Completion is thus a property of distributed representations, and not of local representations. It arises for example in autoassociation (attractor) neuronal networks, which are characterised by recurrent connectivity. It is thought that such networks are important in the hippocampus in enabling incomplete recent episodic memories to be completed, and in the cerebral cortex, where the association fibres between nearby pyramidal cells may help the cells to retrieve a representation which depends on many neurons in the network (Treves and Rolls, 1994; Rolls and Treves, 1998).

6.6. Graceful Degradation or Fault Tolerance

This also arises only if the input patterns have distributed representations, and not if they are local. Local encoding suffers sudden deterioration once the few neurons or synapses carrying the information about a particular stimulus are destroyed.

6.7. Speed of Readout of the Information

The information available in a distributed representation can be decoded by an analyzer more quickly than can the information from a local representation, given comparable firing rates. Within a fraction of an interspike interval, with a distributed representation, much information can be extracted (Treves, 1993; Treves, Rolls and

Tovee, 1996; Rolls, Treves and Tovee, 1997; Treves, Rolls and Simmen, 1997; Panzeri *et al.*, 1999a). In effect, spikes from many different neurons can contribute to calculating the angle between a neuronal population and a synaptic weight vector within an interspike interval. With local encoding, the speed of information readout depends on the exact model considered, but if the rate of firing needs to be taken into account, this will necessarily take time, because of the time needed for several spikes to accumulate in order to estimate the firing rate. It is likely with local encoding that the firing rate of a neuron would need to be measured to some degree of accuracy, for it seems implausible to suppose that a single spike from a single neuron would be sufficient to provide a noise-free representation for the next stage of processing.

7. Invariance in the Neuronal Representation of Stimuli

One of the major problems which must be solved by a visual system is the building of a representation of visual information which allows recognition to occur relatively independently of size, contrast, spatial frequency, position on the retina, angle of view, etc. To investigate whether these neurons in the temporal lobe visual cortex are at a stage of processing where such invariance is being represented in the responses of neurons, the effect of such transforms of the visual image on the responses of the neurons has been investigated.

To investigate whether the responses of these neurons show some of the perceptual properties of recognition including tolerance to isomorphic transforms (*i.e.* in which the shape is constant), the effects of alteration of the size and contrast of an effective face stimulus on the responses of these neurons were analysed quantitatively in macaque monkeys (Rolls and Baylis, 1986). It was shown that the majority of these neurons had responses which were relatively invariant with respect to the size of the stimulus. The median size change tolerated with a response of greater than half the maximal response was 12 times. Also, the neurons typically responded to a face when the information in it had been reduced from 3D to a 2D representation in grey on a monitor, with a response which was on average 0.5 of that to a real face. (This reduction in amplitude does not by itself mean that the point in multidimensional space represented by the ensemble of neurons has moved. The point represented by a facial identity ensemble will move only to the extent that the responses of neurons in the facial identity ensemble are affected differently by this transform. The original data are shown in Rolls and Baylis, 1986.) Another transform over which recognition is relatively invariant is spatial frequency. For example, a face can be identified when it is blurred (when it contains only low spatial frequencies), and when it is high-pass spatial frequency filtered (when it looks like a line drawing). It has been shown that if the face images to which these neurons respond are low-pass filtered in the spatial frequency domain (so that they are blurred), then many of the neurons still respond when the images contain frequencies only up to 8 cycles per face. Similarly, the neurons still respond to high-pass filtered images (with only high spatial frequency edge information) when frequencies down to only 8 cycles per face are included (Rolls *et al.*, 1985). Face recognition shows similar invariance with respect to spatial frequency (see Rolls *et al.*, 1985). Further analysis of these neurons with narrow (octave) bandpass spatial frequency filtered face stimuli shows that the responses of these neurons to an unfiltered face can not be predicted from a linear combination of their responses to the narrow band stimuli (Rolls *et al.*, 1987). This lack of linearity of these neurons, and their responsiveness to a wide range of spatial frequencies, indicate that in at least this part of the primate visual system recognition does not occur using Fourier analysis of the spatial frequency components of images.

To investigate whether neurons in the inferior temporal visual cortex and cortex in the anterior part of the superior temporal sulcus operate with translation invariance in the awake behaving primate (as suggested by earlier findings under anesthesia, see Gross *et al.*, 1985), their responses were measured during a visual fixation (blink) task in which stimuli could be placed in different parts of the receptive field (Tovee, Rolls and Azzopardi, 1994). It was found that in most cases the responses of the neurons were little affected by which part of the face was fixated, and that the neurons responded (with a greater than half-maximal response) even when the monkey fixated 2-5 degrees beyond the edge of a face which subtended 8 - 17 degrees at the retina. Moreover, the stimulus selectivity between faces was maintained this far eccentric within the receptive field. These results held even across the visual midline. It was also shown that these neurons code for identity and not fixation position, in that there was approximately six times more information in the responses of these neurons about which face had been seen than about where the monkey fixated on the face. It is concluded that at least some of these neurons in the temporal lobe visual areas do have considerable translation invariance so that this is a computation which must be performed in the visual system. Ways in which the translation and size invariant representations shown to be present in the brain by these studies could be built are considered below in section 12. It is clearly important that translation invariance in the visual system is made explicit in the neuronal responses, for this simplifies greatly the output of the visual system to memory systems such as the hippocampus and amygdala, which can then remember, or form associations about, objects. The function of these memory systems would be almost impossible if there were no consistent output from the visual system about objects (including faces), for then the memory systems would need to learn about all possible sizes, positions etc of each object, and there would be no easy generalization from one size or position of an object to that object when seen with another retinal size or position (see Rolls and Treves, 1998).

Until now, research on translation invariance has considered the case in which there is only one object in the visual field. The question then arises of how the visual system operates in a cluttered environment. Do all objects that can activate an inferior temporal neuron do so whenever they are anywhere within the large receptive fields of inferior temporal neurons (cf Sato, 1989)? If so, the output of the visual system might be confusing for structures which receive inputs from the temporal cortical visual areas. To investigate this we measured the responses of inferior temporal cortical neurons with face-selective responses of rhesus macaques performing a visual fixation task. We found that the response of neurons to an effective face centred 8.5 degrees from the fovea was decreased to 71% if an ineffective face stimulus for that cell was present at the fovea. If an ineffective stimulus for a cell is introduced parafoveally when an effective stimulus is being fixated, then there was a similar reduction in the responses of neurons. More concretely, the mean firing rate across all cells to a fixated effective face with a non-effective face in the periphery was 34 spikes/s. On the other hand, the average response to a fixated non-effective face with an effective face in the periphery was 22 spikes/s. (These firing rates reflected the fact that in this population of neurons, the mean response for an effective face was 49 spikes/s with the face at the fovea, and 35 spikes/s with the face 8.5 degrees from the fovea.) Thus these cells gave a reliable output about which stimulus is actually present at the fovea, in that their response was larger to a fixated effective face than to a fixated non-effective face, even when there are other parafoveal stimuli ineffective or effective for the cell (Rolls and Tovee, 1995b). Thus the cell provides information biased towards what is present at the fovea, and not equally about what is present anywhere in the visual field. This makes the interface to action simpler, in that what is at the fovea can be interpreted (*e.g.* by an associative memory) partly independently of the surroundings, and choices and actions can be directed if appropriate to what is at the fovea (cf Ballard, 1993). These findings are a step towards understanding how the visual system functions in a normal environment (see also Gallant *et al.*, 1998).

8. A View-independent Representation of Visual Information

For recognizing and learning about objects (including faces), it is important that an output of the visual system should be not only translation and size invariant, but also relatively view invariant. In an investigation of whether there are such neurons, we found that some temporal cortical neurons reliably responded differently to the faces of two different individuals independently of viewing angle (Hasselmo, Rolls, Baylis and Nalwa, 1989), although in most cases (16/18 neurons) the response was not perfectly view-independent. Mixed together in the same cortical regions there are neurons with view-dependent responses (*e.g.* Hasselmo *et al.*, 1989). Such neurons might respond for example to a view of a profile of a monkey but not to a full-face view of the same monkey (Perrett *et al.*, 1985b). These findings, of view-dependent, partially view independent, and view independent representations in the same cortical regions are consistent with the hypothesis discussed below that view-independent representations are being built in these regions by associating together neurons that respond to different views of the same individual. These findings also provide evidence that the outputs of the visual system are likely to include representations of what is being seen, in a view independent way that would be useful for object recognition and for learning associations about objects; and in a view-based way that would be useful in social interactions to determine whether another individual is looking at one, and for selecting details of motor responses, for which the orientation of the object with respect to the viewer is required.

Further evidence that some neurons in the temporal cortical visual areas have object-based rather than view-based responses comes from a study of a population of neurons that responds to moving faces (Hasselmo, Rolls, Baylis and Nalwa, 1989). For example, four neurons responded vigorously to a head undergoing ventral flexion, irrespective of whether the view of the head was full face, of either profile, or even of the back of the head. These different views could only be specified as equivalent in object-based coordinates. Further, for all of the 10 neurons that were tested in this way, the movement specificity was maintained across inversion, responding for example to ventral flexion of the head irrespective of whether the head was upright or inverted. In this procedure, retinally encoded or viewer-centered movement vectors are reversed, but the object-based description remains the same. It was of interest that the neurons tested generalized across different heads performing the same movements.

Also consistent with object-based encoding is the finding of a small number of neurons which respond to images of faces of a given absolute size, irrespective of the retinal image size (Rolls and Baylis, 1986).

To investigate whether view-invariant representations of objects (as well as faces) are also encoded by some neurons in the inferior temporal cortex of the rhesus macaque, the activity of single neurons was recorded while monkeys were shown very different views of 10 objects (Booth and Rolls, 1998). The stimuli were presented for 0.5 s on a colour video monitor while the monkey performed a visual fixation task. The stimuli were images of 10 real plastic objects which had been in the monkey's cage for several weeks, to enable him to build view invariant representations of the objects. Control stimuli were views of objects which had never been seen as real objects. The neurons analyzed were in the TE cortex in and close to the ventral lip of the anterior part of the superior temporal sulcus. Many neurons were found that responded to some views of some objects. However, for a smaller number of neurons, the responses occurred only to a subset of the objects, irrespective of the viewing angle. These neurons thus conveyed information about which object has been seen, independently of view, as confirmed by information theoretic analysis of the neuronal responses. Each neuron did not, in general, respond to only one object, but instead responded to a subset of the objects. They thus showed ensemble, sparse-distributed, encoding. The information available about which object was seen increased approximately linearly with the number of neurons in the ensemble. These experiments provide evidence that there is a view-invariant representation of objects, as well as faces, in the primate temporal cortical visual areas. Further evidence consistent with these findings is that some studies have shown that the responses of some visual neurons in the inferior temporal cortex do not depend on the presence or absence of critical features for maximal activation (*e.g.* Perrett, Rolls and Caan, 1982; see Tanaka, 1993, 1996). For example, Mikami *et al.* (1994) have shown that some TE cells respond to partial views of the same laboratory instrument(s), even when these partial views contain different features. In a different approach, Logothetis *et al.* (1994) have reported that in monkeys extensively trained (over thousands of trials) to treat different views of computer generated wire-frame "objects" as the same, a small population of neurons in the inferior temporal cortex did respond to different views of the same wire-frame object (see also Logothetis and Sheinberg, 1996). The difference in the approach taken by Booth and Rolls (1998) was that no explicit training was given in invariant object recognition, as Rolls= hypothesis (1992b) is that view invariant representations can be learned by associating together the different views of objects as they are moved and inspected naturally in a period that may be in the order of a few seconds.

9. Different Neural Systems Are Specialized for Recognition and for Face Expression Decoding

To investigate whether there are neurons in the cortex in the anterior part of the superior temporal sulcus of the macaque monkey which could provide information about facial expression, neurons were tested with facial stimuli which included examples of the same individual monkey with different facial expressions (Hasselmo, Rolls and Baylis, 1989). The responses of 45 neurons with responses selective for faces were measured to a set of 3 individual monkey faces with three expressions for each monkey, as well as to human expressions. Of these neurons, 15 showed response differences to different identities independently of expression, and 9 neurons showed responses which depended on expression but were independent of identity, as measured by a two-way ANOVA. Multidimensional scaling confirmed this result, by showing that for the first set of neurons the faces of different individuals but not expressions were well separated in the space, whereas for the second group of neurons, different expressions but not the faces of different individuals were well separated in the space. The neurons responsive to expression were found primarily in the cortex in the superior temporal sulcus, while the neurons responsive to identity were found in the inferior temporal gyrus. These results show that there are some neurons in this region the responses of which could be useful in providing information about facial expression, of potential use in social interactions (Rolls, 1981, 1984, 1986a,b, 1990, 1999). Damage to this population may contribute to the deficits in social and emotional behavior which are part of the Kluver-Bucy syndrome produced by temporal lobe damage in monkeys (see Rolls, 1981, 1984, 1986a,b, 1990, 1999; Leonard *et al.*, 1985).

A further way in which some of these neurons may be involved in social interactions is that some of them respond to gestures, *e.g.* to a face undergoing ventral flexion, as described above and by Perrett et. al. (1985a). The interpretation of these neurons as being useful for social interactions is that in some cases these neurons respond not only to ventral head flexion, but also to the eyes lowering and the eyelids closing (Hasselmo *et al.*, 1989). Now these two movements (head lowering and eyelid lowering) often occur together when a monkey is breaking social contact with another, *e.g.* after a challenge, and the information being conveyed by such a neuron could thus reflect the presence of this social gesture. That the same neuron could respond to such different, but normally co-occurrent, visual inputs could be accounted for by the Hebbian competitive self-organization described below. It may also be noted that it is important when decoding facial expression not to move entirely into the object-based domain (in which the description would be in terms of the object itself, and would not contain information about the position and orientation of the object relative to the observer), but to retain some information about the head direction of the face stimulus being seen relative to the observer, for this is very important in determining whether a threat is being made in your direction. The presence of view-dependent representations in some of these cortical regions is consistent with this requirement. Indeed, it may be suggested that the cortex in the superior temporal sulcus, in which neurons are found with responses related to facial expression (Hasselmo, Rolls and Baylis, 1989), head and face movement involved in for example gesture (Hasselmo, Rolls, Baylis and Nalwa, 1989), and eye gaze (Perrett *et al.*, 1985b), may be more related to face expression decoding; whereas the TE areas (more ventral, mainly in the macaque inferior temporal gyrus), in which neurons tuned to face identity (Hasselmo, Rolls and Baylis, 1989) and with view-independent responses (Hasselmo, Rolls, Baylis and Nalwa, 1989) are more likely to be found, may be more related to an object-based representation of identity. Of course, for appropriate social and emotional responses, both types of subsystem would be important, for it is necessary to know both the direction of a social gesture, and the identity of the individual, in order to make the correct social or emotional response.

Outputs from the temporal cortical visual areas reach the amygdala and the orbitofrontal cortex, and evidence is accumulating that these brain areas are involved in social and emotional responses to faces (Rolls, 1990, 1992a-c, 1994, 1999). For example, lesions of the amygdala in monkeys disrupt social and emotional responses to faces, and we have identified a population of neurons with face-selective responses in the primate amygdala (Leonard *et al.*, 1985), some of which may respond to facial and body gesture (Brothers *et al.*, 1990). Rolls, Critchley and Browning (2000) have found a number of face-responsive neurons in the orbitofrontal cortex, and they are also present in adjacent prefrontal cortical areas (Wilson *et al.*, 1993), and also in the ventral striatum, which receives projections from the amygdala and orbitofrontal cortex (Williams, Rolls, Leonard and Stern, 1993).

We have applied this research to the study of humans with frontal lobe damage, to try to develop a better understanding of the social and emotional changes which may occur in these patients. Impairments in the identification of facial and vocal emotional expression were demonstrated in a group of patients with ventral frontal lobe damage who had behavioural problems such as disinhibited or socially inappropriate behaviour (Hornak *et al.*, 1996). A group of patients with lesions outside this brain region, without these behavioural problems, was unimpaired on the expression identification tests. The impairments shown by the frontal patients on these expression identification tests could occur independently of perceptual difficulties. Face expression identification was severely impaired in some patients whose recognition of the identity of faces was normal. Severe impairments on the vocal expression test (which consisted of non-verbal emotional sounds) were found in patients who produced excellent imitations of the sounds they could not identify, and whose identification of environmental sounds was also normal.

These findings suggest that some of the social and emotional problems associated with ventral frontal lobe or amygdala damage may be related to a difficulty in identifying correctly facial (and vocal) expression (Hornak *et al.*, 1996). The question then arises of what functions are performed by the orbitofrontal cortex and amygdala with the face-related outputs they receive from the temporal cortical visual areas. The hypothesis has been developed that these regions are important in emotional and social behaviour because of their role in reward-related learning (Rolls, 1986a,b, 1990, 1995, 1999). The amygdala is especially involved in learning associations between visual stimuli and primary (unlearned) rewards and punishments such as food taste and touch, and the orbitofrontal cortex is especially involved in the rapid reversal (*i.e.* adjustment or relearning) of such stimulus reinforcement associations. According to this hypothesis, the importance of projecting face-related information to the amygdala and orbitofrontal cortex is so that they can learn associations between faces, using information about both face identity and facial expression, and rewards and punishments. Now it is particularly in primate social behaviour that rapid relearning about individuals, identified by their face, and depending on their facial expression, must occur very rapidly and flexibly, to keep up with the continually changing social exchanges between different individuals and groups of individuals. It is crucial to be able to remember recent reinforcement associations of different individuals, and to be able to continually adjust these. It is suggested that these factors have led to the very major development of the orbitofrontal cortex in primates, to receive appropriate inputs (about identity from faces, and about facial expression), and to provide a very rapid and flexible learning mechanism for the current reinforcement associations of these inputs. Consistent with this, the same patients that are impaired in face expression identification are also impaired on stimulus-reinforcement relearning tasks such as visual discrimination reversal and extinction (Rolls, Hornak *et al.*, 1994). Moreover, this learning impairment is highly correlated with the social and behavioural changes found in these patients (Rolls, Hornak *et al.*, 1994).

10. Learning of New Representations in the Temporal Cortical Visual Areas

Given the fundamental importance of a computation which results in relatively finely tuned neurons which across ensembles but not individually specify objects including individual faces in the environment, we have investigated whether experience plays a role in determining the selectivity of single neurons which respond to faces. The hypothesis being tested was that visual experience might guide the formation of the responsiveness of neurons so that they provide an economical and ensemble-encoded representation of items actually present in the environment. To test this, we investigated whether the responses of temporal cortex face-selective neurons were at all altered by the presentation of new faces which the monkey had never seen before. It might be for example that the population would make small adjustments in the responsiveness of its individual neurons, so that neurons would acquire response tuning that would enable the population as a whole to discriminate between the faces actually seen. We thus investigated whether when a set of totally novel faces was introduced, the responses of these neurons were fixed and stable from the first presentation, or instead whether there was some adjustment of responsiveness over repeated presentations of the new faces. First, it was shown for each neuron tested that its responses were stable over 5-15 repetitions of a set of familiar faces. Then a set of new faces was shown in random order (with 1 s for each presentation), and the set was repeated with a new random order over many iterations. Some of the neurons studied in this way altered the relative degree to which they responded to the different members of the set of novel faces over the first few (1-2) presentations of the set (Rolls *et al.*, 1989). If in a different experiment a single novel face was introduced when the responses of a neuron to a set of familiar faces was being recorded, it was found that the responses to the set of familiar faces were not disrupted, while the responses to the novel face became stable within a few presentations. Thus there is now some evidence from these experiments that the response properties of neurons in the temporal lobe visual cortex are modified by experience, and that the modification is such that when novel faces are shown, the relative responses of individual neurons to the new faces alter. It is suggested that alteration of the tuning of individual neurons in this way results in a good discrimination over the population as a whole of the faces known to the monkey. This evidence is consistent with the categorisation being performed by self-organizing competitive neuronal networks, as described below and elsewhere (Rolls, 1989a-c; Rolls and Treves, 1998).

Further evidence that these neurons can learn new representations very rapidly comes from an experiment in which binarized black and white images of faces which blended with the background were used. These did not activate face-selective neurons. Full grey-scale images of the same photographs were then shown for ten 0.5s presentations. It was found that in a number of cases, if the neuron happened to be responsive to that face, that when the binarized version of the same face was shown next, the neurons responded to it (Tovee, Rolls and Ramachandran, 1996). This is a direct parallel to the same phenomenon which is observed psychophysically, and provides dramatic evidence that these neurons are influenced by only a very few seconds (in this case 5 s) of experience with a visual stimulus. We have shown a neural correlate of this effect using similar stimuli and a similar paradigm in a PET (positron emission tomography) neuroimaging study in humans, with a region showing an effect of the learning found in the right temporal lobe, and an area for objects in the left temporal lobe (Dolan *et al.*, 1997).

Such rapid learning of representations of new objects appears to be a major type of learning in which the temporal cortical areas are involved. Ways in which this learning could occur are considered below. It is also the case that there is a much shorter term form of memory in which some of these neurons are involved, for whether a particular visual stimulus (such as a face) has been seen recently, for some of these neurons respond differently to recently seen stimuli in short term visual memory tasks (Baylis and Rolls, 1987; Miller and Desimone, 1994; Xiang and Brown, 1998), and neurons in a more ventral cortical area respond during the delay in a short term memory task (Miyashita, 1993).

11. The Speed of Processing in the Temporal Cortical Visual Areas

Given that there is a whole sequence of visual cortical processing stages including V1, V2, V4, and the posterior inferior temporal cortex to reach the anterior temporal cortical areas, and that the response latencies of neurons in V1 are about 40-50 ms, and in the anterior inferior temporal cortical areas approximately 80-100 ms, each stage may need to perform processing for only 15-30 ms before it has performed sufficient processing to start influencing the next stage. Consistent with this, response latencies between V1 and the inferior temporal cortex increase from stage to stage (Thorpe and Imbert, 1989). This seems to imply very fast computation by each cortical area, and therefore to place constraints on the type of processing performed in each area that is necessary for final object identification. Rapid identification of visual stimuli is important in social and many other situations, and that there must be strong selective pressure for rapid identification. For these reasons, the speed of processing has been investigated quantitatively, as follows.

In a first approach, we measured the information available in short temporal epochs of the responses of temporal cortical face-selective neurons about which face had been seen. We found that if a period of the firing rate of 50 ms was taken, then this contained 84.4% of the information available in a much longer period of 400 ms about which of four faces had been seen. If the epoch was as little as 20 ms, the information was 65% of that available from the firing rate in the 400 ms period (Tovee *et al.*, 1993). These high information yields were obtained with the short epochs taken near the start of the neuronal response, for example in the post-stimulus period 100-120 ms. Moreover, we were able to show that the firing rate in short periods taken near the start of the neuronal response was highly correlated with the firing rate taken over the whole response period, so that the information available was stable over the whole response period of the neurons (Tovee *et al.*, 1993). We were able to extend this finding to the case when a much larger stimulus set, of 20 faces, was used. Again, we found that the information available in short (*e.g.* 50 ms) epochs was a considerable proportion (*e.g.* 65%) of that available in a 400 ms long firing rate analysis period (Tovee and Rolls, 1995). These investigations thus showed that there was considerable information about which stimulus had been seen in short time epochs near the start of the response of temporal cortex neurons.

The next approach was to address the issue of for how long a cortical area must be active to mediate object recognition. This approach used a visual backward masking paradigm. In this paradigm there is a brief presentation of a test stimulus which is rapidly followed (within 1-100 ms) by the presentation of a second stimulus (the mask), which impairs or masks the perception of the test stimulus. This paradigm used psychophysically leaves unanswered for how long visual neurons actually fire under the masking condition at which the subject can just identify an object. Although there has been a great deal of psychophysical investigation with the visual masking paradigm (Turvey, 1973; Breitmeyer, 1980; Humphreys and Bruce, 1989), there is very little direct evidence on the effects of visual masking on neuronal activity. For example, it is possible that if a neuron is well tuned to one class of stimulus, such as faces, that a pattern mask which does not activate the neuron, will leave the cell firing for some time after the onset of the pattern mask. In order to obtain direct neurophysiological evidence on the effects of backward masking of neuronal activity, we analysed the effects of backward masking with a pattern mask on the responses of single neurons to faces (Rolls and Tovee, 1994a). This was performed to clarify both what happens with visual backward masking, and to show how long neurons may respond in a cortical area when perception and identification are just possible. When there was no mask the cell responded to a 16 ms presentation of the test stimulus for 200-300 ms, far longer than the presentation time. It is suggested that this reflects the operation of a short term memory system implemented in cortical circuitry, the importance of which in learning invariant representations is considered below in section 12. If the mask was a stimulus which did not stimulate the cell (either a non-face pattern mask consisting of black and white letters N and O, or a face which was a non-effective stimulus for that cell), then as the interval between the onset of the test stimulus and the onset of the mask stimulus (the stimulus onset asynchrony, SOA) was reduced, the length of time for which the cell fired in response to the test stimulus was reduced. This reflected an abrupt interruption of neuronal activity produced by the effective face stimulus. When the SOA was 20 ms, face-selective neurons in the inferior temporal cortex of macaques responded for a period of 20-30 ms before their firing was interrupted by the mask (Rolls and Tovee, 1994; Rolls *et al.*, 1999a). We went on to show that under these conditions (a test-mask stimulus onset asynchrony of 20 ms), human observers looking at the same displays could just identify which of 6 faces was shown (Rolls *et al.*, 1994b).

These results provide evidence that a cortical area can perform the computation necessary for the recognition of a visual stimulus in 20-30 ms. This provides a fundamental constraint which must be accounted for in any theory of cortical computation. The results emphasise just how rapidly cortical circuitry can operate. This rapidity of operation has obvious adaptive value, and allows the rapid behavioral responses to the faces and face expressions of different individuals which are a feature of primate social and emotional behaviour. Moreover, although this speed of operation does seem fast for a network with recurrent connections (mediated by *e.g.* recurrent collateral or inhibitory interneurons), recent analyses of networks with analog membranes which integrate inputs, and with spontaneously active neurons, shows that such networks can settle very rapidly (Treves, 1993; Treves, Rolls and Tovee, 1996; Rolls and Treves, 1998). This approach has been extended to multilayer networks such as those found in the visual system, and again very rapid propagation (in 40-50 ms) of information through such a 4-layer network with recurrent collaterals operating at each stage has been found (Panzeri *et al.*, 2000).

12. Discussion

The neurophysiological findings described above, and wider considerations on the possible computational properties of the cerebral cortex (Rolls, 1989a,b, 1992b, 1994; Rolls and Treves, 1998), lead to the following outline working hypotheses on object recognition by visual cortical mechanisms described in the next paper. The principles underlying the processing of faces and other objects may be similar, but more neurons may become allocated to represent different aspects of faces because of the need to recognise the faces of many different individuals, that is to identify many individuals within the category faces.

Acknowledgements

The author has worked on some of the investigations described here with P. Azzopardi, G.C. Baylis, P. Foldiak, M. Hasselmo, C.M. Leonard, G. Littlewort, T.J.Milward, D.I. Perrett, M.J. Tovee, A. Treves and G. Wallis, and their collaboration is sincerely acknowledged. Different parts of the research described were supported by the Medical Research Council, PG8513790; by a Human Frontier Science Program grant; by an EC Human Capital and Mobility grant; by the MRC Oxford Interdisciplinary Research Centre in Brain and Behaviour; and by the Oxford McDonnell-Pew Centre in Cognitive Neuroscience.

References

Abbott, L.F., E.T. Rolls and M.J. Tovee (1996) "Representational capacity of face coding in monkeys", *Cerebral Cortex* 6:498-505.

Baddeley, R.J., L.F. Abbott, M.J.A. Booth, F. Sengpiel, T. Freeman, E.A. Wakeman and E.T. Rolls (1997) "Responses of neurons in primary and inferior temporal visual cortices to natural scenes", *Proc. Roy. Soc. B* 264:1775-1783.

Baizer, J.S., L.G. Ungerleider and R. Desimone (1991) "Organization of visual inputs to the inferior temporal and posterior parietal cortex in macaques", *J Neurosci* 11:168-190.

Ballard, D.H. (1993) "Subsymbolic modelling of hand-eye co-ordination", in: *The Simulation of Human Intelligence*, D.E. Broadbent, ed., Oxford: Blackwell, pp.

71-102.

Barlow, H.B. (1972) "Single units and sensation: a neuron doctrine for perceptual psychology?", *Perception* 1:371-394.

Baylis, G.C., E.T. Rolls and C.M. Leonard (1985) "Selectivity between faces in the responses of a population of neurons in the cortex in the superior temporal sulcus of the monkey", *Brain Res.* 342:91-102.

Baylis, G.C., E.T. Rolls and C.M. Leonard (1987) "Functional subdivisions of temporal lobe neocortex", *J. Neurosci* 7:330-342.

Baylis, G.C. and E.T. Rolls (1987) "Responses of neurons in the inferior temporal cortex in short term and serial recognition memory tasks", *Exp. Brain Res.* 65:614-622.

Booth, M.C.A. and E.T. Rolls (1998) "View-invariant representations of familiar objects by neurons in the inferior temporal visual cortex", *Cerebral Cortex* 8:510-523.

Breitmeyer, B.G. (1980) "Unmasking visual masking: a look at the "why" behind the veil of the "how"", *Psychol. Rev.* 87:52-69.

Brothers, L., B. Ring and A.S. Kling (1990) "Response of neurons in the macaque amygdala to complex social stimuli", *Behav. Brain Res.* 41:199-213.

Bruce, C., R. Desimone and C.G. Gross (1981) "Visual properties of neurons in a polysensory area in superior temporal sulcus of the macaque", *J. Neurophys.* 46:369-384.

Desimone, R. (1991) "Face-selective cells in the temporal cortex of monkeys", *J. Cog. Neurosci.* 3:1-8.

Desimone, R. and C.G. Gross (1979) "Visual areas in the temporal lobe of the macaque", *Brain Res.* 178:363-380.

Desimone, R., T.D. Albright, C.G. Gross and C. Bruce (1984) "Stimulus-selective properties of inferior temporal neurons in the macaque", *J. Neurosci.* 4:2051-2062.

Dolan, R.J., G.R. Fink, E.T. Rolls, M. Booth, A. Holmes, R.S.J. Frackowiak and K.J. Friston (1997) "How the brain learns to see objects and faces in an impoverished context", *Nature* 389:596-599.

Elliffe, M.C.M., E.T. Rolls and S.M. Stringer (2000b) "Invariant recognition of feature combinations in the visual system", submitted

Engel, A.K., P. Konig, A.K. Kreiter, T.B. Schillen and W. Singer (1992) "Temporal coding in the visual system: new vistas on integration in the nervous system", *Trends in Neurosci.* 15:218-226.

Gallant, J.L., C.E. Connor and D.C. Van-Essen (1998) "Neural activity in areas V1, V2 and V4 during free viewing of natural scenes compared to controlled viewing", *Neuroreport* 9:85-90.

Gawne, T.J. and B.J. Richmond (1993) "How independent are the messages carried by adjacent inferior temporal cortical neurons?", *J. Neurosci.* 13:2758-2771.

Gross, C.G., R. Desimone, T.D. Albright and E.L. Schwartz (1985) "Inferior temporal cortex and pattern recognition", *Exp. Brain. Res. Suppl.* 11:179-201.

Hasselmo, M.E., E.T. Rolls and G.C. Baylis (1989a) "The role of expression and identity in the face-selective responses of neurons in the temporal visual cortex of the monkey", *Behav. Brain Res.* 32:203-218.

Hasselmo, M.E., E.T. Rolls, G.C. Baylis and V. Nalwa (1989b) "Object-centered encoding by face-selective neurons in the cortex in the superior temporal sulcus of the monkey", *Exp. Brain Res.* 75:417-429.

Hornak, J., E.T. Rolls and D. Wade (1996) "Face and voice expression identification in patients with emotional and behavioural changes following ventral frontal lobe damage", *Neuropsychologia* 34:247-261.

Humphreys, G.W. and V. Bruce (1989) *Visual Cognition*, Hove: Erlbaum.

Leonard, C.M., E.T. Rolls, F.A.W. Wilson and G.C. Baylis (1985) "Neurons in the amygdala of the monkey with responses selective for faces", *Behav. Brain Res.* 15:159-176.

Logothetis, N.K., J. Pauls, H.H. Bulthoff and T. Poggio (1994) "View-dependent object recognition by monkeys", *Current Biol.* 4:401-414.

Logothetis, N.K. and D.L. Sheinberg (1996) "Visual object recognition", *Annu. Rev. Neuroscience* 19:577-621.

Maunsell, J.H.R. and W.T. Newsome (1987) "Visual processing in monkey extrastriate cortex", *Annu. Rev. Neurosci.* 10:363-401.

Mikami, A., K. Nakamura and K. Kubota (1994) "Neuronal responses to photographs in the superior temporal sulcus of the rhesus monkey", *Behav. Brain Res.* 60:1-13.

Miller, E.K. and R. Desimone (1994) "Parallel neuronal mechanisms for short-term memory", *Science* 263:520-522.

Miyashita, Y. (1993) "Inferior temporal cortex: where visual perception meets memory", *Ann. Rev. Neurosci.* 16:245-263.

Panzeri, S., G. Biella, E.T. Rolls W.E. Skaggs and A. Treves (1996) "Speed, noise, information and the graded nature of neuronal responses", *Network* 7:365-370.

Panzeri, S., A. Treves, S.R. Schultz and E.T. Rolls (1999a) "On decoding the responses of a population of neurons from short time epochs", *Neural Computation* 11:1553-1577.

Panzeri, S., S.R. Schultz, A. Treves and E.T. Rolls (1999b) Correlations and the encoding of information in the nervous system. *Proceedings of the Royal Society B* 266:1001-1012.

Panzeri, S., E.T. Rolls, F. Battaglia and R. Lavis (2000) "Speed of information retrieval in multilayer networks of integrate-and-fire neurons", submitted

Perrett, D.I., E.T. Rolls and W. Caan (1982) "Visual neurons responsive to faces in the monkey temporal cortex", *Exp. Brain Res.* 47:329-342.

Perrett, D.I., P.A.J. Smith, A.J. Mistlin, A.J. Chitty, A.S. Head, D.D. Potter, R. Broennimann, A.D. Milner and M.A. Jeeves (1985a) "Visual analysis of body movements by neurons in the temporal cortex of the macaque monkey: a preliminary report", *Behav. Brain Res.* 16:153-170.

Perrett, D.I., P.A.J. Smith, D.D. Potter, A.J. Mistlin, A.S. Head, D. Milner and M.A. Jeeves (1985b) "Visual cells in temporal cortex sensitive to face view and gaze direction", *Proc. Roy. Soc.* 223(B):293-317.

Perrett, D.I., A.J. Mistlin and A.J. Chitty (1987) "Visual neurons responsive to faces", *Trends in Neurosc.* 10:358-364.

Poggio, T. (1990) "A theory of how the brain might work", *Cold Spring Harbor Symposia in Quantitative Biology* 55:899-910.

Rolls, E.T. (1981) "Responses of amygdaloid neurons in the primate", in: *The Amygdaloid Complex*, Y. Ben-Ari, ed., Amsterdam: Elsevier, pp 383-393.

Rolls, E.T. (1984) "Neurons in the cortex of the temporal lobe and in the amygdala of the monkey with responses selective for faces", *Human Neurobiol.* 3:209-222.

Rolls, E.T. (1986a) "A theory of emotion, and its application to understanding the neural basisof emotion", in: *Emotions. Neural and Chemical Control*, Y. Oomura, ed., Tokyo and Karker, Basel: Japan Scientific Societies Press, pp. 325-344.

Rolls, E.T. (1986b) "Neural systems involved in emotion in primates", in: *Emotion: Theory, Research, and Experience*, R. Plutchik and H. Kellerman, eds, Volume 3, Biological Foundations of Emotion. New York: Academic Press, ch 5, pp. 125-143.

Rolls, E.T. (1989a) "Functions of neuronal networks in the hippocampus and neocortex in memory", in: *Neural Models of Plasticity: Experimental and Theoretical Approaches*, J.H. Byrne and W.O. Berry, eds, San Diego: Academic Press, ch 13, pp. 240-265.

Rolls, E.T. (1989b) "The representation and storage of information in neuronal networks in the primate cerebral cortex and hippocampus", in: *The Computing Neuron*, R. Durbin, C Miall and G. Mitchison, eds, Wokingham, England: Addison-Wesley, ch 8, pp. 125-159.

Rolls, E.T. (1989c) "Functions of neuronal networks in the hippocampus and cerebral cortex in memory", in: *Models of Brain Function*, R.M.J. Coterill, ed., Cambridge: Cambridge University Press, pp. 15-33.

Rolls, E.T. (1990) "A theory of emotion, and its application to understanding the neural basis of emotion", *Cog. and Emot.* 4:161-190.

Rolls, E.T. (1991) "Neural organisation of higher visual functions", *Curr. Op. Neurobiol.* 1:274-278.

Rolls, E.T. (1992a) "Neurophysiology and functions of the primate amygdala", in: *The Amygdala*, J.P. Aggleton , ed., New York: Wiley-Liss, ch 5, pp. 143-165.

Rolls, E.T. (1992b) "Neurophysiological mechanisms underlying face processing within and beyond the temporal cortical visual areas", *Phil. Trans. Roy. Soc.* 335:11-21.

Rolls, E.T. (1992c) "The processing of face information in the primate temporal lobe", in: *Processing Images of Faces*, V. Bruce and M. Burton, eds, Norwood,

New Jersey: Ablex, ch 3, pp. 41-68.

Rolls, E.T. (1994) "Brain mechanisms for invariant visual recognition and learning", *Behav. Proc.* 33:113-138.

Rolls, E.T. (1995) "A theory of emotion and consciousness, and its application to understanding the neural basis of emotion", in: *The Cognitive Neurosciences*, M.S. Gazzaniga, ed., Cambridge, Mass: MIT Press, ch 72, pp. 1091-1106.

Rolls, E.T. (1999) *The Brain and Emotion*, Oxford University Press: Oxford.

Rolls, E.T., G.C. Baylis and C.M. Leonard (1985) "Role of low and high spatial frequencies in the face-selective responses of neurons in the cortex in the superior temporal sulcus", *Vis. Res.* 25:1021-1035.

Rolls, E.T. and G.C. Baylis (1986) "Size and contrast have only small effects on the responses to faces of neurons in the cortex of the superior temporal sulcus of the monkey", *Exp. Brain Res.* 65:38-48.

Rolls, E.T., G.C. Baylis, M.E. Hasselmo (1987) "The responses of neurons in the cortex in the superior temporal sulcus of the monkey to band-pass spatial frequency filtered faces", *Vis. Res.* 27:311-326.

Rolls, E.T., G.C. Baylis, M.E. Hasselmo and V. Nalwa (1989) "The effect of learning on the face-selective responses of neurons in the cortex in the superior temporal sulcus of the monkey", *Exp. Brain Res.* 76:153-164.

Rolls, E.T. and A. Treves (1990) "The relative advantages of sparse versus distributed encoding for associative neuronal networks in the brain", *Network* 1:407-421.

Rolls, E.T. and M.J. Tovee (1994) "Processing speed in the cerebral cortex, and the neurophysiology of visual backward masking", *Proc. Roy. Soc.* B 257:9-15.

Rolls, E.T., M.J. Tovee, D.G. Purcell, A.L. Stewart and P. Azzopardi (1994a), "The responses of neurons in the temporal cortex of primates, and face identification and detection", *Exp. Brain Res.* 101:474-484.

Rolls, E.T., J. Hornak, D. Wade and J. McGrath (1994b) "Emotion-related learning in patients with social and emotional changes associated with frontal lobe damage", *J. Neurol., Neurosurg. Psychiat.* 57:1518-1524.

Rolls, E.T. and M.J. Tovee (1995a) "Sparseness of the neuronal representation of stimuli in the primate temporal visual cortex", *J. Neurophys.* 73:713-726.

Rolls, E.T. and M.J. Tovee (1995b) "The responses of single neurons in the temporal visual cortical areas of the macaque when more than one stimulus is present in the visual field", *Exp. Brain Res.* 103:409-420.

Rolls, E.T., H.D. Critchley and A. Treves (1996) "The representation of olfactory information in the primate orbitofrontal cortex", *Journal of Neurophysiology* 75:1982-1996.

Rolls, E.T., A. Treves and M.J. Tovee (1997) "The representational capacity of the distributed encoding of information provided by populations of neurons in the primate temporal visual cortex", *Experimental Brain Research* 114:149-162.

Rolls, E.T., A. Treves, M. Tovee and S. Panzeri (1997) "Information in the neuronal representation of individual stimuli in the primate temporal visual cortex", *Journal of Computational Neuroscience* 4:309-333.

Rolls, E.T. and A. Treves (1998) *Neural Networks and Brain Function*. Oxford University Press: Oxford.

Rolls, E.T., A. Treves, R.G. Robertson, P. Georges-François and S. Panzeri (1998) "Information about spatial view in an ensemble of primate hippocampal cells", *Journal of Neurophysiology* 79:1797-1813.

Rolls, E.T., M.J. Tovee and S. Panzeri (1999a) "The neurophysiology of backward visual masking: information analysis", *Journal of Cognitive Neuroscience* 11:335-346.

Rolls, E.T., M.C.A. Booth, S. Panzeri, S.R. Schultz and A. Treves (1999b) "Information about objects and faces in the responses of simultaneously recorded inferior temporal cortex neurons", *Society for Neuroscience Abstracts* 25.

Rolls, E.T., H.D. Critchley and A.S. Browning (2000) "Face-selective neurons in the primate orbitofrontal cortex",. submitted

Sato,T. (1989) "Interactions of visual stimuli in the receptive fields of inferior temporal neurons in macaque", *Exp. Brain Research* 77:23-30.

Seltzer, B. and D.N. Pandya (1978) "Afferent cortical connections and architectonics of the superior temporal sulcus and surrounding cortex in the rhesus monkey", *Brain Res.* 149:1-24.

Smith, D.V. and J.B. Travers (1979) "A metric for the breadth of tuning of gustatory neurons", *Chem. Sens.* 4:215-229.

Tanaka, K., C. Saito, Y. Fukada and M. Moriya (1990) "Integration of form, texture, and color information in the inferotemporal cortex of the macaque", in: *Vision, Memory and the Temporal Lobe*, E. Iwai and M. Mishkin, eds, New York, Elsevier, ch 10, pp. 101-109.

Tanaka, K. (1993) "Neuronal mechanisms of object recognition", *Science* 262:685-688.

Tanaka, K. (1996) "Inferotemporal cortex and object vision", *Annual Review of Neuroscience* 19:109-139.

Thorpe, S.J. and M. Imbert (1989) "Biological constraints on connectionist models", in: *Connectionism in Perspective*, R. Pfeifer, Z. Schreter and F. Fogelman-Soulie, eds, Amsterdam: Elsevier, pp. 63-92.

Tovee, M.J., E.T. Rolls, A. Treves and R.P. Bellis (1993) "Information encoding and the responses of single neurons in the primate temporal visual cortex", *J. Neurophysiol.* 70:640-654.

Tovee, M.J., E.T. Rolls and P. Azzopardi (1994) "Translation invariance and the responses of neurons in the temporal visual cortical areas of primates", *J. Neurophysiol.* 72:1049-1060.

Tovee, M.J. and E.T. Rolls (1995) "Information encoding in short firing rate epochs by single neurons in the primate temporal visual cortex", *Visual Cognition* 2:35-58.

Tovee, M.J., E.T. Rolls and V.S. Ramachandran (1996) "Rapid visual learning in neurones of the primate temporal visual cortex", *Neuroreport* 7:2757-2760.

Treves, A. (1993) "Mean-field analysis of neuronal spike dynamics", *Network* 4:259-284.

Treves, A. and E.T. Rolls (1991) "What determines the capacity of autoassociative memories in the brain?", *Network* 2:371-397.

Treves, A. and Rolls, E.T. (1994) "A computational analysis of the role of the hippocampus in memory", *Hippocampus* 4:374-391.

Treves, A., E.T. Rolls and M.W.A. Simmen (1995) "Rapid retrieval in an autoassociative network of spiking neurons", in: *Proceedings of the Computation and Neural Systems Meeting,* Monterey, CA, July 11-16, 1995.

Treves, A., E.T. Rolls and M.J. Tovee (1996) "On the time required for recurrent processing in the brain", in: *Neurobiology,* V. Torre and F. Conti, eds, Plenum: New York.

Treves, A., E.T. Rolls and M. Simmen (1997) "Time for retrieval in recurrent associative memories", *Physica* D 107:392-400.

Treves, A., S. Panzeri, E.T. Rolls, M. Booth and E.A. Wakeman (1999) "Firing rate distributions and efficiency of information transmission of inferior temporal cortex neurons to natural visual stimuli", *Neural Computation* 11:611-641.

Turvey, M.T. (1973) "On the peripheral and central processes in vision: inferences from an information processing analysis of masking with patterned stimuli", *Psych. Rev.* 80:1-52.

Williams, G.V., E.T. Rolls, C.M. Leonard and C. Stern (1993) "Neuronal responses in the ventral striatum of the behaving macaque", *Behav. Brain Res.* 55:243-252.

Wilson, F.A.W., S.P. O'Sclaidhe and P.S. Goldman-Rakic (1993) "Dissociation of object and spatial processing domains in primate preforontal cortex", *Science* 260:1955-1958.

Xiang, J.Z. and M.W. Brown (1998) "Differential neuronal encoding of novelty, familiarity and recency in regions of the anterior temporal lobe", *Neuropharmacology* 37:657-76.

FUNCTIONS OF THE PRIMATE TEMPORAL LOBE CORTICAL VISUAL AREAS IN INVARIANT VISUAL OBJECT AND FACE RECOGNITION: COMPUTATIONAL MECHANISMS

EDMUND T. ROLLS

Department of Experimental Psychology, University of Oxford, South Parks Road Oxford, OX1 3UD, England

ABSTRACT

A theory is described of how invariant visual representations may be produced in a hierarchically organized set of visual cortical areas with convergent connectivity. The theory proposes that neurons in these visual areas use a modified Hebb synaptic modification rule with a short term memory trace to capture whatever can be captured at each stage that is invariant about objects as the objects change in retinal view, position, size, and rotation.

1. Introduction and Hypotheses

The neurophysiological findings described in the preceding paper, and wider considerations on the possible computational properties of the cerebral cortex (Rolls, 1989a,b, 1992b, 1994; Rolls and Treves, 1998), lead to the following outline working hypotheses on object and face recognition by visual cortical mechanisms.

Cortical visual processing for object recognition is considered to be organized as a set of hierarchically connected cortical regions consisting at least of V1, V2, V4, posterior inferior temporal cortex (TEO), inferior temporal cortex (*e.g.* TE3, TEa and TEm), and anterior temporal cortical areas (*e.g.* TE2 and TE1). (This stream of processing has many connections with a set of cortical areas in the anterior part of the superior temporal sulcus, including area TPO.) There is convergence from each small part of a region to the succeeding region (or layer in the hierarchy) in such a way that the receptive field sizes of neurons (*e.g.* 1 degree near the fovea in V1) become larger by a factor of approximately 2.5 with each succeeding stage (and the typical parafoveal receptive field sizes found would not be inconsistent with the calculated approximations of *e.g.* 8 degrees in V4, 20 degrees in TEO, and 50 degrees in inferior temporal cortex, Boussaoud *et al.*, 1991) (see Fig. 1). Such zones of convergence would overlap continuously with each other (see Fig. 1). This connectivity would be part of the architecture by which translation invariant representations are computed. Each layer is considered to act partly as a set of local self-organising competitive neuronal networks with overlapping inputs. (The region within which competition would be implemented would depend on the spatial properties of inhibitory interneurons, and might operate over distances of 1-2 mm in the cortex).

These competitive nets operate by a single set of forward inputs leading to (typically non-linear, *e.g.* sigmoid) activation of output neurons; of competition between the output neurons mediated by a set of feedback inhibitory interneurons which receive from many of the principal (in the cortex, pyramidal) cells in the net and project back (via inhibitory interneurons) to many of the principal cells which serves to decrease the firing rates of the less active neurons relative to the rates of the more active neurons; and then of synaptic modification by a modified Hebb rule, such that synapses to strongly activated output neurons from active input axons strengthen, and from inactive input axons weaken (see Rolls and Treves, 1998).

(A biologically plausible form of this learning rule that operates well in such networks is

$$\delta w_{ij} = k \,.\, r_i \,.\, (r'_j - w_{ij})$$

where k is a learning rate constant, r'_j is the jth input to the neuron, r_i is the output of the ith neuron, and w_{ij} is the jth weight on the ith neuron; see Rolls and Treves, 1998).

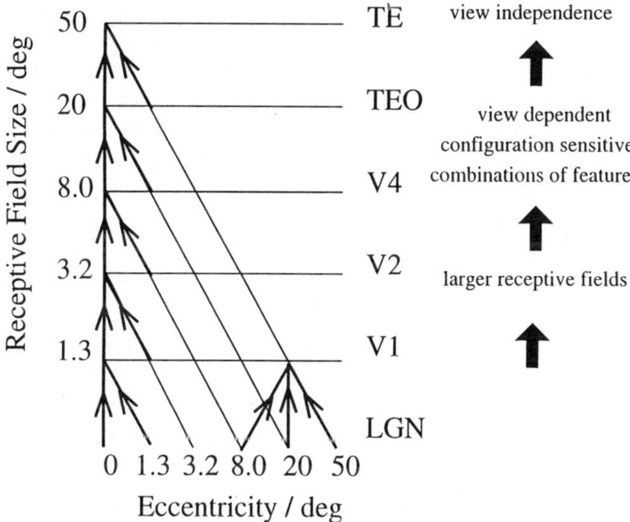

Figure 1. Schematic diagram showing convergence achieved by the forward projections in the visual system, and the types of representation that may be built by competitive networks operating at each stage of the system from the primary visual cortex (V1) to the inferior temporal visual cortex (area TE) (see text). LGN - lateral geniculate nucleus. Area TEO forms the posterior inferior temporal cortex. The receptive fields in the inferior temporal visual cortex (*e.g.* in the TE areas) cross the vertical midline (not shown).

Such competitive networks operate to detect correlations between the activity of the input neurons, and to allocate output neurons to respond to each cluster of such correlated inputs. These networks thus act as categorisers. In relation to visual information processing, they would remove redundancy from the input representation, and would develop low entropy representations of the information (cf. Barlow, 1985; Barlow *et al.*, 1989). Such competitive nets are biologically plausible, in that they utilise Hebb-modifiable forward excitatory connections, with competitive inhibition mediated by cortical inhibitory neurons. The competitive scheme I suggest would not result in the formation of "winner-take-all" or "grandmother" cells, but would instead result in a small ensemble of active neurons representing each input (Rolls, 1989a-c; Rolls and Treves, 1998). The scheme has the advantages that the output neurons learn better to distribute themselves between the input patterns (cf. Bennett, 1990), and that the sparse representations formed have utility in maximising the number of memories that can be stored when, towards the end of the visual system, the visual representation of objects is interfaced to associative memory (Rolls, 1989a,b; Rolls and Treves, 1990, 1998). In that each neuron has graded responses centred about an optimal input, the proposal has some of the advantages with respect to hypersurface reconstruction described by Poggio and Girosi (1990b). However, the system I propose learns differently, in that instead of using perhaps non biologically-plausible algorithms to optimally locate the centres of the receptive fields of the neurons, the neurons use graded competition to spread themselves throughout the input space, depending on the statistics of the inputs received, and perhaps with some guidance from backprojections (see below). The finite width of the response region of each neuron which tapers from a maximum at the centre is important for enabling the system to generalise smoothly from the examples with which it has learned (cf Poggio and Girosi, 1990a,b), to help the system to respond for example with the correct invariances as described below.

Translation invariance would be computed in such a system by utilising competitive learning to detect regularities in inputs when real objects are translated in the physical world. The hypothesis is that because objects have continuous properties in space and time in the world, an object at one place on the retina might activate feature analyzers at the next stage of cortical processing, and when the object was translated to a nearby position, because this would occur in a short period (*e.g.* 0.5 s), the membrane of the postsynaptic neuron would still be in its "Hebb-modifiable" state (caused for example by calcium entry as a result of the voltage dependent activation of NMDA receptors), and the presynaptic afferents activated with the object in its new position would thus become strengthened on the still-activated postsynaptic neuron. It is suggested that the short temporal window (*e.g.* 0.5s) of Hebb-modifiability helps neurons to learn the statistics of objects moving in the physical world, and at the same time to form different representations of different feature combinations or objects, as these are physically discontinuous and present less regular correlations to the visual system. Foldiak (1991) has proposed computing an average activation of the postsynaptic neuron to assist with the same problem. One idea here is that the temporal properties of the biologically implemented learning mechanism are such that it is well suited to detecting the relevant continuities in the world of real objects. Another suggestion is that a memory trace for what has been seen in the last 300 ms appears to be implemented by a mechanism as simple as continued firing of inferior temporal neurons after the stimulus has disappeared, as was found in the masking experiments described above (see also Rolls and Tovee, 1994; Rolls, Tovee *et al.*, 1994a). I also suggest that other invariances, for example size, spatial frequency, and rotation invariance, could be learned by a comparable process. (Early processing in V1 which enables different neurons to represent inputs at different spatial scales would allow combinations of the outputs of such neurons to be formed at later stages. Scale invariance would then result from detecting at a later stage which neurons are almost conjunctively active as the size of an object alters.) It is suggested that this process takes place at each stage of the multiple-layer cortical processing hierarchy, so that invariances are learned first over small regions of space, and then over successively larger regions. This limits the size of the connection space within which correlations must be sought.

Increasing complexity of representations could also be built in such a multiple layer hierarchy by similar mechanisms. At each stage or layer the self-organizing competitive nets would result in combinations of inputs becoming the effective stimuli for neurons. In order to avoid the combinatorial explosion, it is proposed, following Feldman (1985), that low-order combinations of inputs would be what is learned by each neuron. (Each input would not be represented by activity in a single input axon, but instead by activity in a set of active input axons.) Evidence consistent with this suggestion that neurons are responding to combinations of a few variables represented at the preceding stage of cortical processing is that some neurons in V2 and V4 respond to end-stopped lines, to tongues flanked by inhibitory subregions, or to combinations of colours (see references cited by Rolls, 1991); in posterior inferior temporal cortex to stimuli which may require two or more simple features to be present (Tanaka *et al.*, 1990); and in the temporal cortical face processing areas to images that require the presence of several features in a face (such as eyes, hair, and mouth) in order to respond (see above and Yamane *et al.*, 1988). (Precursor cells to face-responsive neurons might, it is suggested, respond to combinations of the outputs of the neurons in V1 that are activated by faces, and might be found in areas such as V4.) It is an important part of this suggestion that some local spatial information would be inherent in the features which were being combined. For example, cells might not respond to the combination of an edge and a small circle unless they were in the correct spatial relation to each other. (This is in fact consistent with the data of Tanaka *et al.*, 1990, and with our data on face neurons, in that some faces neurons require the face features to be in the correct spatial configuration, and not jumbled, Rolls *et al.*, 1994a.) The local spatial information in the features being combined would ensure that the representation at the next level would contain some information about the (local) arrangement of features. Further low-order combinations of such neurons at the next stage would include sufficient local spatial information so that an arbitrary spatial arrangement of the same features would not activate the same neuron, and this is the proposed, and limited, solution which this mechanism would provide for the feature binding problem (Elliffe *et al.*, 2000b; cf.

von der Malsburg, 1990). By this stage of processing a view-dependent representation of objects suitable for view-dependent processes such as behavioural responses to face expression and gesture would be available.

It is suggested that view-independent representations could be formed by the same type of computation, operating to combine a limited set of views of objects. The plausibility of providing view-independent recognition of objects by combining a set of different views of objects has been proposed by a number of investigators (Koenderink and Van Doorn, 1979; Poggio and Edelman, 1990; Logothetis *et al.*, 1994; Ullman, 1996). Consistent with the suggestion that the view-independent representations are formed by combining view-dependent representations in the primate visual system, is the fact that in the temporal cortical areas, neurons with view-independent representations of faces are present in the same cortical areas as neurons with view-dependent representations (from which the view-independent neurons could receive inputs) (Hasselmo *et al.*, 1989b; Perrett *et al.*, 1987; Booth and Rolls, 1998). This solution to "object-based" representations is very different from that traditionally proposed for artificial vision systems, in which the coordinates in 3D-space of objects are stored in a database, and general-purpose algorithms operate on these to perform transforms such as translation, rotation, and scale change in 3D space (*e.g.* Marr, 1982). In the present, much more limited but more biologically plausible scheme, the representation would be suitable for recognition of an object, and for linking associative memories to objects, but would be less good for making actions in 3D-space to particular parts of, or inside, objects, as the 3D coordinates of each part of the object would not be explicitly available. It is therefore proposed that visual fixation is used to locate in foveal vision part of an object to which movements must be made, and that local disparity and other measurements of depth then provide sufficient information for the motor system to make actions relative to the small part of space in which a local, view-dependent, representation of depth would be provided (cf. Ballard, 1990).

The computational processes proposed above operate by an unsupervised learning mechanism, which utilises statistical regularities in the physical environment to enable representations to be built. In some cases it may be advantageous to utilise some form of mild teaching input to the visual system, to enable it to learn for example that rather similar visual inputs have very different consequences in the world, so that different representations of them should be built. In other cases, it might be helpful to bring representations together, if they have identical consequences, in order to use storage capacity efficiently. It is proposed elsewhere (Rolls, 1989a,b; Rolls and Treves, 1998) that the backprojections from each adjacent cortical region in the hierarchy (and from the amygdala and hippocampus to higher regions of the visual system) play such a role by providing guidance to the competitive networks suggested above to be important in each cortical area. This guidance, and also the capability for recall, are it is suggested implemented by Hebb-modifiable connections from the backprojecting neurons to the principal (pyramidal) neurons of the competitive networks in the preceding stages (Rolls, 1989a,b; Rolls and Treves, 1998).

The computational processes outlined above use sparse coding with relatively finely tuned neurons with a graded response region centred about an optimal response achieved when the input stimulus matches the synaptic weight vector on a neuron. The coarse coding and fine tuning would help to limit the combinatorial explosion, to keep the number of neurons within the biological range. The graded response region would be crucial in enabling the system to generalise correctly to solve for example the invariances. However, such a system would need many neurons, each with considerable learning capacity, to solve visual perception in this way. This is fully consistent with the large number of neurons in the visual system, and with the large number of, probably modifiable, synapses on each neuron (*e.g.* 5000). Further, the fact that many neurons are tuned in different ways to faces is consistent with the fact that in such a computational system, many neurons would need to be sensitive (in different ways) to faces, in order to allow recognition of many individual faces when all share a number of common properties.

2. A Computational Model of Invariant Visual Object Recognition

To test and clarify the hypotheses just described about how the visual system may operate to learn invariant object recognition, we have performed a simulation which implements many of the ideas just described, and is consistent and based on much of the neurophysiology summarized above. The network simulated can perform object, including face, recognition in a biologically plausible way, and after training shows for example translation and view invariance (Wallis, Rolls and Foldiak, 1993; Wallis and Rolls, 1997; Rolls and Milward, 1999).

In the four layer network, the successive layers correspond approximately to V2, V4, the posterior temporal cortex, and the anterior temporal cortex. The forward connections to a cell in one layer are derived from a topologically corresponding region of the preceding layer, using a Gaussian distribution of connection probabilities to determine the exact neurons in the preceding layer to which connections are made. This schema is constrained to preclude the repeated connection of any cells. Each cell receives 100 connections from the 32 x 32 cells of the preceding layer, with a 67% probability that a connection comes from within 4 cells of the distribution centre. Fig. 2 shows the general convergent network architecture used, and may be compared with Fig. 1. Within each layer, lateral inhibition between neurons has a radius of effect just greater than the radius of feedforward convergence just defined. The lateral inhibition is simulated via a linear local contrast enhancing filter active on each neuron. (Note that this differs from the global 'winner-take-all' paradigm implemented by Foldiak 1991). The cell activation is then passed through a non-linear cell output activation function, which also produces contrast enhancement of the firing rates.

In order that the results of the simulation might be made particularly relevant to understanding processing in higher cortical visual areas, the inputs to layer 1 come from a separate input layer which provides an approximation to the encoding found in visual area 1 (V1) of the primate visual system. These response characteristics of neurons in the input layer are provided by a series of spatially tuned filters with image contrast sensitivities chosen to accord with the general tuning profiles observed in the simple cells of V1.

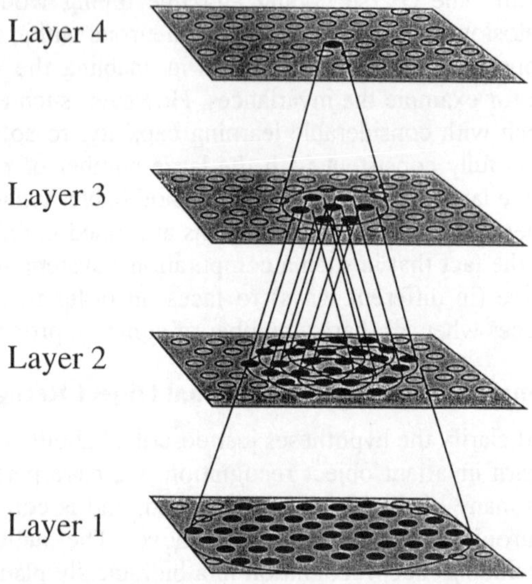

Figure 2. Hierarchical network structure of VisNet.

Currently, only even-symmetric (bar detecting) filter shapes are used. The precise filter shapes were computed by weighting the difference of two Gaussians by a third orthogonal Gaussian (see Wallis and Rolls, 1997). Four filter spatial frequencies (in the range 0.0625 to 0.5 cycles per pixel over four octaves), each with one of four orientations (0° to 135°) were implemented. Cells of layer 1 receive a topologically consistent, localised, random selection of the filter responses in the input layer, under the constraint that each cell samples every filter spatial frequency and receives a constant number of inputs.

The synaptic learning rule used can be summarised as follows:

$$\delta w_{ij} = k \, . \, m_i \, . \, r'_j$$

and

$$m_i^{\mathrm{t}} = (1 - \eta) r_i^{(\mathrm{t})} + \eta m_i^{(\mathrm{t-1})}$$

where r'_j is the jth input to the neuron, r_i is the output of the ith neuron, w_{ij} is the jth weight on the ith neuron, η governs the relative influence of the trace and the new input (typically 0.4 - 0.6), and $m_i^{(\mathrm{t})}$ represents the value of the ith cell's memory trace at time t.

In the simulation the neuronal learning was bounded by normalisation of each cell's dendritic weight vector, as in standard competitive learning (see Rolls and Treves, 1998). An alternative, more biologically relevant implementation, using a local weight bounding operation, has in part been explored using a version of the Oja update rule (Oja 1982; Kohonen 1984). To train the network to produce a translation invariant representation, one stimulus was placed successively in a sequence of 7 positions across the input, then the next stimulus was placed successively in the same sequence of 7 positions across the input, and so on through the set of stimuli. The idea was to enable the network to learn whatever was common at each stage of the network about a stimulus shown in different positions. To train on view invariance, different views of the same object were shown in succession, then different views of the next object were shown in succession, and so on.

One test of the network used a set of three non-orthogonal stimuli, based upon probable 3-D edge cues (such as 'T, L and +' shapes). During training these stimuli were chosen in random sequence to be swept across the 'retina' of the network, a total of 1000 times. In order to assess the characteristics of the cells within the net, a two-way analysis of variance was performed on the set of responses of each cell, with one factor being the stimulus type and the other the position of the stimulus on the 'retina'. A high F ratio for stimulus type (F_s), and low F ratio for stimulus position (F_p) indicates that a cell had learned a position invariant representation of the stimuli. The discrimination factor of a particular cell was thus simply the ratio F_s / F_p (a factor useful for ranking at least the most invariant cells). To assess the utility of the trace learning rule, nets trained with the trace rule were compared with nets trained with standard Hebbian learning without a trace, and with untrained nets (with the initial random weights). The results of the simulations, illustrated in Fig. 3, show that networks trained with the trace learning rule do have neurons with much higher values of the discrimination factor. An example of the responses of one such cell are illustrated in Fig. 4. Similar position invariant encoding has been demonstrated for a stimulus set consisting of 8 faces. View invariant coding has also been demonstrated for a set of 5 faces each shown in 4 views (Wallis, Rolls and Foldiak, 1993; Wallis and Rolls, 1997).

There have been a number of recent investigations to explore this type of learning further. In one investigation, Parga and Rolls (1998) and Elliffe *et al.* (2000a) incorporated the associations between exemplars of the same object in the recurrent synapses of an autoassociative (>attractor=) network, so that the techniques of statistical physics could be used to analyse the storage capacity of a system implementing invariant representations in this way. They showed that such networks did have an >object= phase in which the presentation of any exemplar (*e.g.* view) of an object would result in the same firing state as other exemplars of the same object, and that the number of different objects that could be stored is proportional to the number of synapses per neuron divided by the number of >views= of each object. Rolls and Milward (1999) explored the operation of the trace learning rule used in the VisNet architecture further, and showed that the rule

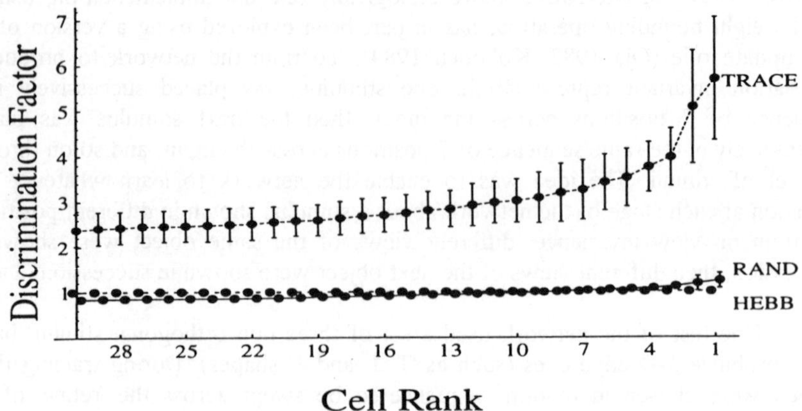

Figure 3. Comparison of network discrimination when trained with the trace learning rule, with a Hebb rule (No trace), and when not trained (Random) on three stimuli, +, T and L, at nine different locations.

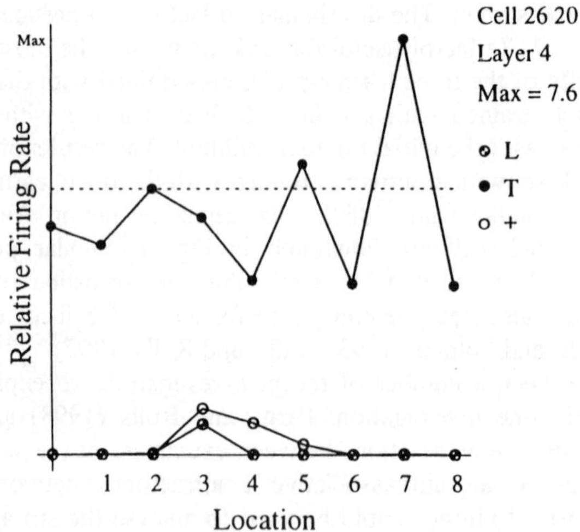

Figure 4. The responses of a layer 4 cell in the simulation shown in Fig. 3. The cell had a translation invariant response to stimulus 1.

operated especially well if the trace incorporated activity from previous presentations of the same object, but no contribution from the current neuronal activity being produced by the current exemplar of the object. The explanation for this is that this temporally asymmetric rule (the presynaptic term from the current exemplar, and the trace from the preceding exemplars) encourages neurons to respond to the current exemplar in the same way as they did to previous exemplars. It is of interest to consider whether intracellular processes related to LTP might implement an approximation of this rule, given that it is somewhat more powerful than the standard trace learning rule described above. Rolls and Stringer (2000) went on to show that part of the power of this type of trace rule can be related to gradient descent and temporal difference (see Sutton and Barto 1998) learning. Elliffe *et al.* (2000b) examined the issue of spatial binding in this general class of hierarchical architecture studied originally by Fukushima (1980, 1989, 1991), and showed how by forming high spatial precision feature combination neurons early in processing, it is possible for later layers to maintain high precision for the relative spatial position of features within an object, yet achieve invariance for the spatial position of the whole object.

These results show that the proposed learning mechanism and neural architecture can produce cells with responses selective for stimulus type with considerable position or view invariance. The ability of the network to be trained with natural scenes may also help to advance our understanding of encoding in the visual system.

Acknowledgements

The author has worked on some of the investigations described here with P. Azzopardi, G.C. Baylis, P. Foldiak, M. Hasselmo, C.M. Leonard, G. Littlewort, T.J.Milward, D.I. Perrett, M.J. Tovee, A. Treves and G. Wallis, and their collaboration is sincerely acknowledged. Different parts of the research described were supported by the Medical Research Council, PG8513790; by a Human Frontier Science Program grant; by an EC Human Capital and Mobility grant; by the MRC Oxford Interdisciplinary Research Centre in Brain and Behaviour; and by the Oxford McDonnell-Pew Centre in Cognitive Neuroscience.

References

Ballard, D.H. (1990) "Animate vision uses object-centred reference frames", in: *Advanced Neural Computers,* R. Eckmiller, ed., Amsterdam: North-Holland, pp. 229-236.

Barlow, H.B. (1985) "Cerebral cortex as model builder" in: *Models of the Visual Cortex*, D. Rose and V.G. Dobson, eds, Chichester: Wiley, pp. 37-46.

Barlow, H.B., T.P. Kaushal, G.J. Mitchison (1989) "Finding minimum entropy codes" *Neural. Computat* 1:412-423.

Bennett, A. (1990) "Large competitive networks", *Network* 1:449-462.

Booth, M.C.A. and E.T. Rolls (1998) "View-invariant representations of familiar objects by neurons in the inferior temporal visual cortex" *Cerebral Cortex* 8:510-523.

Boussaoud, D., R. Desimone and L.G. Ungerleider (1991) "Visual topography of area TEO in the macaque", *J. Comp. Neurol.* 306:554-575.

Elliffe, M.C.M., E.T. Rolls, N. Parga and A. Renart (2000a) "A recurrent model of transformation invariance by association", *Neural Networks* 13:225-237.

Elliffe, M.C.M., E.T. Rolls and S.M. Stringer (2000b) "Invariant recognition of feature combinations in the visual system", submitted.

Feldman, J.A. (1985) "Four frames suffice: a provisional model of vision and space", *Behav. Brain Sci.* 8:265-289 (see p. 279).

Foldiak, P. (1991) "Learning invariance from transformation sequences", *Neural Comp.* 3:193-199.

Fukushima, K. (1980) "Neocognitron: a self-organizing neural network model for a mechanism of pattern recognition unaffected by shift in position" *Biological Cybernetics* 36:193-202.

Fukushima, K. (1989) "Analysis of the process of visual pattern recognition by the neocognitron", *Neural Networks* 2:413-420.

Fukushima, K. (1991) "Neural networks for visual pattern recognition", *IEEE Transactions* E 74:179-190.

Koenderink, J.J. and A.J. Van Doorn (1979) "The internal representation of solid shape with respect to vision", *Biological Cybernetics* 32:211-217.

Kohonen ,T. (1988) *Self-organization and Associative Memory*, New York: Springer-Verlag, 2nd Edition.

Logothetis, N.K., J. Pauls, H.H. Bulthoff and T.Poggio (1994) "View-dependent object recognition by monkeys", *Current Biol.* 4:401-414.

Malsburg, C. von der (1990) "A neural architecture for the representation of scenes" in: *Brain Organization and Memory: Cells, Systems and Circuits*, J.L. McGaugh, N.M. Weinberger and G. Lynch, eds, New York: Oxford University Press, ch 18, pp. 356-372.

Marr, D. (1982) *Vision*, San Francisco: WH Freeman.

Oja, E. (1982) "A simplified neuron model as a principal component analyzer", *J. Math. Biol.* 15:267-73.

Parga, N. and E.T. Rolls (1998) "Transform invariant recognition by association in a recurrent network", *Neural Computation* 10:1507-1525.

Poggio, T. (1990) "A theory of how the brain might work", *Cold Spring Harbor Symposia in Quantitative Biology* 55:899-910.

Poggio, T. and S. Edelman (1990) "A network that learns to recognize three-dimensional objects", *Nature* 343:263-266.

Poggio, T. and F. Girosi (1990a) "Regularization algorithms for learning that are equivalent to multilayer networks", *Science* 247:978-982.

Poggio, T. and F. Girosi (1990b) "Networks for approximation and learning", *Proc. IEEE,* 78:1481-1497.

Rolls, E.T. (1989a) "Functions of neuronal networks in the hippocampus and neocortex in memory", in: *Neural Models of Plasticity: Experimental and Theoretical Approaches*, J.H. Byrne and W.O. Berry, eds, San Diego: Academic Press, ch 13, pp. 240-265.

Rolls, E.T. (1989b) "The representation and storage of information in neuronal networks in the primate cerebral cortex and hippocampus", in: *The Computing Neuron*, R. Durbin, C Miall and G. Mitchison, eds, Wokingham, England: Addison-Wesley, ch 8, pp. 125-159.

Rolls, E.T. (1989c) "Functions of neuronal networks in the hippocampus and cerebral cortex in memory", in: *Models of Brain Function*, R.M.J. Coterill, ed., Cambridge: Cambridge University Press, pp. 15-33.

Rolls, E.T. (1991) "Neural organisation of higher visual functions", *Curr. Op. Neurobiol.* 1:274-278.

Rolls, E.T. (1992b) "Neurophysiological mechanisms underlying face processing within and beyond the temporal cortical visual areas", *Phil. Trans. Roy. Soc.* 335:11-21.

Rolls, E.T. (1994) "Brain mechanisms for invariant visual recognition and learning", *Behav. Proc.* 33:113-138.

Rolls, E.T. and A. Treves (1990) "The relative advantages of sparse versus distributed encoding for associative neuronal networks in the brain", *Network* 1:407-421.

Rolls, E.T. and M.J. Tovee (1994) "Processing speed in the cerebral cortex, and the neurophysiology of visual backward masking", *Proc. Roy. Soc.* B 257:9-15.

Rolls, E.T., M.J. Tovee, D.G. Purcell, A.L. Stewart and P. Azzopardi (1994a), "The responses of neurons in the temporal cortex of primates, and face identification and detection", *Exp. Brain Res.* 101:474-484.

Rolls, E.T. and A Treves (1998) *Neural Networks and Brain Function*, Oxford University Press: Oxford.

Rolls, E.T. and T. Milward (1999) "A model of invariant object recognition in the visual system: learning rules, activation functions, lateral inhibition, and information-based performance measures", *Neural Computation*, in press.

Rolls, E.T. and S.M. Stringer (2000) "Invariant object recognition in the visual system with error correction and temporal difference learning", submitted.

Sutton, R.S. and A.G. Barto (1998) *Reinforcement Learning*, Cambridge, Mass: MIT Press.

Tanaka, K., C. Saito, Y. Fukada and M. Moriya (1990) "Integration of form, texture, and color information in the inferotemporal cortex of the macaque", in: *Vision, Memory and the Temporal Lobe*, E. Iwai and M. Mishkin, eds, New York, Elsevier, ch 10, pp. 101-109.

Tanaka, K. (1993) "Neuronal mechanisms of object recognition", *Science* 262:685-688.

Ullman, S. 1996. *High-Level Vision. Object Recognition and Visual Cognition.*

Bradford/MIT Press: Cambridge, Mass.
Wallis, G., E.T. Rolls and P. Foldiak (1993) "Learning invariant responses to the natural transformations of objects", *Int. Joint Conf. on Neural Net*. 2:1087-1090.
Wallis, G. and E.T. Rolls (1997) "Invariant face and object recognition in the visual system", *Progress in Neurobiology* 51:167-194.
Yamane, S., S. Kaji and K. Kawano (1988) "What facial features activate face neurons in the inferotemporal cortex of the monkey?" *Exp. Brain Res*. 73:209-214.

ORIENTING REFLEX: SELECTIVE HABITUATION

EVGENI N. SOKOLOV
Department of Psychophysiology, Moscow State Lomonosov University
Mokhovaya st. 8, Moscow 103009, Russia

ABSTRACT

Orienting reflex as a part of exploratory behavior is evoked by novel *stimuli*. The repeated presentation of a standard *stimulus* results in orienting reflex gradual decrease (habituation). The process of habituation is selective with respect to parameters of the standard *stimulus*. Such selective habituation is based of formation of an internal representation (a neuronal model) of the standard *stimulus*. The neuronal model is established with participation of hippocampus where novelty-sensitive and familiarity-sensitive neurons are operating. Input excitation vector coming to hippocampal neurons from cortical feature-detectors is selectively blocked by long-term potentiation of plastic synapses constituting a synaptic weight vector. Novelty mismatch signal is widely distributed in neuronal networks being expressed in targeting eye movements, heart rate deceleration, skin galvanic responses, EEG arousal and increase of regional cerebral blood flow.

1. The concept of orienting reflex

Orienting reflex (OR) discovered by I.P. Pavlov (1927) is a set of reactions evoked in humans and animals by a novel *stimulus*. The OR is manifested in an interruption of an ongoing activity (an external inhibition) that parallels turning of eyes, head and body movements toward a novel *stimulus* (targeting response) (Konorski, 1967). Subjectively OR is experienced as an involuntary switching of attention emphasizing a novel *stimulus*.

OR is of an exceptional biological significance as a part of exploratory behavior aimed to extract information from external surrounding. The information can later on be used in different types of specific behaviors such as feeding or avoidance. The exploration is characterized by novelty - seeking drive expressed in attractiveness of complex patterns. In humans exploratory behavior is transformed into scientific investigations.

2. Components of OR

The behavioral manifestations of OR represent only a top of an iceberg of reactions hidden from direct observation but revealed by instrumental research. OR is characterized by a constriction of peripheral vessels and dilation of brain vessels.

The regional cerebral blood flow (rCBF) is of particular interest being related to a local activation of neurons expressed in generation of action potentials (APs). APs are linked with passive inflow of sodium and outflow of potassium ions. To keep a balance of ionic content in neurons an active sodium-potassium pump is operating. It requires an additional energy supplied by oxygen and glucose during intensification of rCBF. The increase of rCBF as an index of OR-related neuronal activation is used in positron emission tomography (PET) and functional magnetic resonance imaging (fMRI) (Posner and Raichle, 1994).

Neuronal networks involved into OR evocation are identified also by electroencephalography (EEG) and magnetoencephalography (MEG). Multichannel EEG and MEG recordings are used to calculate equivalent dipoles localization representing *loci* of neuronal activation. OR is evident as event-related desynchronization (ERD) of a background brain waves in the range of 8-12 Hz (alph*a* waves) that parallels an increase of beta (12-25 Hz) and gamma (25-200 Hz) oscillations (Naatanen, 1992). The ERD is accompanied by a negative shift of a cortical steady potential that is related to an increase of neuronal excitability (Caspers, 1963). The other OR index is nonspecific event-related potentials (ERPs) evoked by novel *stimuli* (Brunia, 1997).

PET and fMRI being characterized by high spatial resolution are suffering by low resolution in time domain. EEG and MEG on the contrary are highly efficient in the time discrimination but have a low spatial resolution. A combination of EEG and MEG dipole analysis with PET and fMRI supplemented by anatomical MRI opens a unique perspective for OR study.

A direct observation of OR related neurons is done by single unit microelectrode recording. The main strategy to study neuronal mechanisms of OR is to look for novelty-sensitive neurons demonstrating novelty responses similar to novelty responses at behavioral level. A number of different OR components evoked by a novel *stimulus* suggests that novelty signal generated in novelty-sensitive neurons spread over wide brain areas.

The OR is expressed also in autonomic indicators: a heart rate deceleration and an increase of skin conductance (Graham, 1979). Heart rate responses are of special significance for differentiation OR from an opponent behavior - namely defense reflex expressed in heart rate acceleration. Sensory OR components refer to an elevation of sensitivity (lowering of thresholds) and to an increase of fusion frequency (rate of discrete *stimuli* building up a continuous percept) (Lindsley, 1968). The sensory OR components are seen not only at psychophysical but also at neuronal level. The thresholds of sensory neurons become lower and fusion frequency increase under ERD known as arousal reaction.

3. Selective habituation of OR

Repeated presentation of a nonsignal standard *stimulus* results in a gradual reduction of OR components (OR habituation). The OR habituation represents a form of negative learning. Negative learning is beneficial eliminating familiar objects from the focus of attention and directing it to events of an informational value. The other side of the negative learning is familiarization constituting a basis of a latent learning. An example of a familiarization and latent learning in rats is constitution of a spatial map during non-reinforced running in a maze. A change of the presented *stimulus* produces OR again demonstrating contribution of novelty. Standard *stimulus* following a novel one results in OR recovery called dishabituation (Groves and Tompson, 1970). The standard *stimulus* characterized by a particular intensity, shape, color, duration and inter*stimulus* intervals builds up in neuronal networks an internal representation of all these parameters - a *stimulus* neuronal model. Any deviation of a *stimulus* from the created neuronal model produces a mismatch signal that determines OR generation. The *stimulus* neuronal model preserves not only simple features, but also their combinations. Thus after habituation of OR to a complex consisting of a light and sound components elimination of either light or sound evokes OR. OR evoked by elimination of a detail from the complex *stimulus* is an evidence that a novel *stimulus* is matched against the trace, induced by the standard one. Such matching takes place by omission of a regularly presented standard *stimulus*. In this case complex *stimulus* is a combination of a sensory event and particular time. By elimination of the sensory event empty time is matched against the complex trace evoking novelty signal.

The *stimulus* neuronal model can be characterized as a multidimensional self-adjustable reject filter. A selective tuning of the filter can be revealed by repeated presentation of the standard *stimulus* and intermittent *stimuli* deviating with respect to a particular parameter from the standard one. The procedure applied to different *stimulus* parameters enables to construct a multidimensional filter characteristic that underlies *stimulus* selective OR habituation. To get the filter characteristic one has to take into consideration dishabituation after a novel test *stimulus* OR to the following standard *stimulus* is recovered. Thus, before application of a next test *stimulus* it is required to repeat standard *stimulus* presentations until dishabituation will completely vanish.

The response to the test *stimuli* increases with their deviation from the standard one so that filter characteristic has a V-like shape. The repeated presentation of a standard *stimulus* reveals generalized and localized ORs. Generalized OR is evident by involvement of a wide variety of components that rapidly habituate. Localized OR is specifically related to the applied *stimulus* modality. Generalized and localized ORs can be demonstrated by event- related desynchronization (ERD) of brain waves. Different brain areas are represented by specific alpha, mu, sigma and tau generators of 8-12 Hz oscillations. Visual area is represented by alpha-rhythm, motor area by mu-rhythm, somatosensory area by sigma-rhythm and auditory area by tau- rhythm. A novel visual *stimulus* evokes a generalized OR composed of generalized ERD, skin galvanic response (SGR) and heart rate deceleration. After 15-20 presentations of the visual *stimulus* only occipital alpha-desynchronization is seen being habituated later on in following trials. Passive hand movement produces initially a generalized ERD that under repetitions is transformed into a local mu-rhythm desynchronization representing a localized OR (Sokolov, 1963).

Thus, in the process of habituation of OR there is a concentration of arousal in the brain areas specific to input *stimulus* and to output reaction. It means that OR is evoked not only by novel *stimuli*, but also to a novel motor reaction. Habituation of OR is occurring in the process of familiarization with sensory events and motor performance. One might say that the process of automatization is accompanied by reduction of OR participation. The degree of automatization of performance parallels degree of OR habituation (Sokolov, 1963).

4. OR and conditioned reflex

The problem of OR and CR integration as a controversial one. On one hand a novel *stimulus* acts as a distracter interrupting CR. I.P. Pavlov studying CRs was concerned with that OR intervention. To minimize interference of OR into CR processing a sound proof "tower of silence" was constructed in St.Petersburg. On the other hand to start CR elaboration the future CS has to evoke OR. Even more when OR to a signal *stimulus* is habituated elaboration of CR meets difficulties.
A combination of neuronal (nonsignal) *stimulus* with an unconditional *stimulus* (US) results in an intensification of OR. The *stimulus* signaling an US is becoming significant and produces more stable OR. In the process of stabilization of the conditioned reflex (CR) the OR gradually habituates. The first elimination of the reinforcement results in on a place of its omission marked appearance of OR. During initial stage of extinction of CR activation of OR is evident by presentation of CSs even no CRs are present. The OR activation during CR extinction can be demonstrated using correlation between skin conductance response (SCR) and fMRI brain activation. Once learning has occurred it may be important to activate an expectancy loop whenever the CS is presented. Such an attention-expectancy circuit involves the frontal lobe. A presentation of a tone results in a gradual decrease of skin conductance. An extinction of vessels after association a tone with electric shocks increases skin conductance. PET compared under habituation and extinction conditions shows an activation of the inferior frontal, orbito-frontal and dorso-frontal prefrontal cortex during extinction of CR. These areas comprise the "anterior attention system" (expectancy circuit). Extinction results in the directing of attention to the CS (Hugdahl, 1998).

An additional *stimulus* that has to be differentiated from the initial CS results in a recovery of OR. The OR is the greater the more difficult is the differentiation task. At initial stage of differential conditioning CS and all differential *stimuli* evoke OR. Such a generalization of OR with respect to the *stimuli* set later on is transformed in more selective OR evocation. OR are elicited only by CS and differential *stimuli* close to it. Differential *stimuli* more distinct from CS are easily discriminated without OR elicitation. Thus such OR intensification is dependent on CR (OR evoked by signal *stimuli*). *Stimuli* very different from CS act as distracters evoking OR due to their novelty. If CSs are presented with regular intervals a time-selective

CR is elaborated that precedes actual presentation of the CS. An anticipatory OR is expressed in saccadic eye movements, ERD and gradual negative shift of study potential (expectancy wave). The anticipatory OR preceding CS reduces CR latency. Irregular intervals between warning and imperative *stimuli* when time conditioning is lacking and anticipatory OR is not operating reaction time significantly increases. The intensification of OR to signal *stimuli* as compared with nonsignal ones suggests a specific neuronal circuit. Thus nonsignal (insignificant) *stimuli* produce rapidly habituatable OR, while signal (significant) *stimuli* trigger more persistent OR.

5. OR and vector encoding

A *stimulus* acting on an ensemble of receptors generates a combination of their excitations - a receptor excitation vector. The receptors overlapping characteristics constitute a non-orthogonal basis. At the next stage of information processing neuronal ensemble performs orthogonalization and normalization so that input *stimuli* are represented in Euclidean space by excitation vectors of a constant length.

The orthogonalization is achieved by a set of excitatory and inhibitory synaptic contacts between receptors and neurons constituting an ensemble of the next neuronal layer. The normalization at the upper stage is due to the addition of a neuron spontaneously active when no stimulation is given. The signals from the responsive neurons are subtracted from the spontaneously active one so that total excitation of the neuronal ensemble remains constant.

An excitation vector evoked by an input *stimulus* is acting in parallel on a set of feature-detectors. Each feature detector is selectively tuned to a particular excitation vector and in this way maximally excited by a respective input *stimulus*. The selective tuning of feature-detectors is assisted by a set of synaptic contacts that are specific for each detector. The set of synaptic contacts of various weights specifying a particular detector can be regarded as a link vector. The link vectors of feature-detectors are of constant length. The response of a feature-detector is equal to an inner (scalar) product of input excitation vector and synaptic link vector of a respective detector. Thus input *stimuli* encoded by excitation vectors are topically represented on a detector map.

The detector map is a hypersphere in a multidimensional space. The output responses are also encoded by means of excitation vectors. A variety of specific responses can be generated by a limited number of channels constituting components of an output excitation vector (Bizzi and Mussa-Ivaldy, 1998). The vector code is operating in conditioning, which is related to the concept of a command neuron. A link between input and output excitation vectors is due to a command neuron having a set of plastic synapses modified in the process of learning so that command neuron become selectively tuned to a CS (Fomin *et al.*, 1979; Sokolov and Vaitkyavicus, 1989).

The set of plastic synapses on a command neuron constitutes a plastic synaptic weight vector composed from single plastic Hebbian synapses (Hebb, 1949) and can be termed "Hebbian link vector". In the process of CR elaboration Hebbian link vector becomes collinear with input excitation vector resulting in a selective tuning of the command neuron to CS. An extinction of CR is related to a process opposite to conditioning. Non-reinforced CS changes Hebbian weight vector so that it become finally orthogonal to the excitation vector and command neuron stops responding. The concept of a command neuron can be applied to OR. The OR habituation as a negative learning is also due to synaptic plasticity of specific novelty related command neurons. The process of OR selective habituation can be explained by formation of a Hebbian link vector opposing an excitation vector evoked by a standard *stimulus*.

The presented model of information processing can be illustrated by color vision. Three types of cones constitute a non-orthogonal basis for receptor excitation vectors. Three types of photopic horizontal cells, red- green, blue-yellow and luminance, are relevant for orthogonal basis where intensity is given by the vector length. At the level of bipolar cells the length of excitation vectors becomes constant. This is achieved by a transition from three to a four-dimensional space with addition of a neuron active under darkness. Such color excitation vector of a constant length is present also in the lateral geniculate body acting in a parallel manner on cortical color detectors. At color specific visual area (V4) different colors are specified by color-selective detectors (Bartels and Zeki, 1998). The color detector map is a hypersphere in the four-dimensional space (Izmailov and Sokolov, 1991). The spherical surface representing colors suggests that different colors are topically separated constituting a colortopic cortical map similar to retinotopic, tonotopic and somatotopic projections. A fMRI used to localize color representations reveals color-selective areas in human occipitotemporal cortex (Beauchamp *et al.*, 1999).

Output vector code is evident in different ballistic movements (Bizzi and Mussa-Ivaldi, 1998). OR vector code is expressed in saccadic eye movements which are selectively triggered by particular combinations of horizontal, vertical and torsion premotor neurons. Vector code refers to autonomic OR components (Sokolov and Cacioppo, 1997).

Vector code in OR can be demonstrated by color-selective habituation of occipital ERD. The magnitude of ERD evoked by presentation of a new color parallels an absolute value of a difference between excitation vectors of the standard and the novel colors. It suggests that in the process of OR habituation the excitation vector evoked by the standard *stimulus* is selectively blocked. The blocking is resulted from establishment of a Hebbian synaptic weight vector opposing to the excitation vector of a standard *stimulus*. The more a novel *stimulus* deviates from the standard one the greater is the novelty response.

6. Neuronal basis of OR

PET and fMRI in combination with EEG and MEG studies in humans have shown that OR circuits differ in passive (involuntary) and active (voluntary) OR. Passive OR is addressed to hippocampus (Tulving, 1994) while active OR involves prefrontal and parietal lobes. Single unit recording in rabbit's hippocampus has demonstrated that neurons under passive OR are separated into two groups. Novelty-sensitive neurons are excited by a novel *stimulus*. Familiarity-sensitive neurons are inhibited by a novel *stimulus*. By repeated *stimulus* presentation responses of excitatory and inhibitory cells are habituated. The habituation selective with respect to a standard *stimulus* is due to a selective decrease of synaptic weights connecting activated feature detectors with novelty-related neurons (Vinogradova, 1976). The mismatch signal evoked by incoincidence of a *stimulus* deviating from the standard one parallels a difference of their excitation vectors. The mismatch signal is addressed to the mesencephalic reticular formation and thalamic reticular nuclei evoking a frequency shift from alpha waves toward gamma oscillations of their pacemaker cells (Moruzzi and Magoun, 1949). Single unit recording from reticular thalamic nucleus in rabbits demonstrates randomization response - a transition from bursting spike activity to a random spiking that parallels ERD of cortical alpha-like oscillations. By repeated *stimulus* presentation novelty neuron spiking, randomization response and ERD selectively habituate in accordance with *stimulus* selective habituation of other OR components (Sokolov, 1975).

The hippocampus plays a distinct role in the orienting reflex and exploratory behavior. The hippocampus is characterized by 1) long-term potentiation of synaptic contacts; 2) hippocampal theta-waves (4-7Hz) during arousal and orienting reaction; 3) novelty and familiarity neurons.

Signals from cortical feature detectors reach the hippocampus via the perforant path contacting apical dendrites of pyramidal cells. Individual fibers make in passing numerous contacts with many pyramidal cells so that excitation spreads over wide area of the hippocampus. A parallel branch the perforant path contact granule cells of the dentate fascia, which sends its mossy fibers to basal dendrites of the pyramidal cells. The potentiation of mossy fibers synapses gradually block direct synaptic contacts so that input signals on pyramidal neurons are selectively diminished. A new *stimulus* acting via synapses not blocked by mossy fibers evoke a novelty responses in pyramidal cells that trigger OR.

A critical role of the dentate fascia in the process of habituation was shown using antibodies against neurons of the dentate fascia. This procedure results in an elimination of novelty responses. The novelty neurons become continuously responsive under habituation procedure (Vinogradova, 1976).

Voluntary OR is supported by prefrontal neurons contributing to working memory. The neurons of working memory are continuously active during period of waiting between warning and imperative *stimuli*. Fixation of eyes on the expected target location is combined with expectancy wave and expectancy related desynchronization of alpha-rhythm. Local damages of frontal lobe produce deficiencies in voluntary OR while passive (involuntary) OR remain intact.

To summarize: OR is a neuronal mechanism emphasizing novel and significant stimuli against familiar and insignificant ones. One might say that OR is a high order contrasting mechanism contributing to long-term memory formation.

7. Conclusion

Single unit recording in animals demonstrated that hippocampal novelty-sensitive neurons have basic characteristics of OR. They selectively habituate to different parameters of the standard stimulus, respond to any change of stimulation and show dishabituation. Parallel to novelty-sensitive neurons in hippocampus familiarity-sensitive neurons are operating.

Recent fMRI data show that human hippocampus also is processing both novelty and familiarity. A left anterior hippocampus responds to perceptual and semantic novelty. By contrast bilateral posterior hippocampus is activated with increasing semantic familiarity. Repeated presentation of semantic stimuli results in topographical spread of hippocampal activity in an anterior-posterior direction. The specific novelty encoding path plays a crucial role in formation of long-term memory within the posterior medial temporal lobe (Strange *et al.*, 1999). PET, fMRI and ERPs show that under active attention task additionally to frontal and parietal areas the anterior cingulate gyrus on the frontal midline is activated demonstrating that active OR to signal stimuli has a specific circuitry different from passive OR (Mountcastle, 1979; Posner, 1995; Posner and Abdullaev, 1999).

Novelty signal is widely distributed among different brain structures including activating reticular formation, autonomic centers and superior colliculus facilitating eyes and body movements as OR components. At the same time novelty signal acts as a distracter of ongoing behavior.

References

Caspers, H. (1963) "Relations of steady potential shift in the cortex to wakefulness-sleep spectrum", in: *Brain function: Cortical excitability and steady potentials*, Vol. 1 M.A.B. Brazier, ed., Berkeley: University of California press, pp. 117-123.

Creutzfeldt, O.D. (1993) "Cortex cerebri. Performance, structural and functional organization of the cortex", Goettingen: M. Creutzfeldt.

Bartles, A. and S. Zeki (1998) "The theory of multistage integration in the visual brain", *Proc. R. Soc. Lond. B. Biol. Sci.* 265:2327-2332.

Beauchamp, M.S., J.V. Haxby, J.E. Jennings and E.A. Deyoe (1999) "An fMRT version of Farnsword-Munsell 100-Hue test reveals multiple color-selective areas in human ventral occipitotemporal cortex", *Cerebral cortex* 257-263.

Bizzi, E. and F.A. Mussa-Ivaldi (1998) "The acquisition of motor behavior", *Daedalus* 127:217-232.

Brunia, C.H.M. (1997) "Gating in Readiness", in: *Attention and Orienting*, P.J. Lang, R.F. Simons and P.M. Balaban, eds, Mahawah, New Jersey, London: Lawrence Erlbaum Associates, pp. 281-306.

Fomin, S.V., E.N. Sokolov and G.G. Vaitkyavicus (1979) "*Artificial Sense Organs. Modeling of Sensory Systems*", in Russian, Moscow: Nauka.

Graham, F.K. (1979) "Distinguishing among orienting, defensive and startle reflexes", in: *The Orienting Reflex in Humans*, A.D. Kimmel, E.H. van Olst and J.F. Orlebeke, eds, Hillsdale, NY: Lawrence Erlbaum Associates, pp. 137-167.

Groves, P.M. and R.F. Tompson (1970) "Habituation: A dual process theory", *Psychological Rev.* 77:419-450.

Hebb, D.O. (1949) "*The Organization of Behavior: A Neuropsychological Theory*", New York: Wiley.

Hugdahl, K. (1998)"Cortical control of human classical conditioning: autonomic and Positron Emission Tomography data", *Psychophysiol.* 35:170-178.

Ishii, R.K., S. Shinosaki, T. Ukai, T. Inouye, T. Ishihara, N. Yoshimine, N. Hirabuki, H. Asada, T. Kihara, S.E. Robinson and M. Takeda (1999) "Medial prefrontal cortex generates frontal midline theta rhythm", *NeuroReport* 10:675-679.

Izmailov, Ch.A. and E.N. Sokolov (1991) "Spherical model of color and brightness discrimination", *Psychol. Sci.* 2:249-259.

Konorski, J. (1967) "*Integrative Activity of the Brain: An Interdisciplinary Approach*", Chicago: Chicago University press.

Lindsley, D.B. (1961) "The reticular activating system and perceptual integration", in: *Electrical Stimulation of the Brain*, D.E. Sheer, ed., Austin, TX: University of Texas press, pp. 331-349.

Moruzzi, G. and H.W. Magoun (1949) "Brain stem reticular formation and activation of the EEG", *Electrophys. and Clin. Neurophysiol.* 1:455-473.

Mountcastle, V.B. (1978) "Brain system for directed attention", *J. Royal Soc. Med.* 71:14-227.

Naatanen, R. (1992) "*Attention and Brain Functions*", Hillsdale, NJ: Lawrence Erlbaum associates.

Pavlov, I.P. (1927) "*Conditioned Reflexes*", Oxford: Oxford University press.

Posner, M.I. and M.E. Raichle (1994) "*Images of Mind*", New York: Scientific American Library.

Posner, M.I. (1995) "Attention in cognitive neuroscience: An over-view", in: *The Cognitive Neurosciences*, M.S. Gazzanige, ed., Cambridge, MA: MII press, pp. 615-624.

Posner, M.I. and Ya.G. Abdullaev (1999) "Neuroanatomy, circuitry and plasticity of word reading", *NeuroReport* 10:R12-R23.

Sokolov, E.N. (1963) "*Perception and the Conditioned Reflex*", Oxford: Pergamon press.

Sokolov, E.N. (1975) "The neuronal mechanisms of the orienting reflex", in: *The Neuronal Mechanisms of the Orienting Reflex*, E.N. Sokolov and O.S. Vinogradova, eds, Hillsdale, NY: Laurence Erlbaum associates.

Sokolov, E.N. and G.G. Vaikyavicus (1989) "*From a Neuron to a Neurocomputer*", in Russian, Moscow: Nauka.

Sokolov, E.N. and J.P. Cacioppo (1997) "Orienting and defense reflexes: Vector coding the cardiac response", in: *Attention and Orienting*, P.J. Lang, R.F. Simons and P.M. Balaban, eds, Mahawah, New Jersey, London: Lawrence Erlbaum associates, pp. 1-22.

Strange, B.A., P.C. Flether, R.N.A. Henson, K.J. Friston and R.J. Dolan (1999) "Segregating the functions of human hippocampus", *Proc. Natl. Acad. Sci.* USA. 96:4034-4039.

Tulving, E., H.J. Markowitsch, S. Kapur, R. Habib and S. Houl (1994) "Novelty encoding networks in the human brain: Positron emission tomography data", *NeuroReport* 5:2525-2528.

Vinogradova, O.S. (1976) "*The Hippocampus and Memory*", in Russian, Moscow: Nauka.

Vinogradova, O.S. (1987) "The hippocampus and the orienting reflex", in: *The Neuronal Mechanisms of the Orienting Reflex*, E.N. Sokolov and O.S. Vinogradova, eds, Hillsdale, NY: Laurence Erlbaum Associates, pp.128-154.

VECTOR CODE IN NEURONAL NETWORKS

EVGENI N. SOKOLOV
Department of Psychophysiology, Moscow State Lomonosov University
Mokhovaya st. 8, Moscow 103009, Russia

ABSTRACT

Modular organization of receptors and neuronal ensembles suggests that input *stimuli* are encoded by excitation vectors. Hierarchy of neuronal ensembles performs their orthogonalization and normalization. The spherical representation of input *stimuli* is achieved with contribution of feature detector maps. Specific behavioral acts are triggered with participation of command neuron output excitation vectors. The output excitation vectors via motor neurons evoke specific behavioral patterns. Conditional reflexes selectively tuned to conditional *stimuli* are established by modifications of plastic synapses on a command neuron in accordance with reinforced input excitation vector. The presented model was tested in the framework of color vision by intracellular recording from color-coding neurons and by factor analysis of confusion matrices found for differential color conditioning.

1. Introduction

I.P. Pavlov (1947) has emphasized in high brain functions two basic concepts: the concept of an analyzer and the concept of conditioned reflex. The progress in this directions achieved during last decades has stressed neuronal mechanisms. At the same time progress in human and animal psychophysics opens new perspectives to integrate subjective, behavioral and neuronal data in an interdisciplinary research suggested by Konorski (1967) and implemented recently in neuroscience.

In the interdisciplinary approach key principles introduced by Hebb (1949) have to be regarded: the principle of a neuronal ensemble and the principle of plastic (Hebbian) synapse. Formalization of a neuronal ensemble excitation leads to excitation vector. Extension of plastic synapse principle on a set of plastic synapses implies introduction of a synaptic weight vector. The formalism suggests a vector code in neuronal network operations (Fomin *et al.*, 1979).

An example of integration of subjective, behavioral and neuronal data is present in color vision with amount of experimental data at neuronal, psychophysical and behavioral levels. The present paper deals with vector encoding in neuronal networks based on results obtained in color vision research.

Multidimentional scaling of subjective differences between color *stimuli* has shown that perceptived colors can be represented by points on a hypersphere in the four-dimentional space. Geometrical distances between color points highly correlate with initial subjective differences between respective colors (Izmailov and Sokolov, 1991). The spherical model of color coding suggests that colors are encoded by vectors represented by excitations of four types of neurons: red-green, blue-yellow, bright and dark. The hypersphere is occupied by color selective color feature detectors. A color *stimulus* generating an excitation vector evokes on the color detector map a local excitation maximum. Modifications of input *stimuli* result in changes of excitation vectors and a translocation of an excitation maximum with respect to the color detector map. Signals from color detectors are transmitted to command neurons responsible for specific behavioral acts. Color detectors are linked with command neurons via plastic synapses, which constitute a synaptic weight vector. Plastic synapses of a command neuron triggering a behavioral act that leads to a positive reinforcement are enchanced.

In the process of learning a synaptic weight vector of the command neuron coincides in orientation with an input location vector. Due to such a synaptic modification the reinforced command neuron become selectively tuned to the conditional *stimulus*. The response probabilities of conditioned reflexes to conditional and differential *stimuli* parallel their excitation vectors. The factor analysis of confusion matrices revealed in trichromatic animals (carp and monkey) the four-dimentional color space supporting vector code hypothesis. Similar vector code was shown for perception of lines, motions, visual stereopsis, emotional expression of faces and cognitive processes in other modalities suggesting universal vector code for perception and learning (Sokolov, 1998). Vector code is operating also in motor and autonomic control. A command neuron excites a limited number of premotor neurons producing an output excitation vector that acts on a population of motor neurons and determines a specific behavior.

The information processing in neuronal networks is characterized by hierarchy and parallelism. Receptor surface can be represented by a mosaic of local receptor ensembles operating in a parallel mode. Hierarchy of neuronal ensembles following receptor mosaic is also organized as parallel branching streams. Command neurons represent a map from which parallel output excitation vectors are initiated. Finally associative learning is a tuning of a command neuron with respect to a conditional *stimulus* based on a parallel distribution of an input excitation vector upon a set of command neurons.

2. Receptor ensemble and receptor excitation vector

Receptor ensemble is a set of neighboring receptors having different but overlapping characteristics. A *stimulus* acting on an ensemble of receptors evokes in each receptor a particular excitation. The set of the excitations constitutes a receptor excitation vector. The overlapping receptor characteristics suggest a nonorthogonal basis of the vector space. Thus input *stimuli* are encoded by receptor excitation vectors having nonorthogonal basis.

The overlapping receptor characteristics of the ensemble are critically important to encode a variety of *stimuli* by limited number of receptors. If receptor characteristics do not overlap such set of receptors would encode only limited number of *stimuli* equal to a number of receptors constituting the receptor ensemble. The overlapping receptor characteristics can encode lots of *stimuli* representing each *stimulus* by a specific combination of receptor excitations - by receptor excitation vector.

An example of receptor ensemble is a set of three types of cones (red, green and blue) in a local area of the vertebrate retina. Theirs wave length characteristics overlap so that light of a particular spectral composition is encoded by specific receptor excitation vector in a non-orthogonal basis. The length of a receptor excitation vector encoded *stimulus* intensity. Its orientation encodes chromatic characteristic of color. Different excitation vectors occupy a region within one octant of the three-dimensional space in a form of color body.

3. Orthogonalization in neuronal networks

The nonorthogonal basis brings about complications for following vector operations. In the process of evolution the receptor ensembles were supplemented by neuronal ensembles constituting next layer of information processing. The characteristics of neurons within such an ensemble were orthogonal against each other. Such an orthogonalization is achieved by adjustment of weights of excitatory and inhibitory synaptic contacts between each receptor and each neuron of a given ensemble.

An example of orthogonalization is present in the vertebrate retina at the level of photopic horizontal cells. Orthogonal opponent characteristics of red-green, blue-yellow and luminosity cells are produced by linear combination of red, green and blue receptor characteristics achieved by natural selection of synaptic contacts between cones and horizontal cells. Excitation vectors after orthogonalization are representing colors within a half of a globe (four octants) constituting in total a color body. The lengths of the excitation vectors represent *stimulus* intensity while their orientations encode different combinations of hue and saturation.

4. Normalization of excitation vectors

After orthogonalization of excitation vectors specific aspects of input *stimuli* are encoded by orientations of vectors, while *stimulus* intensities are represented by lengths of the vectors. Thus *stimulus* intensity is encoded differently as referred to other *stimulus* parameters.

In the process of evolution a universal encoding mechanism was developed. The universal encoding mechanism requires a normalization of excitation vectors. At the next layer of information processing excitation vectors of a neuronal ensemble become of a constant length so that *stimulus* intensity is also encoded by vector orientation. The normalization of excitation vectors is due to an addition to the neuronal ensemble a supplementary neuron spontaneously active under no-stimulation but inhibited in accordance with increase of *stimulus* intensity (Chernorizov, 1999). The next step of the normalization is a replacement of inhibitory phases of opponent neurons by excitatory neurons. Now excitations of neurons are subtracted from the spontaneously active one so that total excitation of all neurons by any *stimulus* presentation remains constant. It means that *stimuli* are encoded by excitation vectors of a constant length equal to the sum of absolute values of excitations of neurons within a given ensemble. This normalization is achieved however in city-block metrics. An approximation of the city-block metric to the Euclidean one is achieved by adaptation reducing extrem values of neuronal excitations.

A concrete example of vector normalization can be found in tonic bipolar cells of the vertebrate retina where an additional neuron spontaneously active under darkness was supplemented to brightness cell. Red-green opponent horizontal cell is represented by two bipolar cells R+G- and G+R-. Blue-yellow opponent horizontal cell is transformed into Y+B- and B+Y- pair of bipolar cells. The number of dimensions is not increased however because either R+G- or G+R- cell is operating. Respectively in operation is either Y+B- or B+Y- bipolar cell. The excitation of R+G- or G+R-; Y+B- or B+Y- cell and of brightness cell were subtracted from darkness neuron producing the four-dimensional color space where color *stimuli* are encoded by excitation vectors of a constant length given in the city-block metric.

5. Representation of *stimuli* on a hypersphere

The normalization in city-block metric is approximated to Euclidean metric by non-linear reduction of extreme values of excitations in respective neurons of the neuronal ensemble due to neuronal adaptation. Due to adaptation input *stimuli* are represented on a hypersphere in the Euclidean space, The dimensionality of the space is equal to the total number of neurons constituting neuronal ensemble including a spontaneously active one.

The change of the input *stimulus* result in a modification of the orientation of excitation vector so that the end of the vector is moving along the hypersphere. Each *locus* of the hypersphere corresponds to a particular input *stimulus*.

The transition from city-block metric to Euclidean metric space can be seen at the level of tonic bipolar cells of the vertebrate retina. The extreme values of excitation are reduced so that the length of excitation vector deviates from city-block metric and approaches to Euclidean metric. Thus at bipolar level solid color body is replaced by a color surface – a kind of colortopic projection. Three angles of the color hypersphere closely correspond with three main subjective characteristics of color perception: hue, lightness and saturation. Subjective differences between colors parallel distances between ends of excitation vectors representing respective colors. The multidimentional scaling returns excitation vectors from a matrix of subjective differences as vectorial differences.

6. Detector map

Representation of input *stimuli* by excitation vectors on hypesphere in Euclidean metric space opens a perspective to reveal mechanisms of sensory discrimination. The discrimination unit is a feature detector. Each feature detector is specified by an ensemble of synapses that can be regarded as synaptic weight vector. A set of feature -detectors relevant to a particular neuronal ensemble constitutes a detector map. The detectors of the detector map are characterized by synaptic weight vectors of a constant length. A particular synapse of the detector multiplies a presynaptic input by a respective postsynaptic weight. The detector sums up all these products "computing" in this way an inner product of two vectors: presynaptic excitation vector and postsynaptic weight vector. Because both vectors are of constant lengths their inner product depends only on the cosine of angle between these vectors. The response of a detector is of a maximal magnitude when both vectors coincide in orientation. The detectors characterized by specific synaptic weight vectors occupy specific *loci* of the hypersphere. A *stimulus* encoded by an excitation vector activates selectively a particular detector on the hypersphere. At the same time an excitation vector acting in parallel on feature-detectors evokes on the detector map profile of excitations - detector population vector. A change of *stimulus* resulting in a change of an excitation vector produces a translocation of an excitation maximum on the detector map. A discrimination threshold between *stimuli* is determined by a distance between neighboring detectors.

A feature-detector map can be demonstrated at color-coding cortical area V4 where color selective detectors are identified. The topological representation of colors on the V4 suggests a colortopic projection similar to retinotopic, tonotopic and somatotopic projections. An input *stimulus* generates on the color detector map a detector population vector equivalent to an excitation vector at the predetector level (Bartles and Zeki, 1998). Particularly interesting are intensity detectors Achromatic *stimuli* activating only brightness and darkness cells constitute a two-dimentional space. Light of different intensity stimulates selectively lightness feature detectors. By an increase of *stimulus* intensity excitation maximum is moving along a semicircle from a particular intensity detector to the next ones. Similar effects refer to color detectors representing different combinations of hue and saturation. Under constant subjective intensity color space is a sphere in the three-dimentional space where the horizontal angle corresponds to hue and the vertical angle represents saturation. By a change of input spectrum excitation maximum is moving along the sphere crossing responsive areas of color selective detectors.

7. Command neuron

The problem arises concerning mechanisms realizing a transition from a detector map encoding input *stimuli* to behavioral acts, which these *stimuli* trigger. To approach a solution command neuron principle has to be considered.

Command neuron is a starting point for response initiation. Different command neurons are responsible for different behavioral patterns. A set of command neurons representing a particular area of behaviors is organized as a map where different behavioral patterns were localized topologically in a form of "mototopic projection".

Each particular command neuron has two types of synapses: plastic Hebbian and nonplastic (stable) ones. A set of Hebbian synapses of a command neuron can be termed "Hebbian weight vector". A detector population vector evoked on a detector map by *stimulus* presentation is distributed among command neurons via Hebbian synapses. In this way *stimulus* is potentially available for any behavioral pattern encoded on command neuron map. A single Hebbian synapse multiplies a presynaptic excitation by a postsynaptic weight. A command neuron summating up these products, "computes" an inner (scalar) product of two vectors: detector population vector and Hebbian weight vector.

The nonplastic synapses on command neurons are reinforcement-specific. Plastic synapses are enhanced only in the case if they were beforehand presynaptically activated. The effect of reinforcement is selectively addressed to a particular command neuron that was activated during performance of a relevant behavioral act. Thus presynaptic *stimulus* and following reinforcement meet on a particular command neuron to produce instrumental associative learning.

Under combination of detector population vector, activation of a command neuron during performing behavioral act and positive reinforcement increase the greater the greater were their presynaptic excitations (Sokolov, 1991). Such presyanptically dependent modification of postsynaptic weight makes Hebbian weight vector to coincide in orientation with detector population vector. The coincidence in orientation of detector population vector and Hebbian weight vector results in a maximal magnitude of their inner (scalar) product and maximal excitatory postsynaptic potential (EPSP) of the command neuron.

The magnitude of the EPSP determines number of action potentials (APs) generated by the command neuron. By fluctuations of the firing threshold spike number varies even under constant stimulation. Command neuron via premotoneurons generates a specific pattern of motoneurons affecting a set of efferents responsible for a particular behavior "gesture". In the process of learning under positive reinforcement of a conditional *stimulus* (CS) when Hebbian weight vector approaches an excitation vector of the CS their inner (scalar) product increases and probability of conditioned reflex (CR) evocation rises. The presented model suggests that probability of CR evocation correspond to inner (scalar) product of excitation vector and Hebbian weight vector. The differential *stimuli* (DSs) not reinforced positively result in decrease of synaptic weights. A differential *stimulus* deviating from a conditional one is characterized by an excitation vector that differ from the excitation vector generated by CS. It means that DS excitation vector deviates also from Hebbian weight vector established by CS. The inner (scalar) product of DS excitation vector and Hebbian weight vector is the smaller the more DS differ from conditional one. With decrease of the inner (scalar) product excitation and response probability of a command neuron diminish. Thus response probabilities to CSs and DSs are measures of inner (scalar) products of their excitation vectors and Hebbian weight vector induced by CS.

8. Vector code in associative learning

The suggested model of vector encoding can be tested using differential conditioning to color *stimuli*. Psychophysiological studies in humans and single unit recordings in animals have shown that in trichromatic visual systems different colors are encoded by excitation vectors of a constant length. Components of the excitation vectors are represented by excitations of four types of cells, red-green, blue-yellow, brightness and darkness ones. If the suggested vector model of associative learning is correct then from response probabilities to CSs a DSs as inner (scalar) products one would find excitation vectors encoding respective *stimuli*.

To perform such a test of the model in a set of color *stimuli* one *stimulus* has to be used as a CS and the other ones as DSs. When in the process of learning response probability to CS approaches a maximum at of plateau level it means that Hebbian weight vector of the command neuron become equal to the excitation vector of respective CS. The response probabilities of DSs are equal to inner (scalar) products of their excitation vectors and the Hebbian weight vector established by CS. A set of response probabilities to CS and DSs constitute a response probability vector specifying the Hebbian weight vector evoked by excitation vector of the CS. Using another CS from the same set of color *stimuli* re-learning is performed to get the next response probability vector specifying the excitation vector of the respective CS. By subsequent re-learning procedures a confusion matrix of response probabilities has to be created. Now each CS is characterized by specific vector of response probabilities. The next step of the testing is to find the basis of the confusion matrix and represent color *stimuli* by their coordinates in the metric space. If the vector model is correct then the coordinates of color *stimuli* found from the confusion matrix have to parallel the excitations of red-green, blue-yellow, brightness and darkness neurons. The inner (scalar) products of excitation vectors of color *stimuli* computed from their coordinates have to correlate positively with respective response probabilities of the confusion matrix. Additionally the excitation vectors of color *stimuli* extracted from the confusion matrix have a constant length suggesting that color *stimuli* are located on a hypersphere in the four-dimentional space.

The predictions of the vector model of associative learning were tested by differential instrumental color conditioning in fishes and monkeys. The confusion matrices were constructed from response probabilities to CSs and DSs found at plateau levels. The factor analysis of the confusion matrices reveals in fishes and monkeys excitation vectors specifying color *stimuli*.

It was found that four coordinates of excitation vectors of color stimuli extracted from confusion matrices closely correspond to excitations of four types of color-coding neurons: red-green, blue-yellow, brightness and darkness ones. The lengths of the excitation vectors are of constant value providing that color stimuli are represented on a hypersphere in the four-dimentional space. Three angles of the hypersphere in accordance with psychophysical data correspond to such subjective aspects of colors as hue, lightness and saturation. The inner (scalar) products of color excitation vectors highly correlate with respective response probabilities of confusion matrix. Thus the predictions of the vector model of associative learning were proved both for fishes and monkeys. The difference between the species with respect to color vision refers to a spectral shift towards long waves in fishes (Latanov *et al.*, 1991, 1997, 1999,. Leonova and Latanov, 1994). The four-dimentional color space suggests that achromatic colors represent a two-dimentional subspace. Differential learning of achromatic stimuli of various intensities was used as an additional test. The confusion matrices reveal two-dimentional space where coordinates of stimuli match excitations of brightness and darkness neurons. The vectors are of a constant length, so that achromatic stimuli were located on a semicircle where angle correspond to lightness (Evtikhin *et al.*, 1995, 1997). The four-dimentional color space predicts also that colors of equal luminosity constitute a three-dimentional subspace. The prediction was studied by color stimuli of equal luminosity obtained by preliminary equalizing procedure. The factor analysis of confusion matrices obtained from differential color learning of equiluminant colors has demonstrated a three-dimentional color space with red-green, blue-yellow and achromatic axes. The color excitation vectors are of a constant length so that colors are located on a sphere in the three-dimentional space. Horizontal angle of the sphere corresponds to hue and vertical angle to saturation.

9. Conclusion

The suggested model of vector encoding in neuronal networks integrates neuronal and cognitive aspects of information processing. An input stimulus is encoded by excitations of neurons constituting an excitation vector. By any change of input stimulus the excitation vector remains of constant length suggesting that stimuli are projected on a spherical surface. The spherical surface is occupied by feature-detectors selectively tuned to particular excitation vectors. The detector map is projected on a set of command neurons via plastic synapses that are modified in the process of associative learning.

Color vision experiments in humans and animals support vector model of information processing at neuronal and psychophysical levels. Direct intracellular recordings from bipolar cells of carp retina demonstrate that their excitations underlie hypersphere in the four-dimentional color space. Behavioral experiments in carp revealed a four-dimentional color space with axes corresponding to color-coding neurons. The correspondence of neuronal and behavioral data suggests that probabilities of conditioned reflexes to conditional and differential *stimuli* implicitly contain information about excitation vectors encoding colors at neuronal level.

I.P. Pavlov (1927) has shown that conditioned reflexes are selective both with respect to conditional *stimuli* and to reflexes evoked by conditional *stimuli*. At the neuronal level selective aspects of conditional *stimuli* are based on detectors selectively tuned to respective *stimuli*. The selective aspects of conditioned reflexes are due to command neurons representing specific unconditioned reflexes. It is assumed that conditioned reflexes are resulted from association between selective detectors and specific command neurons. The detectors activated by a conditional *stimulus* constitute a combination of excitations – an input excitation vector. The detector excitation vector acts on a command neuron via set of plastic synapses – synaptic weight vector. Plastic synapses are modified in the process of learning making command neuron selectively tuned to a specific conditional *stimulus*. The selective tuning of a particular command neuron to a specific conditional *stimulus* is a basis of associative learning selective both to conditional *stimulus* and elicited reaction.

References

Bartles, A. and S. Zeki (1998) "The theory of multistage integration in the visual brain", *Proc. R. Soc. Lond. B. Biol. Sci.* 265:2327-2332.

Chernorizov, A.M. (1999) "*Psychophysiology of Color Vision*", in Russian, Moscow: Moscow State University Press, in press.

Evtikhin, D.V., A.V. Latanov and E.N. Sokolov (1995) "Perceptual brightness space in carp (*Carpio Cyprinus*)", *Zurnal. Vyssh. Nervn. Deyat.* 45:964-975.

Evtikhin, D.V., A.V. Latanov and E.N. Sokolov (1997) "Perceptual brightness space in monkey (*Macaque rhesus*)", *Zurn. Vyssh. Nervn. Deyat.* 47:98-108.

Fomin, S.V., E.N. Sokolov and G.G. Vaitkyavichus (1979) "*Artificial Sensory Organs*", in Russian, Moscow: Nauka.

Hebb, D.O. (1949) "*The Organization of Behavior: A Neuropsychological Theory*", New York: Wiley.

Izmailov, Ch.A. and E.N. Sokolov (1991) "Spherical model of color and brightness discrimination", *Psychol. Sci.* 2:249-259.

Konorski, J. (1967) "*Integrative Activity of The Brain: An Interdisciplinary Approach*", Chicago: Chicago University Press.

Latanov, A.V., V.B. Polyanski and E.N. Sokolov (1991) "Four-dimentional spherical color space in monkey", *Vyssh. Nervn. Deyat.* 41:636-646.

Latanov, A.V., A.Yu. Leonova, D.V. Evtikhin and E.N. Sokolov (1997) "Comparative neurobiology of color vision in humans and animals", *Zurn. Vyssh. Nervn. Deyat.* 47:308-319.

Latanov, A.V., A.Yu. Leonova, D.V. Evtikhin and E.N. Sokolov (1999) "Color spaces in animals –trichromats (*Rhesus* Monkeys and Carps) revealed by instrumental discrimination learning", in: *Conceptual Advances in Russia Neuroscience: Complex Brain Functions*, A.M. Ivanitski and P.M. Balaban, eds, Harvard Academia Publishers, in press.

Leonova, A.Yu. and A.V. Latanov (1994) "Perceptual color space in carp (*Carpio Cyprinus*)", *Zurn. Vyssh. Nervn. Deyat.* 44:1059-1069.

Pavlov, I.P. (1927) "*Conditioned Reflexes*", Oxford: Oxford University Press.

Sokolov, E.N. (1991) "Local plasticity in neuronal learning", in: *Memory: Organization and Locus of Change*, R.L. Squire, N.M. Weinberger, G. Lynch and J.L. McGaugh, eds, New York, Oxford: Oxford University Press., pp. 364-391.

Sokolov, E.N. (1998) "Model of cognitive processes", in: *Advances in Psychological Sciences. Vol. 2: Biological and Cognitive Aspects*, M. Saborin, F. Craik and M. Robert, eds, East Sussex, UK: Psychology Press., pp. 355-379.

VECTOR CODING UNDERLYING INDIVIDUAL TRANSFORMATIONS OF A COLOR SPACE

GALINA V. PARAMEI* and DAVID L. BIMLER°
*Institut für Arbeitsphysiologie an der Universität Dortmund, Ardeystrasse 67, D-44139 Dortmund, Germany
°Massey University, Private Bag 11-222, Palmerston North, New Zealand

ABSTRACT

A color space comprising variations of color vision was reconstructed from large color differences and had a shape of a hypercylinder, with a spherical subspace at each luminance level. Compared to those for normal trichromats, spaces for red-green deficients revealed a compression along the red-green axis, reflected by an axial weighting. In 'private' spaces for protans and deutans the direction of compression differed, which was accomodated by a rotation of the chromatic plane. Vector-coding operations on photoreceptor and post-receptoral outputs are discussed as possible mechanisms underlying the psychophysical findings.

1. Introduction

Traditional color space have three dimensions, corresponding to the red-green (R-G), yellow-blue (Y-B), and brightness subsystems. It is conceived as a geometrical representation in which distances between points are uniformly related to perceived dissimilarities between colors. However, the space reveals a consistent deviation from uniformity in the intervals between adjacent equichroma contours, with the steps at higher chroma levels tending to be smaller. In studies where color spaces were reconstructed from large color differences using multidimensional scaling (MDS) techniques, two approaches of overcoming this discrepancy have been proposed. In the first one, the pattern of deviations becomes less noticeable if color differences are embedded in a 3D elliptic space, being nonlinearly related to distances representing them (Indow, 1999). In the second approach, alternatively, uniformity can be restored by introducing an additional dimension, saturation, with the constraint that equiluminant colors lie on a spherical surface (Izmailov and Sokolov, 1991). We are concerned with elucidating the properties of this psychophysical solution, since an adequate representation is expected to provide a better understanding of the underlying neurophysiological mechanisms.

Another way of probing properties of the color space – opening a second window into the underlying mechanisms – is the inclusion of color-vision deficients (CVDs). MDS studies with observers having congenital red-green abnormalities revealed their individual color spaces to be compressed along a R-G axis, compared to those for normal trichromats (NTs), with the extent of compression reflecting the severity of abnormality (*e.g.* Helm, 1964).

An attempt to reconstruct a color space that covers NTs *and* CVDs have been undertaken in other MDS studies (*e.g.* Chang and Carroll, 1980). These used the weighted-Euclidean model of individual differences, to measure the relative compression along the dimensions in CVDs' spaces. However, the model ignores the fact that color spaces for the two classes of red-green deficients, protans and deutans, are compressed along a slightly different direction, or „axis of confusion", specified by an angular parameter (Farnsworth, 1943). Confirmation of the latter comes from panel diagnostic tests, evaluated qualitatively, and recently quantitatively (Vingrys and King-Smith, 1988).

The present work studies a color space designed to accommodate color normal and abnormal observers applying an individual-difference MDS method extended by the „spherical" approach and varying the direction as well as the extent of compression in subject-specific solutions. Transformations of the color space are considered of interest as indications of underlying vector-coding operations and are addressed in relation to possible neurophysiological mechanisms of color processing.

2. Methods

Dissimilarity between pairs of colors were judged on a 0-9 scale by five NTs, two protans (protanomal, PA; and protanope, P) and four deutans (simple and extreme deuteranomals, DA and EDA; and deuteranopes, D1 and D2), diagnosed using the Ishihara pseudoisochromatic plates, the Nagel anomaloscope, and the FM 100-hue test. The stimulus set included 18 lights, each at 10 and 40 cd/m², displayed on a computer-controlled color monitor as 4° square patches on a dark background. Each pair was exposed for 0.5 sec in quasi-random order, with both luminances interleaved in one session. Resulting values for each subject were means of 10 ratings for a given pair.

3. Results

Color space comprising NTs and CVDs. The data from all observers were analyzed with an individual-differences MDS program to produce a single 3D „group color space" $\mathbf{X}_0$, common for the NTs and CVDs. $\mathbf{X}_0$ consisted of color points covering the surface of a squat cylinder. Two chromatic axes, *Yellow-Blue* (Y-B) and *Red-Green* (R-G), span circular ends at each luminance level (*D1-D2* plane). *D3*, the central axis of the cylinder, is *'Brightness'*[*].

However, this Euclidean 3D solution was found to deviate from the traditional color solid: instead of remaining flat, the ends of the drum bulge away from each other, making the distance between the low- and high-luminance unsaturated stimuli too large to be compatible with the observed dissimilarities. To eliminate this

[*] Since the stimuli used were equiluminant and not equibright at each level, *D3* values are confounded by the interaction of luminance with light composition: the Helmholtz-Kohlrausch effect, hence the label 'Brightness'.

distortion, a 4th dimension was introduced, subject to a 'sphericity' constraint confining points in the *D1-D2-D4* subspace to lie on a (hemi-)spherical surface. *D4* values for stimuli were thus dependent on their distances from *White* in the *D1-D2* plane, warranting an interpretation of *D4* as *Saturation*.

'Private' color spaces $\mathbf{X}_m$ were derived from the group configuration $\mathbf{X}_0$ by compression or elongation along its axes, specified by dimensional weights w_{m1}, w_{m2}, w_{m3} (Table 1)[†]. The best fit (Stress_1 = 0.163) was achieved with non-zero curvature, color planes becoming spherical surfaces having a diameter of 1.62 (in units where the average distance between points in the group configuration is 1).

Type of color vision	*Dimensional weights*			*Angle of rotation* θ_m (°)	*Extent of compression* r_m
	w_{m1}	w_{m2}	w_{m3}		
NT1	1.22	1.22	0.01	undefined	0.00
NT2	1.15	1.20	0.50	123.32	0.04
NT3	1.16	1.29	0.01	128.56	0.10
NT4	0.91	0.96	1.11	62.32	0.05
NT5	0.94	1.05	1.01	41.11	0.11
PA	0.66	1.04	1.21	93.53	0.42
P	0.41	1.27	1.10	97.77	0.81
DA	0.78	1.25	0.90	78.69	0.44
EDA	0.57	1.38	0.87	86.37	0.71
D1	0.39	1.11	1.27	83.35	0.78
D2	0.38	0.99	1.37	88.71	0.74

Table 1. Subject-specific parameters of color spaces

The values of w_{m3} were systematically higher (*i.e. 'Brightness'* was more salient) for the CVDs than for NTs.

The spaces for protans and deutans were found to be compressed but differ slightly in the direction of compression, in line with the prediction of Farnsworth's (1943) model (cf. the „axis of confusion"). We accommodated this distinction by rotating the *D1-D2* 'plane' of $\mathbf{X}_0$ to new orthogonal axes *D1'-D2'* before compressing it, with *D1'* being the individual direction of greatest compression. The *angle of rotation* θ_m ($0° \leq \theta_m \leq 180°$) is one more subject parameter (Table 1).

† In our model, the subject-specific dimensional weights are applied to the 3D $\mathbf{X}_0$ *before* calculating *D4* by projecting the compressed *D1-D2* plane onto a sphere, ensuring that each 4D hypercylindrical $\mathbf{X}_m$, has constant curvature. It should be noted that the simple weighted-Euclidean individual-difference model, with a 4D $\mathbf{X}_0$, is not appropriate: if points are confined to a hemispherical surface in $\mathbf{X}_0$, then the corresponding surface in each compressed $\mathbf{X}_m$ will *not* be spherical. "Spherical MDS" was applied, using program MSPHERE modified by one of the authors to include an additional dimension. MSPHERE uses a simple gradient-descent method to find the weight and angle parameters, and the sphere's curvature (Bimler, 1999).

rotating the *D1-D2* 'plane' of $\mathbf{X}_0$ to new orthogonal axes *D1'-D2'* before compressing it, with *D1'* being the individual direction of greatest compression. The *angle of rotation* θ_m ($0° \le \theta_m \le 180°$) is one more subject parameter (Table 1).

Transformations of the chromatic 'plane' in subject-specific spaces were then investigated by measuring the salience of *R-G* relative to *Y-B* axis: the *extent of compression* $r_m = (w_{m2}^2 - w_{m1}^2)/(w_{m2}^2 + w_{m1}^2)$, which can range from 0 (for an observer who perceives no compression of the color plane relative to the group consensus) to 1 (for observers for whom the color plane is compressed to a single dimension). $\mathbf{X}_0$ has been first normalized to set $w_{m1} = w_{m2}$ for observer NT1.

The indices in Table 1 demonstrate subtle parameter variations among the NTs. For the CVDs the values r_m were systematically higher than for NTs, indicating compression of the color plane along each personal *R-G* axis, with the extent of compression corresponding to the severity of abnormality diagnosed by traditional tools.

Private color spaces of the NT1, P and D2, respectively, are illustrated in Figures 1 and 2a, b.

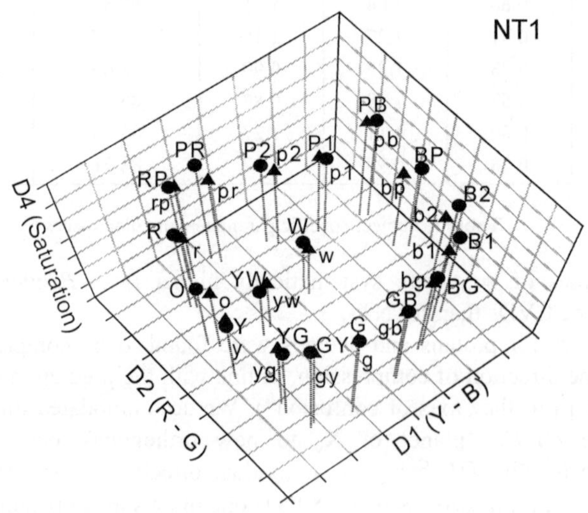

Figure 1. Perspective view of *D1-D2-D4* subspace (without variation in *'Brightness'*) in the color space of a normal trichromat (NT1). Circles designate lights at 40 cd/m² and triangles at 10 cd/m².

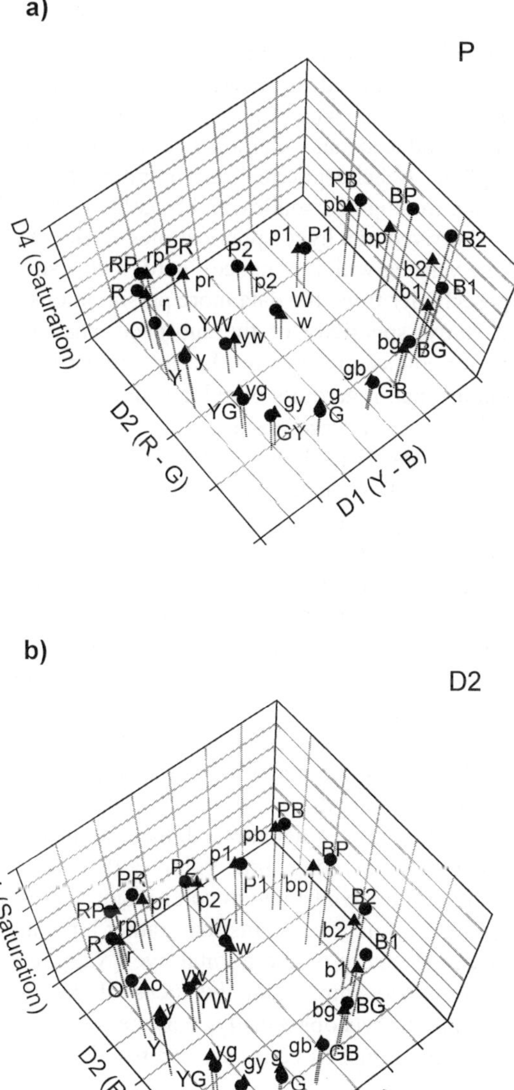

Figure 2. **a)** Perspective view of *D1-D2-D4* subspace in the color space of a protanope P and **b)** of a deuteranope D2.

4. Discussion

The first finding of this study is that the 3D color space representation contains deviations from uniformity related to differences in saturation. To improve the fit between reconstructed differences in the chromatic plane and subjective dissimilarities, we introduced an additional dimension, with a constraint of sphericity for that plane. A hemispherical configuration for the equiluminant-light 'plane' agrees with the finding of Izmailov and Sokolov (1991). That study used six luminance levels and arrived at a hyperspherical model (the average distance within each chromatic plane varying with luminance); here, with a smaller luminance range (two levels), a hypercylinder fitted the data more parsimoniously.

In the present study, sphericity of the subspace was maintained, along with normal trichromats, for color deficients, reinforcing the earlier finding of Paramei *et al.* (1991). This implies that, regardless of color-vision variations, the output values of the chromatic subsystems undergo normalization, a nonlinear form of integrating outputs of underlying channels (Sokolov, 1994, pp.463-476). Partial support for a nonlinear output from the *Y-B* channel comes from another psychophysical study (Fuld, 1991). Recent neurophysiological data provide indications of nonlinearity in color processing beyond the LGN-cell stage (Cottaris and De Valois, 1998), which is considered to be due to the combined activity of a cell network (Werner, 1999).

Another finding, compression of the *R-G* dimension in color spaces of red-green deficients, is in agreement with earlier studies (Helm, 1964; Chang and Carroll, 1980). The extent of compression was found to be proportional to the degree of color vision loss. The spectral proximity hypothesis, recently tested using a molecular genetic analysis (Neitz and Neitz, 1998, pp.101-119), explains the severity of color deficiency by the magnitude of separation in spectral sensitivities of the underlying X-coded pigments.

In addition to the receptor explanation, defects in post-receptoral processing mechanisms are postulated in congenital color deficients. Evidence exists that abnormalities in neural factors contribute to color vision losses: specifically, coding mechanisms are sparse and integrate inputs over rather large areas, in a way similar to that in the extreme periphery of the normal eye (Nagy and Purl, 1987). These receptoral and neural factors imply that the locus of the *R-G* axis compression (weighting), demonstrated psychophysically, may be sought at the post-receptoral stage where outputs of the L and M cones are compared, in L/M opponent ganglion and LGN cells.

Among the present findings, last but not least, is the rotation of the chromatic plane in abnormal spaces. Whereas compression is conceived as reflecting the operation of weighting photoreceptor *inputs* at the L/M chromatic differencing stage, the rotation might indicate an additional weighting of these units' *outputs* at a later processing stage. This operation may be presumably attributed to perceptual opponent color subsystems, postulated by De Valois and De Valois (1993), at a cortical level. A further assumption concerns the opposite sign of the rotation direction in the two classes of red-green deficients which might be related to differences in integrative characteristics of the color-opponent functions of protans and deutans - relative magnitudes of positive and negative lobes and crossover points.

Presently neurophysiological evidence about late mechanisms of color processing in normal trichromats is scarce and that about post-receptoral mechanisms in color deficients is very scanty. With this in mind, indications of vector-coding operations gained psychophysically from normal and abnormal observers may serve as a basis for simulations modeling transformations in the color-vision processing and as an impetus for neurophysiological studies of different loci of the system.

Acknowledgements

Supported by DFG grant Ca 126/3-1 and Ca 126/3-2. We thank C. Richard Cavonius for his helpful comments on an earlier manuscript and John Mollon and John Werner for critical discussion of the study. Technical assistance of Ute Lobisch is gratefully acknowledged.

References

Bimler, D.L. (1999) "A multidimensional scaling comparison of color metrics for response times and rated dissimilarities", *Percept. and Psychophys.* 61:1675-1680.

Chang, J.-J. and J.D. Carroll (1980) "Three are not enough: An INDSCAL analysis suggesting that color space has seven (± 1) dimensions", *Col. Res. Appl.* 5:193-206.

Cottaris, N.P. and R.L. De Valois (1998) "Temporal dynamics of chromatic tuning in macaque primary visual cortex", *Nature* 395:896-900.

De Valois, R.L. and K.K. De Valois (1993) "A multi-stage color model", *Vision Res.* 33:1053-1065.

Farnsworth, D. (1943) "The Farnsworth-Munsell 100-Hue and dichotomous tests for color vision", *J. Opt. Soc. Am.* 33:568-578.

Fuld, K. (1991) "The contribution of chromatic and achromatic valence to spectral saturation", *Vision Res.* 31:237-246.

Helm, C.E. (1964) "A multidimensional ratio scaling analysis of perceived color relations", *J. Opt. Soc. Am.* 54:256-262.

Indow, T. (1999) "Predictions based on Munsell notation. I. Perceptual color differences", *Col. Res. Appl.* 24:10-18.

Izmailov, Ch.A. and E.N. Sokolov (1991) "Spherical model of color and brightness discrimination", *Psychol. Sci.* 2:249-259.

Nagy, A.L. and K.F. Purl (1987) "Color discrimination and neural coding in color deficients", *Vision Res.* 27:483-489.

Neitz, M. and J. Neitz (1998) "Molecular genetics and the biological basis of color vision", in: *Color Vision. Perspectives from Different Disciplines*, W.G.K. Backhaus, R. Kliegl and J.S. Werner, eds, Berlin-New York: Walter de Gruyter, pp.101-119.

Paramei, G.V., Ch.A. Izmailov and E.N. Sokolov (1991) "Multidimensional scaling of large chromatic differences by normal and color-deficient subjects", *Psychol. Sci.* 2:244-248.

Sokolov, E.N. (1994) "Vector coding in neuronal nets: Color vision", in: *Origins: Brain and Self Organization*, K. Pribram, ed., Hillsdale, NJ: Erlbaum, pp. 463-476.

Vingrys, A.J. and P.E. King-Smith (1988) "A quantitative scoring technique for panel tests of color vision", *Invest. Ophthalmol. Vis. Sci.* 29:50-63.

Werner, J. (1999) "Human colour vision: 2. Colour Appearance and cortical transformation", in: *From Neuronal Coding to Consciousness. Series of Biophysics and Biocybernetics, V. 9.* W. Backhaus, ed., Singapore: World Scientific Publishers; 1999, in press.

GABOR POPULATION CODES FOR ORIENTATION SELECTION

NIKLAS LÜDTKE and EDWIN HANCOCK
Department of Computer Science, University of York, York Y01 5DD, UK

ABSTRACT

In this paper we adopt the principle of *population vector coding*, which has been used to describe neural representations of limb movements and properties of stimuli of various sensory modalities for orientation measurement in computer vision. In analogy to orientation sensitive units found in a cortical hypercolumn, a bank of Complex Gabor filters is used, since the moduli of their responses resemble receptive field properties of complex cells in primary visual cortex. Vectorial combination of units with very broad orientation tuning allows a precise and reliable estimate of stimulus orientation. By investigating the performance of the orientation measurement we demonstrate the applicability of a biologically inspired method in machine vision.

1. Introduction

Population coding is considered a neural coding scheme of general importance. Stimulus properties are not represented by single specialized neurons ("grandmother cells"), but by the activities of an ensemble of neurons. Although each individual neuron provides only little information about the stimulus, the whole population can implicitly characterize the stimulus at a very high precision.

Georgopoulos *et al.* (1986) introduced the concept of the *population vector* to decode the internal representation of limb movements by direction sensitive neurons in motor cortex. Each neuron is assigned a vector component with a magnitude equal to the strength of its response (spike frequency) and a direction according to its preferred orientation. The population vector is the sum of these components.

Population vector decoding has also been suggested to understand the representation of *visual* information. Vogels (1990) examined a model of population vector coding of visual stimulus orientation by striate cortical cells. Based on an ensemble of broadly orientation-tuned units, the model explains the high accuracy of orientation discrimination in the mammalian visual system. The aim in this paper is to show the technical applicability of population coding for the measurement of edge orientation in computer vision.

2. Orientation Measurement with Gabor Filters

2.1. Orientation Tuning

Like complex cells in V1, Gabor filters have a rather broad orientation tuning. Each filter is characterized by a size parameter, σ_e (the width of its Gaussian envelope). A value of 3 × σe can be interpreted as the *radius of the "receptive field"*. To examine their tuning properties we have convolved Gabor filters with 256×256-grey-scale images of single straight lines with varying orientations (0° - 170°). Figure 1 shows the tuning curves for three filters which had a preferred orientation θ = 90°, a wavelength λ = 8 pixels, and a size parameter σe = 0.6×λ (4.8 pixels), 1.0×λ (8 pixels), and 2.0×λ (16 pixels), respectively. The estimated half-widths are w = 16.7°, 9.7°, and 5.2°. The first two values are comparable to typical orientation tuning half-widths of striate cortical cells (Vogels 1990).

Obviously the tuning width depends on the size of the receptive field. In figure 2 the tuning width is plotted as a function of the size of the receptive field σe in a log-log diagram. The tuning width and σe are inversely proportional to each other.

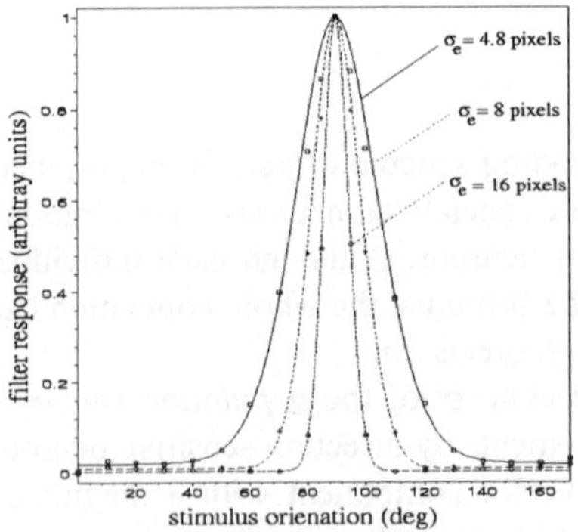

Figure 1. Tuning curve for the moduli of three Gabor-filters with wavelength of eight pixels and preferred vertical orientation.

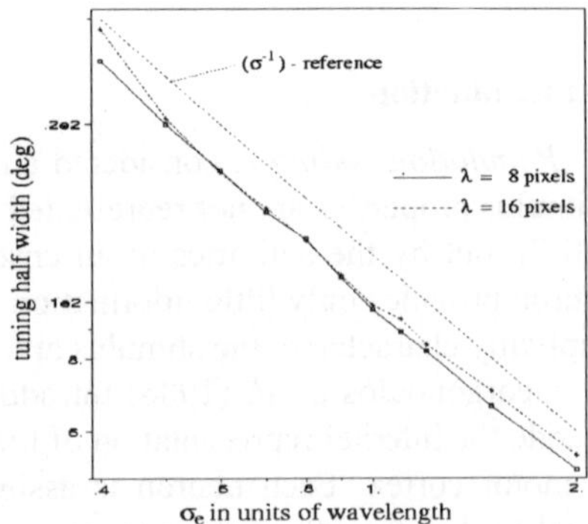

Figure 2. The log-log-plot of the tuning width depen-ding on the size parameter σe of the receptive field reveals their antiproportionality: $w \propto (\sigma e)^{-1}$.

2.2. The Population Vector of a Gabor Filter Bank

The individual Gabor filters only yield a very coarse estimate of the local orienation. However, by adopting the concept of population vector coding a much more precise value of the stimulus orientation can be calculated from the filter bank.

According to Georgopoulos *et al.* (1986), we define the population vector for a set of n Gabor-filters as follows. Consider a wavelength λ. Let $G(x,y;\theta_i,\lambda)$ be the modulus response of a complex Gabor filter of preferred orientation θ_i. Let $\boldsymbol{e}_i$ = $(cos\theta_i, sin\theta_i)^T$ be the unit vector in the direction θ_i.

Then the population vector $\boldsymbol{p}$ is defined as:

$$\vec{p}(x,y) = \sum_{i=1}^{n} G(x,y;\theta_i,\lambda)\, \vec{e}_i \,. \qquad (1)$$

Each filter is represented by a component vector. The vector orientation and length are given by the preferred orientation θ_i and the response magnitude $G(x,y;\theta_i)$ of the filter at location (x,y). The population vector is the vector sum of the n components. The decoding of the stimulus orientation is achieved by determining the orientation of the population vector: $\theta_p = \arctan(p_y / p_x)$.

The magnitude of the population vector $\|\boldsymbol{p}(x,y)\|$ characterizes the overall response ``energy" of the filter bank at position (x,y). If evaluated at contours, i.e., local maxima of $\|\mathbf{p}\|$, θ_p gives an estimate of the local tangent angle. Theoretically, the coding error decreases with an increasing number of filters. However, computational cost and discretization errors of digital images place limitations on the optimal number of filters.

3. Performance

3.1. Accuracy of Orientation Measurement in Grey-scale Images

The same stimuli and filter parameters as in Section 2.1 were used to determine the accuracy of the orientation estimate from population coding for 8, 16, and 32 Gabor filters. Figure 3 shows how the root mean square error of the population coding $\langle\delta\theta_p\rangle_{rms}$ depends on the tuning (half-)width of the applied Gabor filters.

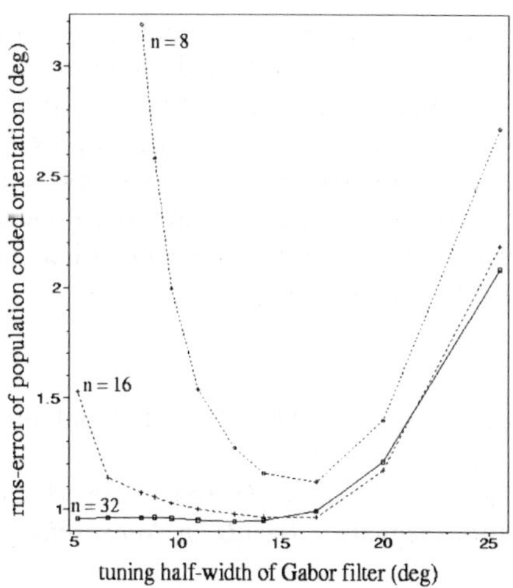

Figure 3. Dependence of the rms-error of the population-coded orientation on the tuning width.

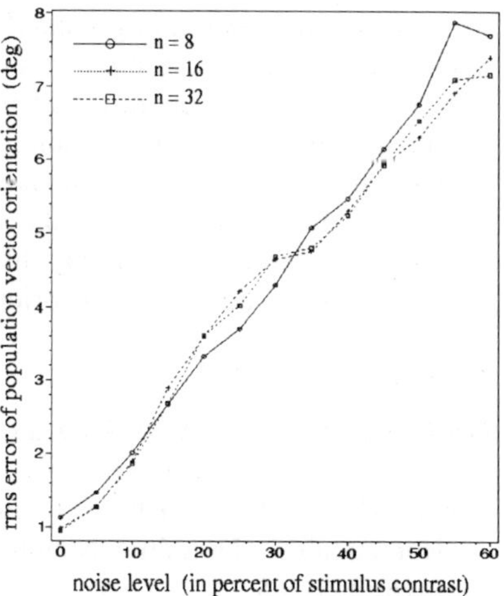

Figure 4. The rms-error as a function of the noise variance for additive Gaussian noise.

The optimal choice for the tuning width would be simultaneous minimization of σ_e and $\langle \delta\theta_p \rangle_{rms}$ because small σ_e means better spatial resolution, as the filters extend less far in the image. With 32 filters the rms error decreases even further for decreasing tuning width because the filters are still dense enough to have sufficient overlap to cover the whole range of 180 degrees. With 16 or 8 filters this no longer the case. However, the spatial resolution then becomes poor for small tuning widths, as σ_e is rather large. Therefore, a filter bank of 16 filters seems to yield the best results. The rms-deviation of the angle of the population vector from the stimulus orientation is only $\langle \delta\theta_p \rangle_{rms} \approx 1°$, which is very accurate compared to the half-width of the tuning curve for the most suitable filter ($w \approx 17°$). The error of the population-coded orientation estimate consists of two components: the coding error, due to the limited number of filters, and the discretization errors of the digital images. The measured rms-error is consistent with simulations by Vogels (1990).

3.2. Robustness to Noise

Finally, we experiment with the sensitivity to additive Gaussian noise. Figure 4 shows the rms-error as a function of the noise variance for different numbers of filters (8, 16 and 32). The dependence is roughly linear for all three filter banks with no significant difference in noise sensitivity.

4. Conclusion and Discussion

Although a population vector interpretation allows a readout of the information encoded by a neural ensemble, it is rather unlikely that such decoding is actually performed in the brain (Pouget and Zhang 1997). Instead, distributed coding must be maintained in order to secure robustness against neural mortality and limited signal to noise ratio. However, in a technical system the decoding is justified since the intention here is not to model biological information processing. For the purpose of local orientation estimation with Gabor filters, the population vector is a simple and efficient tool. Alternative methods, such as Bayesian analysis, would require more data and the accuracy would only differ significantly for a large number of filters or a non-even distribution of filters orientations (Oram *et al.* 1998).

References

Georgopoulos, A.P., A.B. Schwarz and R.E. Kettner (1986) "Neural Population Coding of Movement Direction", *Science* 233:1416-1419.

Oram, M.W., P. Földiàk, D.I. Perrett and F. Sengpiel (1998) "The 'Ideal Homunculus': Decoding Neural Population Signals", *Trends in Neuroscience* 21(6):259-265.

Pouget, A. and K. Zhang (1997) "Statistically Efficient Estimation Using Cortical Lateral Connections", *Adv. Neur. Inform. Proc. Syst.* 9:97-103.

Vogels, R. (1990) "Population Coding of Stimulus Orientation by Striate Cortical Cells", *Biol. Cybern.* 64:25-31.

ANALYSIS AND INTERPRETATION OF ESCHER'S IMPOSSIBLE BUILDINGS BY MEANS OF A SPACE-VARIANT RETINAL MODEL

PAOLO GUALTIERI and LUIGI TAIBI
Istituto di Biofisica, CNR, Via S. Lorenzo 26, 56127 Pisa, Italy

ABSTRACT

The use of a discrete model of the human retina structure based on a non-uniform sampling of the visual field is proposed for the analysis and explanation of Cornelius Escher's (1898-1972) paintings such as Belvedere (1958), Ascending and Descending (1960), and Waterfall (1961), based on impossible figures (*i.e.*, impossible Penrose triangle, Penrose infinite staircase and Necker cube). These figures give the impression of spatial interpretation of a picture, and are more relevant to psychology of vision than to geometry. In the following we will demonstrate that the perception of these impossible figures as real 3D objects is due to the impossibility of our visual system to acquire the entire image at once and with the highest resolution. A two-stage mechanism for spatial interpretation is identified.

1. Introduction

An impossible figure gives the observer the impression of looking at some three-dimensional objects, even though those objects cannot exist, since any attempt to its construction leads to geometrical contradiction (Kupla, 1983). For example a figure such as the Penrose triangle (Penrose and Penrose, 1958) looks like a real, 3D object; however, a deeper examination reveals that this figure cannot be a representations of a real object, since its parts are spatially non-connectable. Why have impossible figures drawn so much attention? The answer is: impossible figures causes a mental conflict in the observer, hence, they create emotions.

Looking at the figure, the observer appropriately interprets different local and global depth clues to arrive at some over-all model of a three-dimensional structure of the object present in the figure. This process of interpretation is mostly unconscious and depends substantially on a huge set of memorized "most likely" models of various familiar shapes (Marr and Nishihara, 1978). The identification of the scene by the observer is immediate. The attention then turns from one point to another of the scene, guiding eye movements: the observer fix his/her eyes on a detail of the scene (fixation pause), then rapidly takes them off to fix them again on another detail, and the process goes on until the entire scene has been explored. The field of view (30°) is split into a homogeneous high resolution area, the fovea (1°), and a periphery with decreasing resolution (Baily, 1981; Fiorentini, 1989); therefore, during every instant of a fixation pause, only a small region of the figure, namely that surrounding the fixation point, appears well defined and detailed.

The rest of the figure is less perceivable, since its image is formed onto the peripheral part of the retina, where vision is less distinct. However, the overall impression of the observer is that the entire picture is at the same time clear and in detail. This is the result of a mnemonic process of re-assembling the sub-images acquired during the fixation pauses. This process gives the idea of a figure belonging to a well-defined category of objects; however, since this process is merely mnemonic, it does not consider spatial relationship among the acquired sub-images. This means that the observer gathers only an idea of the image, without three-dimensional consistency (Treisman and Schmidt 1982; Watt, 1988). When we analyze an impossible figure, our visual system is not able to work out the problem, and it gives us both the idea of a real image and its spatial oddness. Therefore, we look at something that we do know cannot exist. To solve the problem, we need a further detailed analysis of the image, based on our ability of abstraction.

Several impossible figures and Escher's impossible buildings based on the Penrose triangle, the Penrose infinite staircase and the Necker cube were used to verify this hypothesis. A discrete model of the retina (Sandini and Tagliasco, 1980) that simulates the spatial sampling of retinal images was used to represent these images at the retinic level.

2. Material and Methods

Escher's paintings were acquired by a scanner. The acquisition procedure uses 256 gray levels per pixel. The 600 x 600 dots per inch sampling step was chosen to guarantee the best compromise between the need of storing image data of reasonable size (about 600 x 400 pixels) and the requirement of resolution; sensitivity was adjusted and input gamma corrected for the radiometric accuracy of the generated image. For the handling of the digital b/w images we present, we used a Pentium personal computer (IBM, USA) with a driver direct color display (24 bits) as hardware platform. A fast DRAM (Diamond, USA) based on 64-bit graphics technology with accelerated video playback and a 64-bit graphics accelerator chip was plugged in the Intel PCI bus. We assembled user friendly interfaces using libraries and objects that are implemented in MS-Windows environment. Our algorithms were implemented using C++ (Borland, USA) that provides a development system and a visual language for image processing.

The spatial sampling performed on the retinal image can be modeled by a discrete distribution of elements whose radii increase linearly with eccentricity. Our model covers the visual field with circular elements with variable overlapping and fits well the density of retinal cells. In order to minimize the overlapping, the intersections of the circular elements' contours are vertex of hexagons (Figure 1a).

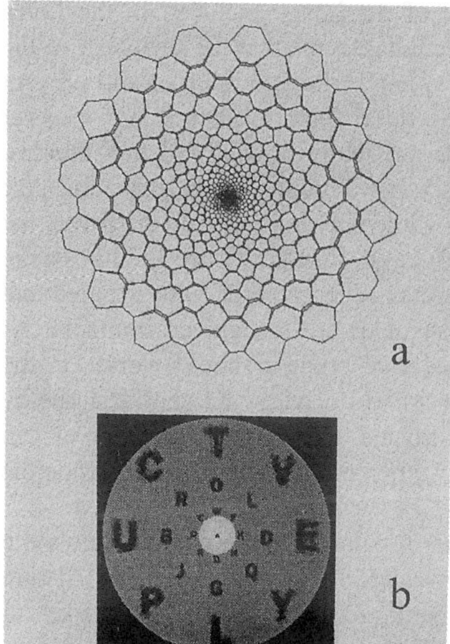

Figure 1

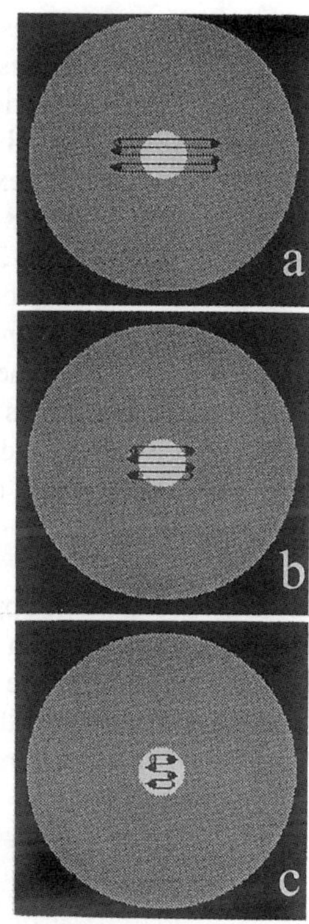

Figure 2

The linear relationship between radius R and eccentricity E can be expressed, in term of number N of equal size elements at a given eccentricity, by the equation: $R=(2p/(3*N))*E$. The retinic resulting image has pixels whose dimension varies as the model. The value of each pixel is obtained by computing the average of the light intensity distribution of the original image in the correspondent filter area. Figure 1b shows how big should be the letters to be detected moving from the fovea to the periphery of the visual field.

3. Results and Discussion

A first example is constituted by a generated image representing an impossible figure (Figure 2a). The first impression the figure produces is the drawing of four pencils. Since pencil shapes are very familiar, it seems very easy to recognize them. But this image is built in such a way that more than one fixation pause are needed to explore the entire subject (three in our case). The light zone in Figure 2a represents that part of the image included in the fovea during the three fixation pauses; details of the image remain at the periphery of the visual field, where the resolution is very low, hence they are blurred and prevent the recognition of the true object. After the three fixation pauses, an observer perceives that only the two external pencils are really drawn, and that the two internal ones are only lines. As a matter of fact, if we progressively reduce the size of the four pencils (Figure 2b, 2c), when the whole drawing is inside the fovea, the observer perceive, with only one pause of fixation, that the figure is impossible. With this arrangement, the figure as a whole has been gathered with detailed and uniform perception and geometrical inconsistencies are recognized immediately. This experiment demonstrates how the perceptive process should work, *i.e.* the zones of the image corresponding to the fovea field of view are analyzed one by one so that the whole image can be reconstructed. Otherwise, the observer can have the conscious perception of a scene only when the figure is entirely inside the fovea.

This analysis has been applied to Escher's painting representing impossible buildings (Figure 3, for reason of room only Waterfall example is shown). Also in this case, the conclusion is the same: the structure of our visual field is the key to understand these paintings. All Escher's buildings seem well built and intriguing, but they are based on hidden impossible figures and therefore "impossible".

Figure 3

The Waterfall lithography is based on the Penrose's triangle: the water falling down from the mill wheel goes back to its starting point. The Belvedere lithography is based on Necker's cube: there are two terraces one over the other, but a staircase connects the inside of the lower terrace to the outside of the upper terrace. The Ascending and Descending lithography is based on the Penrose's infinite staircase: the monk going up the stair comes down the steps to his starting point. To demonstrate that the impossibility of these images is related to the non-uniform sampling of our visual field, for each lithography we simulated the retinic acquisition. Figure 4 shows the result of the process on the Waterfall lithography. On the left, the fovea (lighter zone) of an observer looking at the picture 2 meter away is placed on a vertex of the hidden impossible figure; on the right, the Escher's figure elaborated by our filter. It could be noticed that in the fovea field there exist a 3D consistency, while at the periphery of the visual field there is a loss in resolution so dramatic to prevent the recognition of the objects.

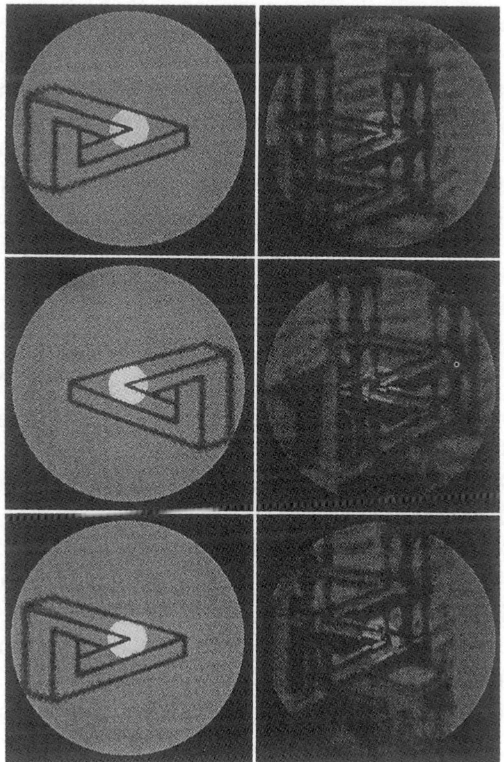

Figure 4

This process is equivalent to removing an increasing amount of high-frequency component from the center to periphery. Our elaboration introduced noise artifact represented by sharp discontinuities between filter elements; from a perceptive point of view, these discontinuities are not true, but are without influence in our demonstration. Since in these paintings the 3D consistency is not limited to the area of the fovea, wherever an observer fixes his/her eyes, during each pause of fixation, there exist spatial consistency; therefore, an observer can understand with his/her abstraction capacity the false relations between the different zones in these paintings only following sequentially and with many pause of fixation the impossible hidden figure. This demonstration should clarify that the visual process is a two instant mechanism: first, object perception is made comparing it with abstract categories stored in the memory, second, the visual system reconstructs the scene associating the several sub-images acquired by the fovea during fixation. Impossible figure are "impossible" since in the first phase are recognized as object belonging to an existing category. However, the second phase recognizes the inconsistency of this belonging. In the case of a space-invariant structure, we could acquire the entire image with a very high accuracy. This would allow us to gather the true image with only a fixation pause also in the case of an impossible figure. We have simulated this acquisition process by reducing the image to the size of the fovea. However, this would produce a very high increase of the information to process, due to the increased number of photoreceptors and connections involved. As a consequence, the vision system could not be able to provide a real time response in a dynamically changing environment.

4. References

Baily, C.H. (1981) "Visual System. I. The Retina", in: *Principles of Neural Science*, E. R. Kandel and J. H. Schwartz, eds, Elsevier, NewYork.

Biederman, I. (1985) "Human image understanding: Recent research and theory", *CVGIP* 32:29-73.

Fiorentini, A. (1989) "Differences between fovea and parafovea in visual search processes", *Vis. Res.* 29:1153-1164.

Kupla, Z. (1983) "Are impossible figures possible?", *Signal Process.* 5:201-220.

Marr, D. and H.K. Nishihara (1978) "Representation and recognition of thee dimensional shapes", *Proc. Roy. Soc. London B* 200:269-294.

Noton, D. and L. Stark (1971) "Scan paths in eye movements during pattern perception", *Science* 171:308-311.

Penrose, L.S. and R. Penrose (1958) "Impossible objects: A special type of illusion", *Brit. J. Psychol.* 49:31-33.

Sandini, G. and V. Tagliasco (1980) "An anthtropomorphic retina-like structure for scene analysis", *CGIP* 14:365-372.

Treisman, A.M. and H. Schmidt (1982) "Illusory conjunctions in the perception of objects", *Cognit. Psychol.* 14:107-141.

Watt, R.J. (1988) *Visual Processing: Computational, Psychophysical and Cognitive Research,* Erlbaum, London.

A REALISTIC NEURAL NETWORK SIMULATING FUNCTIONS OF A VISUAL CORTICAL MODULE

INNA Z. KREMEN
Laboratory for Neurobiology of Action Programming, Institute of Human Brain, Acad. Pavlov Street, 9a, St.-Petersburg, 197376, Russia

ABSTRACT

The paper presents a neural network based on neurophysiological data and simulating some functions of mammalian visual cortical module, such as orientational selectivity, spatial frequency analysis, texture discrimination and noise filtering.

1. Structure of the network

1.1. Network architecture

The network presented in the paper has been developed in the Laboratory for Neurobiology of Action Programming and is being studies during last years. General architecture of the cortical module is shown at Fig.1, *a*.

The first layer (R) represents the retina, the LGB-ON and LGB-OFF layers simulate two opponent layers in the lateral geniculate body and the other layers represent striate cortex, which is divided into two independent modules.

Each module has four layers of inhibitory interneurons (In_i, i =1..4), four layers of simple cortical neurons (S_i, i=1..4) and one layer of complex cortical neurons (C). The response of complex neuronal layer is considered as response of the whole network.

The two modules differ by spatial patterns of two types of the interneurons in the In_i layers. One module has a pinwheel interneuronal pattern and the other has a concentric circular one (Fig.1,*b*); these patterns were suggested from studying experimental data on distribution of neurons in different areas of mammalian visual cortex (Bauer et al., 1989; Bonhoeffer & Grinvald, 1991; Ghose & Tso, 1997).

The density of neurons in all the layers, except In_i, is equal to 1. The densities of interneurons' distribution in the planes In_i (i = 1..4) are represented as

for pinwheel: for circular:

$$n_{In(i)on} = 1 - \sin(\omega\psi + \pi i/2) \qquad n_{In(i)on} = 1 - \sin(2\pi\nu r + \pi i/2)$$
$$n_{In(i)off} = 1 + \sin(\omega\psi + \pi i/2) \qquad n_{In(i)off} = 1 + \sin(2\pi\nu r + \pi i/2) \quad (1)$$

Here, ω and ν – circular and radial frequencies determining the number of "pins" or "rings" in the mosaic; r and ψ – polar coordinates of inhibitory neuron in the plane In_i.

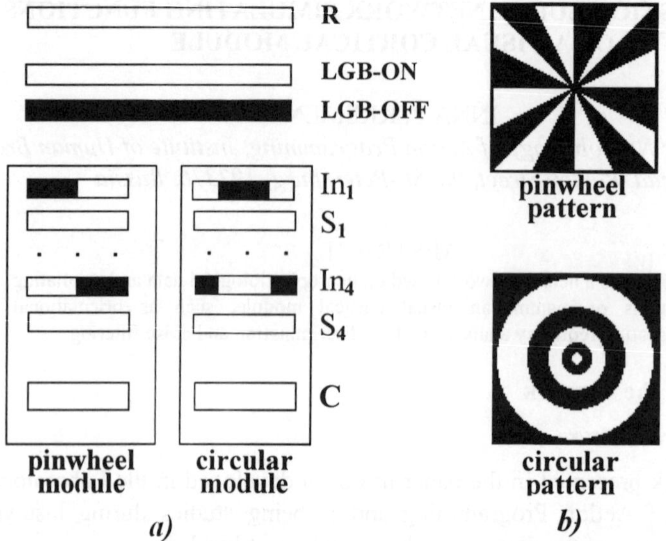

Figure 1. a) General scheme of the network layers; b) Two patterns of inhibitory interneurons' spatial distribution. Two types of model neurons – "on" and "off" – are indicated by white and black.

1.2. Network output signal

If input signal is $s(r,\psi)$, then a simple model cortical neuron input I_{Si} is derived as follows (here $^{(p)}$ superscript is used for pinwheel, $^{(c)}$ – for circular submodule):

$$I_{Si}^{(c)}(R,\varphi)=\iint \exp\left(-\alpha^{(c)2}[R^2+r^2+2Rr\cos(\varphi-\psi)]/R^2\right)\sin(2\pi vr+\pi i/2)S(r,\psi)rdrd\psi,$$

$$I_{Si}^{(p)}(R,\varphi)=\iint \exp\left(-\alpha^{(p)2}[R^2+r^2+2Rr\cos(\varphi-\psi)]/R^2\right)\sin(\omega\psi+\pi i/2)S(r,\psi)rdrd\psi \quad (2)$$

Here $\alpha^{(p)}$ and $\alpha^{(c)}$ – parameters determining the simple neuron receptive field width (for pinwheel and circular module respectively); $S(r,\psi) = s(r,\psi) - sv$, where sv - input averaged over the network field of view.

Outputs for model simple and complex neurons (O_S and O_C) are derived as

$$O_C^{(p)}(R,\varphi) = \sum_{i=1}^{4} O_{Si}^{(p)}(R,\varphi) = \sum_{i=1}^{4} I_{Si}^{(p)}(R,\varphi)\, H[I_{Si}^{(p)}(R,\varphi)] \quad (3)$$

$$O_C^{(c)}(R,\varphi) = \sum_{i=1}^{4} O_{Si}^{(c)}(R,\varphi) = \sum_{i=1}^{4} I_{Si}^{(c)}(R,\varphi)\, H[I_{Si}^{(c)}(R,\varphi)] \quad (4)$$

where $H(z) = 1$ if $z > 0$ and $H(z) = 0$ in the opposite case.

2. Results of the computer investigations of the network

2.1. Orientational and spatial frequency selectivity

Different functions and abilities of the network have been described in several papers (Kropotov et al., 1998; Norseen et al., 1999). Here we concern the capability of the network to perform sort of harmonic analysis similar to one carried out in real visual cortex (Glezer et al., 1989).

To study this question, sine-wave gratings were chosen as input signals:

$$s(r,\psi) = A(1+\sin[\Omega\, r \sin(\psi-\Phi)+\Theta]) \quad (A>0) \tag{5}$$

Substituting Eq. 5 into Eqs. 2-4, we obtain orientational and spatial frequency characteristics of the network. Complex neurons' outputs were shown to be invariant to spatial phase (Kropotov et al., 1998), like real complex cortical cells are (Glezer et al., 1989).

A pinwheel module neuron was found to give maximal response to gratings *collinear* to its radius vector in the module: $\Phi_{max}{}^{(p)}(R,\varphi) = \varphi$, while a circular module neuron – to *orthogonal* gratings: $\Phi_{max}{}^{(c)}(R,\varphi) = \varphi + \pi/2$. This property simulates orientational selectivity of real cortical neurons (Bonhoeffer & Grinvald, 1991; Ghose & Tso, 1997).

Fig. 2 illustrates spatial frequency selectivity of complex neurons. In the circular module all the neurons give maximal response to the same spatial frequency $\Omega_{max}{}^{(c)}(R,\varphi) = 2\pi\nu$, while a pinwheel module neuron gives maximal response to the frequency depending on the pinwheel pattern frequency and the polar radius of the neuron: $\Omega_{max}{}^{(p)}(R,\varphi) = \omega/R$. So we can say that a circular module has a characteristic *resonant* spatial frequency, while a pinwheel module *decomposes* the input signal by mapping different spatial frequencies to different module areas.

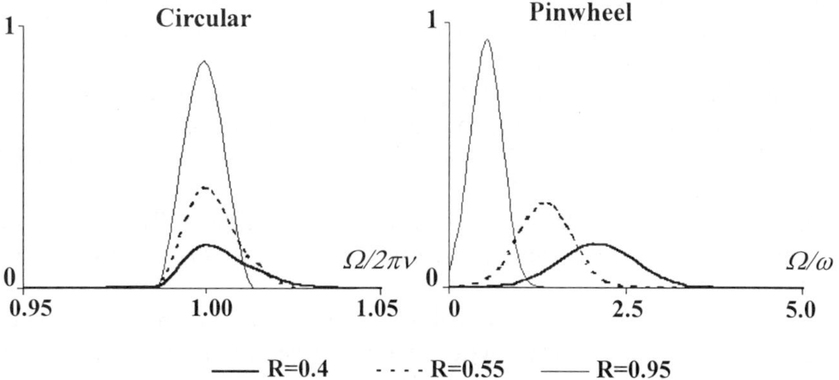

Figure 2. Spatial frequency characteristics of complex neurons.

2.4. Texture discrimination and noise filtering

A *texture* can be considered as a set of local orientation/frequency characteristics of visual pattern. So different textures produce activation of local patches located in different parts of the pinwheel module. Profiles of complex neurons' activities might be used for texture discrimination, interpreting a texture as a sum of several sine-wave gratings.

A pinwheel module can also discriminate textures in presence of additive white spatial *noise* (if signal-to-noise ratio (SNR) exceeds a certain level). Fig. 3 compares responses of a pinwheel module and a Gaussian filter (with the same radius) to a chess-like texture both without noise and with it (SNR = 4). It shows that the Gaussian filter *loses* texture information in the presence of noise, while the pinwheel module *keeps* it.

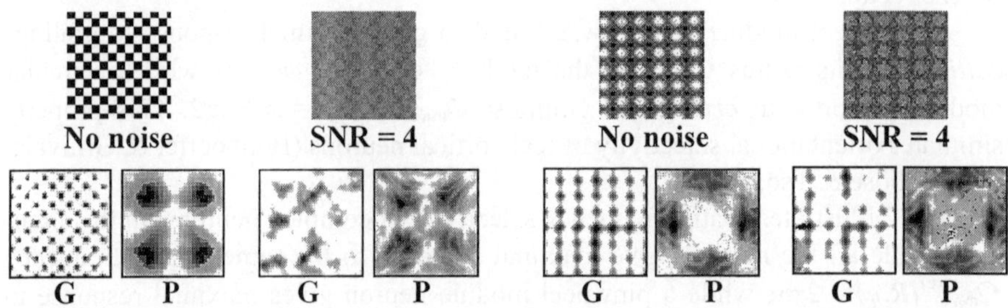

Figure 3. Responses of a Gaussian filter (G) and a pinwheel module (P) to different input textures in absence and in presence of additive noise.

References

Bauer, R., R. Eckhorn and W. Jordan (1989) "Iso- and cross-orientation columns in cat striate cortex: a re-examination with simultaneous single- and multi-unit recordings", *Neuroscience* 30:733-740.

Bonhoeffer, T. and A. Grinvald (1991)"Orientational columns in cat striate cortex are organized in pinwheel-like patterns", *Nature* 353:429-431.

Ghose, G.M. and D.Y.Tso (1997) "Form processing modules in primate area V4", *J. Neurophysiol.* 77:2191-2196.

Glezer,V.D., V.V.Yakovlev and V.E.Gauzelman (1989) "Harmonic basis functions for spatial coding in the cat striate cortex", *Vis. Neurosci.* 3:351-363.

Kropotov, J.D., V.A. Ponomarev and I.Z. Kremen (1998) "A realistic neural network simulating a visual system and its usage in tasks of invariant image description", *Optical Journal* 65:40-45.

Norseen, J.D., J.D. Kropotov and I.Z. Kremen (1999) "Bio-fusion for intelligent systems control", *Proceedings of the SPIE* 3719:410-417.

SYNCHRONIZATION IN THE VISUAL CORTEX: A BIOPHYSICAL APPROACH

A. DI GARBO, M. BARBI, S. CHILLEMI
Istituto di Biofisica CNR, Via S. Lorenzo 26, 56127 Pisa, Italia
E-mail: digarbo@ib.pi.cnr.it

ABSTRACT

A single functional column in the visual cortex is modelled as a FitzHugh-Nagumo oscillator. We report a numerical study of the synchronization phenomena occurring in a network with all-to-all synaptic coupling described by α function. A variety of states with different degree of synchronization (either in amplitude, or phase, or frequency) is found to occur, depending on the number of oscillators and on the coupling parameters.

1. Introduction

The information about the visual world arrives to the visual cortex. Complex information-coding processes occur in the cells of this area of the cerebral cortex. Current theories suggest that the brain processes the visual input by splitting it into its component features (colour, motion, depth, etc.). A very interesting question is how the brain binds these individual features of the visual scene. Recently it was proposed that this binding problem could be solved by synchronizing the firing time of the neurons that encode the features of the visual scene (Malsburg and Schneider, 1986). A striking characteristic of the visual cortex is its columnar organization: each cell, within a given functional column, will respond only to stimuli encoding a given orientation. Recent studies on the visual cortex of cats revealed oscillatory responses of the neurons in a given functional column induced by external stimuli (Gray and Singer, 1989). Moreover, stimulus dependent synchronization between spatially separate columns was also observed (Gray *et al.*, 1989). These experimental results suggest that the coherency between oscillatory activities in the visual cortex could be used to link together different features of a scene. How synchronization between different columns is achieved is not yet clear. However, the tangential intracortical excitatory connections between columns seem to be good candidates (Singer, 1990). As emphasized by Singer (1990), both the strength of the coupling and the characteristics of the synapses can modify the degree of synchronization between different functional columns.

In this paper we investigate the synchronization properties in a network of all-to-all coupled FitzHugh-Nagumo (FHN) oscillators (FitzHugh, 1961; Nagumo, 1962). The synaptic coupling is described by α function, $\alpha\, t\, e^{-\alpha t}$ (α^{-1} represents the duration of the synaptic current). In this context we mean that each unit of the network represents a synchronized neural population within a given functional column of the visual cortex. In particular we will study how the features of the excitatory synaptic currents, as the current duration and the coupling intensity, influence the synchronization properties of the network.

2. Network description

The FHN model is defined by the two first order differential equations:

$$\varepsilon \frac{dv}{dt} = v(v-a)(1-v) - w \ , \qquad (2.1a)$$

$$\frac{dw}{dt} = v - d\,w - b, \qquad (2.1b)$$

where v(t) is the fast voltage variable, while w(t) represents the slower recovery one. ε, a, b, d are the model parameters. The firing regime is reached through a supercritical Hopf bifurcation. The usual bifurcation parameter is b and, for the parameter values $\varepsilon = 0.005$, $a = 0.5$, $d = 1$ used here, the bifurcation occurs at $b_H = 0.264$. In the following, the value $b = 0.265$ will be used.

Our network consists of N identical FHN units each one being coupled to all the others. The network equations are thus:

$$\varepsilon \frac{dv_i}{dt} = F(v_i, w_i) + \sigma\, C_i(t), \qquad (2.2a)$$

$$\frac{dw_i}{dt} = G(v_i, w_i), \qquad (2.2b)$$

$$C_i(t) = \frac{\alpha^2}{N-1} \sum_{j \neq i}^{N} \sum_{k=1}^{n_j} (t - t_{j,k})\, e^{\alpha(t_{j,k} - t)}, \qquad (2.2c)$$

where the functions $F(v_i, w_i)$ and $G(v_i, w_i)$ are, respectively, the expressions in the right hands of equation (2.1a,b), $t_{j,k}$ (j = 1, 2, .., N) are the times when the j-th oscillator fires and n_j is the index of its last firing, $\sigma > 0$ represents the strength of the excitatory coupling and α characterizes the time scale of the "synaptic current". In our numerical simulations the firing condition is that v_j crosses from below a threshold set at 0.6. In order to investigate the synchronization properties of the network, we will use three different indicators, as defined in (Chillemi *et al.*, 1999) and describing the amplitude, phase and frequency coherences (for each indicator the value 0 means complete incoherence and the value 1 complete synchronization). The numerical integration of equations (2.2) was performed by using a 4th-order Runge-Kutta method (Mascagni, 1989) with integration step Δt = 0.002.

3. Results

First we investigated how the network dynamics depends on the number of oscillators. In Fig. 1 the three synchronization indicators are plotted against the number of oscillators for three different α values. For the highest α the network exhibits a complete synchronization independent of N while, for low and intermediate α values, the indicators, as the number of oscillators is increased, exhibit (except for the frequency indicator in the case $\alpha = 2$) an initial decrease and then reach a level approximately constant.

To study both the effect of the coupling strength σ and that of the duration of the synaptic current (defined as α^{-1}) on the synchronization of the network, three FHN units were simulated. The result are plotted in Fig. 2. Inspection of the plots shows that for low and high α values the coherence in frequency is always complete. All indicators exhibit a very slight dependence on σ. For the intermediate α values the plots are more complex and with different levels of synchronization for the three indicators. Moreover, this region widens as the coupling strength increases.

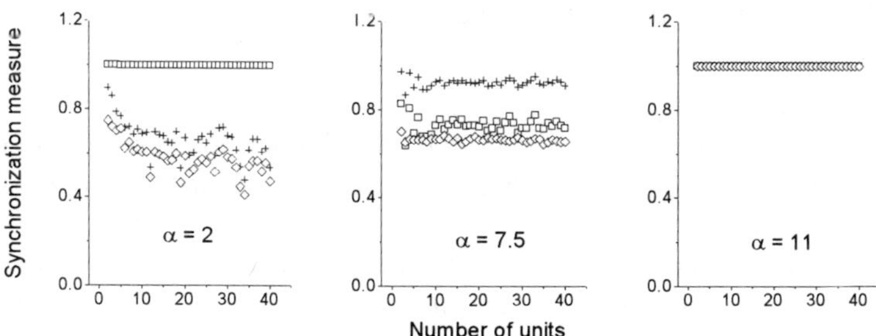

Figure 1. Synchronization measures against the number of units for different durations of the synaptic current (defined as α^{-1}). Strength of the excitatory coupling: $\sigma = 0.002$. Open squares: frequency synchronization; crosses: phase synchronization; open diamonds: amplitude synchronization.

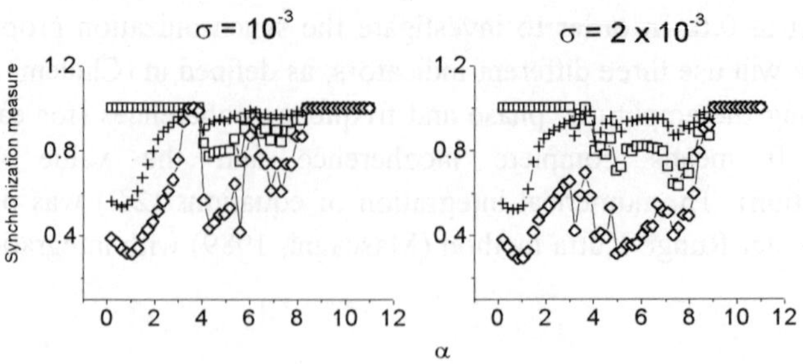

Figure 2. Synchronization measures against α (the duration of the synaptic current is defined as α^{-1}) for different strengths of the excitatory coupling, σ. Number of units: N = 3. Open squares: frequency synchronization; crosses: phase synchronization; open diamonds: amplitude synchronization.

4. Conclusions

In this study we modelled the functional columns of the visual cortex by FHN oscillators to investigate how the synchronization behaviour of a network of such units is influenced by the number of units and the features of their coupling. Our results suggest that the dynamics of large ensembles of model columns can be predicted from the knowledge of the small ones. The dependence of the network dynamics on the coupling strength is weak. Moreover, we found that a crucial role for the synchronization of the network is played by the duration α^{-1} of the synaptic coupling. In fact, the significative variation of the network synchronization obtained by simply varying this parameter could suggest relevant physiological mechanisms underlying the features binding of a visual scene.

References

Chillemi, S., M. Barbi, A. Di Garbo (1999) "Synchronization in a network of FHN units with synaptic-like coupling", *Lecture Notes in Computer Science* 1606:230-239.

FitzHugh, R.A. (1961) "Impulses and physiological states in theoretical models of nerve membrane", *Biophys. J.* 1:445-466.

Gray, C.M., P. Koning, A.K. Engel and W. Singer (1989) "Oscillatory responses in cat visual cortex exhibit inter-columnar synchronization which reflects global stimulus properties", *Nature* 338:334-337.

Gray, C.M. and W. Singer (1989) "Stimulus-specific neuronal oscillations in orientation columns of cat visual cortex", *Proc. Natl. Acad. Sci. USA* 86:1698-1702.

Mascagni, M.V. (1989) "Numerical methods for neuronal modeling" in: *Methods in Neuronal Modeling*, C. Koch and I. Segev, eds, MIT Press, Cambridge, MA, Massachusetts, pp. 439-485.

Malsburg, C. von der and W. Schneider (1986) "A neural cocktail-party", *Biol. Cybern.* 54:29-40.

Nagumo, J.S., S. Arimoto, S. Yoshizawa (1962) "An active pulse transmission line simulating nerve axon", *Proc. I.R.E.* 50:2061-2070.

Singer, W. (1990) "Search for coherence: A basic principle of cortical self-organization", *Concepts in Neuroscience* 1:1-26.

THE INTERPOLATION BETWEEN UNSIMILAR VIEWS OF A 3-D OBJECT INCREASES THE SIMILARITY AND DECREASES THE SIGNIFICANCE OF LOCAL PHASE

GABRIELE PETERS

Institut für Neuroinformatik, Ruhr-Universität Bochum,
Universitätsstraße 150, D-44780 Bochum, Germany

ABSTRACT

For 3-D object recognition it is useful to partition the viewing hemisphere of an object into areas of similar views, termed *view bubbles*. I generate view bubbles by tracking local object features in the form of Gabor wavelet responses to the east, west, north, and south direction until a termination condition is met. I examine two termination conditions: First, the starting view is compared *directly* to other views on the hemisphere, and the boundaries of the view bubble are reached if a similarity threshold is reached. Second, the starting view is compared to a view *interpolated* from other views. I use two different similarity functions. One takes the phases of the Gabor responses into account, the other does not. For each condition the areas of the resulting view bubbles are calculated. Two major results are obtained: 1. The similarity of a view to an interpolated view is larger than to either of both views from which the interpolation was calculated. 2. The influence of the phases on the similarity is diminished by interpolation.

1. Introduction

For the purpose of 3-D object recognition the viewing hemisphere of an object can be partitioned into areas of similar views, termed *view bubbles*. The preprocessing and the generation of view bubbles is described in (Peters *et al.*, 1999). Each image of the hemisphere is preprocessed by a gray level segmentation described in (Vorbrüggen, 1995). Then a grid graph is put on the object. Each vertex of the graph is labeled with Gabor wavelet responses, which provide a local description of the surroundings of the vertex. Such a local descriptor J (called *jet*) consists of amplitudes a_i and phases ϕ_i for each wavelet i and allows the tracking of the vertex from one view to a neighbouring view by a phase-based disparity estimation. This is described in Maurer and Malsburg, 1995.

2. Generation of View Bubbles

For each view of the hemisphere I create the affiliated view bubble by tracking the grid graph to the east, west, north, and south direction on the viewing hemisphere until a termination condition is met. If *(i, j)* is the view for which I want to create the view bubble I track the graph representing view *(i, j)* to the views *(i-1, j)* and *(i+1, j)* and examine the termination condition for them. If it is not yet fulfilled, I track the graph to the views *(i-2, j)* and *(i+2, j)*. I stop this procedure until the termination condition is met. By doing the same for the north-south direction I obtain four views *(i-n, j)*, *(i+n, j)*, *(i, j-m)*, and *(i, j+m)* on the hemisphere, which define the view bubble for the starting view. To depict a view bubble I draw an ellipse through these four views (see figure 1).

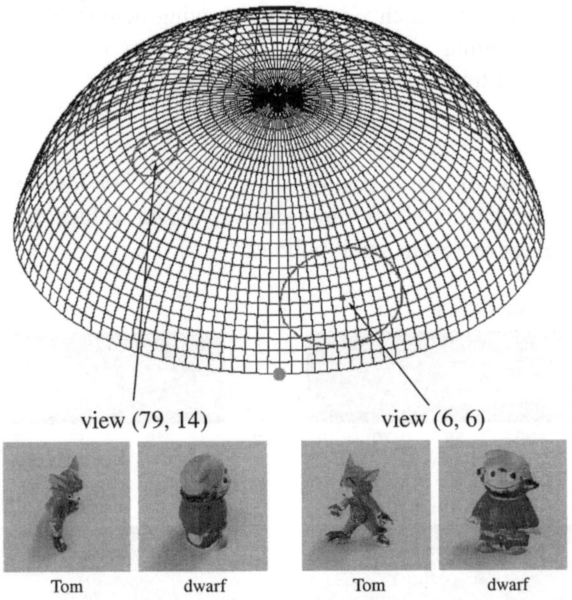

Figure 1. Viewing Hemisphere with Examples of View Bubbles. Each crossing of the grid stands for one view. Two examples of view bubbles are depicted on the hemisphere grid. Neigbouring views have a distance of 3.6 degrees on longitudes as well as on latitudes.

2.1. Termination Conditions

I examine two termination conditions: In the first case, the boundaries of the view bubbles are reached if the *similarities* of the current tracked east and west view (resp. north and south view) to the starting view are below a preset threshold (condition SIM). In the second case, I *interpolate* the Gabor responses of the east and west view (resp. north and south view) and stop tracking if the similarity of the interpolated view to the starting view is below the threshold (condition INT).

I use two different similarity functions for either condition: $S(J,J') = 0.5 \cdot \left(\left(\sum_i a_i a_i' \cos(\phi_i - \phi_i')\right) / \left(\sqrt{\sum_i a_i^2 \sum_i a_i'^2}\right) + 1\right)$ takes the phases of the complex wavelet responses into account, $\tilde{S}(J,J') = \left(\sum_i a_i a_i'\right) / \left(\sqrt{\sum_i a_i^2 \sum_i a_i'^2}\right)$ does not.

This results in four conditions: SIM/PHASE, SIM/NO_PHASE, INT/PHASE, and INT/NO_PHASE. For each condition I choose the same similarity threshold. I calculate view bubbles for each view of the viewing hemisphere. Then I compare the areas of the resulting view bubbles (i.e., the areas of the ellipses described above) for each condition.

3. Results

The distribution of the areas of view bubbles are depicted for all four conditions for the object "Tom" in figure 2. Light values encode large view bubbles, dark values represent small bubbles.

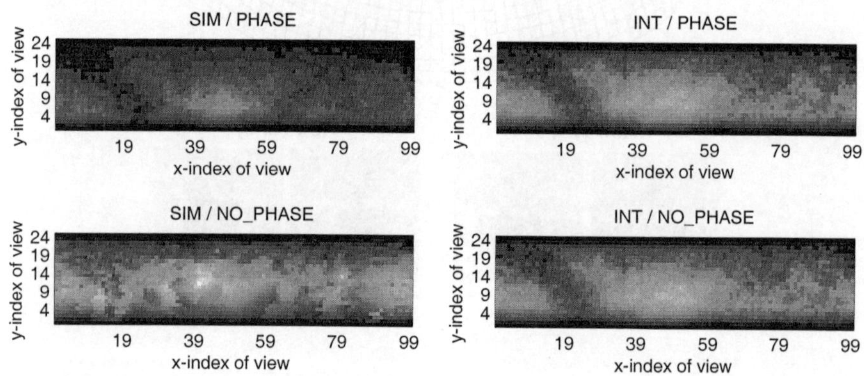

Figure 2. Areas of View Bubbles. Left diagrams: SIM condition. Right diagrams: INT condition. Upper diagrams: PHASE condition. Lower diargams: NO_PHASE condition.

I obtain two main results:

- The INT/PHASE condition provides larger bubbles than the SIM/PHASE condition for the majority of views. The interpolation from two unsimilar views retrieves similarity to the center view, i.e., the similarity of the center view to the interpolated view is larger than the similarity to either of both views from which the interpolated view was calculated (see figure 3).
- The INT/NO_PHASE condition provides exactly the same view bubbles as the INT/PHASE condition, which means that the phases are useless for the generation of view bubbles. The influence of the phases on the similarity is diminished by interpolation.

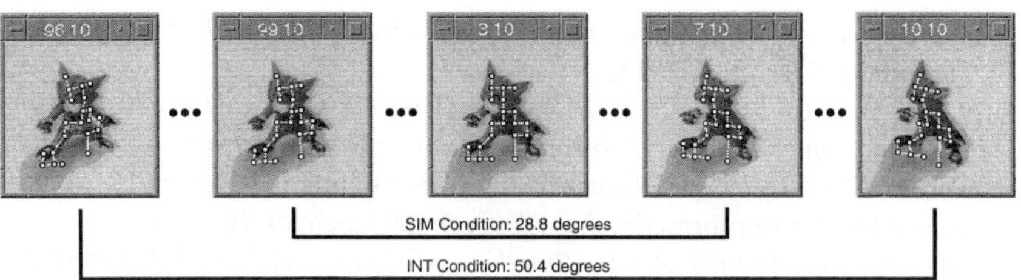

Figure 3. Size of Example View Bubble in East-West-Direction. The size of the view bubble for the example view (3, 10) is larger for the INT condition than for the SIM condition. The view interpolated from the *unsimilar* views (96, 10) and (10, 10) is *similar* to the center view (3, 10).

The condition SIM/NO_PHASE provides the largest view bubbles of all conditions, but the correspondences of the tracked object features become poor at the periphery of the view bubbles, in contrast to the other conditions where the vertices of the tracked graphs remain at corresponding object features even at the boundaries of the view bubbles (see figure 3 again).

As a side-effect centers of areas of large view bubbles can be regarded as *canonical views* (see figure 4).

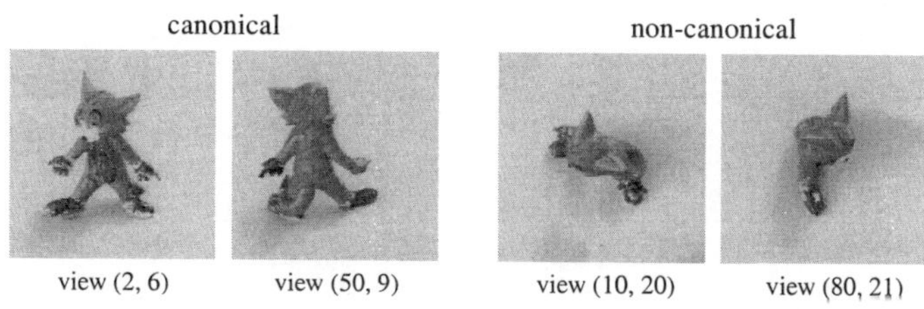

Figure 4. Canonical and Non-Canonical Views. Compare with figure 2 for view numbers.

Acknowledgements

I want to express my thanks to Prof. Christoph von der Malsburg for supporting this work, Michael Neef for system administration, Uta Schwalm for office administration and Pervez Mirza for inspiring talks and for proof-reading this paper.

References

Maurer, T. and C.v.d. Malsburg (1996) "Tracking and Learning Graphs and Pose on Image Sequences of Faces", *Proceedings of the 2nd International*

Conference on Automatic Face- and Gesture- Recognition, Killington, Vermont (USA), pp. 176-181.

Peters, G., B. Zitova and C.v.d. Malsburg (1999) *A Comparative Evaluation of Matching and Tracking Object Features for the Purpose of Estimating Similar-View-Areas of 3-Dimensional Objects*, Internal Report IRINI 99-06, Institut für Neuroinformatik, Ruhr-Universität Bochum, Bochum.

Vorbrüggen, J.C. (1995) *Zwei Modelle zur datengetriebenen Segmentierung visueller Daten*, volume 47 of *Reihe Physik*, Verlag Harri Deutsch, Thun, Frankfurt am Main.

SIMPLIFYING RAW IMAGES

CARLO ARCELLI and LUCA SERINO
Istituto di Cibernetica, CNR, 80072 Arco Felice, Napoli, Italy

ABSTRACT

We describe a process which transforms a grey-tone digital image constituted by a large number of elementary regions, each one consisting of a connected set of pixels with the same grey-value, into a simplified image where only a reduced number of connected sets of pixels with the same grey-value, called components, is present. Components are expected, as a whole, to give an impression of what is seen by an observer, only on the basis of the information related to the grey-value associated to every pixel and without any specific knowledge of the contents of the image. Morphological operators, including also reduction operators and reconstruction operators, are employed to carry out the process.

1. Introduction

An initial step in digital image processing is to find an organisation of the contents of the image; for instance, by identifying a number of regions among which an appropriate procedure could select the ones constituting the image subset of interest for the specific task at hand (Everat *et al.*, 1997; Lowe, 1985; Ullman, 1998).

A grey-tone image may be understood as a set of elementary regions, each one consisting of a connected set of pixels with the same grey-value. Scope of the paper is to describe a process which removes most of the elementary regions and expands the remaining ones so as to obtain a new image where only a reduced number of connected sets of pixels with the same grey-value (called components) is present. Components are found in correspondence with some of the elementary regions the grey-value of each of which is a local extreme of intensity, *i.e.*, they are locally lighter or locally darker connected sets of pixels. Components are expected, as a whole, to give an impression of what is seen by an observer, only on the basis of the information related to the grey-value associated to every pixel and without any specific knowledge of the contents of the image.

We deal with images where darker subsets and lighter subsets are respectively understood as constituting foreground and background. In particular, we take into account images where the foreground is perceived as union of elongated shapes. In fact, the components are obtained by using reconstruction operators applied to an input image modified in accordance with the watershed transformation (Beucher and Meyer, 1993), which is particularly suitable to highlight elongated image subsets. The resulting image is characterised by a high difference in grey-value between adjacent components and, besides being a simplified version of the input, also constitutes a perceptually meaningful sketch of it.

2. Outline of the transformation

Consider a digital image with pixels having one of the grey-values g_k, $k = 0, 1, \ldots, N$, and assume that darker areas correspond to sets of pixels with higher grey-value. We regard the image as a mountainous relief (the grey-value of a pixel being interpreted as its height) and call a top (a bottom) any region whose adjacent regions have all lower (higher) grey-values. The watersheds are digital lines, generally dividing adjacent bottoms, and are constituted by pixels the neighbourhood of each of which has a structure in agreement with topographical features such as ridges, peaks and saddles. Watersheds are found by applying reduction operators to the image, in order to expand bottoms, while simultaneously eroding darker regions, until only a one-pixel-thick line remains to separate adjacent bottoms or to identify a promontory (Arcelli and Serino, 1998; Arcelli and Serino, 2000).

To avoid the creation of a too busy network of watersheds, it is necessary that the image be suitably pre-processed by filtering out regions with negligible size. To this purpose, we resort to the consecutive use of morphological opening and morphological closing (Serra, 1982) to remove narrow peaks and pits, and to the construction of multi-valued regions (Wang and Bhattacharya, 1996) to obtain only a number of significant (*i.e.*, deep enough) bottoms.

Once the set of watersheds has been obtained, we modify it into a number of connected subsets (branches), by removing any pixel which either does not constitute a high separation between its adjacent bottoms or belongs to a promontory sticking out not significantly from the surrounding bottom.

In detail, for every watershed the grey-value $g(q)$ of any pixel q is compared with the maximum grey-value, say M, present in the image and with the grey-values, say g_r and g_s, of the adjacent bottoms. Particularly, q is lowered to $\min(g_r, g_s)$ if any of the following occurs:

$$g(q) - \max(g_r, g_s) < t_1 \cdot (g(q) - \min(g_r, g_s))$$
$$g(q) - \min(g_r, g_s) < t_2 \cdot (M - \min(g_r, g_s))$$

Where t_1, t_2 are suitable thresholds, respectively set to 0.25 and 0.6 in this paper.

Bottoms which have become adjacent because of the lowering of the previous pixels are assigned a common grey-value equal to the minimum of the grey-values possessed by the bottoms. Then, the i-th branch is labelled with a grey-value M_i which is the maximum among the grey-values characterising its pixels.

To obtain the desired image components, we apply a reconstruction operator to the image resulting after watershed modification. Image reconstruction is achieved by regarding each branch as a seed which propagates its label M_i over the neighbouring bottoms and assigns new grey-values to the pixels of the bottoms.

The pixels of all the branches iteratively propagate parallelwise their grey-value to the pixels of the adjacent bottoms. Any pixel reached during propagation, say *p'*, assumes one out of two possible grey-values depending on M_i, on the grey-value of the bottom *p'* belongs to, and on the grey-value *g(p)* of the pixel *p* placed in the corresponding position of the input image.

If g_k denotes the grey-value of the k-th bottom and if, for every pixel *p'* of that bottom, the corresponding pixel *p* in the input image has grey-value *g(p)*, we consider the reconstruction operator which assigns the new grey-value *g(p')* as follows:

$$g(p') = M_i \text{ if } g(p) > t_2 \cdot (M_i - g_k) + g_k$$
$$g(p') = g_k \text{ otherwise.}$$

Since seed propagation occurs parallelwise, it may happen that a bottom pixel be reached by more than one grey-value M_i (each grey-value M_i being propagated from a different branch). In this case, we select as new grey-value *g(p')* the one related to the lowest of the propagated grey-values reaching *p'*.

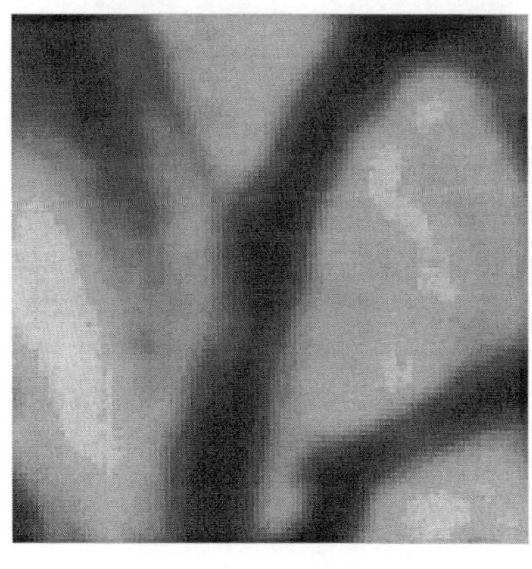

a)

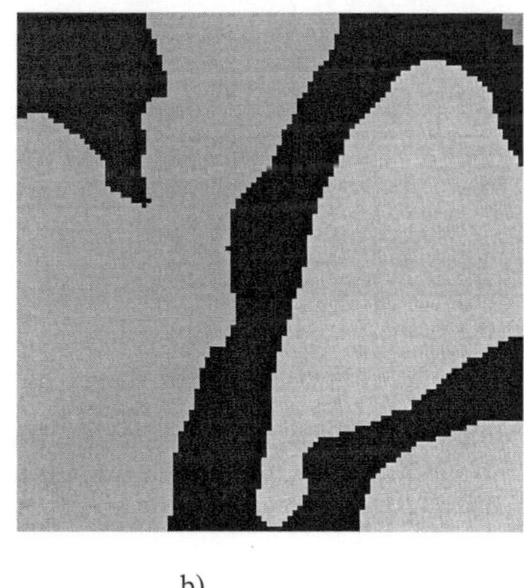

b)

Figure 1. a) Input image; b) Output image.

3. Concluding remarks

We have illustrated an algorithm to transform a grey-tone image into one constituted by a reduced number of components and grey-values. Three main steps have been accomplished during the process. The first is a pre-processing step, during which the input image is cleaned by filtering out regions with negligible size.

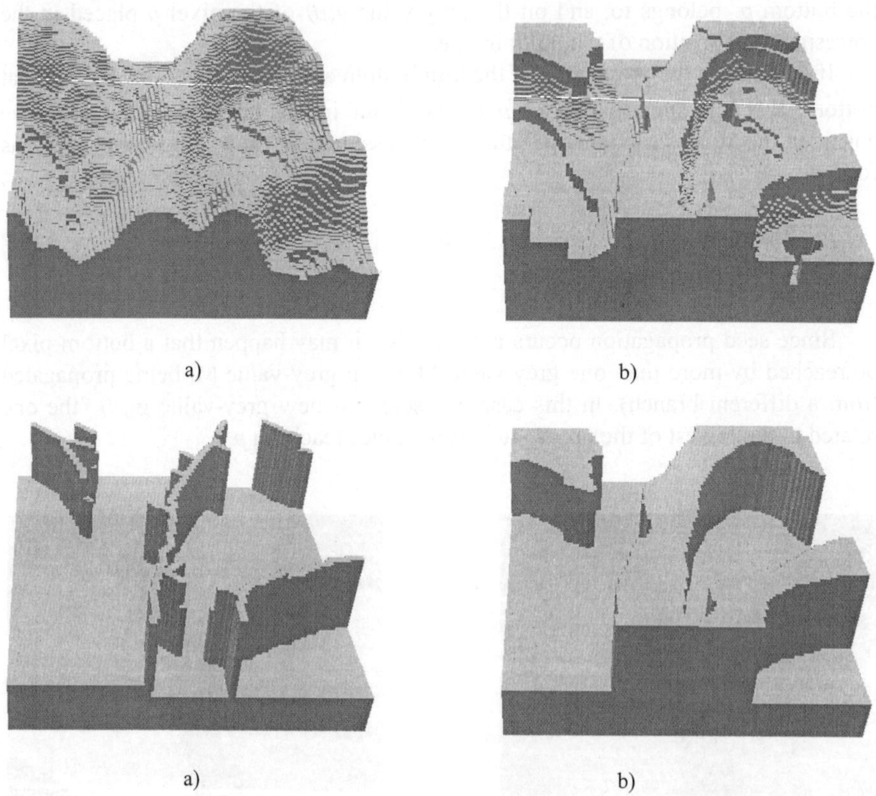

Figure 2. Topographic representation: a) Input image; b) Pre-processed image; c) Watersheds and expanded bottoms; d) Output image.

The second step computes the watershed transformation and finds out the seeds from which the desired components will be originated. Finally, during the third step, the components constituting the simplified image become available by growing the seeds and applying reconstruction operators.

Work regarding the performance of the algorithm has been carried out by analysing a number of test images constituted by parts of photos of magnified biological material (neurons) scanned at 300 d.p.i., 256 grey-levels. As an example, see Figures 1, 2. It is straightforward to observe that the described approach greatly reduces the number of components and of grey-values present in the image, and aims at producing a sketched version of the input, where foreground and background are more clearly distinguishable. In particular, we may note that the shown example is characterised by 6130 regions and 111 grey levels in the input image, 1303 regions and 66 grey levels in the pre-processed image and 7 regions and 6 grey levels in the output image.

References

Arcelli, C. and L. Serino (1998) "Parallel lowering of digital pictures by topology preserving operations", *Proc. 14th International Conference on Pattern Recognition*, Brisbane, pp.1601-1603.

Arcelli, C. and L. Serino (2000) "Parallel reduction operators for grey-tone pictures", *Int. Journal of Pattern Recognition and Artificial Intelligence* 14: in press.

Beucher, S. and F. Meyer (1993) "The morphological approach to segmentation: the watershed transformation", in *Mathematical Morphology in Image Processing*, E.R. Dougherty, ed., New York: Marcel Dekker, pp. 433-481.

Everat, J.C., G. Bertrand and M. Couprie (1997) "Reconstruction operators for image segmentation", in: *Advances in Visual Form Analysis*, C. Arcelli, L.P. Cordella and G. Sanniti di Baja, eds., Singapore: World Scientific, pp.188-197.

Lowe, D.G. (1985) *Perceptual Organization and Visual Recognition*, Boston: Kluwer.

Serra, J (1982) *Image Analysis and Mathematical Morphology*, London: Academic Press.

Ullman, S. (1998) "Aspects of segmentation and object recognition", in: *Downward Processes in the Perception Representation Mechanisms*, C. Taddei-Ferretti and C. Musio, eds., Singapore: World Scientific, pp. 129-144.

Wang, Y. and P. Bhattacharya (1996) "On parameter-dependent connected components of gray images", *Pattern Recognition*, 29, pp. 1359-1368.

PRESERVING PATTERN FEATURES AT DIFFERENT SCALES

GIULIANA RAMELLA and GABRIELLA SANNITI DI BAJA
Image Analysis Department, Istituto di Cibernetica, CNR
Via Toiano 6, 80072 Arco Felice, Naples, Italy

ABSTRACT

Resolution (digital) pyramids can be used to simulate the effect produced on the shape of a pattern by a modification of the observation distance in the real world. Binary OR- and AND-pyramids are easy to build, but distort the shape significantly. Here a grey level pyramid is built, which produces significantly better results.

1. Introduction

Images in the real world appear different when seen from different distances. There is a connection between distance and size. Some image features can be interpreted as having a finer structure and tend to disappear as soon as the distance from which the image is seen increases. Other features, the most significant ones, are preserved longer, when the observation distance increases, and are often enough to roughly represent the image. When this is the case, human beings can still perform recognition regardless of the observation distance.

In general, a real image can be interpreted as a continuous function of grey level along the image. Digitisation and quantization transform the real image into a digital picture. To build representations at different resolutions one should not necessarily resort to repeated digitization and quantization of the real image. Starting from the highest available resolution picture, all desired lower resolution pictures are created by means of a decimation process that associates with suitably identified subsets of pixels, called the *children*, a single pixel, called the *parent*, in the lower resolution picture. In this way, a pyramid structure is created. This communication is the follow up of previous papers in this field, see all papers in section References. Here, we build a resolution pyramid and focus on the task of preserving shape as much as possible when resolution decreases.

2. The Multi-Scale Representation

For the sake of simplicity we will refer to images that in the real world can be perceived as black (pattern) and white (background) and have, as their digital counterpart, binary images including pixels with values 1 and 0, respectively. In this way, we will restrict ourselves to space resolution only. The most straightforward example of a multi-resolution picture is the 2×2 binary pyramid. Let us suppose that the original, highest resolution picture be $2^n \times 2^n$. Otherwise, white rows and columns are added as needed. The next, lower, resolution level is built by partitioning the picture in 2×2 blocks of *children*, and associating to each block the relative *parent*. The colour of the parent pixel is determined according to an a priori fixed *rule*. The partitioning process is repeated onto the previously computed lower resolution representation, and then further iterated to build all possible resolution levels, ending in a single pixel in the lowest resolution level.

The logical OR and AND operations are the two rules most commonly used to build binary pyramids, but they are not adequate to preserve shape information when resolution decreases. The pattern is transformed into a sort of amorphous blob, if the OR operation is used; in fact, one black child in a 2×2 block is enough to cause the parent pixel to be black. If the AND operation is preferred, the pattern is rapidly shrunk and its narrower regions tend to completely vanish or become disconnected; in fact, all children in a 2×2 block should be black to originate a black parent. In any case, the shape of the pattern is not adequately preserved in OR- and AND-pyramids.

As an example, see Figure 1, where an AND-pyramid is built starting from a 64×64 picture. Only four levels are shown, as lower resolution pictures consisting of 4×4, 2×2 and 1×1 pixels do not include enough pixels to provide an adequate shape representation of the pattern, or are totally empty.

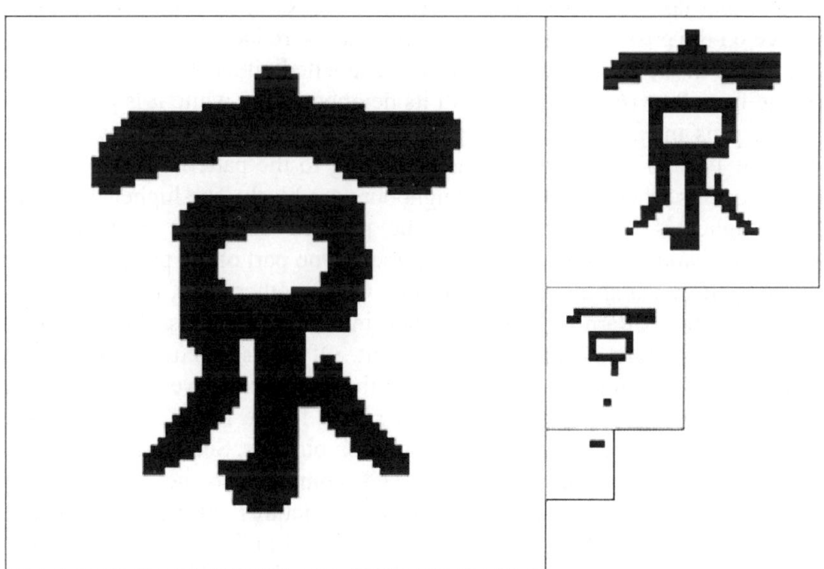

Figure 1. A binary AND-pyramid.

Besides the rule used to determine the colour of the parent pixels, also the *position* of the grid used to partition the picture into 2×2 blocks influences the shape of the resulting lower resolution pattern. The grid can be shifted in four different positions, originating four different pictures at the next lower level. From each picture, four lower resolution pictures are possible, and so on. Better results can be obtained by using a combination of the four possible pictures resulting when the grid is shifted, and by using grey values to take into account the local configurations from which the parent pixels originate. To this aim, the decimation rule should be based on the examination of a set of pixels larger than a 2×2 block.

3. The Grey Level Pyramid

Let us consider a pixel p in a given resolution picture and its eight neighbours. The partition grid can be shifted in four positions, so originating four 2×2 blocks of children, each of which including p. Edge neighbours of p belong to a pair of the above four 2×2 blocks, while point neighbours belong to only one block. The occurrences of p and its neighbours in the four blocks containing p can be used to form a 3×3 mask of weights, with central value 4, edge-neighbour value 2 and point-neighbour value 1. By using this mask on every other pixel in the highest resolution picture, we compute a grey value for the pixel p', parent of a 2×2 block including p, that takes into account all four 2×2 blocks including p. In this way, dependence on the position of the partition grid can be reduced.

Since pixels in the initial binary picture are labelled either 0 or 1, grey values for p' range from 0 to 16. When p and all its neighbours are white it is p' = 0, and in this case p' is interpreted as belonging to the background. Pixels with value ranging from 1 to 16 are interpreted as belonging to the pattern. The maximum value for p' occurs when p and all its neighbours are black. The higher their grey level, the higher is their relevance in the pattern. Since the basic pattern-background decisions are those of the OR-pyramid, no part of the pattern is lost as it would happen by using the AND operation. On the other hand, the use of grey values prevents the pattern to be transformed into an amorphous blob. The next lower resolution picture is then similarly built. Of course, pattern grey values would now be in the range {1, 256}. To limit the complexity, we linearly rescale the values assigned to the non-zero pattern pixels to the range {1, 16} before computing similarly all remaining lower resolution pictures. See Figure 2.

Pattern shape is noticeably better preserved, compared to the AND-pyramid case, even at the lowest resolution. The obtained pictures can be interpreted as discrete elevation maps, where the grey level of a pixel indicates the height of the region that is the continuous counterpart of the pixel itself. The original binary picture can be interpreted as a landscape of flat *islands*, some of which possibly including one or more *lakes*. In the pictures with lower resolution, the lakes (as well as all other pattern and background components) are shrunk, but, even when they do not include any white pixel, they can still be identified - as *craters* - in the elevation map. Narrow *peninsulas*, that in a binary AND-pyramid would have soon disappeared or become disconnected, are detectable also when resolution decreases, although their height diminishes.

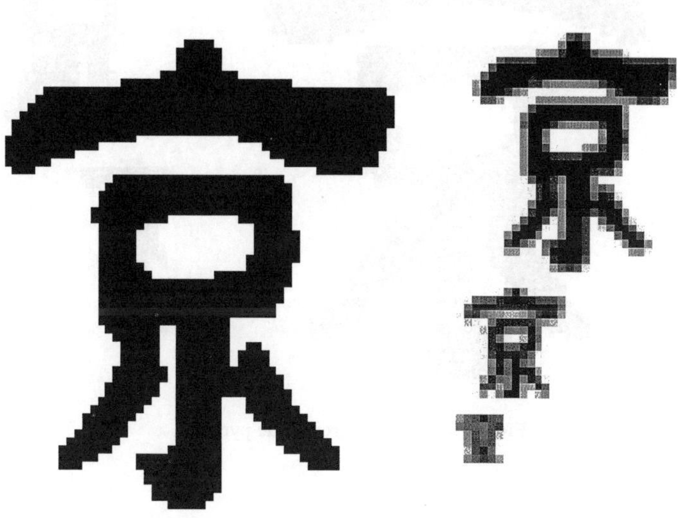

Figure 2. The grey value pyramid.

Unfortunately, also some strips of land emerge, when the resolution decreases, and link regions that were sufficiently close to each other in the previous resolution picture. Due to this unwanted side effect, *bays* are likely to be transformed into craters and regions that were not connected at higher resolution are possibly merged into a unique connected component. To solve these problems, the pyramid has to be analysed to infer geometrical and topological information from the highest resolution level onto the lower levels. Only craters corresponding to lakes in the input binary picture can originate (true) craters at lower resolution. All other craters, having no descendants belonging to lakes in the highest resolution picture, are spurious craters and as such are removed. To this purpose, a lowering process that suitably reduces the grey value of some pixels along the border of the spurious craters is used. Lowering is iterated until all spurious craters are transformed into bays. To avoid unwanted merging of distinct regions, connected component labelling is accomplished to ascribe a different label to every connected component in the binary picture preliminarily. The label is then taken into account when computing the grey level pyramid and allows us to separate regions whose merging was not desired. The resulting grey level picture is shown in Figure 3. Shape is now very well preserved. Also topology is fairly well preserved through pyramid levels.

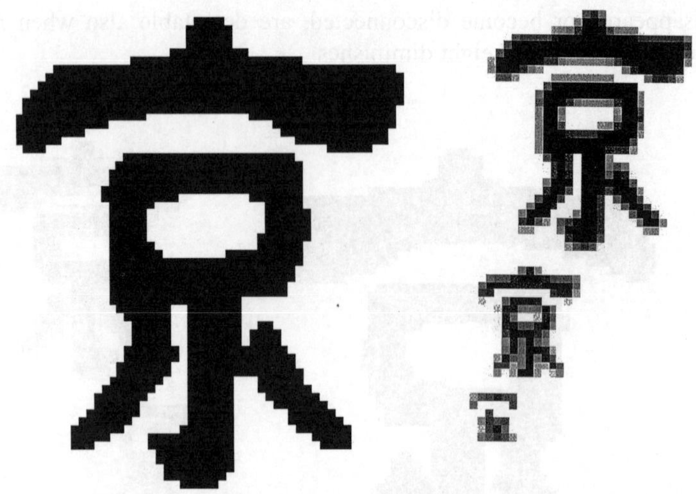

Figure 3. The final grey value pyramid.

4. Conclusion

Starting from a binary picture a grey level pyramid is built in such a way to preserve shape (and topology) as much as possible, when resolution decreases. The grey level pictures constituting the pyramid levels can be interpreted as discrete elevation maps, where the grey level of a pixel indicates the height of the region that is the continuous counterpart of the pixel itself. Our method is designed to preserve perceived shape, more than to maintain topology. However, topology *is* fairly well preserved. Changes of binary topology are, in any case, unavoidable when image resolution is decreased. Maintaining the connectedness in the pattern and avoiding the creation of holes (i.e., maintaining the connectedness of the background) are dual problems that can never be solved simultaneously. Our approach favours maintenance of pattern connectedness.

References

Borgefors, G. and G. Sanniti di Baja (1996) "Shape preserving binary pyramids", *Proc. RECPAD'96*, Guimarães (P), pp. 197-203.

Borgefors, G., G. Ramella and G. Sanniti di Baja (1996) "Multiresolution representation of shape in binary images", in: *Lecture Notes in Computer Science*, vol. 1176, Berlin: Springer-Verlag, pp. 51-58.

Borgefors, G., G. Ramella, G. Sanniti di Baja and S. Svensson, (1999) "On the multi-scale representation of 2D and 3D Shapes", *GMIP* 61:44-62.

PICTORIAL AND VERBAL COMPONENTS IN ARTIFICIAL INTELLIGENCE EXPLANATIONS

E. BURATTINI*, M. DE GREGORIO* and G. TAMBURRINI°
Istituto di Cibernetica, C.N.R., I-80072 Arco Felice (NA), Italy
°Dipartimento di Filosofia, Università di Pisa, 56126 Pisa, Italy
e-mail: {ernb, massimo, gugt}@sole.cib.na.cnr.it

ABSTRACT

Neurosymbolic computational systems exploit largely complementary advantages of both neural and symbolic information processing (Sun and Bookman, 1995). This broad feature of neurosymbolic systems is clearly appealing for modelling tasks that involve both verbal and visual information: in view of their learning and associative capacities, neural networks are known to perform well in visual pattern recognition, whereas symbol manipulation algorithms are tailored for "higher-level" reasoning. This paper investigates the integration of verbal and visual information in expert system explanation, from a hybrid neurosymbolic perspective. This sort of integration is a desirable feature for many expert system domains, as the effectiveness of explanations provided by human experts is known to be often enhanced by multiple representation techniques, using both words and pictures (Tabachneck-Schijf *et al.*, 1997).

1. Introduction

The explanation task computationally modelled here is relative to a visual classification process, schematised in Burattini *et al.* (1998; 2000) as a process involving (a) detection of visual cues from an input image (b) rule-based abductive reasoning, enabling one to advance classification hypotheses about a visually presented object, (c) prediction and testing of new visual cues from the selected hypotheses, in order to arrive at a final classification, if any.

The visually presented objects that we selected for instantiating this classification process are arches of various kinds (lancet, round, lintel, polygonal, segmental, polycentric), as shown in actual photographs of house portals. Arches are a classical example of computational learning theory, first discussed by Winston (1975) in the framework of a simplified block world. More recently, arches are a chief example used in Glasgow's discussion of visual (more specifically, diagrammatic) imagery in problem solving (Glasgow, 1993). Both learning from examples and visual imagery play a significant role in the present neurosymbolic approach to expert explanation, exemplified by means of the hybrid neurosymbolic system ARCS (Arches Recall and Classification System - part of an expert system analysing data concerning buildings with unknown construction records in historical centres of Naples province, Italy). ARCS, however, does not inhabit a block world, since its input is formed by photographs of actual building portals, and the sorts of recalled images it makes use of in explanation are more "pictorial" than the diagrams in Glasgow's proposal.

The information stored during training phases enables the neural component of ARCS to recognise and to generate characteristic visual features for arch shapes (Burattini *et al.*, 1998; 2000). The capability of detecting such visual features is used in classification. The generation of visual features is crucial for a vivid explanation, when coupled to a verbal account of the reasoning process.

2. Justification and teacher-mode explanation

There are two different explanation modes in ARCS: justification and teacher-mode explanation.

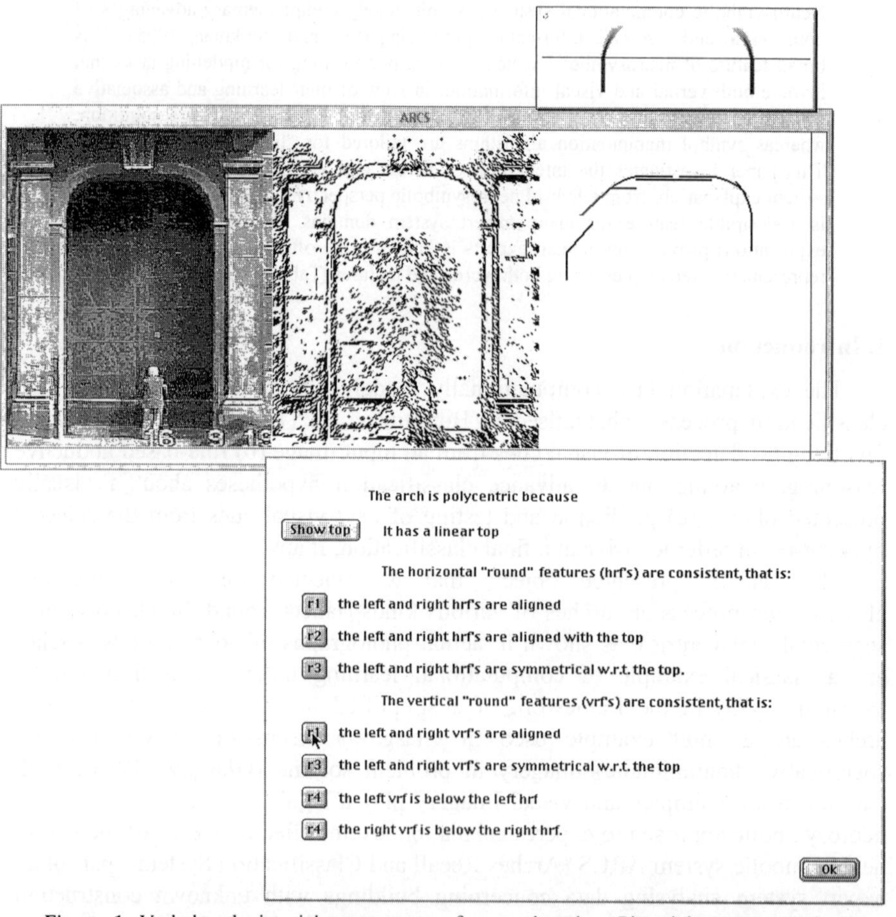

Figure 1. Verbal and pictorial components of an explanation. Pictorial component: input image (left), filtered input (centre), detected features in input image (right), generated visual cues corresponding to the preferred interpretation with proper alignment (top right). The verbal component (bottom) tells the users why the detected features allow the system to converge on the given classification hypothesis.

In the justification mode, ARCS has to explain the results of the classification process, if any. Justification starts with the neural component of ARCS generating the initially detected visual features; thereafter, its symbolic component traces the rule-based, abductive reasoning steps leading from such visual cues to classification hypotheses. The latter are selected from a fixed set of stored hypotheses. Then, neural and symbolic components of ARCS cooperate to trace the hypothesis-driven search for additional visual cues: for each abduced hypothesis h_i, the symbolic component lists the visual features that should or may be detected, if h_i were correct, and their arrangement in the input image; the neural component generates these visual features, and the ones that were actually detected are singled out. (It is worth noticing that this stage of the justification process may correspond to one or more cycles of the abduction-prediction-test loop in the classification process). Subsequently, the symbolic component indicates how the actually detected features induce a partial order between the initially abduced hypotheses. At this point, if the classification process was successful, there is a further and final step of the corresponding justification process. The system declares that the detected visual cues make the selection of the higher-ranked classification hypothesis possible, and the visual features associated to this higher-ranked classification hypothesis are generated and assembled together so as to satisfy the spatial relationship that were verified to hold between them (see Fig. 1).

In teacher-mode explanations (see Fig. 2), examples of the various classes are provided, and misclassifications of arch shapes by users are corrected by the system, exploiting both the neural generation of visual cues and the abduction-prediction-test cycle briefly described above.

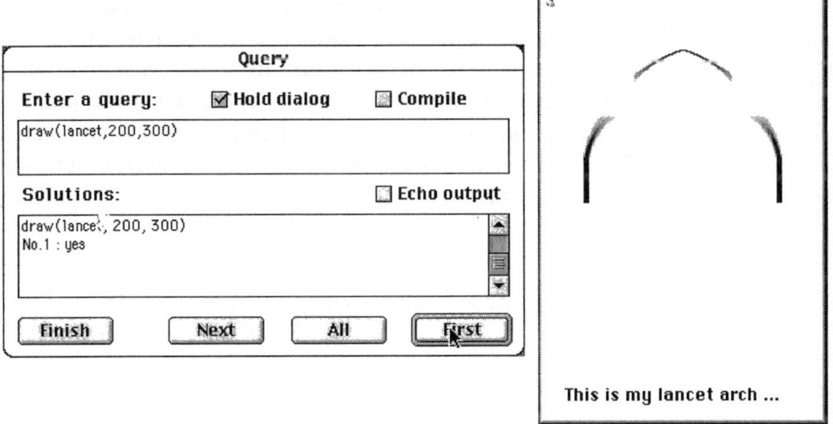

Figure 2. Teacher-mode explanation: generation of a lancet arch from query specifying required class of arches and dimensions of the desired instance.

3. System's architecture

Let us now turn to briefly describe the system's architecture. Aleksander's RAM-based weightless neural discriminators form the neural component of ARCS (Aleksander and Morton, 1990). Clearly, these digital neural systems can be trained to classify visual patterns. But in addition, if certain training procedures are adopted, they can store information which can be later used to generate an example of a given class of visual features. The particular procedure we have adopted enables one to record the relative frequency of appearance of a subpattern formed by a fixed number of pixels in the training set. The example generated starting from this information can be qualitatively described as obtained by "superimposing" the members of the training set for that class (see Burattini *et al.*, 1998 for technical details). It should be clear that this sort of bidirectional behaviour in RAM-based weightless neural discriminators is not identical to the I/O reversal exhibited by the Bidirectional Associative Memories (BAMs) introduced in (Kosko, 1988).

4. Conclusion

It is a widely shared view that visual perception and imagery share various underlying mechanisms (Kosslyn, 1994, pp. 54-60; Ishai and Sagi, 1995). This basic view is reflected in the fact that visual object classification and generation (the latter one is a sort of imagery function performed by the system, since a picture of the object is generated from a verbal stimulus only) are performed by the same modules of ARCS, because these are capable of appropriate forms of bidirectional behaviour. Equally important, ARCS is internally organised in ways that closely reflect Kosslyn's "protomodel" of visual object identification and imagery. In fact, ARCS might even be regarded, relative to its limited domain of application, as a computational rendering of basic mechanisms postulated in Kosslyn's protomodel. The hybrid neurosymbolic nature of ARCS raises, from the viewpoint of its possible interpretation as a cognitive model, interesting epistemological questions. In this framework, a unifying interpretation for neural and symbolic computations can be found at an I/O level, capturing the functional organisation of the main perception/imagery subsystems of Kosslyn's protomodel.

Acknowledgements

The authors gratefully acknowledge support from MURST, Project 9911263337_004.

References

Aleksander, I. and E. Morton (1990) *An Introduction to Neural Computing*, Chapman & Hall, London.

Burattini, E., M. De Gregorio and G. Tamburrini (1998) "Generating and classifying recall images by neurosymbolic computation", *Proceedings of the*

2nd European Conference on Cognitive Modelling, F.E. Ritter and R.M. Young, eds, Nottingham: Nottingham University Press, pp. 127-134.

Burattini, E., M. De Gregorio and G. Tamburrini (2000) "Hybrid expert systems: An approach to combining neural computation and rule-based reasoning", in: *Expert Systems, Techniques and Applications*, C.D. Leondes, ed, London: Gordon and Breach, in press.

Glasgow, J.I. (1993) "The imagery debate revisited: a computational perspective", *Computational Intelligence* 9:309-333.

Ishai, A. and D. Sagi. (1995) "Common mechanisms of visual imagery and perception", *Science* 268:1772-1774.

Kosko, B. (1988) "Bidirectional Associative Memories", *IEEE Transactions on Systems, Man and Cybernetics* 18:49-60.

Kosslyn, S.M. (1994) *Image and Brain*, Cambridge, MA: MIT Press.

Sun, R. and L.A. Bookman (1995) *Computational Architectures Integrating Neural and Symbolic Processes*, Dordrecht: Kluwer Academic Pub.

Tabachneck-Schijf, H.J.M., A.M. Leonardo and H.A. Simon (1997) "CaMeRa: A computational model of multiple representations", *Cognitive Science* 21:305-350.

Winston, P.H. (1975) "Learning structural descriptions from examples", in: *The Psychology of Computer Vision*, P.H. Winston, ed., New York: McGraw-Hill, pp. 157-209.

A MATHEMATICAL MODEL OF DEPTH DISPLACEMENT OF CONTRACTING 2-D FIGURES. PART A: RECTANGLES OF CONSTANT WIDTH UNDERGOING LATERAL DISPLACEMENTS

E. XAUSA*, L. BEGHI** and M. ZANFORLIN*

* *Department of General Psychology, University of Padua, Via Venezia 8, Padova 35131, Italy*

** *Department of Pure and Applied Mathematics, University of Padua, Via Belzoni 7, Padova 35131, Italy*

ABSTRACT

In the present paper we examine the stereokinetic phenomena generated by rectangles with constant width and periodically contracting height, undergoing simultaneous lateral displacements from left to right on the frontal plane.The observer perceives rectangles rotating around their horizontal symmetry axis and simultaneously undergoing a translation in depth. Slight elastic deformations of the figures (periodic changes in width) can also become apparent. These apparent movements differ from those performed by rigid 2-D figures having the same orthogonal projections on the frontal plane. Our mathematical model is based on a *principle of velocity differences minimization*, which states that the visual system performs a transformation of the velocity field of the stimulus in such a way as to minimize the differences between the lengths of the velocity vectors of the points of the moving image. A velocity component "in depth" (*i.e.* along a direction orthogonal to the frontal plane) is added to each velocity vector in order to fulfill the requirements of the minimality condition. As a result a rotational component around the horizontal symmetry axis of the figure and a translatory one become apparent. The theoretical predictions are in a fairly good agreement with the experimental results.

1. Introduction

In a previous paper (Xausa *et al.*, 1999) we presented a mathematical model of the stereokinetic phenomena generated by a vertical bar undergoing periodic contractions and simultaneous lateral displacements (from left to right) on the frontal plane (see Fig. 1): the observer perceives a line segment of (slightly) increasing length, undergoing a displacement in depth and a simultaneous rotation around its midpoint, in a sagittal plane (see Fig. 2).

Our mathematical model was based on a *principle of velocity differences minimization* (Zanforlin, 1988; Beghi *et al.*, 1991, 1999; Xausa *et al.*, 1999), stating that the visual system performs a transformation of the velocity field of the stimulus in such a way as to minimize the differences between the lengths of the velocity vectors of the points of the moving image.

With reference to Fig. 1, assuming that M is the midpoint of the contracting line segment of extremes, P, P' and Q is a (generic) point of the segment PP', satisfying the condition:

$$Q-M = \lambda \cdot (P-M) \ , \quad -1 \leq \lambda \leq 1 \tag{1}$$

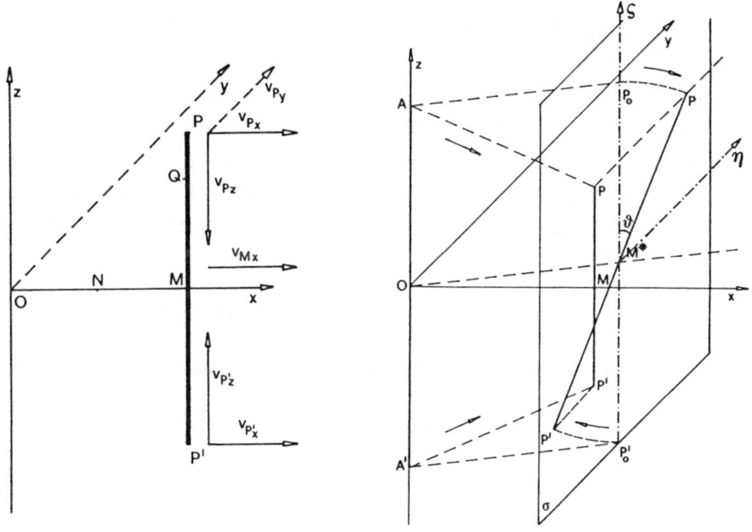

Figure 1 **Figure 2**

and assuming that at a generic time t ($0 \leq t \leq T$, where T is the period of the phenomenon) the velocity components of point Q along the horizontal and vertical direction respectively are given by the relations:

$$v_{Qx}(t) = \lambda \cdot v_{Px}(t) \quad , \qquad v_{Qz}(t) = \lambda \cdot v_{Pz}(t) \tag{2}$$

(where $v_{Px}(t)$, $v_{Pz}(t)$ are assigned functions), then, according to this minimality principle, a third velocity component $v_{Qy}(t)$, oriented "in depth" is associated to Q, which is defined by the relation :

$$v_{Qy}(t) = \sqrt{v_{max}^{\ 2} - v_{Qx}(t)^2 - v_{Qz}(t)^2} \ , \tag{3}$$

where

$$v_{max}^{\ 2} = (v_{x\,max}^{\ 2} + 4\, v_{z\,max}^{\ 2}) \ , \tag{4}$$

$$v_{x\,max} = \max_{0 \leq t \leq T} |\, v_{Px}(t)\,| , \quad v_{z\,max} = \max_{0 \leq t \leq T} |\, v_{Pz}(t)\,| .$$

When $\lambda = 0$, point Q coincides with the midpoint M and the velocity components $v_{Qx}(t)$, $v_{Qy}(t)$, $v_{Qz}(t)$ become:

$$v_{Mx}(t) = v_{Px}(t) \quad , \quad v_{My}(t) = \sqrt{v_{max}{}^2 - v_{Px}(t)^2} \quad , \quad v_{Mz}(t) = 0 \quad . \qquad (5)$$

They define the motion, on the Oxy plane, of a virtual point M*, according to the time law :

$$x_{M*}(t) = \int_0^t v_{Mx}(\tau)\, d\tau \quad , y_{M*}(t) = \int_0^t v_{My}(\tau) d\tau \quad , z_{M*}(t) = 0 \quad . \qquad (6)$$

Taking now into consideration the relative velocity components $v_{QMx}(t)$,$v_{QMy}(t)$, $v_{QMz}(t)$ of point Q with respect to point M, we obtain the relations :

$$v_{QMx}(t) = 0 \quad , v_{QMy}(t) = |\ v_{Qy}(t) - v_{My}(t)\ | \cdot \mathrm{Sign}(\lambda) \quad , \quad v_{QMz}(t) = v_{Pz}(t) \ .$$

They define the motion, on a $M^*\eta\zeta$ plane (where the $M^*\zeta$ axis is parallel to the Oz axis and the $M^*\eta$ axis is parallel to the Oy axis) of a virtual point Q* (see Fig. 2), according to the time law :

$$\eta_{Q*}(t) = \int_0^t v_{QMy}(\tau)\, d\tau \quad , \zeta_{M*}(t) = \int_0^t v_{QMz}(\tau)\, d\tau \qquad . \qquad (7)$$

A rough estimate of the rotational angle $\Delta\vartheta$ can be obtained by the relation:

$$\Delta\vartheta \cong \frac{1}{h_o} \cdot \int_0^T \sqrt{v_{QMy}(\tau)^2 + v_{QMz}(t)^2}\, d\tau \quad , \qquad (8)$$

where h_o is the initial half-heigth of the contracting line segment on the frontal plane.

2. Stereokinetic phenomena generated by rectangles of constant width undergoing lateral displacements

In the presence of a rectangle of constant width b_o, initial height h_o, final height h_T, which periodically contracts (with period T) and simultaneously undergoes a lateral displacement (from left to right) on the frontal plane (see Fig. 3), an observer

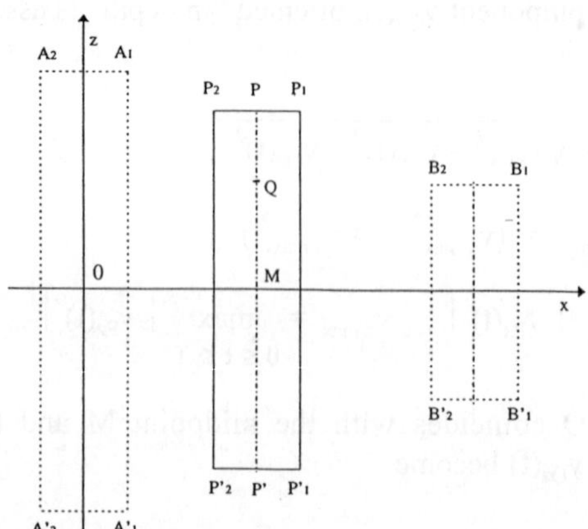

Figure 3

perceives a rectangular figure undergoing a displacement in depth which is less accentuated than in the case of a contracting vertical line segment of same heigth; at the same time, this rectangle appears to rotate around its horizontal symmetry axis, with a rotational angle which is always greater than in the case of a line segment .

In order to obtain a quantitative model of the phenomenon, we slightly modified the described in the previous section and referring to vertical line segments with the same height and the lateral displacement of our rectangles.In order to obtain a quantitative model of the phenomenon, we slightly modified the described in the previous section and referring to vertical line segments with the same height and the lateral displacement of our rectangles.

Assuming that the velocity components $v_{Px}(t)$, $v_{Pz}(t)$, $v_{P'x}(t) = v_{Px}(t)$, $v_{P'z}(t) = -v_{Pz}(t)$ of points P, P' lying at the intersection between the vertical symmetry axis of rectangle and the two bases of the rectangle are assigned, and indicating by M the midpoint of the rectangle moving on the frontal plane (and by M* the midpoint of the apparent rectangle diplacing itself in depth) and by Q the generic point of the PP' line segment, we introduced the following relations for the velocity component $v_{My}(t)$ of point M* along the Oy direction and for the relative velocity component $v_{QMy}(t)$ of point Q along the same direction :

$$v^*_{My}(t) \cong \alpha_y \cdot v_{My}(t) \quad , v^*_{QMy}(t) = \left| v_{Qy}(t) - v^*_{My}(t) \right| \cdot \mathrm{Sign}(\lambda) \quad , \tag{9}$$

where

$$\alpha_y = \sqrt{1 - \frac{b_o}{h_o} \cdot \frac{h_T}{h_o}} \quad , \tag{10}$$

and $v_{Qy}(t)$, $v_{My}(t)$ are given in formulas (3), (5) respectively .

According to the procedure illustrated in the previous section, the total amount Δy_{M*} of depth displacement of the apparent midpoint M* of the rectangle after a period is given by the formula :

$$\Delta y_{M*} = \int_0^T v^*_{My}(\tau)\, d\tau \quad , \tag{11}$$

while a rough estimate of rotational angle $\Delta\vartheta$ is given by the formula :

$$\Delta\vartheta \cong \frac{1}{h_o} \cdot \int_0^T \sqrt{v^*_{QMy}(\tau)^2 + v_{QMz}(t)^2}\, d\tau \quad . \tag{12}$$

3. Experimental results and conclusions

We examined the following cases :

a) rectangles with periodically decreasing height according to the hyperbolic time law

$$z_P(t) = h_o \cdot \frac{1}{1 + [\frac{h_o}{h_T} - 1] . \frac{t}{T}} \;,\; 0 \le t \le T$$

and with a lateral displacement according to the hyperbolic time law :

$$x_P(t) = b\,\frac{h_o}{h_T} \cdot \frac{t}{T} \cdot \frac{1}{1 + [\frac{h_o}{h_T} - 1] . \frac{t}{T}} \;,\; 0 \le t \le T$$

in such a way that

$$z_P\,(x_P) = h_o \,\{\, 1 - [\, 1 - \frac{h_T}{h_o}] \cdot \frac{x_P}{b} \,\} \;,\; 0 \le x_P \le b$$

b) rectangles with periodically decreasing height according to the uniform time law

$$z_P(t) = h_o \,\{\, 1 - [\, 1 - \frac{h_T}{h_o}] \cdot \frac{t}{T} \} \;,\; 0 \le t \le T$$

and with a lateral displacement according to the uniform time law :

$$x_P(t) = b \cdot \frac{t}{T} \;,\; 0 \le t \le T$$

in such a way that

$$z_P\,(x_P) = h_o \,\{\, 1 - [\, 1 - \frac{h_T}{h_o}] \cdot \frac{x_P}{b} \,\} \;,\; 0 \le x_P \le b$$

c) rectangles with periodically decreasing height according to the harmonic time law :

$$z_P(t) = h_o \cos [\, \Phi \cdot \frac{t}{T}] \;,\; 0 \le t \le T \;,\; \Phi = \cos^{-1}(\frac{h_T}{h_o})$$

and with a lateral displacement according to the uniform time law :

$$x_P(t) = b \cdot \frac{t}{T} \;,\; 0 \le t \le T$$

in such a way that

$$z_P\,(x_P) = h_o \cos [\, \Phi \cdot \frac{x_P}{b}] \;,\; 0 \le x_P \le b$$

The theoretical estimates (solide lines) and the experimental results (dashed lines) are illustrated in Figure 4, for different values of side ratio $\frac{b_o}{h_o}$, in the hyperbolic–hyperbolic case. In all examined cases the lateral displacement ratio was $\frac{b}{h_o} = 0.67$.

Slight elastic deformations of the figures (periodic changes in width) can also become apparent. They are due to the fact that, according to projective geometry, a rigid rectangular figure displacing itself in depth should evidence a progressive reduction in width as far as the distance of the figure from the frontal plane of increases .

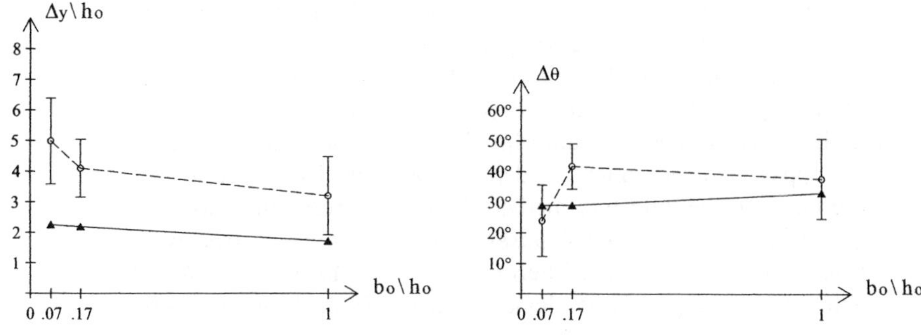

Figure 4

Slight elastic deformations of the figures (periodic changes in width) can also become apparent. They are due to the fact that, according to projective geometry, a rigid rectangular figure displacing itself in depth should evidence a progressive reduction in width as far as the distance of the figure from the frontal plane of increases .This reduction in width does not take place in our case (rectangle of constant width): as a consequence, the figure does not appear as a rigid one.

The model presented here is an attempt to describe in mathematical terms the logics of human visual processing, as far as the 3–D recovering of moving images is concerned. This is a legitimated goal of biological cybernetics. The fairly good agreement between the theoretical estimates and the experimental results seems to indicated that our approach is correct.

References

Beghi, L., E. Xausa and M. Zanforlin (1991) "Analytic determination of the depth effect in stereokinetic phenomena without a rigidity assumption", *Biol. Cyber.* 65:425-432.

Beghi, L., E. Xausa, C. De Biasio and M. Zanforlin (1991) "Quantitative determination of the 3-D appearances of a rotating ellipse without a rigidity assumption", *Biol. Cyber.* 65:433-440.

Beghi, L., E. Xausa and M. Zanforlin (1999) "Mathematical model of the apparent displacement in depth of a vertical contracting bar, according to the trajectory method. Part A: hyperbolic decrements", in: *Neuronal Bases and Psychological Aspects of Consciousness*,

C. Taddei-Ferretti and C. Musio, eds, World Scientific, Singapore, New Jersey, London, Hong Kong, pp. 399-407.

Beghi, L., E. Xausa and M. Zanforlin (1997) "The depth effect of an oscillating tilted bar", *J. Math. Psych.* 41:11-18.

Xausa, E., L. Beghi and M. Zanforlin (1999) "Mathematical model of the apparent displacement in depth of a vertical contracting bar, according to the trajectory method. Part B: Uniform and harmonic decrements", in: *Neuronal Bases and Psychological Aspects of Consciousness*, C. Taddei-Ferretti and C. Musio, eds, World Scientific, Singapore, New Jersey, London, Hong Kong, pp. 408-414.

Xausa, E., L. Beghi and M. Zanforlin (1999) "A mathematical model of depth displacement with a contracting bar", submitted to *J. Math. Psych.*

Zanforlin, M. (1988) "Stereokinetic phenomena as good Gestalts. The minimum principle applied to circles and ellipses in rotation: a quantitative analysis and theoretical Discussions", *Gestalt Theory* 10:187-214.

A MATHEMATICAL MODEL OF DEPTH DISPLACEMENT OF CONTRACTING 2-D FIGURES . PART B: CONTRACTING RECTANGLES WITH INVARIANT SIDE RATIO

L. BEGHI*, E. XAUSA ** and M. ZANFORLIN**

* *Department of Pure and Applied Mathematics, University of Padua, Via Belzoni, 7, Padova 35131, Italy*

** *Department of General Psychology , University of Padua, Via Venezia 8, Padova 35131, Italy*

ABSTRACT

In the present paper we examine the stereokinetic phenomena generated by contracting rectangles keeping constant their side ratio. When the height of a figure is greater than its width, the observer perceives a contracting rectangle, simultaneously undergoing a translatory movement in depth (*i.e.*, along the direction orthogonal to the frontal plane). The amount of this translatory displacement is smaller than what we should expect according to the laws of perspective in case of a rigid figure. Our mathematical model is based on a *principle of velocity differences minimization*, which states that the visual system performs a transformation of the velocity field of the stimulus in such a way as to minimize the differences between the lengths of the velocity vectors of the points of the moving image. A velocity component along the direction orthogonal to the frontal plane is added to each velocity vector in order to fullfil the requirements of the minimality condition, thus producing the apparent displacement in depth of the figure.

1. Introduction

When a rectangular figure undergoes periodic contractions on the frontal plane of an observer , while keeping constant its side ratio (see Fig. 1), the observer's visual system perceives a contracting rectangular figure which displaces itself in depth. The same is true in case of periodically contracting squares.

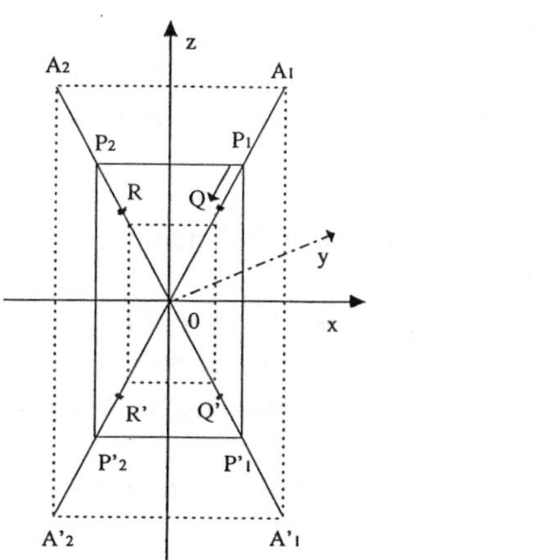

Figure 1

2. Mathematical model

With reference to Fig.1, let us consider a rectangular figure of vertices P_1, P_2, P'_1, P'_2 belonging to the frontal plane π of an observer; let us assume that this figure undergoes periodical homothetic transformations of period T, moving from the initial position A_1, A_2, A'_1, A'_2 to the final position B_1, B_2, B'_1, B'_2.
At any time t ($0 \leq t \leq T$) the half-heigth h(t) and the half-width b(t) of the rectangle satisfy the invariance relation :

$$\frac{h(t)}{b(t)} = \frac{h_o}{b_o} = k \text{ (constant)}, \quad k \in \Re^+ \tag{1}$$

where h_o, b_o are the initial half-heigth and half-width of the rectangle.

Let us introduce on π a cartesian reference system Oxy, where O is the (stationary) centre of the contracting rectangle, Ox is an axis which is parallel to the horizontal plane and is oriented from left to right, Oz is an axis which is orthogonal to the Ox axis and is oriented "upwards".
An additional reference axis Oy, orthogonal to the frontal plane and oriented "in depth" is finally introduced .

Let $\{v_{1x}(t), v_{1z}(t)\}$, $\{v'_{1x}(t), v'_{1z}(t)\}$, $\{v_{2x}(t), v_{2z}(t)\}$, $\{v'_{2x}(t), v'_{2z}(t)\}$ be the velocity components of vertices P_1, P'_1, P_2, P'_2; they satisfy the relations :

$$v_{1x}(t) = v'_{1x}(t) = v_x(t) \;,\; v_{2x}(t) = v'_{2x}(t) = -v_x(t) \;, \tag{2}$$

$$v_{1z}(t) = v_{2z}(t) = v_z(t) \;,\; v'_{1z}(t) = v'_{2z}(t) = -v_z(t) \;, \tag{3}$$

$$\frac{v_z(t)}{v_x(t)} = k \quad , \quad k \in \Re^+ \tag{4}$$

where $v_x(t)$ is an assigned, negative time function over the interval $0 \leq t \leq T$.

With reference to Fig.1, let us now consider a generic point Q of the half-diagonal OP_1 of our rectangle, i.e. a point such that the (oriented) segment Q–O satisfies the following relation :

$$Q - O = \lambda \cdot (P_1 - O) \;,\; 0 \leq \lambda \leq 1 \tag{5}$$

In the same way , we could consider a generic point R of the half-diagonal OP_2, i.e. a point such that the (oriented) segment R–O satisfies the following relation:

$$R - O = \mu \cdot (P_2 - O) \;,\; 0 \leq \mu \leq 1 \;. \tag{6}$$

Similarly, we could consider points Q', R' belonging to the half diagonals OP'_1, OP'_2, satisfying the relations:

$Q' - O = \lambda' \cdot (P'_1 - O) \;,\; 0 \leq \lambda' \leq 1 \;,\; R' - O = \mu' \cdot (P'_2 - O) \;,\; 0 \leq \mu' \leq 1$.

For the sake of simplicity, let us now confine ourselves to point Q. Under the given hypotheses, the velocity components $\{v_{Qx}(t), v_{Qz}(t)\}$ of point Q will satisfy the relations :

$$v_{Qx}(t) = \lambda \cdot v_x(t) \;,\; v_{Qz}(t) = \lambda \cdot v_z(t) = \lambda \cdot k \cdot v_x(t) \tag{7}$$

In order to develop a mathematical model explaining the apparent depth displacement of our rectangular figure, we make now use of the so-called *principle of velocity differences minimization* (Zanforlin, 1988, Beghi *et al.*, 1991, 1997; Xausa *et al.*, 1997), stating that under the given conditions the visual system performs a transformation of the velocity field of the stimulus in such a way as to minimize the differences between the lengths of the velocity vectors of the points of the moving image.
This uniformization of the velocity field could be achieved by associating to point Q an additional velocity component $v_{Qy}(t)$ oriented "in depth" and satifying the following relation :

$$\sqrt{v_{max}{}^2 - v_{Qx}(t)^2 - v_{Qz}(t)^2} = \sqrt{v_{max}{}^2 - (1+k^2)\,\lambda^2 \cdot v_x(t)^2}\ , \qquad (8)$$

where

$$v_{max}{}^2 = 4\,(1+k^2)\cdot v_{x\,max}{}^2 \quad , \qquad v_{x\,max} = \max_{0 \le t \le T} |\, v_x(t)\,| \qquad (9)$$

The velocity components { $v_{Qx}(t)$, $v_{Qy}(t)$, $v_{Qz}(t)$ } can be finally attributed to a virtual point Q* displacing itself in depth according to the parametric equations :

$$x_Q(t) = b + \lambda \int_0^t v_x(\tau)\, d\tau\ ,\ y_Q(t) = \int_0^t v_{Qy}(\tau)\, d\tau\ ,\ z_Q(t) = h_o + \lambda\, k \int_0^t v_z(\tau)\, d\tau \qquad (10)$$

At time t = T the amount of depth displacement Δy will be given by the relation :

$$\Delta y\ = y_Q(t) = \int_0^T v_{Qy}(\tau)\, d\tau\ . \qquad (11)$$

3. Theoretical estimates

We took into consideration the following cases :

a) side contraction of hyperbolic type

$$v_x\,(t) = -\ v_o \cdot [\frac{h_o}{h_T} - 1\,] \cdot \frac{1}{\{\,1 + [\frac{h_o}{h_T} - 1\,]\,.\,\frac{t}{T}\,\}^2}\ , \qquad (12)$$

where b_o is the initial half-width of the contracting rectangle, h_o is its initial half-heigth , h_T is its final half-heigth and $v_o = \frac{b_o}{T}$;

b) side contraction of uniform type

$$v_x(t) = -\,v_o\,.\,[1 - \frac{h_T}{h_o}\,] \quad . \qquad (13)$$

c) side contraction of harmonic type

$$v_x\,(t) = -\,v_o \cdot \Phi \cdot \sin\,[\Phi\,.\,\frac{t}{T}\,] \quad ,\ \Phi = \cos^{-1}(\frac{h_T}{h_o}) \qquad (14)$$

In Figs 2a, 2b, 2c the numerical values of the ratio $\frac{\Delta y}{h_o}$ (as a function of the ratio $\frac{h_T}{h_o}$) are plotted, for different values of the ratio $\frac{b_o}{h_o}$ (namely $\frac{b_o}{h_o} = 0.25$: solide lines; $\frac{b_o}{h_o} = 0.5$: dashed lines; $\frac{b_o}{h_o} = 1$: dot-dashed lines) in case of hyperbolic, uniform, harmonic decrements respectively .

It is worthwhile to be recalled that the case k = 1 refers to contracting squares.

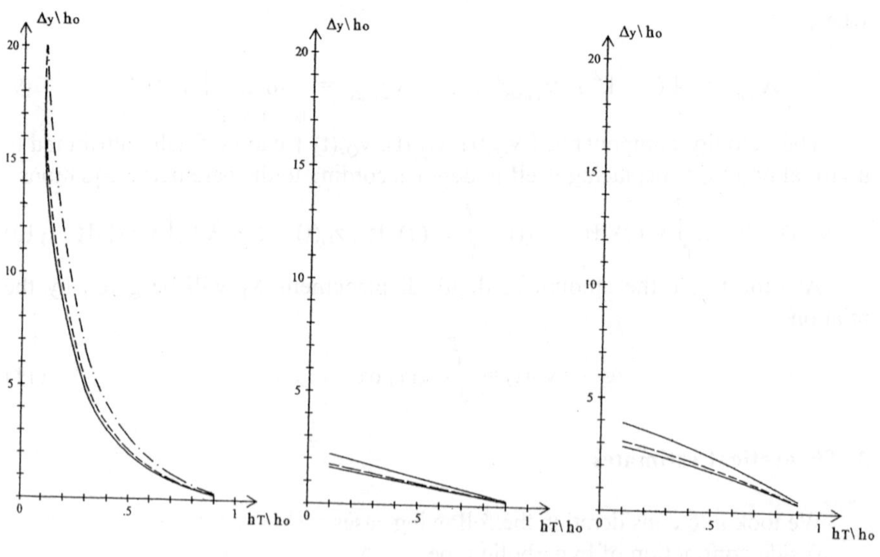

Figure 2a **Figure 2b** **Figure 2c**

From these results it appears that in case a), i.e. in case of side decrements according to an hyperbolic time law the apparent displacement in depth is much higher than in cases b) and c) for low values of the height ratio $\frac{h_T}{h_o}$. In any examined case the apparent displacement in depth is heigher when the figures are squares and decreases to a well defined lower bound when the side ratio $\frac{b_o}{h_o}$ tends to zero, i.e. when the vertical rectangles tend to become vertical line segments.

The model illustrated in this paper, which continues our studies in the field of the 3–D recovery of 2–D moving objects by the human system (as presented in Part A of this work), is a mathematical one, which still requires an experimental validation. The fairly good agreement between the theoretical estimates and the experimental results shown in Part A, incourages the authors in continuing their research in this direction.

References

Beghi, L., E. Xausa and M. Zanforlin (1991) "Analytic determination of the depth effect in stereokinetic phenomena without a rigidity assumption", *Biol. Cyber.* 65:425-432.

Beghi, L., E. Xausa, C. De Biasio and M. Zanforlin (1991) "Quantitative determination of the 3-D appearances of a rotating ellipse without a rigidity assumption", *Biol. Cyber.* 65:433-440.

Beghi, L., E. Xausa and M. Zanforlin (1997) "Mathematical model of the depth effect in the translatory alternating movement. Part B: swinging gate phenomenon", in: *Biocybernetics of Vision. Integrative Mechanisms and Cognitive Processes*, C. Taddei-Ferretti ed., World Scientific, Singapore, New Jersey, London, Hong Kong, pp. 297-300.

Beghi, L., E. Xausa and M. Zanforlin (1997) "The depth effect of an oscillating tilted bar", *J. Math. Psych.* 41:11-18.

Xausa, E., L. Beghi and M. Zanforlin (1997) "Mathematical model of the depth effect in the translatory alternating movement. Part A: 3–D perception of length amplification", in: *Biocybernetics of Vision. Integrative Mechanisms and Cognitive Processes*, C. Taddei-Ferretti ed., World Scientific, Singapore, New Jersey, London, Hong Kong, pp. 293-296.

Xausa, E., L. Beghi and M. Zanforlin (1999) "A mathematical model of depth displacement with a contracting bar", submitted to *J. Math. Psych.*

Zanforlin, M. (1988) "Stereokinetic phenomena as good Gestalts. The minimum principle applied to circles and ellipses in rotation: a quantitative analysis and theoretical Discussions", *Gestalt Theory* 10:187-214.

Zanforlin, M., L. Beghi, E. Xausa and L. Tomat (1997) "The depth effect in the stereokinetic phenomenon of swinging gate", in: *"Biocybernetics of Vision. Integrative Mechanisms and Cognitive Processe,* C. Taddei-Ferretti, ed., World Scientific, Singapore, New Jersey, London, Hong Kong, pp. 301-304.

PARTICIPANTS

Members of the Advisory Board, Lecturers and members of the Organizing Committee are indicated respectively by three, two and one asterisks

Angotzi Anna Rita
International Marine Centre
Loc. Sa Mardini
I 09072 Torregrande (Oristano)
Italy
tel: +39 0783 22027
fax: +39 0783 22002
e-mail: imceco@tin.it

*Aprile Silvana
Istituto Italiano per gli Studi Filosofici
Via Gennaro Serra, 75
I 80132 Napoli
Italy
tel: +39 081 2452086
fax: +39 081 7645058

Arcelli Carlo
Istituto di Cibernetica, CNR
Via Toiano, 6
I 80072 Arco Felice (Napoli)
Italy
tel: +39 081 8534206
fax: +39 081 5267654
e-mail: car@imagm.cib.na.cnr.it

**Arikawa Kentaro
Graduate School of Integrated Science
Yokohama City University,
22-2 Seto, Kanazawa-ku
J 236-0027 Yokohama
Japan
tel/fax: +81 45 787 2212
e-mail: arikawa@yokohama-cu.ac.jp

Beghi Luigi
Dipartimento di Matematica Pura e Applicata
Università di Padova
Via Belzoni, 7
I 35131 Padova
Italy
tel +39 049 8275967
fax: +39 049 8758596
e-mail: beghi@galileo.math.unipd.it

**Berardi Nicoletta
Istituto di Neurofisiologia, CNR
Area della Ricerca di Pisa – S. Cataldo
Via V. Alfieri, 1
I 56010 Ghezzano (Pisa)
Italy
tel: +39 050 3153164
fax: +39 050 3153220
e-mail: berardi@in.pi.cnr.it

**Borg-Graham Lyle
Equipe Cognisciences, Institut Alfred Fessard
CNRS UPR 2212 Bâtiment 33
Avenue de la Terrasse
F 91198 Gif-sur-Yvette
France
tel: +33 1 69 82 34 13
fax: +33 1 69 82 34 27
e-mail: lyle@cogni.iaf.cnrs-gif.fr
http://www.cnrs-gif.fr/iaf/iaf9/surf-hippo.html

Campana Gianluca
Centro Interdipartimentale di Scienze Cognitive
Università di Padova
Via Beltrame, 1b
I 35100 Padova
Italy
tel: +39 049 8715751
fax: +39 049 8276600
e-mail: campana@snoopy.psy.unipd.it

**Chalupa Leo
Section of Neurobiology Physiology & Behavior
Division of Biological Sciences
University of California Davis
Davis, CA, 95616
USA
tel: +1 530 752 2559
fax: +1 530 752 5582
e-mail: lmchalupa@ucdavis.edu

Chillemi Santi
Istituto di Biofisica, CNR
Area della Ricerca di Pisa – S. Cataldo
Via V. Alfieri, 1

I 56010 Ghezzano (Pisa)
Italy
tel: +39 050 3152578
fax: +39 050 3152760
e-mail: santi.chillemi@ ib.pi.cnr.it

Chukova Svetlana V.
Pavlov Institute of Physiology
Emb. Makarov 6
CSI 197376 St.-Petersburg
Russia
tel: +7 812 328 4571
fax: +7 812 328 0501
e-mail: svch@infran.ru

*Cotugno Antonio
Istituto di Cibernetica, CNR
Via Toiano, 6
I 80072 Arco Felice (Napoli)
Italy
tel: +39 081 8534113
fax: +39 081 5267654
e-mail: neurosistemi@sole.cib.na.cnr.it

De Gregorio Massimo
Istituto di Cibernetica, CNR
Via Toiano, 6
I 80072 Arco Felice (Napoli)
Italy
tel: +39 0818534228
fax: +39 0815267654
e-mail: massimo@sole.cib.na.cnr.it

Di Garbo Angelo
Istituto di Biofisica, CNR
Area della Ricerca di Pisa – S. Cataldo
Via V. Alfieri, 1
I 56010 Ghezzano (Pisa)
Italy
tel: +39 050 3152578
fax: +39 050 3152760
e-mail: angelo.digarbo@ib.pi.cnr.it

Di Maio Vito
Istituto di Cibernetica, CNR
Via Toiano, 6
I 80072 Arco Felice (Napoli)
Italy
tel: +39 0818534118
fax: +39 0815267654
e-mail: vdm@biocib.cib.na.cnr.it

***Dowling John E.
Department of Molecular and Cellular Biology
Harvard University
Cambridge, MA 02138
USA
tel: +1 617 495 2245
fax.: +1 617 496 3321
e-mail: dowling@fas.harvard.edu

***Fiorentini Adriana
Istituto di Neurofisiologia, CNR
Area della Ricerca di Pisa – S. Cataldo
Via V. Alfieri, 1
I 56010 Ghezzano (Pisa)
Italy
tel: +39 050 3153176
fax: +39 050 3153220
e-mail: adri@in.pi.cnr.it

**Gualtieri Paolo
Istituto di Biofisica, CNR
Area della Ricerca di Pisa – S. Cataldo
Via V. Alfieri, 1
I 56010 Ghezzano (Pisa)
Italy
tel: +39 050 3153026
fax: +39 050 3152760
e-mail: Paolo.Gualtieri@ib.pi.cnr.it

Guglielmotti Vittorio
Istituto di Cibernetica, CNR
Via Toiano, 6
I 80072 Arco Felice (Napoli)
Italy
tel: +39 0818534117/114
fax: +39 0815267654
e-mail: vigu@biocib.cib.na.cnr.it

***Karten Harvey J.
Department of Neurosciences
University of California San Diego
La Jolla CA 92093-0662
USA
tel: +1 619 534 4938

fax: +1 619 534 6602
e-mail: hjkarten@ucsd.edu

Kremen Inna
Laboratory of Action Programming
Institute of Human Brain
Acad.Pavlov Street, 9a
CSI197376 St.-Petersburg
Russia
tel: +7 812 2341314
fax: +7 812 2343247
e-mail: action@brain.nw.ru

**Land Michael F.
Sussex Centre for Neuroscience
School of Biological Sciences
University of Sussex
Brighton BN1 9QG
UK
tel: +44 0 1273 678505
fax: +44 0 1273 678535
e-mail: m.f.land@sussex.ac.uk

***Lagnado Leon
Medical Research Council
Laboratory of Molecular Biology
Hills Road
Cambridge, CB2 2QH
UK
tel +44 0 1223 402177/2453
fax: +44 0 1223 213556
e-mail: ll1@mrc-lmb.cam.ac.uk

***Laughlin Simon Barry
Department of Zoology
University of Cambridge
Downing St.
Cambridge CB2 3EJ
UK
tel: +44 223 336 608
fax: +44 223 336 676
e-mail: SL104@cam.ac.uk
http://www.zoo.cam.ac.uk/zoostaff/laughlin/

Levy Hanna
The Bruce Rappaport Faculty of Medicine,
Technion-Israel Institute of Technology,
Rappaport Institute,
PO Box 9697
IL 31096 Haifa
Israel
tel: +972 4 8295279
fax: +972 4 8535969
e-mail: levyh@techunix.technion.ac.il

Lüdtke Niklas
Department of Computer Science
University of York
Heslington
York YO10 5DD
UK
tel: +44 1904 432774
e-mail: niklu@cs.york.ac.uk

**Moya Kenneth L.
CNRS URA 2210
Service Hospitalier Frédéric Joliot, CEA
4, place du Général Leclerc
F 91401 Orsay
France
tel: +33 1 69867711
fax +33 1 69867816
e-mail: moya@shfj.cea.fr

**Musio Carlo
Istituto di Cibernetica, CNR
Via Toiano, 6
I 80072 Arco Felice (Napoli)
Italy
tel: +39 081 8534131/113
fax: +39 081 5267654
e-mail: carlom@biocib.cib.na.cnr.it

Musumeci Daniela
Dipartimento di Fisiologia e Biochimica
Università di Pisa
Via S. Zeno, 31
I 56127 Pisa
Italy
tel: +39 050 553517
fax: +39 050 552183
e-mail: dmusumeci@dfb.unipi.it

Neu Georgia
Department of Cell- and Neurobiology
Institute of Zoology I

University of Karlsruhe
Kornblumenstrasse13
D 76128 Karlsruhe
Germany
tel.: +49 721 6083354
fax:+49 721 6084848
e-mail: dc08@rz.uni-karlsruhe.de

Orioli Mauro
Dipartimento di Psicologia Generale
Università di Padova
Via Venezia, 8
I 35131 Padova
Italy
tel +39 049 8276920
e-mail: orioli@mail.psy.unipd.it

Paramei Galina
Institut für Arbeitsphysiologie
Universität Dortmund
Ardeystrasse 67
D 44139 Dortmund
Germany
tel: +49 231 1084276
fax: +49 231 1084401
e-mail: paramei@arb-phys.uni-dortmund.de

**Paulsen Reinhard
Department of Cell- and Neurobiology
Institute of Zoology I
University of Karlsruhe
Kornblumenstrasse13
D 76128 Karlsruhe
Germany
tel: +49 721 608 2218
fax: +49 721 608 4848
e-mail: Reinhard.Paulsen@bio-geo.uni-karlsruhe.de

**Perlman Ido
The Bruce Rappaport Faculty of Medicine,
Technion-Israel Institute of Technology,
Rappaport Institute,
PO Box 9697
IL 31096 Haifa
Israel
tel: +972 4 8295279
fax: +972 4 8535969
e-mail: iperlman@techunix.technion.ac.il

Peters Gabi
ND 03/72
Ruhr-Universität Bochum
Institut für Neuroinformatik
Lehrstuhl fuer Systembiophysik
Universitaetsstrasse 150
D 44780 Bochum
Germany
tel: +49 234 700 7971
fax: +49 234 7094 210
e-mail: gpeters@neuroinformatik.ruhr-uni-bochum.de

Ramella Giuliana
Istituto di Cibernetica, CNR
Via Toiano, 6
I 80072 Arco Felice (Napoli)
Italy
tel: +39 081 8534213
fax: +39 081 5267654237
e-mail: gr@imagm.cib.na.cnr.it

Raynauld Jean-Pierre
Département de physiologie
Université de Montréal
CP 6128 Succ. Centre-Ville
Montréal (Québec) H3C 3J7
Canada
tel 514 343-2032
fax 514 343-2111
e-mail: raynauld@ERE.UMontreal.CA

Rispoli Giorgio
Dipartimento di Biologia
Sezione di Fisiologia Generale
Universita' di Ferrara
Via Borsari, 46
I 44100 Ferrara
Italy
tel: +39 0532 291472/462
fax: +39 0532 207143
E-mail: RSG@DNS.UNIFE.IT

**Rolls Edmund T.
Department of Experimental Psychology
Oxford University

South Parks Road
Oxford, OX1 3UD
UK
tel: +44 0 1865 271348/271323
fax: +44 0 1865 310447
e-mail: Edmund.Rolls@psy.ox.ac.uk

Sanniti di Baja Gabriella
Istituto di Cibernetica, CNR
Via Toiano, 6
I 80072 Arco Felice (Napoli)
Italy
tel: +39 081 8534237
fax: +39 081 5267654
e-mail: gsdb@imagm.cib.na.cnr.it

*Santillo Silvia
Istituto di Cibernetica, CNR
Via Toiano, 6
I 80072 Arco Felice (Napoli)
Italy
tel: +39 081 8534113
fax: +39 081 5267654
e-mail: silvia@sole.cib.nc.cnr.it

Schulz Simone
Department of Cell- and Neurobiology
Institute of Zoology I
University of Karlsruhe
Kornblumenstrasse13
D 76128 Karlsruhe
Germany
tel.: +49 721 6083354
fax: +49 721 6084848
e-mail: dc28@rz.uni-karlsruhe.de

Serino Luca
Istituto di Cibernetica, CNR
Via Toiano, 6
I 80072 Arco Felice (Napoli)
Italy
tel: +39 081 8534206
fax: +39 081 5267654
e-mail: ls@imagm.cib.na.cnr.it

**Sokolov Evgenij N.
Department of Psychophysiology
Faculty of Psychology
Lomonosov State University
Mokhovaya, 8/5,
Moscow, 103009
Russia
fax: +7 95 2033607
e-mail: SOKOLOV@cogsci.msu.su

Subbarao V. Veluvali
Department of Physiology
Mamata Medical College
507002 Khammam, Andhra Pradesh
India
fax +91 40 3329402
e-mail: vvsubbarao@hotmail.com

***Taddei-Ferretti Cloe
Istituto di Cibernetica, CNR
Via Toiano, 6
I 80072 Arco Felice (Napoli)
Italy
tel: +39 081 8534113
fax: +39 081 5267654
e-mail: cloe@sole.cib.na.cnr.it

Twig Gilad
The Bruce Rappaport Faculty of Medicine,
Technion-Israel Institute of Technology,
Rappaport Institute,
PO Box 9697
IL 31096 Haifa
Israel
tel. +972 4 8295279
fax: +972 4 8535969
e-mail: giladt@techunix.technion.ac.il

Ukhanov Kyrill
Institute for Zoophysiology
University of Potsdam
Lennestrasse 7a
D 14471 Potsdam
Germany
tel: +49 331 9774859
fax: +49 331 9774861
e-mail: ukhanov@rz.uni-potsdam.de

Vestri Alec
Dipartimento di Psicologia dello Sviluppo e
della Socializzazione

Università di Padova
via Venezia, 8
I 35131 Padova
Italy
tel. +44 049 8276616
fax +44 049 8276611
e-mail: avestri@psico.unipd.it

Xausa Elisabetta
Dipartimento di Matematica Pura e Applicata
Università di Padova
Via Belzoni, 7
I 35131 Padova
Italy
tel: +39 049 8275905
fax: +39 049 8758596
e-mail: xausa@galileo.math.unipd.it

**Yau King-Wai
Howard Hughes Medical Institute
Department of Neurosciences
John Hopkins University School of Medicine
MD 21205 Baltimore
USA
tel: +1 410 955 1260
fax: +1 410 614 3579/955 4897
e-mail: kwyau@welchlink.welch.jhu.edu
kwyau@welch.jhu.edu